Remote Sensing in Hydrology and Water Resources Management

Remote Sensing in Hydrology and Water Resources Management

Editors

Weili Duan
Shreedhar Maskey
Pedro Luiz Borges Chaffe
Pingping Luo
Bin He
Yiping Wu
Jingming Hou

MDPI • Basel • Beijing • Wuhan • Barcelona • Belgrade • Manchester • Tokyo • Cluj • Tianjin

Editors

Weili Duan
State Key Laboratory of Desert
and Oasis Ecology
Xinjiang Institute of Ecology
and Geography
Urumqi
China

Shreedhar Maskey
Water Resources and Ecosystems
IHE Delft Institute for Water
Education
Delft
The Netherlands

Pedro Luiz Borges Chaffe
Department of Sanitary and
Environmental Engineering
Universidade Federal
de Santa Catarina
Florianopolis
Brazil

Pingping Luo
School of Water and
Environment
Chang'an Unviersity
Xi'an
China

Bin He
Water Resources Centre
Institute of Eco-environmental
and Soil Sciences
Guangzhou
China

Yiping Wu
Department of Earth and
Environmental Science
Xi'an Jiaotong University
Xi'an
China

Jingming Hou
School of Water Resources and
Hydro-electric Engineering
Xi'an University of Technology
Xi'an
China

Editorial Office
MDPI
St. Alban-Anlage 66
4052 Basel, Switzerland

This is a reprint of articles from the Special Issue published online in the open access journal *Remote Sensing* (ISSN 2072-4292) (available at: www.mdpi.com/journal/remotesensing/special_issues/RS_Hydrology_and_Water_Resources_Management).

For citation purposes, cite each article independently as indicated on the article page online and as indicated below:

LastName, A.A.; LastName, B.B.; LastName, C.C. Article Title. *Journal Name* **Year**, *Volume Number*, Page Range.

ISBN 978-3-0365-2701-7 (Hbk)
ISBN 978-3-0365-2700-0 (PDF)

© 2021 by the authors. Articles in this book are Open Access and distributed under the Creative Commons Attribution (CC BY) license, which allows users to download, copy and build upon published articles, as long as the author and publisher are properly credited, which ensures maximum dissemination and a wider impact of our publications.

The book as a whole is distributed by MDPI under the terms and conditions of the Creative Commons license CC BY-NC-ND.

Contents

About the Editors ix

Preface to "Remote Sensing in Hydrology and Water Resources Management" xi

Weili Duan, Shreedhar Maskey, Pedro L. B. Chaffe, Pingping Luo, Bin He, Yiping Wu and Jingming Hou
Recent Advancement in Remote Sensing Technology for Hydrology Analysis and Water Resources Management
Reprinted from: *Remote Sens.* **2021**, *13*, 1097, doi:10.3390/rs13061097 1

Nur Atirah Muhadi, Ahmad Fikri Abdullah, Siti Khairunniza Bejo, Muhammad Razif Mahadi and Ana Mijic
The Use of LiDAR-Derived DEM in Flood Applications: A Review
Reprinted from: *Remote Sens.* **2020**, *12*, 2308, doi:10.3390/rs12142308 5

Aifeng Lv, Zhilin Zhang and Hongchun Zhu
A Neural-Network Based Spatial Resolution Downscaling Method for Soil Moisture: Case Study of Qinghai Province
Reprinted from: *Remote Sens.* **2021**, *13*, 1583, doi:10.3390/rs13081583 25

Yuan Zhang, Xiaoming Feng, Bojie Fu, Yongzhe Chen and Xiaofeng Wang
Satellite-Observed Global Terrestrial Vegetation Production in Response to Water Availability
Reprinted from: *Remote Sens.* **2021**, *13*, 1289, doi:10.3390/rs13071289 47

Zhenjie Zhu, Bingjun Liu, Hailong Wang and Maochuan Hu
Analysis of the Spatiotemporal Changes in Watershed Landscape Pattern and Its Influencing Factors in Rapidly Urbanizing Areas Using Satellite Data
Reprinted from: *Remote Sens.* **2021**, *13*, 1168, doi:10.3390/rs13061168 67

Yuanyuan Zhou, Nianxiu Qin, Qiuhong Tang, Huabin Shi and Liang Gao
Assimilation of Multi-Source Precipitation Data over Southeast China Using a Nonparametric Framework
Reprinted from: *Remote Sens.* **2021**, *13*, 1057, doi:10.3390/rs13061057 93

Fangdi Sun, Ronghua Ma, Caixia Liu and Bin He
Comparison of the Hydrological Dynamics of Poyang Lake in the Wet and Dry Seasons
Reprinted from: *Remote Sens.* **2021**, *13*, 985, doi:10.3390/rs13050985 115

Jéssica G. Nascimento, Daniel Althoff, Helizani C. Bazame, Christopher M. U. Neale, Sergio N. Duarte, Anderson L. Ruhoff and Ivo Z. Gonçalves
Evaluating the Latest IMERG Products in a Subtropical Climate: The Case of Paraná State, Brazil
Reprinted from: *Remote Sens.* **2021**, *13*, 906, doi:10.3390/rs13050906 135

Fan Sun, Yi Wang, Yaning Chen, Yupeng Li, Qifei Zhang, Jingxiu Qin and Patient Mindje Kayumba
Historic and Simulated Desert-Oasis Ecotone Changes in the Arid Tarim River Basin, China
Reprinted from: *Remote Sens.* **2021**, *13*, 647, doi:10.3390/rs13040647 153

Xingxing Han, Wei Chen, Bo Ping and Yong Hu
Implementation of an Improved Water Change Tracking (IWCT) Algorithm: Monitoring the Water Changes in Tianjin over 1984–2019 Using Landsat Time-Series Data
Reprinted from: *Remote Sens.* **2021**, *13*, 493, doi:10.3390/rs13030493 169

Zonghan Ma, Bingfang Wu, Nana Yan, Weiwei Zhu, Hongwei Zeng and Jiaming Xu
Spatial Allocation Method from Coarse Evapotranspiration Data to Agricultural Fields by Quantifying Variations in Crop Cover and Soil Moisture
Reprinted from: *Remote Sens.* **2021**, *13*, 343, doi:10.3390/rs13030343 . . . . **187**

Xiongpeng Tang, Jianyun Zhang, Guoqing Wang, Gebdang Biangbalbe Ruben, Zhenxin Bao, Yanli Liu, Cuishan Liu and Junliang Jin
Error Correction of Multi-Source Weighted-Ensemble Precipitation (MSWEP) over the Lancang-Mekong River Basin
Reprinted from: *Remote Sens.* **2021**, *13*, 312, doi:10.3390/rs13020312 . . . . **205**

Jiachao Chen, Zhaoli Wang, Xushu Wu, Chengguang Lai and Xiaohong Chen
Evaluation of TMPA 3B42-V7 Product on Extreme Precipitation Estimates
Reprinted from: *Remote Sens.* **2021**, *13*, 209, doi:10.3390/rs13020209 . . . . **229**

Yonghua Zhu, Pingping Luo, Sheng Zhang and Biao Sun
Spatiotemporal Analysis of Hydrological Variations and Their Impacts on Vegetation in Semiarid Areas from Multiple Satellite Data
Reprinted from: *Remote Sens.* **2020**, *12*, 4177, doi:10.3390/rs12244177 . . . . **245**

Yanfen Yang, Jing Wu, Lei Bai and Bing Wang
Reliability of Gridded Precipitation Products in the Yellow River Basin, China
Reprinted from: *Remote Sens.* **2020**, *12*, 374, doi:10.3390/rs12030374 . . . . **265**

Xianghu Li, Zhen Li and Yaling Lin
Suitability of TRMM Products with Different Temporal Resolution (3-Hourly, Daily, and Monthly) for Rainfall Erosivity Estimation
Reprinted from: *Remote Sens.* **2020**, *12*, 3924, doi:10.3390/rs12233924 . . . . **289**

Fangdi Sun, Ronghua Ma, Bin He, Xiaoli Zhao, Yuchao Zeng, Siyi Zhang and Shilin Tang
Changing Patterns of Lakes on The Southern Tibetan Plateau Based on Multi-Source Satellite Data
Reprinted from: *Remote Sens.* **2020**, *12*, 3450, doi:10.3390/rs12203450 . . . . **311**

Michael Kalua, Anna M. Rallings, Lorenzo Booth, Josué Medellín-Azuara, Stefano Carpin and Joshua H. Viers
sUAS Remote Sensing of Vineyard Evapotranspiration Quantifies Spatiotemporal Uncertainty in Satellite-Borne ET Estimates
Reprinted from: *Remote Sens.* **2020**, *12*, 3251, doi:10.3390/rs12193251 . . . . **331**

Christian Schwatke, Denise Dettmering and Florian Seitz
Volume Variations of Small Inland Water Bodies from a Combination of Satellite Altimetry and Optical Imagery
Reprinted from: *Remote Sens.* **2020**, *12*, 1606, doi:10.3390/rs12101606 . . . . **349**

Wenqing Xu, Like Ning and Yong Luo
Applying Satellite Data Assimilation to Wind Simulation of Coastal Wind Farms in Guangdong, China
Reprinted from: *Remote Sens.* **2020**, *12*, 973, doi:10.3390/rs12060973 . . . . **381**

Jinyu Hui, Yiping Wu, Fubo Zhao, Xiaohui Lei, Pengcheng Sun, Shailesh Kumar Singh, Weihong Liao, Linjing Qiu and Jiguang Li
Parameter Optimization for Uncertainty Reduction and Simulation Improvement of Hydrological Modeling
Reprinted from: *Remote Sens.* **2020**, *12*, 4069, doi:10.3390/rs12244069 . . . . **409**

Jiabin Peng, Tie Liu, Yue Huang, Yunan Ling, Zhengyang Li, Anming Bao, Xi Chen, Alishir Kurban and Philippe De Maeyer
Satellite-Based Precipitation Datasets Evaluation Using Gauge Observation and Hydrological Modeling in a Typical Arid Land Watershed of Central Asia
Reprinted from: *Remote Sens.* **2021**, *13*, 221, doi:10.3390/rs13020221 . . . . **433**

Shi Hu and Xingguo Mo
Attribution of Long-Term Evapotranspiration Trends in the Mekong River Basin with a Remote Sensing-Based Process Model
Reprinted from: *Remote Sens.* **2021**, *13*, 303, doi:10.3390/rs13020303 . . . . **457**

T. A. Jeewanthi G. Sirisena, Shreedhar Maskey and Roshanka Ranasinghe
Hydrological Model Calibration with Streamflow and Remote Sensing Based Evapotranspiration Data in a Data Poor Basin
Reprinted from: *Remote Sens.* **2020**, *12*, 3768, doi:10.3390/rs12223768 . . . . **475**

About the Editors

Weili Duan

Weili Duan is a Professor in water resources and climate change at the State Key Laboratory of Desert and Oasis Ecology, Xinjiang Institute of Ecology and Geography, Chinese Academy of Sciences. Weili pursued a PhD degree at the Department of Civil and Earth Resources Engineering, Kyoto University, Japan, in September 2014. His research interests are water for sustainable development, climate change and water-related disasters management. He has obtained about 10 research projects including 2 Talent projects and 3 NSFC projects and published more than 70 papers and 1 monograph.

Shreedhar Maskey

Shreedhar Maskey, PhD, is currently an Associate Professor of Hydrology and Water Resources at IHE Delft Institute for Water Education, The Netherlands. He holds a BSc degree in civil engineering and MSc in hydroinformatics, and has obtained a PhD degree from Delft University of Technology. He has more than 25 years of experience ranging from post-graduate and graduate level teaching, research supervision, international projects on floods and droughts modelling and forecasting, climate change impacts, and water resources assessment. His hydrological modelling experience includes many major river basins, such as the Yellow River, Red River, Mekong, Irrawaddy, Ganges basins (Koshi, Bagmati, Narayani and West Rapti), Upper Indus, Karkheh, Aral Sea basins (Amu Darya and Syr Darya), Loire, Blue Nile, Limpopo, Niger, St. Paul River, and several other river catchments in Africa, Asia and Europe. He has co-authored over 50 peer-reviewed journal papers and book chapters and over 35 conference papers. He is also an associate editor of the Journal of Hydrology, and has served as associate editor of the Journal of Flood Risk Management.

Pedro Luiz Borges Chaffe

Pedro Luiz Borges Chaffe is an Associate Professor in Hydrology and Water Resource at the Federal University of Santa Catarina (UFSC), Brazil. He obtained his Bachelor (2007) and Masters (2009) in Environmental Engineering at UFSC, and PhD degree at Kyoto University (2012). Currently, he is the Supervisor of the Hydrology Lab at UFSC, hoping to understand the climatic and physiographic controls of streamflow from small to large catchments in Brazil.

Pingping Luo

Pingping Luo is a full professor in School of Water and Environment, Chang'an University. He graduated from Disaster Prevention Research Institute (DPRI), Kyoto University in 2012. He was a postdoc researcher in DPRI, Kyoto University and IAS, Unitied Nations University from 2012 to 2016. Prof. Luo was appointed as full professor in Chang'an University in 2016. He has published more than 60 papers and 5 book chapters. His major is focused on Hydrological modelling, Flood disaster, Water resource management and Environmental Engineering.

Bin He

Prof. Bin He is working in the Institute of Eco-environmental and Soil Sciences, Guangdong Academy of Sciences,China. He has more than 20 years experience in hydrological engineering and water pollution control. He holds the Ph.D. of "Hydrology for Environmental Engineering" from Ehime University in Japan, and then obtained the Research Fellow Award from Japan Society for the Promotion of Science (JSPS). After that, he worked as an Assistant Professor in the University of Tokyo (Japan), an Associate Professor in Kyoto University (Japan), and a Professor in Chinese Academy of Sciences. His main interests include GIS-based hydrological modeling for nonpoint source pollution (NPS), developing engineering techniques for water pollution control. He has published more than 100 papers in journals, 3 books and 30 patents in environmental engineering. He is the leader of more 30 domestic and international projects, an Associate Editor-in-Chief of Journal "Geoenvironmental Disasters", and an Editor of the journals "Landslides", "Hydrological Research Letter".

Yiping Wu

Dr. Yiping Wu is a professor in the Institute of Global Environmental Change, Xi'an Jiaotong University and engaged in watershed eco-hydrological study. He received his B.E. degree from Xi'an University of Architecture and Technology and Ph.D. degree from The University of Hong Kong, and served as a senior scientist at ASRC Federal, USGS EROS. He gained the talent plans of the National Youth Talent Program and Shaanxi Talent Program. He also serves the Education Instruction Committee of Nature Conservation and Environmental Ecology as a member, Agricultural and Forestry Branch of Education Instruction Committee of Shaanxi Province as deputy director, Asia Oceania Geosciences Society (AOGS) as a Secretary in Hydrological Section, and of China Ecological Society as a committee member in Eco-hydrological Section. He acts as an associate editor or editorial board member in the international renowned SCI journals including Engineering, Stochastic Environmental Research and Risk Assessment, Carbon Balance and Management, and Geoscience Letters.

Jingming Hou

Prof. Jingming Hou serves as the Executive Deputy Director of State Key Laboratory of Eco-hydraulics in Northwest Arid Region of China, Xi'an University of Technology. He graduated from the Technical University of Berlin, Germany, and won the Tiburtius Outstanding Doctoral Thesis Award. In 2016, he worked in Xi'an University of Technology and was selected into the Shaanxi Hundred Talents Program, and selected into the national talent plan in 2018. In 2020, he was elected as the vice chairman of the Urban Water Conservancy Special Committee of the Chinese Hydraulic Engineering Society. His research areas cover the fields in theoretical development and practical application of mathematical and numerical models of surface water flows and associated transport processes, Urban and catchment flood management, High resolution numerical modeling of coupled hydrodynamic and pollutant transport processes and application of deep learning techniques in urban flood inundation rapid forecasting.

Preface to "Remote Sensing in Hydrology and Water Resources Management"

In the last few decades, remote sensing (RS) technology has developed rapidly, which provides a means of observing hydrological and hydraulic state variables including precipitation, temperature, soil moisture, water levels, evapotranspiration, flood extent, flow velocity, river discharge, and land water storage over regional/global areas. All these variables could be the input files for integrated hydrodynamics or hydrological or hydrometeorological models to simulate and assess water resources and water-related issues, contributing to fully understand global- and regional-scale hydrological processes under climate change and human activities, which could be useful for improving sustainable water management. The objective of this book is to present reviews and recent advances of general interest that make use of remote sensing techniques in hydrology and water resources management. In general, remote sensing technology can improve land-surface and hydrologic modeling from three aspects, including model inputs (watershed information, atmospheric information, boundary conditions, etc.), state estimation (data assimilation), and model calibration and parameter estimation. This book aspires to stimulate further research into the remote sensing technology of hydrology analysis and water resources management.

This book is jointly supported by the National Natural Science Foundation of China (Grant No. 42122004), the Project of Tianshan Innovation Team in Xinjiang (202113050), and the Program for Foreign High-Level Talents Introduction in Xinjiang Uygur Autonomous Region (Y941181).

Weili Duan, Shreedhar Maskey, Pedro Luiz Borges Chaffe, Pingping Luo, Bin He, Yiping Wu, Jingming Hou

Editors

Editorial

Recent Advancement in Remote Sensing Technology for Hydrology Analysis and Water Resources Management

Weili Duan [1,2,*], Shreedhar Maskey [3], Pedro L. B. Chaffe [4], Pingping Luo [5,6], Bin He [7,8], Yiping Wu [9] and Jingming Hou [10]

1. State Key Laboratory of Desert and Oasis Ecology, Xinjiang Institute of Ecology and Geography, Chinese Academy of Sciences, Urumqi 830011, China
2. College of Resources and Environment, University of Chinese Academy of Sciences, Beijing 100049, China
3. Department of Water Resources and Ecosystems, IHE Delft Institute for Water Education, 2611 AX Delft, The Netherlands; s.maskey@un-ihe.org
4. Key Laboratory of Subsurface Hydrology and Ecological Effects in Arid Region, Ministry of Education, Chang'an University, Xi'an 710054, China; pedro.chaffe@ufsc.br
5. Department of Sanitary and Environmental Engineering, Federal University of Santa Catarina, Florianópolis 88040-900, Brazil; lpp@chd.edu.cn
6. School of Water and Environment, Chang'an University, Xi'an 710054, China
7. Guangdong Key Laboratory of Integrated Agro-Environmental Pollution Control and Management, Institute of Eco-Environmental and Soil Sciences, Guangdong Academy of Sciences, Guangzhou 510650, China; bhe@soil.gd.cn
8. National-Regional Joint Engineering Research Center for Soil Pollution Control and Remediation in South China, Guangzhou 510650, China
9. Department of Earth and Environmental Science, School of Human Settlements and Civil Engineering, Xi'an Jiaotong University, Xi'an 710049, China; yipingwu@xjtu.edu.cn
10. State Key Laboratory of Eco-Hydraulics in Northwest Arid Region of China, Xi'an University of Technology, Xi'an 710048, China; jingming.hou@xaut.edu.cn
* Correspondence: duanweili@ms.xjb.ac.cn

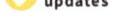

Citation: Duan, W.; Maskey, S.; Chaffe, P.L.B.; Luo, P.; He, B.; Wu, Y.; Hou, J. Recent Advancement in Remote Sensing Technology for Hydrology Analysis and Water Resources Management. *Remote Sens.* **2021**, *13*, 1097. https://doi.org/10.3390/rs13061097

Received: 1 March 2021
Accepted: 11 March 2021
Published: 13 March 2021

Publisher's Note: MDPI stays neutral with regard to jurisdictional claims in published maps and institutional affiliations.

Copyright: © 2021 by the authors. Licensee MDPI, Basel, Switzerland. This article is an open access article distributed under the terms and conditions of the Creative Commons Attribution (CC BY) license (https://creativecommons.org/licenses/by/4.0/).

1. Introduction

Water is undoubtedly the most valuable resource of human society and an essential component of the ecosystem. Under climate change and human activities, water resources management for sustainable socio-economic development presents many challenges around the world, especially in arid regions and high-altitude regions, with sparse in situ hydro-meteorological monitoring networks. Rare and uncertain hydrological information generally causes uncertainty in hydrologic modeling and eventually impedes continuous water resources management.

In recent decades, remote sensing (RS) technology has been developed rapidly to obtain sufficient information on hydrological state variables including precipitation, temperature, soil moisture, water levels, evapotranspiration, flood extent, flow velocity, river discharge, and land water storage over regional/global areas, especially in those remote regions where measurements are not feasible or can only be carried out under very difficult circumstances with high costs. This remote sensing information could be the input files for integrated hydrodynamics or hydrological or hydrometeorological models to simulate and assess water resources and water-related issues, largely supporting the development of more efficient hydrological models and water resources management.

Therefore, in order to fully understand recent advancements in remote sensing technology for hydrology analysis and water resources management, this special issue hosts 18 papers devoted to remote sensing in hydrology and water resources management. The volume includes studies on satellite remote sensing for water resources management, water quality monitoring and evaluation using remote-sensing data, remote sensing for detecting the global impact of climate extremes, the use of remote sensing data for improved calibration of hydrological models, and so on.

The following section concludes the individual articles hosted in this Special Issue in alphabetical order according to the first author's name.

2. Overview of Contributions

Based on the fuzzy C-means algorithm and L-moment-based regional frequency analysis method, Chen et al. [1] evaluated the performance of 3B42-V7 satellite-based precipitation product on extreme precipitation estimates compared with the China Gauge-based Daily Precipitation Analysis (CGDPA) product in China.

An improved water change tracking (IWCT) algorithm was proposed by Han et al. [2], which could remove built-up shade noise and correct omitted water pixels by taking the time-series data into consideration and was applied to evaluate water changes in Tianjin during the period 1984–2019.

Using the Global Land Surface Satellite (GLASS) leaf area index (LAI), Hu and Mo [3] used a remote sensing-based process to model spatial-temporal patterns of the actual evapotranspiration (ETa) and available water resources in the Mekong River Basin from 1981 to 2012.

In order to reduce and quantify parameter uncertainty in hydrological simulations, Hui et al. [4] attempted to introduce additional remotely sensed data (such as evapotranspiration (ET)) into the Soil Water Assessment Tool in the Guijiang River Basin (GRB) in China and found that the simulation accuracy of ET was substantially improved when adding remotely sensed ET data.

Kulua et al.'s [5] study compares measurements of evapotranspiration (ET) from a commercial vineyard in California using data collected from the Small Unmanned Aerial Systems (sUAS) and the Earth Observation System (EOS) sources for 10 events over a growing season using multiple ET estimation methods. This study indicates that limited deployment of sUAS can provide important estimates of uncertainty in EOS ET estimations for larger areas and also improve irrigation management at a local scale.

Based on the Tropical Rainfall Measuring Mission (TRMM) 3B42 3-hourly, daily, and 3B43 monthly rainfall data, the rainfall erosivity (RE) was quantified by Li et al. [6]. They found that all three TRMM rainfall products can generally capture the overall spatial patterns of RE and could be used to assess the risk of soil erosion.

By developing an allocation factor, Ma et al. [7] proposed a new field-scale ET estimation method to quantitatively evaluate field-level ET variations and allocate coarse ET to the field scale. Results from the new method can fully meet the demands of wide application for controlling regional water consumption, which is beneficial to effective management and control of water resources.

Muhadi et al. [8] reviewed the potential and the applications of light detection and ranging (LiDAR) technology in flood studies, and pointed out that LiDAR-derived data are very useful in flood mapping and risk assessment, especially in the future assessment of flood-related problems.

In developing countries, generally, water resources management is highly restricted because of the lack of high-precision measurement of precipitation in large areas. Nowadays, the Integrated Multi-satellite Retrievals for GPM (IMERG) can offer a new source of precipitation data with high spatial and temporal resolution. Therefore, Nascimento et al. [9] evaluated the performance of the GPM products in the state of Paraná, Brazil, from June 2000 to December 2018.

Peng et al. [10] evaluated the performances of six gauge-adjusted-version satellite precipitation datasets in the Bosten Lake Basin, a typical arid land watershed of Central Asia, which provides a reference for the hydrological and meteorological application of satellite precipitation datasets in Central Asia with sparse in situ hydro-meteorological monitoring networks.

Based on water levels from satellite altimetry and surface areas from optical imagery, Schwatke et al. [11] developed a new approach for estimating water volume variations of lakes and reservoirs. The method was applied to investigate volume changes in 28 lakes and reservoirs located in Texas.

The study by Sirisena et al. [12] evaluated the use of measured streamflow and RS-based ET data to calibrate a Soil and Water Assessment Tool (SWAT) and evaluate the performances for Chindwin Basin, Myanmar. The results indicated the advantage of remote-sensing-based and multiple data sources for calibration of hydrological modelling in data poor basins.

By interpreting satellite imagery from 1990, 2000, and 2015, Sun et al. [13] investigated the dynamic evolution of the desert-oasis ecotone in the Tarim River Basin and then predicted the near-future land-use change based on the cellular automata-Markov (CA-Markov) model.

After collecting high frequency (moderate-resolution imaging spectroradiometer) MODIS images, altimetry, and data from the Hydroweb database, Sun et al. [14] delineated the detailed hydrological changes of 15 lakes in three basins—Inner Basin, Indus Basin, and Brahmaputra Basin—on the southern Tibetan Plateau.

Because remote sensing and reanalysis quantitative precipitation products are inevitably subject to errors, Tang et al. [15] proposed a novel daily-scale precipitation bias correction framework to combine these precipitation products from various institutions. The framework was applied to do error correction for multi-source weighted-ensemble precipitation in the Lancang-Mekong River Basin.

Through combining the Weather Research and Forecast (WRF) model with the three-dimensional variation (3DVar) data assimilation system, Xu et al. [16] put the satellite data assimilation to wind speed simulation in wind resource assessments in Guangdong, China.

In the paper by Yang et al. [17], five gridded precipitation products including Multi-Source Weighted-Ensemble Precipitation (MSWEP), CPC Morphing Technique (CMORPH), Global Satellite Mapping of Precipitation (GSMaP), Tropical Rainfall Measuring Mission (TRMM) Multi-Satellite Precipitation Analysis 3B42, and Precipitation Estimation from Remotely Sensed Information using Artificial Neural Networks (PERSIANN) were evaluated against observations in the Yellow River Basin, China, at daily, monthly, and annual scales during 2001–2014.

Using multiple satellite data, Zhou et al [18] analyzed the spatiotemporal changes of hydrological elements in semiarid areas from 2002 to 2014 and their effects on vegetation, which demonstrated that the application of satellite data could significantly improve the water assessment capability in semiarid areas.

3. Conclusions

This Special Issue aimed to summarize the recent advancement in remote sensing technology for hydrology analysis and water resources management. In general, remote sensing technology can improve land-surface and hydrologic modeling from three aspects including model inputs (watershed information, atmospheric information, boundary conditions, etc.), state estimation (data assimilation), and model calibration and parameter estimation. This special collection of papers aspires to stimulate further research into the remote sensing technology of hydrology analysis and water resources management.

Funding: This research was funded by the National Natural Science Foundation of China (Grant No. 41971149), the Tianshan Xuesong Program (Grant No. 2019XS10), and the Program for High-Level Talents Introduction in Xinjiang Uygur Autonomous Region (Grant No. Y941181).

Acknowledgments: As the Guest Editor of this Special Issue entitled "Remote Sensing in Hydrology and Water Resources Management", we would like to thank all the authors who contributed towards this volume. We are also thankful to the reviewers of the submitted manuscripts who added value to the volume by providing timely and thorough reviews with comments and very constructive feedback to the authors. Last, but not least, we wish to express our gratitude to the editorial staff of Remote Sensing for their collaboration and prompt efforts in completing this task.

Conflicts of Interest: The authors declare no conflict of interest.

References

1. Chen, J.; Wang, Z.; Wu, X.; Lai, C.; Chen, X. Evaluation of TMPA 3B42-V7 Product on Extreme Precipitation Estimates. *Remote Sens.* **2021**, *13*, 209. [CrossRef]
2. Han, X.; Chen, W.; Ping, B.; Hu, Y. Implementation of an Improved Water Change Tracking (IWCT) Algorithm: Monitoring the Water Changes in Tianjin over 1984–2019 Using Landsat Time-Series Data. *Remote Sens.* **2021**, *13*, 493. [CrossRef]
3. Hu, S.; Mo, X. Attribution of Long-Term Evapotranspiration Trends in the Mekong River Basin with a Remote Sensing-Based Process Model. *Remote Sens.* **2021**, *13*, 303. [CrossRef]
4. Hui, J.; Wu, Y.; Zhao, F.; Lei, X.; Sun, P.; Singh, S.K.; Liao, W.; Qiu, L.; Li, J. Parameter Optimization for Uncertainty Reduction and Simulation Improvement of Hydrological Modeling. *Remote Sensing*. **2020**, *12*, 4069. [CrossRef]
5. Kalua, M.; Rallings, A.M.; Booth, L.; Medellín-Azuara, J.; Carpin, S.; Viers, J.H. sUAS Remote Sensing of Vineyard Evapotranspiration Quantifies Spatiotemporal Uncertainty in Satellite-Borne ET Estimates. *Remote Sens.* **2020**, *12*, 3251. [CrossRef]
6. Li, X.; Li, Z.; Lin, Y. Suitability of TRMM Products with Different Temporal Resolution (3-Hourly, Daily, and Monthly) for Rainfall Erosivity Estimation. *Remote Sens.* **2020**, *12*, 3924. [CrossRef]
7. Ma, Z.; Wu, B.; Yan, N.; Zhu, W.; Zeng, H.; Xu, J. Spatial Allocation Method from Coarse Evapotranspiration Data to Agricultural Fields by Quantifying Variations in Crop Cover and Soil Moisture. *Remote Sens.* **2021**, *13*, 343. [CrossRef]
8. Muhadi, N.A.; Abdullah, A.F.; Bejo, S.K.; Mahadi, M.R.; Mijic, A. The Use of LiDAR-Derived DEM in Flood Applications: A Review. *Remote Sens.* **2020**, *12*, 2308. [CrossRef]
9. Nascimento, J.G.; Althoff, D.; Bazame, H.C.; Neale, C.M.U.; Duarte, S.N.; Ruhoff, A.L.; Gonçalves, I.Z. Evaluating the Latest IMERG Products in a Subtropical Climate: The Case of Paraná State, Brazil. *Remote Sens.* **2021**, *13*, 906. [CrossRef]
10. Peng, J.; Liu, T.; Huang, Y.; Ling, Y.; Li, Z.; Bao, A.; Chen, X.; Kurban, A.; De Maeyer, P. Satellite-Based Precipitation Datasets Evaluation Using Gauge Observation and Hydrological Modeling in a Typical Arid Land Watershed of Central Asia. *Remote Sens.* **2021**, *13*, 221. [CrossRef]
11. Schwatke, C.; Dettmering, D.; Seitz, F. Volume Variations of Small Inland Water Bodies from a Combination of Satellite Altimetry and Optical Imagery. *Remote Sens.* **2020**, *12*, 1606. [CrossRef]
12. Sirisena, T.A.; Maskey, S.; Ranasinghe, R. Hydrological Model Calibration with Streamflow and Remote Sensing Based Evapotranspiration Data in a Data Poor Basin. *Remote Sens.* **2020**, *12*, 3768. [CrossRef]
13. Sun, F.; Wang, Y.; Chen, Y.; Li, Y.; Zhang, Q.; Qin, J.; Kayumba, P.M. Historic and Simulated Desert-Oasis Ecotone Changes in the Arid Tarim River Basin, China. *Remote Sens.* **2021**, *13*, 647. [CrossRef]
14. Sun, F.; Ma, R.; He, B.; Zhao, X.; Zeng, Y.; Zhang, S.; Tang, S. Changing Patterns of Lakes on The Southern Tibetan Plateau Based on Multi-Source Satellite Data. *Remote Sens.* **2020**, *12*, 3450. [CrossRef]
15. Tang, X.; Zhang, J.; Wang, G.; Ruben, G.B.; Bao, Z.; Liu, Y.; Liu, C.; Jin, J. Error Correction of Multi-Source Weighted-Ensemble Precipitation (MSWEP) over the Lancang-Mekong River Basin. *Remote Sens.* **2021**, *13*, 312. [CrossRef]
16. Xu, W.; Ning, L.; Luo, Y. Applying Satellite Data Assimilation to Wind Simulation of Coastal Wind Farms in Guangdong, China. *Remote Sens.* **2020**, *12*, 973. [CrossRef]
17. Yang, Y.; Wu, J.; Bai, L.; Wang, B. Reliability of gridded precipitation products in the Yellow River Basin, China. *Remote Sens.* **2020**, *12*, 374. [CrossRef]
18. Zhu, Y.; Luo, P.; Zhang, S.; Sun, B. Spatiotemporal Analysis of Hydrological Variations and Their Impacts on Vegetation in Semiarid Areas from Multiple Satellite Data. *Remote Sens.* **2020**, *12*, 4177. [CrossRef]

Review

The Use of LiDAR-Derived DEM in Flood Applications: A Review

Nur Atirah Muhadi [1,*], Ahmad Fikri Abdullah [1,2], Siti Khairunniza Bejo [1], Muhammad Razif Mahadi [1] and Ana Mijic [3]

[1] Department of Biological and Agricultural Engineering, Faculty of Engineering, Universiti Putra Malaysia, Serdang 43400, Selangor, Malaysia; ahmadfikri@upm.edu.my (A.F.A.); skbejo@upm.edu.my (S.K.B.); razifman@upm.edu.my (M.R.M.)
[2] International Institute of Aquaculture and Aquatic Sciences, Batu 7, Jalan Kemang 6, Teluk Kemang, Si Rusa 71050, Port Dickson, Negeri Sembilan, Malaysia
[3] Department of Civil and Environmental Engineering, Skempton Building, Imperial College London, South Kensington Campus, London SW7 2AZ, UK; ana.mijic@imperial.ac.uk
* Correspondence: gs53332@student.upm.edu.my; Tel.: +603-97694337

Received: 29 April 2020; Accepted: 21 June 2020; Published: 18 July 2020

Abstract: Flood occurrence is increasing due to escalated urbanization and extreme climate change; hence, various studies on this issue and methods of flood monitoring and mapping are also increasing to reduce the severe impacts of flood disasters. The advancement of current technologies such as light detection and ranging (LiDAR) systems facilitated and improved flood applications. In a LiDAR system, a laser emits light that travels to the ground and reflects off objects like buildings and trees. The reflected light energy returns to the sensor, whereby the time interval is recorded. Since the conventional methods cannot produce high-resolution digital elevation model (DEM) data, which results in low accuracy of flood simulation results, LiDAR data are extensively used as an alternative. This review aims to study the potential and the applications of LiDAR-derived DEM in flood studies. It also provides insight into the operating principles of different LiDAR systems, system components, and advantages and disadvantages of each system. This paper discusses several topics relevant to flood studies from a LiDAR-derived DEM perspective. Furthermore, the challenges and future perspectives regarding DEM LiDAR data for flood mapping and assessment are also reviewed. This study demonstrates that LiDAR-derived data are useful in flood risk management, especially in the future assessment of flood-related problems.

Keywords: airborne LiDAR; DEM; flood inundation; flood map; flood model; LiDAR; terrestrial LiDAR

1. Introduction

Floods are a major severe natural catastrophe experienced in many countries around the world including Malaysia. In Malaysia, flooding is the most frequent danger among all disasters, and it can be considered as an annual disaster due to its consistent occurrence [1,2]. The flood issue is gaining attention globally with significant efforts made to develop effective flood prevention and monitoring solutions. Preparation of flood hazard and floodplain maps is one of the examples of the preparedness phase in a disaster management cycle, which is widely used to reduce the impact of disasters, to react during the event, and to take action to recover after a disaster occurs, including flood disasters [3].

Information on how far the floodwater inundates and how deep the area is flooded at what velocity is required in floodplain management and flood damage estimation. To obtain such information, elevation data that represent the earth's surface represent one of the primary components for flood studies. Accurate elevation information is crucial to both the input and the output of flood hydraulic analysis, as well as to producing floodplain maps [4]. A flood hydraulic model requires several input parameters that can be derived from the digital elevation data. The output of the flood model simulation is then mapped onto a digital elevation surface to determine the flood hazard zone and to further analyze the products to estimate probable flood damage in terms of flood inundation and flood depth.

A digital elevation model (DEM) is a predominant source of elevation data due to its simplicity and easy-to-use data [4,5]. A DEM provides gridded elevation data in a raster structure that represents the terrain's surface. It contains x-, y-, and z-values, which represent x- and y-coordinates and elevation information, respectively. A DEM is commonly generated by extracting surface features from a digital surface model (DSM). DEMs can be generated from many sources such as ground surveys, digitizing existing hardcopy topographic maps, or remotely sensed technology. DEM sources range from no cost to high-cost data sources depending on their accuracy. With the rapid development of remote sensing technology, DEMs generated from this technology are a preferred choice using photogrammetry, interferometric synthetic aperture radar (IfSAR), or light detection and ranging (LiDAR). Photogrammetry is the science of measuring features without physical contact with the features from photographs [6]. Aerial photographs are widely adopted due to their ability to provide high-resolution and high-accuracy DEMs. However, this emerging technology can only be acquired during cloud-free and low-haze conditions, which is not the finest condition during flood events.

On the other hand, IfSAR uses a microwave sensor to send signals to the earth's surface and records the scattered signal from the surface. It is an active microwave radar system that can obtain imagery over a vast area at night or in cloud cover. Spaceborne IfSAR such as shuttle radar topography mission (SRTM) is the most commonly used global DEM because it is an open-access DEM with acceptable resolution and accuracy. In contrast to spaceborne IfSAR, airborne IfSAR systems have more flexible system deployment and provide higher spatial resolution. However, both spaceborne and airborne IfSAR have limitations in urban areas due to complex scattering environments [7]. Furthermore, using this technology in densely vegetated areas is challenging because the radar cannot penetrate the ground surface beneath vegetation canopy.

Another emerging technology is LiDAR. A LiDAR sensor that is mounted on platforms such as aircraft and helicopters is known as airborne LiDAR. Meanwhile, LiDAR systems that collect data from the ground are referred to as ground-based LiDAR or terrestrial LiDAR. The generation of DEM data using the LiDAR system has several advantages over other sources. LiDAR data can be acquired during daylight or night time, as well as during cloudy conditions [7,8]. Moreover, it has the ability to penetrate the ground surface in vegetated and urban areas more reliably than either photogrammetry or IfSAR. Due to these reasons, it became a recent solution for flood-related problems. The ability of LiDAR systems to provide higher-resolution and centimeter-accuracy outcomes diversified their use in wide-ranging applications of flood studies.

In studies involving the application of remote sensing in floodplain and flood risk assessment, DEMs are used to visualize the interface of floodwater with the elevation of the ground surface. Moreover, a DEM is an important indicator in determining the flood inundation and flood depth [4,9,10]. The accuracy of a DEM is critical in hydrological modeling as it can affect the discharge values, water depth, and the extent of flood inundation maps [11,12]. In a flat floodplain, a vertical error of 1 m in the DEM leads to an error of 100 km^2 in the estimated flood inundation [10]. Hence, accurate and high-resolution DEM data are needed to produce reliable flood mapping, especially in the context of flood simulation modeling.

In Malaysia, the Department of Irrigation and Drainage (DID) is responsible for providing flood forecasts and flood hazard maps. A higher level of accuracy of the DEM such as LiDAR data will improve the accuracy and reliability of the flood maps. Hence, the DID provides LiDAR-derived DEM as the backbone of the hydrological model. Nevertheless, the existing LiDAR data coverage is minimal; thus, IFSAR data are used to cover the rest of the potentially flooded area. Furthermore, LiDAR-derived

DEM was also applied for flood risk assessments in some regions of the Philippines, as well as using IFSAR and SAR DEM as other alternatives in the absence of LiDAR in certain areas [13,14].

In flood inundation modeling, DEM resolution and accuracy play an important role in terms of modeling resolution and accuracy. For instance, a low-resolution DEM allows quick model simulations but it simplifies the topographic information that may affect the flood propagation. High-resolution DEM is needed, especially for urban areas, due to the presence of small features such as road curbs and dykes; thus, it is likely that the accuracy of flood simulations can be affected by the resolution of the DEM. Therefore, many researchers carried out studies to see whether coarser DEM resolution decreases the accuracy of the predicted flood inundation extent.

Tamiru and Rientjes [15] investigated the effects of LiDAR-derived DEM resolution in flood modeling. In this study, several DEMs with different resolutions were used as the input to the flood model. The authors concluded that the DEM resolution has a significant effect on simulation results. The affected flood simulation characteristics are inundation extent, flow depth, and flood velocity. In short, a coarser-resolution DEM results in a larger loss of information, while a high-resolution DEM results in excessive computational time. Furthermore, Casas et al. [16] conducted a study on different topographic data sources and resolution on flood modeling. The differences between each DEM were measured in terms of flood model outputs such as water discharge, water level, and flood inundation. The authors emphasized that the flood modeling results are majorly dependent on the DEM accuracy, whereby a LiDAR-derived DEM has the least root-mean-square error (RMSE) in terms of elevation accuracy and estimated flood inundation.

Vaze et al. [17] carried out a study on the accuracy and resolution of LiDAR DEM to improve the quality of hydrological features extracted from DEMs. The authors also investigated the effect of re-sampling DEM data into coarser resolution. The results obtained demonstrate that the accuracy and resolution of input DEM have a significant impact on the values of the hydrologically important spatial indices derived from the DEM. Hsu et al. [18] conducted a case study on the influence of DEM resolutions on the simulation of flood inundation in Tainan City, Taiwan. Five different grid sizes of DEM from 1×1 m to 40×40 m were used as the input of the flood models. The results showed that coarser DEM may cause losses of important small-scale features. Therefore, the inundation area may increase with a coarser DEM, resulting in a reduction in the accuracy of flood inundation models.

Ozdemir et al. [19] investigated the impact of using different high-resolution terrestrial LiDAR data on water depth, inundation extent, arrival time, and velocity predicted by the flood simulations. It was found that increasing the terrain resolution significantly affected the flood simulation results. The finding demonstrated that fine-resolution DEM can lead to significant differences in the dynamics of flood inundation. On the other hand, a coarser resolution reduced the performance of flood inundation prediction due to changes in flow paths at coarser resolution caused by losses of feature representation [20]. Furthermore, de Almeida et al. [21] investigated the influence of fine-resolution DEM on the flood inundation model over urban areas. The authors performed four different scenarios with small-scale modifications to analyze the influence of the decametric-scale changes. The findings from this study confirmed that flood hazard prediction was sensitive to decimetric-scale features, and they had an impact on the dynamic and distribution of flooded areas.

In summary, the findings from previous studies confirmed that accurate terrain data had a big impact on flood hazard prediction. Results of flood simulations varied in response to different DEM resolutions, which could be associated with the degree of topography representation. It was found that high-resolution DEM can provide relevant and reliable flood modeling results [22]. In contrast, coarser resolutions deteriorate the performance of flood models [20]. The previous studies confirmed that accurate terrain data have a big impact on flood hazard prediction and that the inundation area evaluation increases with coarser DEMs. Hence, a finer model resolution is necessary if the decision-maker is interested in local-scale inundation predictions [23]. Based on previous studies, the researchers suggested that LiDAR data offer high-quality data as an essential input of flood modeling.

Therefore, this review highlights the basic principles of LiDAR systems, system components, and their applications in flood studies. It also presents a brief discussion on the number of papers published in the Scopus database that focused on LiDAR data in flood studies over the past 10 years. Moreover, this paper reviews the challenges and future directions of the technology when using LiDAR data for flood modeling.

2. Principles of LiDAR Systems

LiDAR is an active remote sensing system, which allows the system to operate during the day and at night. LiDAR systems are used in various applications, and their advantages are well noted by researchers and practitioners all around the world. The development of laser scanning differs by the position of a sensor, i.e., whether an airborne-based LiDAR system or a ground-based LiDAR system. These two systems vary in terms of data acquisition modes, scanning mechanisms, and product accuracy and resolution, with several similarities. One similarity is that both systems can capture point cloud data and simultaneously acquire imagery [24].

2.1. Airborne LiDAR: Fundamentals and System Components

An airborne LiDAR is a multi-sensor system [25] that consists of several components which are the platform, laser scanner, positioning hardware, photographic or video recording equipment, computer, and data storage. For airborne LiDAR, the platform to mount the laser scanner can either be a fixed-wing aircraft or a helicopter, which is used to fly the laser sensor over a region of interest.

A LiDAR sensor with a wavelength of 1000–1600 nm emits laser pulses toward the ground, and the signal is backscattered by different objects, such as man-made structures, vegetation, and the ground surface [26]. The reflected light energy returns to the sensor, whereby the sensor logs the returning signal. The time of travel of the return pulse is used to measure the distance traveled. The measurements of distance and orientation are done by utilizing positioning systems, including a global positioning system (GPS) and inertial measurement unit (IMU).

LiDAR can produce high-resolution and high accuracy data by relying on the accuracy of GPS and IMU components [27]. IMUs are used to measure the accurate position, trajectory, and orientation of the aircraft. Meanwhile, the purpose of the GPS is to identify the X, Y, and Z location. The GPS is responsible for providing the precise location of the sensor; hence, differential GPS is adopted by setting up a ground GPS station to achieve a required position accuracy of better than 10 cm in the airborne LiDAR [28,29].

The camera or video recording equipment flies along with the LiDAR sensor to provide color information to represent the real-world color. The process is carried out by mapping red, green, and blue values onto the georeferenced point location [30]. Other components in the airborne LiDAR system are the control and data recording unit and onboard computer. The control and data recording unit stores raw data collected by the scanner, IMU, and GPS. Laser scanners can produce about 20 gigabytes of ranging data per hour as compared to the summation of GPS and IMU data, which only produce about 0.1 gigabytes per hour [31].

2.2. Terrestrial LiDAR: Fundamentals and System Components

Terrestrial LiDAR, also known as terrestrial laser scanning, is a ground-based version of the airborne LiDAR, which is frequently used for terrain and topographic mapping. Terrestrial LiDAR includes stationary laser scanning, whereby the sensor is mounted on a tripod for fixed positions and mobile laser scanning, while the sensor is mounted on a mobile ground-based platform such as a vehicle. The term terrestrial LiDAR usually refers to static laser scanning. Because static and mobile laser scanning differ in terms of components and mechanisms, these two categories are discussed separately in this section.

Nevertheless, both systems still have several similarities. For instance, the main component of both terrestrial LiDAR systems is a laser scanner. Lasers with a wavelength of 500–600 nm are typically used in ground-based LiDAR systems [26]. Furthermore, terrestrial LiDAR, just like airborne LiDAR systems, utilizes an integrated digital camera or video recording, which is responsible for colorizing point clouds and three-dimensional (3D) models to represent color in the real world.

2.2.1. Static Laser Scanner

Static laser scanning is performed from the top of a fixed surveying tripod. Static terrestrial laser scanning needs a two-dimensional (2D) scanning pattern to complete a scan survey, which is why static terrestrial LiDAR integrates with one or two mirrors that can change the direction of laser pulses [32]. Therefore, this LiDAR system can scan and measure the distances of the surrounding objects. Terrestrial LiDAR systems identify the range between the sensor and the targets by measuring the time required for the laser pulse to travel to the target and return to the sensor. The basic components for this system include the ranging unit, scanning mechanisms, laser controller, data recorder, and (optionally) a digital camera.

Theoretically, the laser scanner operates by emitting an infrared laser beam to the center of a rotating mirror, which deflects the laser beam around the scanning area. Once the scattered light hits the objects, it reflects onto the scanner. The digital camera can be mounted on the scanner rotating axis to provide images of the surroundings [33] The recorded time it takes is divided by two and multiplied by the speed of light to get the distance. The coverage of terrestrial laser scanning usually ranges from 100 m to 300 m [24]. Because static terrestrial LiDAR scans are provided from a stable position and orientation, point clouds with good geometric quality are obtained [31].

2.2.2. Mobile Laser Scanner

On the other hand, mobile laser scanning has similar data collection modes to airborne LiDAR. Showing many similarities with airborne LiDAR, mobile laser scanning requires only one scanning (1D) direction, whereas the other is performed by the moving platform. In mobile laser scanning systems, the laser scanner is mounted on a moving vehicle such as a car or van. Due to the continuous motion of the scanner, positioning systems based on GPS and IMU technologies are required to precisely measure the respective positions and orientations. The systems perform as the vehicle moves around, while the positioning systems track the trajectory and attitude of the vehicle for producing a 3D point cloud from the range of data collected. Figure 1 illustrates the operating principles of all types of laser scanning.

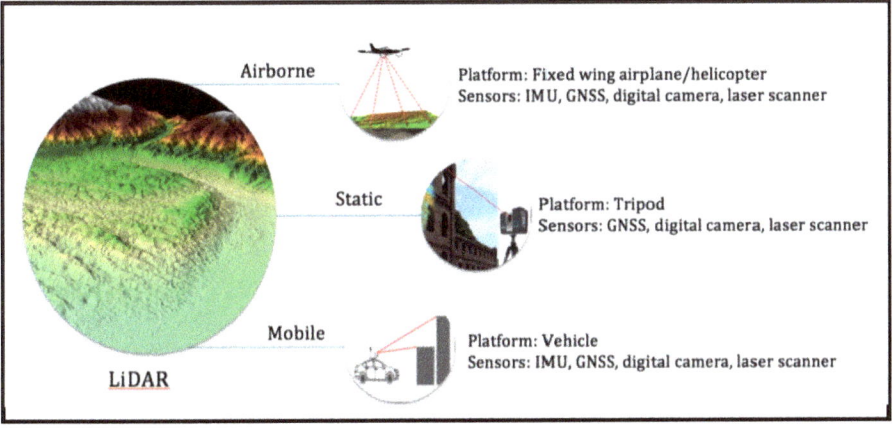

Figure 1. Light detection and ranging (LiDAR) survey data acquisition.

2.3. Advantages and Disadvantages of Airborne and Terrestrial LiDAR

Both LiDAR systems have their advantages and drawbacks. Airborne LiDAR data offer rapid data acquisition capability and a high degree of automation. Data are captured with a speed of up to about 50 km^2/h [34]. Due to this, it is considered to be a fast method of generating accurate DEM. Furthermore, because airborne LiDAR captures data from above, it gives a direct and clearer view of roads and rooftops of buildings as compared to terrestrial LiDAR.

In contrast, terrestrial LiDAR is preferable compared to airborne LiDAR in certain situations because terrestrial LiDAR is more cost-effective for small-scale areas and can be portable, while it produces high-resolution of terrain data. The main advantages of terrestrial LiDAR data are the high measurement density and high data accuracy. It can collect higher point density, typically 100 points/m^2 [35]. It can also provide scan rates up to half a million points per second for 100 m to 300 m, depending on the distance range of the scanner [31]; thus, it provides detailed terrain description and high-resolution surface roughness [15,36–39]. Terrestrial LiDAR yields high-resolution digital elevation models (DEMs) with pixel sizes on the scale of centimeters rather than the 1–3-m-resolution DEMs derived from airborne LiDAR [40].

Even though terrestrial LiDAR data can be used to provide information in small-scale areas, it cannot provide data in certain areas such as private lands and steep slope areas. Therefore, several researchers suggest fusing terrestrial LiDAR with airborne sources to cover topographic features in inaccessible areas [38,41].

2.4. Overview

In summary, there are different ways of achieving data acquisition using LiDAR data, including airborne LiDAR and terrestrial LiDAR. Previous studies demonstrated that terrestrial LiDAR data held an advantage over other DEM sources, including airborne LiDAR, as they responded to small-scale topographic features, which were important factors that influenced the flood prediction results. Airborne LiDAR has difficulty in detecting small-scale features which are often not well represented in DEMs, which is the reason why many researchers opted for terrestrial LiDAR to generate a high-resolution DEM. Nevertheless, both LiDAR sensors proved to be able to maintain high accuracy and produce high-resolution data due to their high scanning rates [42] compared to other DEM sources.

3. Applications of LiDAR System in Flood Monitoring

The application of LiDAR in supporting many science research activities such as geologic mapping, landslide hazards, and flood risk management cannot be disputed [43]. The number of publications of peer-reviewed research literature recorded in the Scopus database for the past 10 years, from 2010 to 2019, which discussed LiDAR data in flood studies, was determined. The related papers were searched using the boolean "AND" to combine the words "LiDAR" and "flood", and we sought these words in the abstract, title, and keywords of the documents. This study decided to focus only on research articles and conference proceedings of the related topic to be counted, as presented in Figure 2.

The graph shows that there was a rapid increase in the rate of publishing papers on LiDAR and flood applications in early 2010, and this increasing trend remained until 2019. There may be various reasons for the increase such as the availability and accessibility of LiDAR technology and the occurrence of flood disasters in the world. For instance, Duan et al. [44] suggested that flood disasters became more severe in China in recent years based on flood variations from 1950 to 2013. Furthermore, the potential of LiDAR technology to provide high-quality data may receive attention from researchers and practitioners, which leads to an increment in LiDAR data applications in multidisciplinary studies, especially in flood applications.

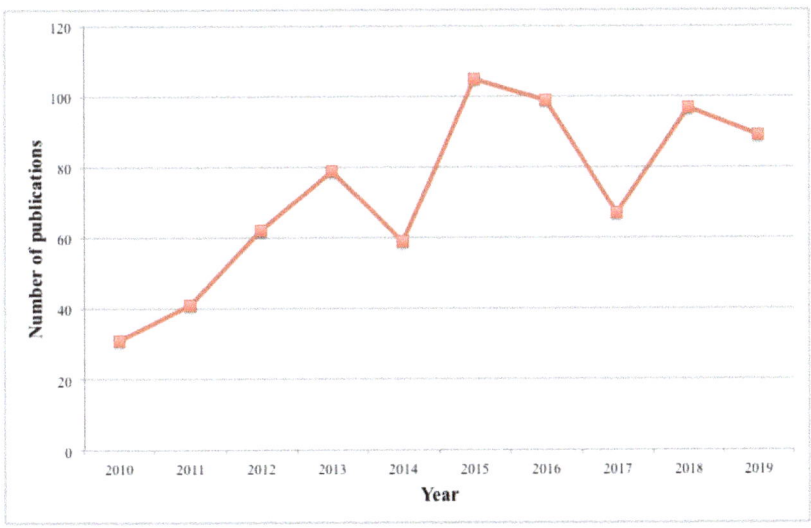

Figure 2. The number of papers that discussed on LiDAR and flood applications by year in the Scopus database.

LiDAR provides detailed information on the elevation of the ground surface for predicting flood inundation from rivers. Detailed LiDAR measurements not only offer higher-resolution elevation data for floodplain modeling, but they also provide a source of high-resolution surface roughness information. High resolution and high accuracy of a topographic dataset are also important in predicting the flood inundation [45].

Due to this, LiDAR technology is in demand for creating DEMs for flood-prone areas, especially in urban areas. The application of high-resolution LiDAR data is increasing in developed countries [46] such as in the United Kingdom, as well as in the United States. This section reviews the previous studies on the application of the airborne and terrestrial LiDAR in flood mapping and monitoring applications. This paper discusses several topics relevant to flood studies from a LiDAR data perspective.

3.1. Development of Flood Models Using DEM LiDAR

One of the important factors in producing reliable flood inundation maps is the availability of high-accuracy topographic data [47]. Detailed and accurate DEMs are needed, to represent specific properties that may obstruct and conduct the flow of water in the real world. Inaccurate topographic representation in a small-scale area would affect the simulation results [15], especially in urban areas; hence, researchers tend to use high-resolution input data for flood simulation in urban areas and floodplains to collect important small-scale features. Many studies were carried out to demonstrate the effectiveness of using LiDAR-derived DEMs in developing flood models. Results from the flood model simulations were compared with the observed water level during previous flood events to validate the simulated results.

Priestnell et al. [36] discussed the methods of extracting surface features from DSMs generated by airborne LiDAR. The extraction of features could help in many applications, including flood inundation modeling. This study explained the way in which the DEM and surface roughness layer could be generated from the original DSM from LiDAR by using a simple filtering procedure and an artificial neural network. The findings were illustrated in the case of flood inundation modeling. Furthermore, Webster et al. [48] investigated the coastal impacts due to climate change and sea-level rise in Charlottetown, Canada. Detailed topographic data were derived from airborne LiDAR for flood risk mapping, and they were used to define flood risk hazards. This finding demonstrates the effectiveness of airborne LiDAR for identifying the impact of climate change and storm surge in coastal areas.

Webster et al. [49] generated a flood risk map by using airborne LiDAR and geographical information system (GIS) processing to study flood inundation in Southeast New Brunswick, Canada. The flood inundation and flood depth of the proposed approach were validated by comparing the results with water levels observed during the flood event in January 2000. It was found that the flood extent and flood depth were accurate within 10–20 cm. Bales et al. [50] also carried out a study on flood inundation maps derived from LiDAR data for real-time flood mapping applications in Tar River Basin, North Carolina. This study used airborne LiDAR data with a vertical accuracy of about 20 cm to produce topographic data for the inundation maps. The difference between the measured and simulated water levels to high-water marks was less than 25 cm.

For terrestrial LiDAR, the first attempt at developing an urban hydraulic model using terrestrial LiDAR was done by Fewtrell et al. [51]. The performance of the flood model was analyzed by comparing the simulation results with the 50-cm-resolution model as a benchmark. This study found that errors in coarse-scale topographic datasets were significantly high. Moreover, the authors concluded that terrestrial LiDAR data can be used to provide information in small-scale flood risk management and suggested fusing airborne and terrestrial LiDAR to cover topographic features in inaccessible areas. Sampson et al. [52] investigated the capability of terrestrial LiDAR to provide high accuracy of DEMs for improving flood inundation models in urban areas. The study found that small features such as curbs and dykes, which had a significant impact on the flood propagation, could be represented from the terrestrial LiDAR data. The authors concluded that terrestrial LiDAR could be employed when an accurate representation of surface features is required, especially in urban inundation studies.

Furthermore, Poppenga and Worstell [53] demonstrated the need for hydrologic information derived from airborne LiDAR elevation surfaces for flood inundation monitoring in coastal regions. The study demonstrated how inland areas are hydrologically disconnected to ocean water due to bridge decks or culverts. Next, Yin et al. [54] used LiDAR-derived DEM in a high-resolution 2D hydraulic model to study the impact of land subsidence on urban pluvial flooding. The authors concluded that land subsidence could lead to moderate impacts on flood extent and flood depth in the urban areas. Chen et al. [55] assessed the accuracy of airborne LiDAR-derived flood extent by evaluating the data during the 2008 Iowa flood in the United States (US) with field measurements collected by the US Geological Survey (USGS) and Federal Emergency Management Agency (FEMA). The root-mean-square error (RMSE) of the floodwater surface profile from LiDAR to field measurement was 30 cm. The finding showed that LiDAR surveys could be used in measuring floodwater heights with reliable quality.

After the devastating flood in 2008, the Iowa Flood Center (IFC) was established to improve the availability of flood-relevant information to the community. Krajewski et al. [56] discussed several projects conducted by the IFC that were related to flood disasters. One of the projects was flood mapping that could be accessed online, working as flood inundation map libraries. Hydrodynamic modeling was used to simulate river and floodplain flows by using the best approach to describe river and floodplain topography, which was LiDAR data. In summary, these findings demonstrated the effectiveness of airborne LiDAR for identifying the impact of flooding.

In the past few years, LiDAR technology was widely used in flood inundation research due to its high potential of providing inundation models with detailed elevation data. Based on these studies, it was found that LiDAR data produce high-resolution DEMs for flood simulation modeling, which can be an efficient tool in floodplain inundation management.

3.2. Generation of Surface Roughness Maps Using LiDAR Data

One of the essential input parameters in a flood model is surface roughness, which is useful for boundary conditions. Roughness maps can be derived from different sources such as orthophoto, LiDAR data, and land-use data. The surface roughness has a significant effect on the output of hydrodynamic modeling. The roughness parameter is often defined through Manning's formula [57]. Meanwhile, the roughness values are mostly derived from a look-up table based on land-use/land-cover classification (LULC).

According to Straatsma and Baptist [58], roughness values have to be estimated accurately to reduce the variation of input parameters during calibration. The authors carried out a study to derive roughness parameterization using multispectral and airborne LiDAR data. After that, the results of the proposed method were compared with a traditional roughness parameterization approach, which was a manual interpretation of aerial photographs and a look-up table. This approach led to a high-resolution roughness map.

Vetter et al. [59] used airborne LiDAR to derive hydraulic surface roughness estimations based on geometry data by using vertical vegetation structure analysis. The effects of different roughness coefficient values were quantified by calculating the inundated depth maps. The results showed that the roughness values derived from airborne LiDAR represented the area in detail as compared to the traditionally derived map.

High-resolution data are recommended as the best option for damage assessment applications. Joyce et al. [60] recommended using airborne LiDAR data to generate DEMs and surface roughness layers to be included in hazard models. Moreover, the high density of LiDAR data provides a high-resolution surface roughness of floodplains. This information is very useful for boundary conditions in flood simulations [61].

Dorn et al. [57] derived roughness maps based on several different datasets, including LiDAR. This study aimed to analyze the effect of the roughness maps on flood simulations. LiDAR point clouds were used to derive surface roughness by using a voxel structure, an approach developed by Vetter et al. [59]. The results based on different roughness maps differed in terms of inundation area, water depth, and flood intensity. The authors suggested using LiDAR data to derive a roughness map for estimating the consequences of floods. Moreover, the authors mentioned that the use of the same LiDAR data in producing the DEM data and the roughness maps is beneficial, as there is no issue of temporal difference.

In short, in addition to topography, surface roughness has a great influence on hydrodynamic models as it affects the flow regime [62–64]. Hence, appropriate roughness maps should be generated for the use of hydrodynamic models for predicting the reliable consequences of flood disasters. Based on previous research, it was concluded that LiDAR data provide high-resolution surface roughness, which will increase the accuracy of the flood extent simulated by the hydrodynamic model. Hence, laser scanning technology is able to produce a roughness map with a high level of spatial detail [58].

3.3. Comparisons of LiDAR-Derived DEMs with Other DEM Sources

Due to the significant impacts of DEM accuracy on the flood model outputs, it is important to know which DEM sources could provide higher accuracy and spatial resolution before the selected DEM is used for the assessment of flood hazard risk. Therefore, many comparative studies were carried out using different DEM sources to understand the importance of the accuracy of DEM on the flood model. The results were analyzed based on significant differences in the model output. This section discusses the comparisons between LiDAR-derived DEM and other DEM sources, as well as their significant characteristics in flood applications.

Casas et al. [16] evaluated the effects of DEM sources on the hydraulic modeling of floods in terms of the hydraulic model outputs such as flood inundation and water surface elevation. The results of this study demonstrated that the flood model output was highly dependent on the DEM quality with LiDAR data, showing a high potential source for the parameterization of channel and floodplain topography.

Schumann et al. [65] carried out a comparison of DEMs generated from airborne LiDAR, contours, and SRTM in terms of the effect on a flood inundation model. The results were compared with inundation maps from a model calibrated with ground-surveyed maximum watermarks. As expected, the authors found that LiDAR had the lowest RMSE, followed by contour DEM and SRTM. The estimated inundated area for LiDAR was the largest area and the nearest to the reference value. It was concluded that LiDAR is the most reliable source of topographic data for flood hazard estimation.

Moreover, Wang and Zheng [66] compared LiDAR-derived DEM with United States Geologic Survey (USGS) national elevation data (NED) on floodplains in North Carolina. Sanders [67] extended the scope by evaluating the difference between LiDAR-derived DEM and NED with airborne IfSAR and SRTM for flood inundation modeling. Sanders found that flood model predictions were highly dependent on the DEM resolution. The author also concluded that LiDAR-derived DEM was more accurate than other DEM sources which overestimated the flood extent. The need for LiDAR data is now a fundamental input to hydrologic and hydraulic models, especially in flood inundation models.

Furthermore, Coveney and Fotheringham [68] examined the impact of DEM data sources on flood risk prediction in the coastal areas. The authors used national-coverage DEM known as Ordnance Survey Ireland, two GPS-derived DEMs generated at low and medium resolution, and terrestrial LiDAR-derived DEM to model flood risk. The findings demonstrated that the DEM generated from terrestrial LiDAR was more advantageous than other DEM data sources, especially in representing small topographical features in a local flood.

Additionally, Papaioannou et al. [69] investigated the influence of different DEM sources used in a hydraulic model for flood analysis. The results of the flood models were compared with historical flood records. According to this study, DEMs derived from terrestrial LiDAR were best, as they generated the closest values to the historical data. This finding indicated that the high accuracy of DEMs helped improve the flood risk analysis task. This study concluded that the accuracy of DEMs is the major factor that affects flood modeling results.

In addition, Li and Wong [70] studied the effects of different DEM sources on flood simulation results. The authors concluded that different DEM sources have major impacts on inundation areas from flood prediction results as compared to DEM spatial resolution. Based on the experimental results, it was found that inundation areas from LIDAR-derived DEM were the closest to reality. Furthermore, this study also highlighted that the reliability of the DEM source significantly affected the flood simulation results.

Jakovljevic and Govedarica [71] simulated flood inundation by selecting the grid cell of a DEM lower than the projected water level, connected to an adjacent flooded grid cell. In this study, the authors used LiDAR-derived DEM and the Advanced Spaceborne Thermal Emission and Reflection Radiometer Global Digital Elevation Model (ASTER GDEM) to study the difference in the estimated flood extents. It was found that land elevation from ASTER GDEM was overestimated, which directly resulted in an underestimation of flood inundation risk. The inundation map generated from ASTER GDEM indicated that the inundation area was two times smaller than that generated from LiDAR-derived DEM. Figure 3 shows the visual comparison of the flood extent from LiDAR and ASTER GDEM.

Based on previous research, LiDAR proved to be an efficient method to provide terrain data with high resolution as compared to other DEM sources [72]. According to Sampson et al. [73], LiDAR-derived DEMs are considered the most reliable DEMs for flood modeling to date. In summary, hydrological modeling studies showed that the vertical accuracy of DEMs does affect the accuracy of hydrologic predictions [70].

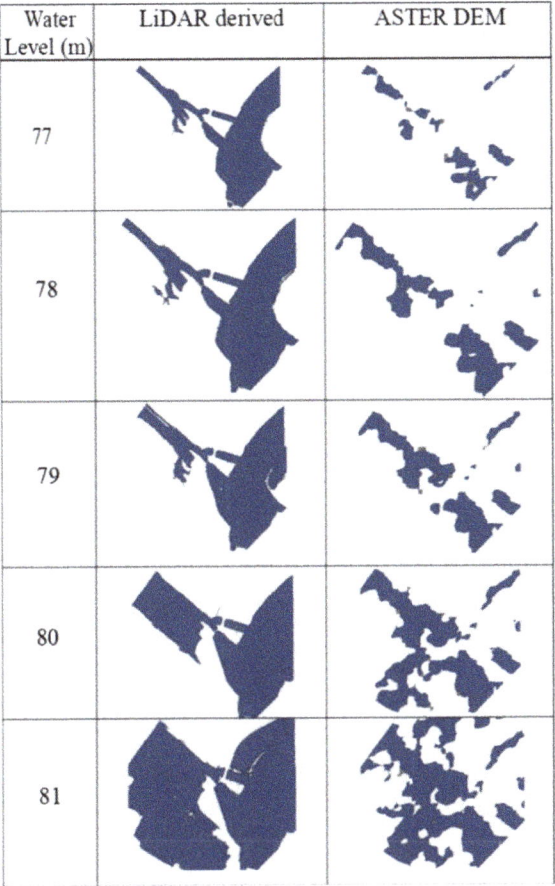

Figure 3. A comparison of flood extent derived from LiDAR and the Advanced Spaceborne Thermal Emission and Reflection Radiometer Global Digital Elevation Model (ASTER GDEM) according to water level [71].

3.4. LIDAR as a Source of Information for Hydrodynamic Model Verification

Based on previous studies, it was found that LiDAR data are capable of producing high-resolution DEMs for flood simulation modeling, which can be an efficient tool in floodplain inundation management. Hence, they are commonly used for hydrodynamic model verification. Courty et al. [74] mentioned that inundation areas from LIDAR-derived DEM were the closest to reality as reported by Li and Wong [69]; therefore, they used LiDAR-derived DEM as a reference when comparing DEMs generated from Advanced Land Observing Satellite (ALOS) World 3D-30m (AW3D30), SRTM, and Advanced Spaceborne Thermal Emission and Reflection Radiometer (ASTER) for flood modeling purposes. Based on the flood simulation results, AW3D30 performed better than SRTM, while ASTER was the worst performer of all global DEMs.

Hashemi et al. [75] also used LiDAR-derived DEMs as a reference when investigating the quality of DEMs generated from an unmanned aerial vehicle (UAV) used in flood modeling. These studies concluded that the reliability of floodplain maps is dependent on the quality of DEM. Van de Sande et al. [76] adopted LiDAR DEM data as ground truth referring to the terrain elevation. Hence, the flood risk assessment of publicly available DEMs such as ASTER and SRTM DEM was compared with flood risk based on LiDAR DEM. The inundation maps of these publicly available DEMs were smaller than inundation maps produced using LiDAR DEM. The underestimations of the flood risk influence the credibility when making appropriate decisions regarding flood risk management and mitigation.

Furthermore, most small river basins in many countries are not characterized by high-quality DEMs such as LiDAR data [77]. Hence, aerial photographs or globally available DEMs such as ASTER and SRTM are commonly used, which leads to low accuracy of flood prediction due to the significant effect of low-accuracy DEMs. Therefore, this study proposed using corrected DEMs generated from aerial photographs as an option in flood modeling. The correction of DEM was performed based on field measurements to determine vertical errors. Then, a reference DEM that was developed from LiDAR data was used to validate the performance of the original and corrected DEM. The impact of DEM accuracy was evaluated using the flood model. The results from the model indicated that the flood prediction of corrected DEM was better than that of the original DEM when compared with the simulated result of the reference DEM, as shown in Figure 4. However, the authors suggested that the proposed method was not suitable for urban areas.

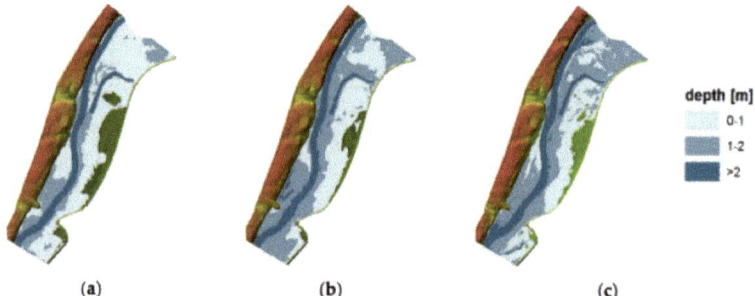

Figure 4. Comparison of flood extent and flood depth obtained from flood model simulation: (**a**) original DEM; (**b**) corrected DEM; (**c**) reference DEM [77].

3.5. LiDAR DEM for Flood Hazard and Flood Risk Mapping

Flood risk assessment and management rely on the accuracy of flood extent simulated using a flood model. Most flood risk mapping is based on a conceptual risk approach that uses DEMs to predict the flood hazard according to the projected water levels and to indicate the vulnerability of areas to flood events with damage to properties and livelihood. Hazard mapping is an important element in assessing risk and designing mitigation measures for flood-prone areas.

Flood hazard is usually generated based on the outcome of hydrological models that simulate the water movement across the floodplain like flood extent, water velocity, or water depth [11,37,78]. In addition, flood hazards can also be produced using a statistical or machine-learning approach integrated with GIS technology by using fluvial stage records and topographic data [79,80]. Flood hazard and flood risk maps indicate the flood-prone area with possible destructive impact, which is used for flood planning purposes.

For instance, existing digital mapping was not sufficient enough to provide a high accuracy of flood risk maps for Annapolis Royal, Nova Scotia, Canada, an area that is vulnerable to coastal flooding [81]. Hence, the need for a high-resolution DEM was studied to produce accurate inundation maps based on sea level and climate change. As the sea level rises, water inundates the nearby lands; thus, it is important to define the extent of the flood inundation. The predicted results were compared with the benchmark of a past storm event to test the model. Based on the prediction results, mitigation structures such as dykes could be suggested if coastal development is planned to take place in any of the risk areas.

Puno et al. [13] conducted flood simulations at different return periods with LiDAR-derived DEMs as a primary source of elevation data in the hydrologic model. The model was calibrated by comparing the predicted flood simulation with a real flood event in 2016. Flood hazard maps were generated from the simulated flood events using GIS and LiDAR-derived DEM. The generated maps were validated through an interview with the affected localities. The authors found that using LiDAR data in the hydrologic model could produce high-resolution flood hazard maps that can offer more accurate decisions and actions in disaster management and mitigation.

Ogania et al. [14] evaluated the effect of DEM resolutions on generating flood hazard maps using hydraulic modeling software for disaster preparedness and mitigation. This study presented the performance of three different DEM resolutions, which were LiDAR, IfSAR, and SAR DEMs in flood modeling studies. The accuracy of each generated flood map was evaluated using a confusion matrix approach by comparing the generated maps with the actual flood data. This paper revealed that LiDAR-derived DEMs provide a more defined flood extent and clear distribution of flood hazards. Furthermore, they offer more accurate flood maps compared to other DEM data sources, which aligned with the findings from previous researchers such as Hailes and Rientjes [82] and Schumann et al. [65].

Mihu-Pintilie et al. [83] used high-density LiDAR data with 2D hydraulic modeling to improve urban flood hazard maps. This study simulated four different multi-scenarios at different discharge values. Because LiDAR data provide a precise representation of the hydraulic conditions such as channels and roads, the combination of 2D hydraulic and LiDAR DEMs produced accurate information regarding flood hazard vulnerability. Flood hazard maps were generated based on flood depth classification according to the Japanese criteria of the Ministry of Land Infrastructure and Transport (MLIT). The criteria suggested five hazard classes of very low, low, medium, high, and extreme classified as H1, H2, H3, H4, and H5, respectively. Figure 5 shows that all hazard classes were encountered according to scenario 1 (s1). However, most of the affected areas were assigned with the very low or low class of hazard (H1 and H2).

LiDAR datasets were implemented in a new procedure of flood hazard estimation proposed by Guerriero et al. [84]. The authors developed algorithms of interpolation of multiple probability models of hydrometric time-series data combined with topography derived from LiDAR data for the production of flood hazard maps. Flood hazard maps produced from this method were compared with a flood event observation in 2015 for validation. This suggested method can be considered as another option for hydraulic simulations to provide flood hazard analysis.

In conclusion, high-resolution DEMs have great influence on producing accurate and reliable maps in the field of flood simulations. Using these maps helps in disaster risk reduction and management, especially in identifying specific areasthat need to be prioritized for providing appropriate flood risk management measures to be taken to combat flood disaster. Previous studies implied that LiDAR-derived DEMs improve the accuracy of flood parameters; hence, they can help in producing high-quality flood hazard and flood risk mapping.

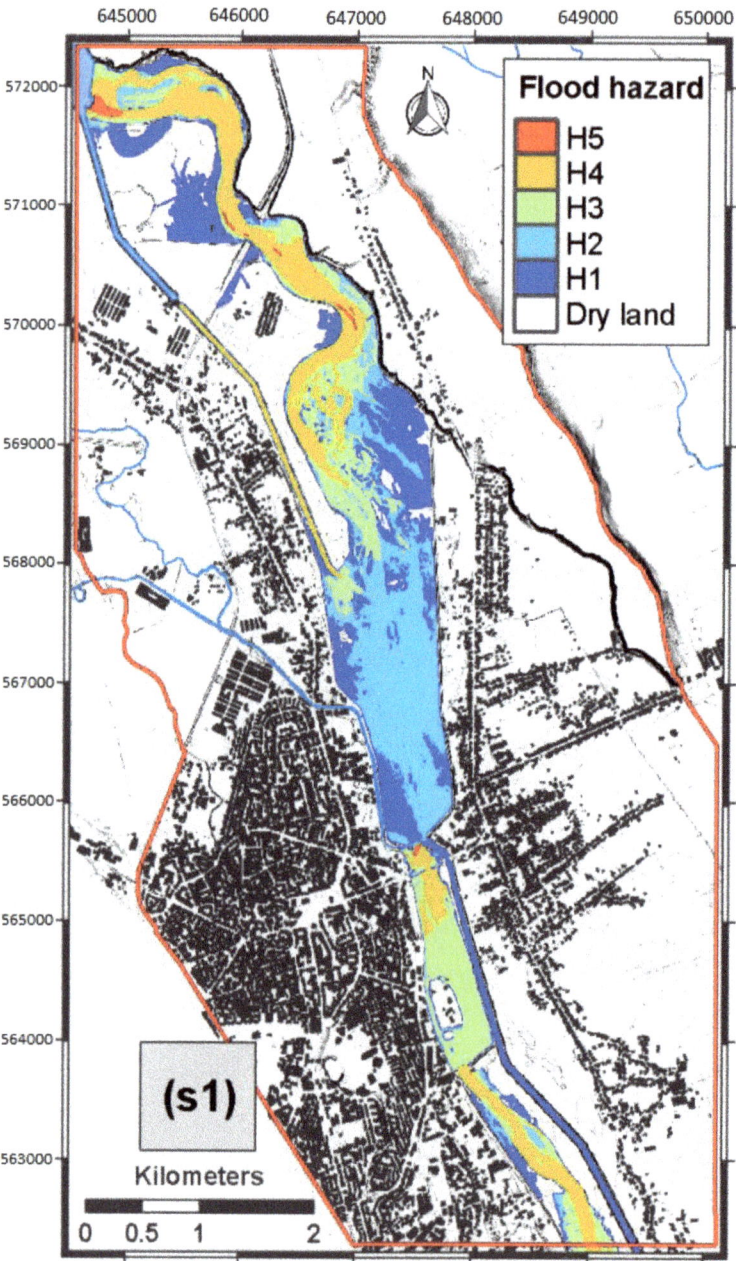

Figure 5. Flood hazard map based on flood depth classification according to the Ministry of Land Infrastructure and Transport (MLIT) [83].

4. Challenges and Future Perspectives

The frequency of flood disasters all over the world is increasing due to climate change and rapid urbanization. Future climate projections could provide an additional understanding of extreme climate changes, including the risk of flood events [85]. Furthermore, studies on flood mapping and monitoring increased with the advancement of current technologies to reduce the impact of flood disasters. LiDAR data acquisitions seem to be a promising approach to solve the problems associated with the inadequate representation of topographic data. Both airborne and terrestrial LiDAR systems are active imaging techniques operating with light that allow the systems to collect data during daylight or nighttime. Previous studies revealed that LiDAR technology has many advantages, which makes it suitable for flood modeling, particularly in flat areas and complex urban environments. Depending on the spatial scale, LiDAR data offer different advantages for accurate terrain mapping compared to other sources. Moreover, LiDAR could be advantageous to provide information in small-scale flood risk management by having small important topographic features such as dykes, ditches, and levees [15,51,86,87]. Furthermore, the integration of LiDAR technology with any remotely sensed products may be used to increase the effectiveness of this technology, especially in flood modeling.

However, there are some restrictions in using LiDAR-derived DEM in the context of flood applications. The main drawback of both LiDAR systems is the process of classifying ground from non-ground data for DEM generation, which is needed in simulations of the flood model. Ground surface information is not easily extracted, especially in areas with complex terrain surface and features such as buildings and vegetation [88]. The ground filtering process proves to be a challenging task as it can affect the accuracy of the LiDAR products [8,49,89]. Several filtering algorithms were developed by previous researchers to process LiDAR data. However, the LiDAR data must be correctly processed because they could influence the outcomes of flood mapping [69]. The algorithms perform differently depending on the specific surface conditions. This means that not all algorithms are competent in producing high-quality LiDAR-derived DEM data [90]. A filtering algorithm should be selected based on its ability to produce the desired result [91]. Common filtering algorithms used in LiDAR data processing include elevation threshold with expand window (ETEW), maximum local slope, adaptive triangulated irregular network (TIN), and progressive morphology [90,92,93]. Filtering problems are expected to be better solved with the evolution of machine learning [88].

In addition, the sensitive response of flood inundation to small changes in topography representation gives rise to several challenges [21]. Collecting small-scale features needs a high resolution of DEM data, but the data are rarely available, especially for developing countries. Not all countries can afford to use LiDAR data due to economic constraints. The high cost and the difficulty of processing huge LiDAR datasets could be the main reason why LiDAR data are not used in some developing countries. Even developed countries like the United States and the United Kingdom do not have LiDAR data available for the entire country. Another challenge when using LiDAR data is the need for huge data storage due to the high-point-density data. High-point-density data need a longer computational time to process [94,95]. Between airborne and terrestrial LiDAR systems, the time required for flood model simulations using terrestrial LiDAR is 10 times longer than that required for airborne LiDAR [96].

Furthermore, even though high-resolution DEMs offer detailed information topography, they take a longer time to process or analyze the data. Abucay and Tseng [97] carried out a visibility analysis that could be used in identifying flood-prone areas using various DEM sources. The authors reported that the LiDAR-derived DEM required 28 min to complete the visibility analysis, followed by the SAR DEM that took 19 s, while ALOS and ASTER GDEM both required only 3 s to complete the process. Nevertheless, the computational time problem may be solved with future advancements in computer technology. Moreover, the LiDAR system cannot penetrate water bodies as its laser beam is absorbed by the water. Therefore, the inaccurate elevation measurement of water-covered areas influences cross-section attributes, leading to inaccuracies in hydrodynamic simulations [98].

5. Conclusions

Detailed topographic information is a crucial input parameter for flood modeling and monitoring. The performance of flood modeling is highly dependent on the DEM accuracy [10], especially in small-scale flood modeling studies. Flood model simulation results show differences in water depth and inundation when using detailed DEMs, proving that DEM accuracy has a significant impact on flood hazard estimation [21,41]. Therefore, the need for high-resolution DEM explains the interest in exploring new technology to generate detailed elevation data. In this review, the promising applications in numerous flood studies demonstrate that the LiDAR system is capable of offering high-density and high-resolution DEM data to improve the flood model input, thus resulting in a higher accuracy of flood modeling results. However, LiDAR data also face several difficulties that need to be addressed in the future regarding the filtering process for DEM generation and enormous point density data that need huge data storage, resulting in a longer computational time to simulate flood models. Additionally, integration between terrestrial and airborne LiDAR or any remotely sensed products seems to be a promising approach to solve the problems associated with the inadequate representation of topographic data in topographically complex areas [99]; hence, more investigation and research work for the expansion of LiDAR systems can be foreseen in upcoming applications of flood detection and monitoring.

Author Contributions: N.A.M. executed the manuscript writing, coordinated the paper revisions, and contributed to the workflow implementation. A.F.A. contributed to the workflow implementation, as well as the manuscript compilation and revisions. S.K.B. proposed the research idea. M.R.M. and A.M. supervised the final manuscript. All authors have read and agreed to the published version of the manuscript.

Funding: This research was funded by Universiti Putra Malaysia, grant number UPM/800-3/3/1/GPB/2019/9678700.

Acknowledgments: The authors wish to acknowledge the assistance of the Department of Biological and Agricultural Engineering, Faculty of Engineering, Universiti Putra Malaysia for supplying the facilities for this study. The authors would also like to appreciate the support for this study from the Institute of Aquaculture and Aquatic Sciences.

Conflicts of Interest: The authors declare no conflict of interest.

References

1. Chan, N.W. *Impacts of Disasters and Disaster Risk Management in Malaysia: The Case of Floods BT—Resilience and Recovery in Asian Disasters: Community Ties, Market Mechanisms, and Governance*; Aldrich, D.P., Oum, S., Sawada, Y., Eds.; Springer: Tokyo, Japan, 2015; pp. 239–265. ISBN 978-4-431-55022-8.
2. DID. *Department of Irrigation and Drainage (DID) Manual (Volume 1–Flood Management)*; Department of Irrigation and Drainage: Kuala Lumpur, Malaysia, 2009; Volume 1.
3. Khan, H.; Vasilescu, L.G.; Khan, A. Disaster management cycle—A theoretical approach. *J. Manag. Mark.* **2008**, *6*, 43–50.
4. National Research Council. *Elevation Data for Floodplain Mapping*; National Academies Press: Washington, DC, USA, 2007; ISBN 0309104092.
5. Wilson, J.P.; Gallant, J.C. *Terrain Analysis: Principles and Applications*; John Wiley & Sons: Hoboken, NJ, USA, 2000; ISBN 0471321885.
6. Schenk, T. *Introduction to Photogrammetry*; Department of Civil and Environmental Engineering and Geodetic Science The Ohio State University: Columbus, OH, USA, 2005; pp. 79–95.
7. Dowman, I. Integration of LIDAR and IFSAR for mapping. *Int. Arch. Photogramm. Remote Sens.* **2004**, *35*, 90–100.
8. Hodgson, M.E.; Jensen, J.R.; Schmidt, L.; Schill, S.; Davis, B. An evaluation of LIDAR- and IFSAR-derived digital elevation models in leaf-on conditions with USGS Level 1 and Level 2 DEMs. *Remote Sens. Environ.* **2003**, *84*, 295–308. [CrossRef]
9. Merwade, V.; Olivera, F.; Arabi, M.; Edleman, S. Uncertainty in Flood Inundation Mapping: Current Issues and Future Directions. *J. Hydrol. Eng.* **2008**, 608–620. [CrossRef]
10. Sanyal, J.; Lu, X.X. Application of Remote Sensing in Flood Management with Special Reference to Monsoon Asia: A Review. *Nat. Hazards* **2004**, 283–301. [CrossRef]
11. Bates, P.D.; De Roo, A.P.J. A simple raster-based model for flood inundation simulation. *J. Hydrol.* **2000**, *236*, 54–77. [CrossRef]

12. Chaubey, I.; Cotter, A.S.; Costello, T.A.; Soerens, T.S. Effect of DEM data resolution on SWAT output uncertainty. *Hydrol. Process.* **2005**, *19*, 621–628. [CrossRef]
13. Puno, G.R.; Angelica Amper, R.L.; Allan Talisay, B.M. Flood simulation using geospatial and hydrologic models in Manupali Watershed, Bukidnon, Philippines. *J. Biodivers. Environ. Sci.* **2018**, *294*, 294–303.
14. Ogania, J.L.; Puno, G.R.; Alivio, M.B.T.; Taylaran, J.M.G. Effect of digital elevation model's resolution in producing flood hazard maps. *Glob. J. Environ. Sci. Manag.* **2019**, *5*, 95–106. [CrossRef]
15. Tamiru, A.; Rientjes, T.H.M. Effects of Lidar Dem Resolution in Flood Modelling: A Model Sensitivity Study for the City of Tegucigalpa, Honduras. In Proceedings of the ISPRS WG III/3, III/4, V/3 Workshop, Enschede, The Netherlands, 12–14 September 2005; pp. 168–173.
16. Casas, A.; Benito, G.; Thorndycraft, V.R.; Rico, M. The topographic data source of digital terrain models as a key element in the accuracy of hydraulic flood modelling. *Earth Surf. Process. Landf.* **2006**, *31*, 444–456. [CrossRef]
17. Vaze, J.; Teng, J.; Spencer, G. Impact of DEM accuracy and resolution on topographic indices. *Environ. Model. Softw.* **2010**, *25*, 1086–1098. [CrossRef]
18. Hsu, Y.C.; Prinsen, G.; Bouaziz, L.; Lin, Y.J.; Dahm, R. An Investigation of DEM Resolution Influence on Flood Inundation Simulation. *Procedia Eng.* **2016**, *154*, 826–834. [CrossRef]
19. Ozdemir, H.; Sampson, C.C.; De Almeida, G.A.M.; Bates, P.D. Evaluating scale and roughness effects in urban flood modelling using terrestrial LIDAR data. *Hydrol. Earth Syst. Sci.* **2013**, *17*, 4015–4030. [CrossRef]
20. Savage, J.T.S.; Bates, P.; Freer, J.; Neal, J.; Aronica, G. When does spatial resolution become spurious in probabilistic flood inundation predictions? *Hydrol. Process.* **2016**, *30*, 2014–2032. [CrossRef]
21. de Almeida, G.A.M.; Bates, P.; Ozdemir, H. Modelling urban floods at submetre resolution: Challenges or opportunities for flood risk management? *J. Flood Risk Manag.* **2018**, *11*, S855–S865. [CrossRef]
22. Erpicum, S.; Dewals, B.; Archambeau, P.; Detrembleur, S.; Pirotton, M. Detailed Inundation Modelling Using High Resolution DEMs. *Eng. Appl. Comput. Fluid Mech.* **2010**, *4*, 196–208. [CrossRef]
23. Savage, J.T.S.; Pianosi, F.; Bates, P.; Freer, J.; Wagener, T. Quantifying the importance of spatial resolution and other factors through global sensitivity analysis of a flood inundation model. *Water Resour. Res.* **2015**, *52*, 9146–9163. [CrossRef]
24. Olsen, M.J. Methodology for Assessing Coastal Change Using Terrestrial Laser Scanning. Ph.D. Thesis, University of California, San Diego, CA, USA, 2009.
25. Lemmens, M. *Geo-Information: Technologies, Applications and the Environment*; Springer: London, UK, 2011; ISBN 9789400716667.
26. Dong, P.; Chen, Q. *LiDAR Remote Sensing and Applications*; CRC Press: Boca Raton, FL, USA, 2017; ISBN 1351233343.
27. Chen, Z. *The Application of Airborne Lidar Data in the Modelling of 3D Urban Landscape Ecology*; Cambridge Scholars Publishing: Cambridge, UK, 2016; ISBN 1443857602.
28. Cracknell, A.P. *Introduction to Remote Sensing*; CRC press: Boca Raton, FL, USA, 2007; ISBN 1420008978.
29. Li, Z.; Chen, J.; Baltsavias, E. *Advances in Photogrammetry, Remote Sensing and Spatial Information Sciences: 2008 ISPRS Congress Book*; CRC Press: Boca Raton, FL, USA, 2008; Volume 7, ISBN 0203888448.
30. Pradhan, B. *Laser Scanning Applications in Landslide Assessment*; Springer: Berlin/Heidelberg, Germany, 2017; ISBN 3319553429.
31. Vosselman, G.; Maas, H.-G. *Airborne and Terrestrial Laser Scanning*; Whittles Publishing: Scotland, UK, 2010; ISBN 9781904445876.
32. Thenkabail, P. *Remote Sensing Handbook-Three Volume Set*; CRC Press: Boca Raton, FL, USA, 2018; ISBN 1482282674.
33. Dassot, M.; Constant, T.; Fournier, M. The use of terrestrial LiDAR technology in forest science: Application fields, benefits and challenges. *Ann. For. Sci.* **2011**, *68*, 959–974. [CrossRef]
34. National Research Council. *Earth Materials and Health: Research Priorities for Earth Science and Public Health*; National Academies Press: Washington, DC, USA, 2007; ISBN 030910470X.
35. Olsen, M.J. *Guidelines for the Use of Mobile LIDAR in Transportation Applications*; Transportation Research Board: Washington, DC, USA, 2013; Volume 748, ISBN 0309259142.
36. Priestnall, G.; Jaafar, J.; Duncan, A. Extracting urban features from LiDAR digital surface models. *Comput. Environ. Urban Syst.* **2000**, *24*, 65–78. [CrossRef]

37. Teng, J.; Jakeman, A.J.; Vaze, J.; Croke, B.F.W.; Dutta, D.; Kim, S. Flood inundation modelling: A review of methods, recent advances and uncertainty analysis. *Environ. Model. Softw.* **2017**, *90*, 201–216. [CrossRef]
38. Turner, A.B.; Colby, J.D.; Csontos, R.M.; Batten, M. Flood modeling using a synthesis of multi-platform LiDAR data. *Water (Switzerland)* **2013**, *5*, 1533–1560. [CrossRef]
39. Kerle, N.; Heuel, S.; Pfeifer, N. Real-time data collection and information generation using airborne sensors. In *Geospatial Information Technology for Emergency Response*; CRC Press: Boca Raton, FL, USA, 2008; pp. 59–90. ISBN 9780415422475.
40. Lim, S.; Thatcher, C.A.; Brock, J.C.; Kimbrow, D.R.; Danielson, J.J.; Reynolds, B.J. Accuracy assessment of a mobile terrestrial lidar survey at Padre Island National Seashore. *Int. J. Remote Sens.* **2013**, *34*, 6355–6366. [CrossRef]
41. Leitão, J.P.; de Sousa, L.M. Towards the optimal fusion of high-resolution Digital Elevation Models for detailed urban flood assessment. *J. Hydrol.* **2018**, *561*, 651–661. [CrossRef]
42. Awange, J. *GNSS Environmental Sensing*; Springer International Publishers: New York, NY, USA, 2018; Volume 10, pp. 973–978.
43. Stoker, J.M.; Brock, J.C.; Soulard, C.E.; Ries, K.G.; Sugarbaker, L.J.; Newton, W.E.; Haggerty, P.K.; Lee, K.E.; Young, J.A. *USGS Lidar Science Strategy—Mapping the Technology to the Science*; US Department of the Interior, US Geological Survey: Reston, VA, USA, 2016.
44. Duan, W.; He, B.; Nover, D.; Fan, J.; Yang, G.; Chen, W.; Meng, H.; Liu, C. Floods and associated socioeconomic damages in China over the last century. *Nat. Hazards* **2016**, *82*, 401–413. [CrossRef]
45. Brasington, J.; Vericat, D.; Rychkov, I. Modeling river bed morphology, roughness, and surface sedimentology using high resolution terrestrial laser scanning. *Water Resour. Res.* **2012**, *48*, 1–18. [CrossRef]
46. Mason, D.C.; Trigg, M.; Garcia-Pintado, J.; Cloke, H.L.; Neal, J.C.; Bates, P.D. Improving the TanDEM-X Digital Elevation Model for flood modelling using flood extents from Synthetic Aperture Radar images. *Remote Sens. Environ.* **2016**, *173*, 15–28. [CrossRef]
47. Hawker, L.; Bates, P.; Neal, J.; Rougier, J. Perspectives on Digital Elevation Model (DEM) Simulation for Flood Modeling in the Absence of a High-Accuracy Open Access Global DEM. *Front. Earth Sci.* **2018**, *6*, 233. [CrossRef]
48. Webster, T.L.; Forbes, D.L.; Dickie, S.; Shreenan, R. Using topographic lidar to map flood risk from storm-surge events for Charlottetown, Prince Edward Island, Canada. *Can. J. Remote Sens.* **2004**, *30*, 64–76. [CrossRef]
49. Webster, T.L.; Forbes, D.L.; Mac Kinnon, E.; Roberts, D. Flood-risk mapping for storm-surge events and sea-level rise using lidar for southeast New Brunswick. *Can. J. Remote Sens.* **2006**, *32*, 194–211. [CrossRef]
50. Bales, J.D.; Wagner, C.R.; Tighe, K.C.; Terziotti, S. *LiDAR-Derived Flood-Inundation Maps for Real-Time Flood-Mapping Applications, Tar River Basin, North Carolina*; Geological Survey (US): Reston, VA, USA, 2007.
51. Fewtrell, T.J.; Duncan, A.; Sampson, C.C.; Neal, J.C.; Bates, P.D. Benchmarking urban flood models of varying complexity and scale using high resolution terrestrial LiDAR data. *Phys. Chem. Earth* **2011**, *36*, 281–291. [CrossRef]
52. Sampson, C.C.; Fewtrell, T.J.; Duncan, A.; Shaad, K.; Horritt, M.S.; Bates, P.D. Use of terrestrial laser scanning data to drive decimetric resolution urban inundation models. *Adv. Water Resour.* **2012**, *41*, 1–17. [CrossRef]
53. Poppenga, S.; Worstell, B. Evaluation of airborne lidar elevation surfaces for propagation of coastal inundation: The importance of hydrologic connectivity. *Remote Sens.* **2015**, *7*, 11695–11711. [CrossRef]
54. Yin, J.; Yu, D.; Wilby, R. Modelling the impact of land subsidence on urban pluvial flooding: A case study of downtown Shanghai, China. *Sci. Total Environ.* **2016**, *544*, 744–753. [CrossRef] [PubMed]
55. Chen, B.; Krajewski, W.F.; Goska, R.; Young, N. Using LiDAR surveys to document floods: A case study of the 2008 Iowa flood. *J. Hydrol.* **2017**, *553*, 338–349. [CrossRef]
56. Krajewski, W.F.; Ceynar, D.; Demir, I.; Goska, R.; Kruger, A.; Langel, C.; Mantilllla, R.; Niemeier, J.; Quintero, F.; Seo, B.C.; et al. Real-time flood forecasting and information system for the state of Iowa. *Bull. Am. Meteorol. Soc.* **2017**, *98*, 539–554. [CrossRef]
57. Dorn, H.; Vetter, M.; Höfle, B. GIS-based roughness derivation for flood simulations: A comparison of orthophotos, LiDAR and Crowdsourced Geodata. *Remote Sens.* **2014**, *6*, 1739–1759. [CrossRef]
58. Straatsma, M.W.; Baptist, M.J. Floodplain roughness parameterization using airborne laser scanning and spectral remote sensing. *Remote Sens. Environ.* **2008**, *112*, 1062–1080. [CrossRef]

59. Vetter, M.; Höfle, B.; Hollaus, M.; Gschöpf, C.; Mandlburger, G.; Pfeifer, N.; Wagner, W. Vertical Vegetation Structure Analysis and Hydraulic Roughness Determination Using Dense Als Point Cloud Data—A Voxel Based Approach. *ISPRS-Int. Arch. Photogramm. Remote Sens. Spat. Inf. Sci.* **2012**, *XXXVIII-5/W12*, 265–270. [CrossRef]
60. Joyce, K.E.; Samsonov, S.V.; Levick, S.R.; Engelbrecht, J.; Belliss, S. Mapping and monitoring geological hazards using optical, LiDAR, and synthetic aperture RADAR image data. *Nat. Hazards* **2014**, *73*, 137–163. [CrossRef]
61. Brown, J.D.; Damery, S.L. Managing flood risk in the UK: Towards an integration of social and technical perspectives. *Trans. Inst. Br. Geogr.* **2002**, *27*, 412–426. [CrossRef]
62. Aronica, G.; Hankin, B.; Beven, K. Uncertainty and equifinality in calibrating distributed roughness coefficients in a flood propagation model with limited data. *Adv. Water Resour.* **1998**, *22*, 349–365. [CrossRef]
63. Fu, J.C.; Hsu, M.H.; Duann, Y. Development of roughness updating based on artificial neural network in a river hydraulic model for flash flood forecasting. *J. Earth Syst. Sci.* **2016**, *125*, 115–128. [CrossRef]
64. Liu, Z.; Merwade, V.; Jafarzadegan, K. Investigating the role of model structure and surface roughness in generating flood inundation extents using one- and two-dimensional hydraulic models. *J. Flood Risk Manag.* **2019**, *12*, e12347. [CrossRef]
65. Schumann, G.; Matgen, P.; Cutler, M.E.J.; Black, A.; Hoffmann, L.; Pfister, L. Comparison of remotely sensed water stages from LiDAR, topographic contours and SRTM. *ISPRS J. Photogramm. Remote Sens.* **2008**, *63*, 283–296. [CrossRef]
66. Wang, Y.; Zheng, T. Comparison of light detection and ranging and national elevation dataset digital elevation model on floodplains of North Carolina. *Nat. Hazards Rev.* **2005**, *6*, 34–40. [CrossRef]
67. Sanders, B.F. Evaluation of on-line DEMs for flood inundation modeling. *Adv. Water Resour.* **2007**, *30*, 1831–1843. [CrossRef]
68. Coveney, S.; Fotheringham, A.S. The impact of DEM data source on prediction of flooding and erosion risk due to sea-level rise. *Int. J. Geogr. Inf. Sci.* **2011**, *25*, 1191–1211. [CrossRef]
69. Papaioannou, G.; Loukas, A.; Vasiliades, L.; Aronica, G.T. Flood inundation mapping sensitivity to riverine spatial resolution and modelling approach. *Nat. Hazards* **2016**, *83*, 117–132. [CrossRef]
70. Li, J.; Wong, D.W.S. Effects of DEM sources on hydrologic applications. *Comput. Environ. Urban Syst.* **2010**, *34*, 251–261. [CrossRef]
71. Jakovljevic, G.; Govedarica, M.; Alvarez-Taboada, F.; Pajic, V. Accuracy assessment of deep learning based classification of LiDAR and UAV points clouds for DTM creation and flood risk mapping. *Geosciences* **2019**, *9*, 323. [CrossRef]
72. Hsieh, Y.C.; Chan, Y.C.; Hu, J.C. Digital elevation model differencing and error estimation from multiple sources: A case study from the Meiyuan Shan landslide in Taiwan. *Remote Sens.* **2016**, *8*, 199. [CrossRef]
73. Sampson, C.C.; Smith, A.M.; Bates, P.D.; Neal, J.C.; Trigg, M.A. Perspectives on open access high resolution digital elevation models to produce global flood hazard layers. *Front. Earth Sci.* **2016**, *3*, 85. [CrossRef]
74. Courty, L.G.; Soriano-Monzalvo, J.C.; Pedrozo-Acuña, A. Evaluation of open-access global digital elevation models (AW3D30, SRTM, and ATSER) for flood modelling purposes. *J. Flood Risk Manag.* **2019**, *12*, e12550. [CrossRef]
75. Hashemi-Beni, L.; Jones, J.; Thompson, G.; Johnson, C.; Gebrehiwot, A. Challenges and opportunities for UAV-based digital elevation model generation for flood-risk management: A case of princeville, north carolina. *Sensors (Switzerland)* **2018**, *18*, 3842. [CrossRef]
76. Van de Sande, B.; Lansen, J.; Hoyng, C. Sensitivity of coastal flood risk assessments to digital elevation models. *Water (Switzerland)* **2012**, *4*, 568–579. [CrossRef]
77. Laks, I.; Sojka, M.; Walczak, Z.; Wrózyński, R. Possibilities of using low quality digital elevation models of floodplains in hydraulic numerical models. *Water (Switzerland)* **2017**, *9*, 283. [CrossRef]
78. Toda, L.L.; Yokingco, J.C.E.; Paringit, E.C.; Lasco, R.D. A LiDAR-based flood modelling approach for mapping rice cultivation areas in Apalit, Pampanga. *Appl. Geogr.* **2017**, *80*, 34–47. [CrossRef]
79. Wang, Z.; Lai, C.; Chen, X.; Yang, B.; Zhao, S.; Bai, X. Flood hazard risk assessment model based on random forest. *J. Hydrol.* **2015**, *527*, 1130–1141. [CrossRef]
80. Alfonso, L.; Mukolwe, M.M.; Di Baldassarre, G. Probabilistic Flood Maps to support decision-making: Mapping the Value of Information. *Water Resour. Res.* **2016**, *52*, 1026–1043. [CrossRef]

81. Webster, T.L. Flood risk mapping using LiDAR for annapolis Royal, Nova Scotia, Canada. *Remote Sens.* **2010**, *2*, 2060–2082. [CrossRef]
82. Haile, A.T.; Rientjes, T.H.M. Effects of LiDAR DEM resolution in flood modelling: A model sensitivity study for the city of Tegucigalpa, Honduras. In Proceedings of the ISPRS WG III/3, III/4, Enschede, The Netherlands, 12–14 September 2005; pp. 12–14.
83. Mihu-Pintilie, A.; Cîmpianu, C.I.; Stoleriu, C.C.; Pérez, M.N.; Paveluc, L.E. Using high-density LiDAR data and 2D streamflow hydraulic modeling to improve urban flood hazard maps: A HEC-RAS multi-scenario approach. *Water (Switzerland)* **2019**, *11*, 1832. [CrossRef]
84. Guerriero, L.; Ruzza, G.; Guadagno, F.M.; Revellino, P. Flood hazard mapping incorporating multiple probability models. *J. Hydrol.* **2020**, *587*, 125020. [CrossRef]
85. Duan, W.; Hanasaki, N.; Shiogama, H.; Chen, Y.; Zou, S.; Nover, D.; Zhou, B.; Wang, Y. Evaluation and future projection of Chinese precipitation extremes using large ensemble high-resolution climate simulations. *J. Clim.* **2019**, *32*, 2169–2183. [CrossRef]
86. Thatcher, C.; Lim, S.; Palaseanu-Lovejoy, M.; Danielson, J.; Kimbrow, D. Lidar-based mapping of flood control levees in South Louisiana. *Int. J. Remote Sens.* **2016**, *37*, 5708–5725. [CrossRef]
87. Horritt, M.S.; Bates, P.D. Effects of spatial resolution on a raster based model of flood flow. *J. Hydrol.* **2001**, *253*, 239–249. [CrossRef]
88. Podobnikar, T.; Vrečko, A. Digital Elevation Model from the Best Results of Different Filtering of a LiDAR Point Cloud. *Trans. GIS* **2012**, *16*, 603–617. [CrossRef]
89. Zhang, K.; Chen, S.C.; Singh, P.; Saleem, K.; Zhao, N. A 3D visualization system for hurricane storm-surge flooding. *IEEE Comput. Graph. Appl.* **2006**, *26*, 18–25. [CrossRef]
90. Abdullah, A.; Rahman, A.; Vojinovic, Z. LiDAR filtering algorithms for urban flood application: Review on current algorithms and filters test. *Int. Arch. Photogramm. Remote Sens. Spat. Inf. Sci.* **2009**, *38*, 30–36.
91. Muhadi, N.A.; Abdullah, A.F.; Kassim, M.S.M. Quantification of terrestrial laser scanner (TLS) elevation accuracy in oil palm plantation for IFSAR improvement. *IOP Conf. Ser. Earth Environ. Sci.* **2016**, *37*, 012042. [CrossRef]
92. Meng, X.; Currit, N.; Zhao, K. Ground filtering algorithms for airborne LiDAR data: A review of critical issues. *Remote Sens.* **2010**, *2*, 833–860. [CrossRef]
93. Zhang, K.; Chen, S.C.; Whitman, D.; Shyu, M.L.; Yan, J.; Zhang, C. A progressive morphological filter for removing nonground measurements from airborne LIDAR data. *IEEE Trans. Geosci. Remote Sens.* **2003**, *41*, 872–882. [CrossRef]
94. De Santis, R.; Macchione, F.; Costabile, P.; Costanzo, C. A comparative analysis of 3-D representations of urban flood map in virtual environments for hazard communication purposes. *E3S Web Conf.* **2018**, *40*, 8. [CrossRef]
95. Dutta, D.; Teng, J.; Vaze, J.; Lerat, J.; Hughes, J.; Marvanek, S. Storage-based approaches to build floodplain inundation modelling capability in river system models for water resources planning and accounting. *J. Hydrol.* **2013**, *504*, 12–28. [CrossRef]
96. Bates, P.D. Integrating remote sensing data with flood inundation models: How far have we got? *Hydrol. Process.* **2012**, *26*, 2515–2521. [CrossRef]
97. Abucay, E.R.; Tseng, Y.-H. Assessing landscape visibility using LiDAR, SAR DEM and globally available elevation data: The case of Bongabong, Oriental Mindoro, Philippines. In Proceedings of the 40th Asian Conference on Remote Sensing: Progress of Remote Sensing Technology for Smart Future, Daejeon, Korea, 14–18 October 2019.
98. Podhoranyi, M.; Fedorcak, D. Inaccuracy introduced by LiDAR-generated cross sections and its impact on 1D hydrodynamic simulations. *Environ. Earth Sci.* **2014**, *73*, 1–11. [CrossRef]
99. Muhadi, N.A.; Mohd Kassim, M.S.; Abdullah, A.F. Improvement of Digital Elevation Model (DEM) using data fusion technique for oil palm replanting phase. *Int. J. Image Data Fusion* **2019**, *10*, 232–243. [CrossRef]

© 2020 by the authors. Licensee MDPI, Basel, Switzerland. This article is an open access article distributed under the terms and conditions of the Creative Commons Attribution (CC BY) license (http://creativecommons.org/licenses/by/4.0/).

Article

A Neural-Network Based Spatial Resolution Downscaling Method for Soil Moisture: Case Study of Qinghai Province

Aifeng Lv [1,2,*], Zhilin Zhang [3] and Hongchun Zhu [3]

1. Key Laboratory of Water Cycle and Related Land Surface Processes, Institute of Geographic Sciences and Natural Resources Research, CAS, Beijing 100101, China
2. University of Chinese Academy of Sciences, Beijing 100049, China
3. School of Surveying and Mapping Science and Engineering, Shandong University of Science and Technology, Qingdao 266000, China; ZZL_ZY_1995@163.com (Z.Z.); hongchun@sdust.edu.cn (H.Z.)
* Correspondence: lvaf@igsnrr.ac.cn

Citation: Lv, A.; Zhang, Z.; Zhu, H. A Neural-Network Based Spatial Resolution Downscaling Method for Soil Moisture: Case Study of Qinghai Province. *Remote Sens.* **2021**, *13*, 1583. https://doi.org/10.3390/rs13081583

Academic Editor: Nicolas Baghdadi

Received: 29 March 2021
Accepted: 16 April 2021
Published: 19 April 2021

Publisher's Note: MDPI stays neutral with regard to jurisdictional claims in published maps and institutional affiliations.

Copyright: © 2021 by the authors. Licensee MDPI, Basel, Switzerland. This article is an open access article distributed under the terms and conditions of the Creative Commons Attribution (CC BY) license (https://creativecommons.org/licenses/by/4.0/).

Abstract: Currently, soil-moisture data extracted from microwave data suffer from poor spatial resolution. To overcome this problem, this study proposes a method to downscale the soil moisture spatial resolution. The proposed method establishes a statistical relationship between low-spatial-resolution input data and soil-moisture data from a land-surface model based on a neural network (NN). This statistical relationship is then applied to high-spatial-resolution input data to obtain high-spatial-resolution soil-moisture data. The input data include passive microwave data (SMAP, AMSR2), active microwave data (ASCAT), MODIS data, and terrain data. The target soil moisture data were collected from CLDAS dataset. The results show that the addition of data such as the land-surface temperature (LST), the normalized difference vegetation index (NDVI), the normalized shortwave-infrared difference bare soil moisture indices (NSDSI), the digital elevation model (DEM), and calculated slope data (SLOPE) to active and passive microwave data improves the retrieval accuracy of the model. Taking the CLDAS soil moisture data as a benchmark, the spatial correlation increases from 0.597 to 0.669, the temporal correlation increases from 0.401 to 0.475, the root mean square error decreases from 0.051 to 0.046, and the mean absolute error decreases from 0.041 to 0.036. Triple collocation was applied in the form of [NN, FY3C, GEOS-5] based on the extracted retrieved soil-moisture data to obtain the error variance and correlation coefficient between each product and the actual soil-moisture data. Therefore, we conclude that NN data, which have the lowest error variance (0.00003) and the highest correlation coefficient (0.811), are the most applicable to Qinghai Province. The high-spatial-resolution data obtained from the NN, CLDAS data, SMAP data, and AMSR2 data were correlated with the ground-station data respectively, and the result of better NN data quality was obtained. This analysis demonstrates that the NN-based method is a promising approach for obtaining high-spatial-resolution soil-moisture data.

Keywords: soil moisture; neural network; downscaling; microwave data; MODIS data

1. Introduction

Moisture stored in surface soil accounts for less than 0.001% of total global freshwater by volume but plays an important role in connecting global terrestrial water, energy, and carbon cycling processes [1]. By influencing soil evaporation and transpiration, soil moisture (SM) strongly affects the interaction between the land surface and the atmosphere [2]. Thus, a thorough understanding of SM can contribute to efficient monitoring of the climate and environmental changes and provide valuable guidance for drought monitoring and flood forecasting in agriculture and forestry [3]. In addition, SM determines the distribution of precipitation infiltration and surface runoff, which controls plant growth [4]. Therefore, high-quality SM data is crucial in multiple technological fields, such as hydrology, meteorology, climatology, and water-resources management.

Traditional methods to monitor SM usually rely on automatic or manual collection methods, which have the advantages of temporal continuity and guaranteed accuracy. However, these methods are unsatisfactory because, for starters, there are insufficient observation stations, which is especially serious because SM results are representative only of the soil near the given station. In addition to poor spatial representation, these methods are time-consuming and labor-intensive [5].

However, recent developments in remote-sensing methods have created the possibility to obtain large-scale, long-term soil-moisture data. In this field, microwave radiometers have become the most important source of global SM data due to their better temporal sampling features. In particular, microwave bands such as the L (0.5–1.5 GHz), C (4–8 GHz), and X (8–12 GHz) bands have been widely used to measure SM [6]. Currently, four passive microwave satellites and one active microwave satellite monitor SM globally. Four passive microwave sensors are currently in orbit: the microwave radiation imager (MWRI), which operates in the X-band, onboard the Fengyun-3 (FY3) satellite launched by the China National Space Administration (2008–present) [7], the Advanced Microwave Scanning Radiometer (AMSR2), which operates in the X and C bands, onboard the Global Change Observation Mission-Water (GCOM-W) satellite launched by the Japan Aerospace Exploration Agency (JAXA) (2012–present) [8], and two dedicated satellites equipped with L-band radiometers: the Soil Moisture and Ocean Salinity (SMOS) (2009–present) instrument launched by the European Space Agency (2010–present) [9] and the Soil Moisture Active Passive (SMAP) instrument launched by the National Aeronautics and Space Administration (NASA) (2015–present) [10]. Another contributor is the ASCAT (2007–present) instrument, which monitors active scatterer in the C band from the MEOP satellite launched by the ESA and is an important source of active microwave data [11]. These microwave radiometers have the advantages of providing a complete observation of the global land surface within two to three days and providing surface soil-moisture information on a large scale. Their major disadvantage, however, is the poor spatial resolution of the microwave radiometer, which is typically about 25–40 km. However, SM is subject to complex interactions between topography, soil, vegetation, and other meteorological factors, which leads to high spatial variability. Therefore, many regional hydrological and agricultural applications require SM data with a spatial resolution of several kilometers or even tens of meters. It is thus vital to develop techniques to obtain accurate, high-precision, soil-moisture data with high coverage.

The low spatial resolution of soil-moisture data extracted from passive microwave data is typically downscaled by combining it with other high-spatial-resolution data. Based on the combined data type, the following two categories emerge: (i) combinations of active and passive microwave data and (ii) combinations of visible, infrared, and microwave data. In previous work, Njoku et al. combined radar (active) and radiometer (passive) data to study SM under vegetated-terrain cover and analyzed the sensitivity with which multi-channel low-frequency passive and active measurements can detect SM under different vegetation conditions [12]. In other work, Das et al. obtained a linear relationship between radar backscatter and soil-moisture data by merging coarse-scale radiometer SMAP SM data with the fine-scale backscatter coefficient to produce high-spatial-resolution (9 km) SM data [13]. Zhan et al. used a Bayesian method to merge relatively accurate 36-km radiometer brightness temperature with the relatively noisy 3-km radar backscatter coefficient and explored the potential for retrieving SM from these results. Their results prove that the Bayesian method produces better data than direct extraction of either the brightness temperature or radar backscatter [14]. To combine visible and infrared remote sensing with passive microwave data, Wilson et al. combined and weighted terrain maps and other spatial attributes according to the correlations to generate SM data [15]. Srivastava et al. used artificial neural networks (NN), support vector machines, relevance vector machines, and generalized linear models to combine MODIS surface temperature with SM retrieved by SMOS to conclude that the artificial NN produced better results than other methods [16]. Yang et al. estimated soil parameters by assimilating the brightness temperature data simulated by the land surface model and the radiative transfer model. By minimizing the brightness temperature errors of AMSR2, they estimated the SM [17]. In researching SM downscaling, Chen et al. used dual Kalman filters to assimilate the brightness temperature of AMSR-E with the MODIS surface temperature [18]. Finally, Chauhan et al. used the universal triangle approach to link the high-resolution normalized difference vegetation index (NDVI), surface albedo, and land-surface temperature to SM data, thereby disaggregating low-spatial-resolution microwave SM into high-spatial-resolution SM [19]. The common idea behind these methods is to establish a statistical correlation or physical model between SM and auxiliary variables.

Qinghai Province is in the northeastern part of the Qinghai-Tibet Plateau, which is the source of the Yangtze River, the Yellow River, and the Lancang-Mekong River, and is an important water-conserving area in China and Asia [20]. In recent years, under the influence of global warming, the climate of Qinghai has been warming and humidifying, glaciers and snowfields are shrinking year by year, rivers, lakes, and wetlands are shrinking, soil erosion is expanding, and the water-conserving function is deteriorating seriously. Soil moisture is an important surface characteristic parameter and has an irreplaceable role in related land degradation, drought monitoring, and water conservation monitoring [21]. Therefore, an urgent need exists to systematically monitor the soil moisture information in Qinghai Province, which is an area that seriously lacks ground truth data, making the use of remote sensing data to retrieve SM in Qinghai Province of significant potential value.

The quality of remote sensing data largely determines the results of remote-sensing retrieval of soil moisture. Existing studies, such as those mentioned above, usually involve only a single passive microwave radiometer or a single passive microwave radiometer combined with a single active microwave radar for SM retrieval and do not involve the three bands L, C, and X simultaneously [22]. In this paper, we use the powerful multivariate and nonlinear fitting capability of NN to analyze the single-band as well as multi-band synergy in detecting soil moisture in the region of Qinghai Province for the three bands L, C, and X and select multiple microwave sensors (SMAP, AMSR2, FY3C, ASCAT) as data sources. The ability to detect SM information in Qinghai Province through multi-band synergy compensates for the shortcomings of insufficient information from a single sensor. At the same time, elevation and slope data are introduced to treat the complex topography of Qinghai Province to make the algorithm more universal [23]. Finally, we use MODIS data with high spatial resolution and topographic data to downscale SM experiments with the NN model trained with low-resolution data.

The paper is organized as follows: Section 2 explains the data and methods used in this study. Section 3 presents and discusses the main findings. Finally, Section 4 gives the main conclusions of the study.

2. Materials and Methods

2.1. Data

2.1.1. Microwave Data

This study relies mainly on the passive microwave soil-moisture datasets SMAP, FY3C, and AMSR2 and the active microwave soil-moisture dataset of ASCAT as input microwave data. The datasets FY3C (10.65, 18.7, 23.8, 36.8, 89 GHz) and AMSR2 (6.925, 7.3, 10.65, 18.7, 23.8, 36.5, 89 GHz) have multiple frequencies, whereas SMAP (1.41 GHz) and ASCAT (5.3 GHz) have fixed frequencies; see Table 1 for details. In general, the L, C, and X bands are sensitive to SM, whereas other bands in other frequency ranges are not sensitive to SM. Therefore, this study uses the brightness temperature from the 1.41 GHz channel from SMAP, from the 6.9, 7.3, and 10.65 GHz channels of AMSR2, and from the 10.65 GHz channel of FY3C. Specifically, the SMAP radiometer uses a 24 MHz bandwidth centered at 1.41 GHz. The AMSR2 radiometer uses 0.35, 0.35, 0.10 GHz bandwidths centered at 6.9, 7.3, and 10.65 GHz, respectively. The FY3C radiometer uses a 180% ± 10% MHz bandwidth centered at 10.65 GHz. In addition, we use the backscatter coefficient (σ40) from the 5.3 GHz channel of the ASCAT active microwave data as the input variables for the NN. The specific spatial distribution of soil moisture is shown in Figure 1.

Table 1. Comparison of global soil-moisture products.

Satellite	Unit	Ascending/Descending	Spatial Resolution	Time Series
SMAP	m^3/m^3	18:00/6:00	36 km	2015–present
AMSR2	m^3/m^3	13:30/1:30	0.25°	2012–present
FY3C	m^3/m^3	13:40/1:40	25 km	2014–present
ASCAT	m^3/m^3	——	25 km	2007–present
GEOS-5	m^3/m^3	——	0.25° × 0.3125°	2014–present
CLDAS	m^3/m^3	——	0.0625° × 0.0625°	2017–present

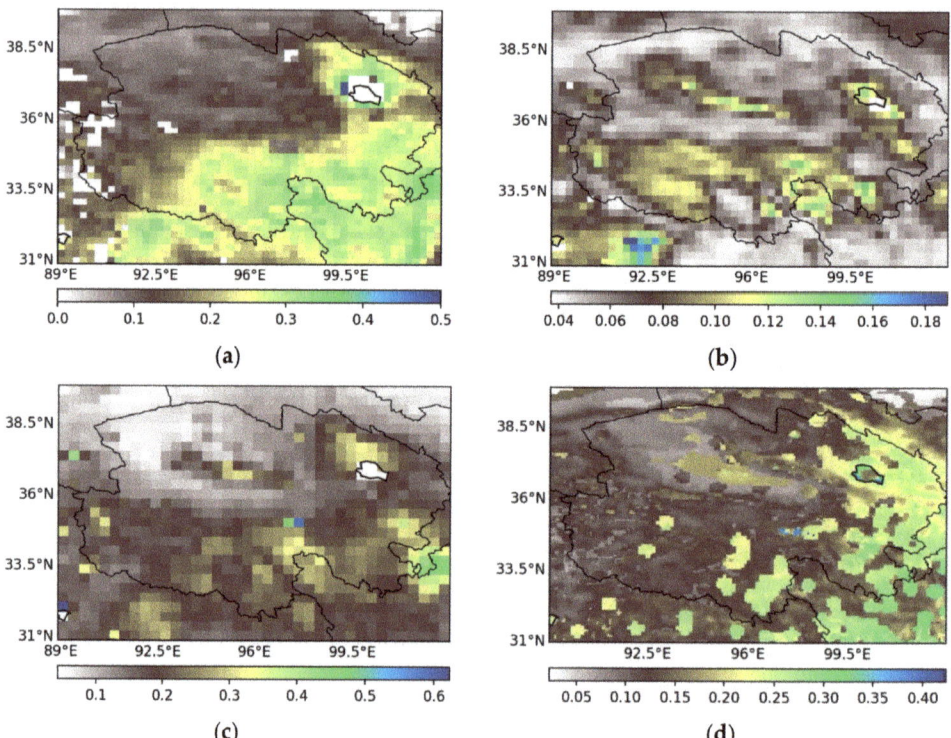

Figure 1. Daily average soil-moisture maps from (a) FY3C; (b) AMSR2; (c) SMAP; (d) CLDAS.

2.1.2. Data from Land Surface Model

As target data, we use the CLDAS soil volumetric water content analysis product, which is published by China Meteorological Data Service Centre. Comparing the quality-controlled SM observation data from automated monitoring stations in China with the CLDAS-V2.0 soil volumetric water content data shows that the CLDAS soil volumetric water content product fits the actual ground observation data [24], with a national regional average correlation coefficient of 0.89, a root mean square error (RMSE) of 0.02 m^3/m^3, and a deviation of 0.01 m^3/m^3. Therefore, CLDAS SM products are considered of higher quality than similar international products (such as the GLDAS and NLDAS products), and they also offer better spatial and temporal resolution. In addition, we use the SM data from the SM ground model GEOS-5 as the input product for triple collocation (TC). The GEOS-5 model provides hourly products, and this study uses the time of 05:30 for the surface SM dataset to represent the SM between 0 and 7 cm within the surface layer [25].

2.1.3. MODIS Data and Terrain Data

The auxiliary input data included land surface temperature (LST) and vegetation index (VI), and the VI data further included the enhanced VI (EVI) and the NDVI [26]. The daily LST data were provided by the MODIS-Terra LST product (MOD11A1) with a spatial resolution of 1 km. The EVI and NDVI data were provided by the MODIS-Terra VI products (MOD13A2) and have a temporal resolution of 16 days and a spatial resolution of 1 km. This study also uses data from the MODIS-Terra surface reflectance product (MOD09A1), with a temporal resolution of 8 days and a spatial resolution of 500 m. The annual land cover data (MCD12Q1) were also used here. The specific spatial distribution of vegetation cover is shown in Figure 2. In addition, we use the results of the SRTM 90-m digital elevation model (DEM) data and the calculated slope data (SLOPE) based on the DEM data for the study area. Table 2 summarizes the datasets used. The above data were obtained from the land processes distributed active archive center (https://lpdaac.usgs.gov/, accessed on 29 March 2021).

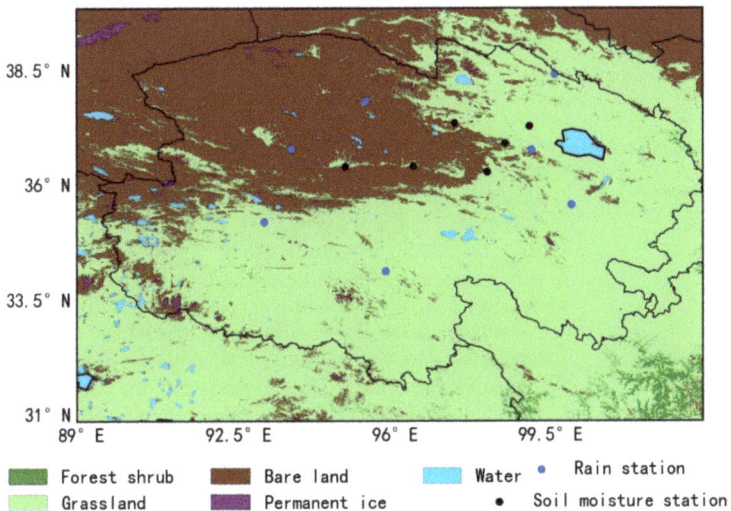

Figure 2. Map of vegetation cover (MCD12Q1).

Table 2. Summary of auxiliary input variables.

Input Variables	Dataset	Spatial Resolution	Temporal Resolutions	Time Series
NDVI	MOD13A2	1 km	16 days	2000–present
EVI	MOD13A2	1 km	16 days	2000–present
LST	MOD11A1	1 km	daily	2000–present
DEM	SRTM	90 m	-	2000
SLOPE	SRTM	90 m	-	2000
NSDSI	MOD09A1	500 m	8 days	2000–present

2.1.4. In Situ Observations

The measured data used in this paper are mainly divided into SM data and precipitation data from stations in Qinghai area. SM data are obtained from a soil depth of 10 cm and in time intervals of hours from six automated SM stations (Delingha, Dulan,

Golmud, Nomuhong, Tianjun, and Wulan). The time interval for precipitation data is one day, and the areas contain the seven stations Yeniugou, Xiaozaohuo, Dachaidan, Chaka, Wudaoliang, Xinghai, Qumarai.

2.1.5. Data Preprocessing

Due to snow and ice coverage in winter, active and passive microwave data were not available from December to March in the study area because snow cover and frozen soil typically cover the land surface, which might introduce large biases in satellite-retrieved products such as SM [27]. Therefore, this study uses the satellite data and ground station data for Qinghai Province (31°–40° N, 89°–103° E) from 1 April 2017 to 30 November 2017 and from 1 April 2018 to 30 November 2018. To apply satellite observation data and SM data in the NN model, all data were resampled to a grid of 0.25° × 0.3125°. SMAP and AMSR2 data were treated by using bilinear interpolation, ASCAT and FY3C data by the inverse distance weighted method, and CLDAS, MODIS, and terrain data by simple average aggregation. Since the passive microwave datasets (SMAP, AMSR2, and FY3C) include both ascending and descending orbits, we processed the data from these orbits separately. In addition, the brightness temperature data of the passive microwave data could be divided into vertical and horizontal polarization channels, based on which the microwave polarization difference index (MPDI) is calculated as [28]

$$MPDI = (Tbv - Tbh)/(Tbv + Tbh), \qquad (1)$$

where Tbv and Tbh are the brightness temperature of the vertical and horizontal polarizations, respectively. Based on the assumption that the microwave channel is not subject to strong atmospheric attenuation, the MPDI is designed to eliminate the influence of surface temperature on microwave signals. In addition, it is a normalized polarization difference, which can serve as an indicator of SM status as a function of incident angle. In addition, the MPDI is sensitive to the dielectric properties of soil, and even more so to the surface roughness. Therefore, the MPDI is high for flat surfaces but relatively low for rough surfaces, such as areas with vegetation cover. In the preprocessing of the ASCAT data, the $\sigma 40$ time series of each grid was renormalized to the range of [0 1], which means that the highest (lowest) backscatter value measured in this study is assigned the value 1 (0). The backscatter time index obtained from this preprocessed ASCAT data is abbreviated "BTI" [29]. This processing method emphasizes the time mode of the ASCAT signal and has been shown to reduce the retrieval time. However, since the processing is performed on the grid, it might reduce the spatial information provided by the radar.

2.2. Method

2.2.1. Triple Collocation Method

The traditional error estimation method typically compares retrieved SM data with actual observation data obtained from ground stations. However, such comparisons are usually limited in number and location of instrument verification points, which makes it difficult to ensure robust datasets. In addition, the spatial mismatch between the ground data and the remote-sensing satellite data, as well as the heterogeneity of the ground surface, lead to representative errors and scale-conversion errors. Therefore, we used TC analysis [30–32] to estimate SM error. Compared with the traditional method to estimate SM error, it (1) does not require a high-quality reference dataset, which means that it can verify the three different SM data in the study area without ground measurement data. (2) Triple collocation simultaneously obtains the error variances of the three different SM data and (3) avoids the representative error caused by the spatial mismatch between the ground measurement data and the remote-sensing satellite data in the traditional estimation method. (4) Finally, the improved extended TC method [33] detects correlations between the retrieved SM data and the actual surface layer SM data. The error variance is expressed by [33]

$$\sigma^2_{\varepsilon_Y} = \sigma^2_Y - \frac{\sigma_{YX}\sigma_{YZ}}{\sigma_{XZ}},$$
$$\sigma^2_{\varepsilon_Y} = \sigma^2_Y - \frac{\sigma_{YX}\sigma_{YZ}}{\sigma_{XZ}},$$
$$\sigma^2_{\varepsilon_Z} = \sigma^2_Z - \frac{\sigma_{ZX}\sigma_{ZY}}{\sigma_{XY}},$$
(2)

where σ^2_X is the variance of X, and σ_{XY} is the covariance of X and Y. The correlation coefficient is [33]

$$R_X = \sqrt{\frac{\sigma_{XY}\sigma_{XZ}}{\sigma^2_X \sigma_{YZ}}},$$
(3)

where R_X is the correlation between X and the unknown true SM state.

In this study, the SMs retrieved by the NN method, collected from satellite data, and obtained from the ground model GEOS-5 are triple-matched in the form [NN data, satellite data, reanalysis data] to estimate the error variance and correlation coefficient of the TC. The SM obtained from the NN method is evaluated based on these results.

2.2.2. Evaluation Index

This study uses spatial correlation, temporal correlation, root mean square error, and mean absolute error to quantitatively analyze the aspects that differentiate SM retrieved by the NN and SM obtained from a model, as well as aspects that differentiate SM retrieved by the NN and SM collected from ground stations.

(1) Spatial correlation: ρspatial

Spatial correlation serves to evaluate the accuracy with which the spatial model retrieves SM. It is obtained by calculating the Pearson correlation coefficient between the retrieved SM, which produces a daily correlation value between the whole area and the simulated SM map. For a better comparison, the average spatial correlation is calculated as the average of all daily spatial correlations with a significance greater than 95%.

(2) Temporal correlation: ρtemporal

Temporal correlation is used to evaluate how well the retrieved SM matches the temporal variations in the SM. It is a location-related metric calculated at the pixel level. The Pearson correlation between the retrieved time series and the modeled SM is calculated for each pixel, which gives a correlation map. The mean temporal correlation is the mean value of all the pixels in the temporal correlation map.

(3) Root mean square error: RMSE

The RMSE is calculated based on the unit error and the deviation from a reference of the unit error. Therefore, it provides a comprehensive assessment of recalculation, including the accuracy and precision of data retrieval. The RMSE is calculated at the pixel level by using the original SM time series, and a map of the RMSE is obtained for each retrieval. The mean RMSE is the mean value of all the pixels in the RMSE map.

(4) Mean absolute error: MAE

The MAE is the absolute error between the retrieved SM and the simulated SM. It is calculated at the pixel level, and each search generates a map. The MAE correlation is the mean value of all the pixels in the MAE map.2.2.3. Downscaling scheme based on neural network.

A NN [34–36] is essentially a system to do nonlinear mathematical calculations and can represent any complex nonlinear process. The multivariable nature and nonlinear ability of NN fully exploit the synergy between different data. The NN used in this study has three parts: (1) an input layer, which receives the satellite observation data and auxiliary variable inputs; (2) a hidden layer; and (3) an output layer, which provides the SM. This structure suffices to fit any continuous function.

The NN was trained with satellite observation data as input data and the corresponding ground model data as target data. The training dataset must represent the entire range of expected scenarios, which means that it must include all climate regimes and seasons. If the training data are well selected, the NN's performance when applied to the training data should differ little from its performance when applied to the entire dataset. Similarly, a NN should perform in the same way when applied to two sufficiently representative but completely different target datasets, meaning that any potential local or regional bias in the target data is corrected. These characteristics can be traced to the fact that the estimated spatial-temporal structure of the NN is determined by satellite observations instead of by target data [37]. In addition, the NN correlates the satellite observations with the most common SM among the input values in the target data, regardless of the location or acquisition time of the data [29].

The NN constructed in this study uses the Levenberg-Marquardt (LM) [38,39] training algorithm and applies error backpropagation [40] to update the weights. Since the LM algorithm stops when it finds a local minimum, the error surface is not fully explored. Therefore, in this study, the NN training was repeated four times, each time using random initial NN weights to ensure different starting points on the error surface; the optimal NN was selected for retrieving SM products.

The key step in downscaling SM in this study is to build a statistical relationship using low spatial-resolution data, and then input high-spatial-resolution data into the statistical relationships to obtain the downscaled SM. [26,41,42] The spatial scale of different data is unified and scaled by different resampling methods, as shown in Section 2.1.5. In particular, the low-spatial-resolution microwave data are resampled to 1 km spatial resolution by replication expansion, without changing the specific values, to make them consistent with the spatial resolution of MODIS and other auxiliary data. Figure 3 shows a flow chart of this process, which is described as follows:

- Aggregate auxiliary data (NDVI, EVI, LST, NSDSI, DEM, and SLOPE) into a grid with a resolution of 0.25° × 0.3125°, which is consistent with the spatial resolution of the resampling microwave data (Tbh, Tbv, σ40, and BTI). Specific bands of microwave data are available in Section 2.1.1. The relationship between these different input variables and the ground model CLDAS SM was established through the NN, and the quality of the NN SM was evaluated by comparing it with the CLDAS SM.
- Evaluate the SM dataset obtained from the NN model by using the TC method.
- Input the 1 km medium-resolution data from 2017 to 2018 into the verified NN model to obtain 1-km-resolution SM data.
- Use the data collected from the ground station to verify the downscaled NN SM data.

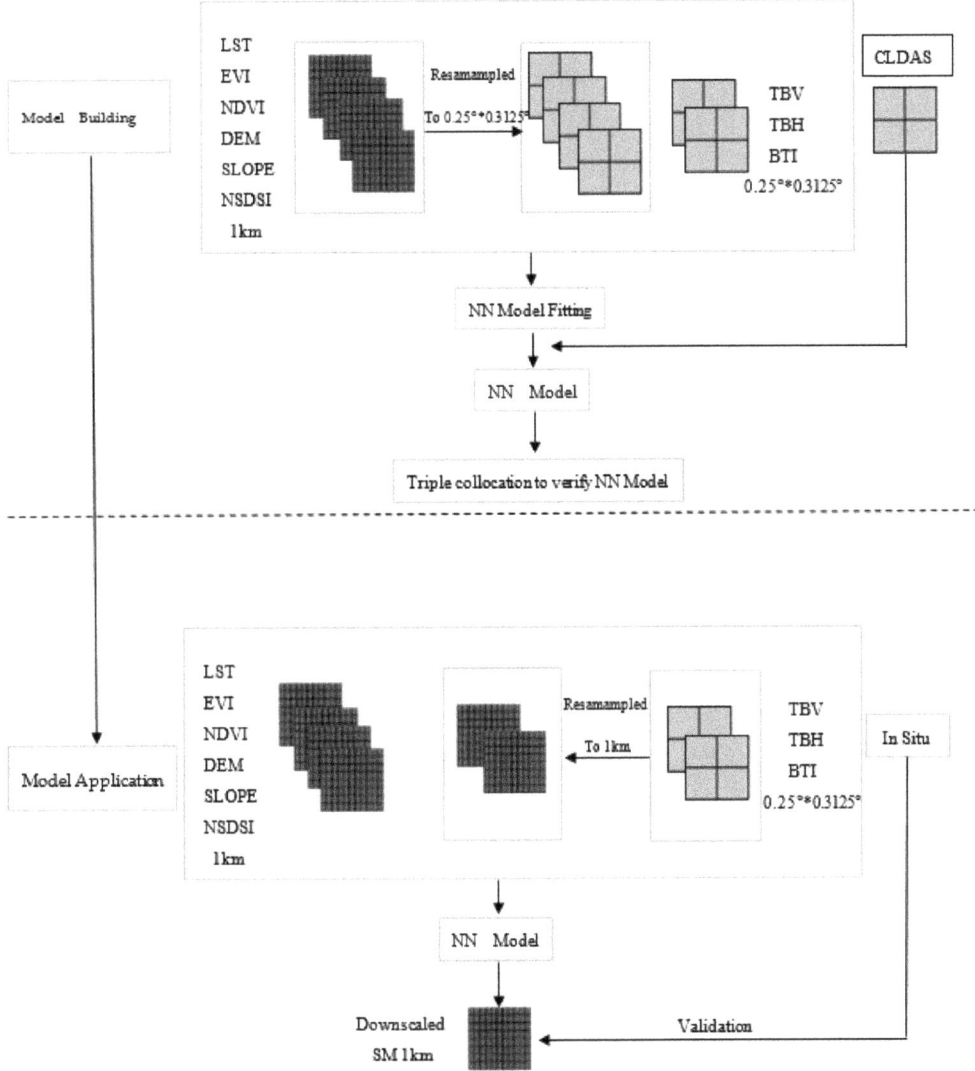

Figure 3. Schematic of the proposed SM downscaling method.

3. Results and Discussion

3.1. Selection Microwave Band

By studying the quality of data retrieved from various satellites and in different bands, we identified the SM data from different sensors. This exercise was done using the CLDAS SM dataset as reference data. Although the NN model was trained on a small subset of the available dataset, the entire dataset was used for retrieval and evaluation. Table 3 summarizes the average quality index of the SM calculated by comparing a single microwave input dataset with the target SM dataset (CLDAS). Table 4 summarizes the average quality index of the SM calculated by comparing a combination microwave input dataset with the target SM dataset (CLDAS). Below, we discuss in detail the results of using different satellites and different bands.

Table 3. Average quality index for soil moisture calculated by comparing a single microwave input dataset with data from CLDAS.

Ascending	ρspatial	ρtemporal	RMSE	MAE	Descending	ρspatial	ρtemporal	RMSE	MAE
SMAP					SMAP				
1.41 Tbv	0.388	0.322	0.065	0.053	1.41 Tbv	0.374	0.235	0.066	0.055
1.41 Tbh	0.254	0.333	0.067	0.055	1.41 Tbh	0.306	0.260	0.068	0.056
1.41 MPDI	0.186	0.291	0.069	0.057	1.41 MPDI	0.121	0.259	0.070	0.058
AMSR2					AMSR2				
6.9 Tbv	0.159	0.158	0.070	0.058	6.9 Tbv	0.250	0.182	0.068	0.056
6.9 Tbh	0.396	0.187	0.064	0.052	6.9 Tbh	0.474	0.182	0.062	0.050
6.9 MPDI	0.486	0.186	0.063	0.051	6.9 MPDI	0.509	0.214	0.061	0.050
7.3 Tbv	0.171	0.176	0.069	0.057	7.3 Tbv	0.275	0.166	0.067	0.056
7.3 Tbh	0.406	0.191	0.063	0.052	7.3 Tbh	0.488	0.194	0.061	0.050
7.3 MPDI	0.491	0.186	0.062	0.051	7.3 MPDI	0.511	0.217	0.061	0.050
10.7 Tbv	0.186	0.201	0.069	0.057	10.7 Tbv	0.312	0.216	0.066	0.054
10.7 Tbh	0.459	0.202	0.062	0.050	10.7 Tbh	0.525	0.205	0.059	0.048
10.7 MPDI	0.513	0.182	0.061	0.050	10.7 MPDI	0.529	0.214	0.060	0.049
FY3C					FY3C				
10.7 Tbv	0.249	0.183	0.067	0.055	10.7 Tbv	0.182	0.232	0.067	0.055
10.7 Tbh	0.471	0.188	0.061	0.049	10.7 Tbh	0.428	0.233	0.060	0.049
10.7 MPDI	0.511	0.177	0.060	0.049	10.7 MPDI	0.470	0.185	0.060	0.049
ASCAT					ASCAT				
σ40	0.259	0.334	0.066	0.054	BTI	0.269	0.336	0.064	0.053

Table 4. Average quality index for soil moisture calculated by comparing different microwave observation combinations with data from CLDAS.

Input Variable	ρspatial	ρtemporal	RMSE	MAE
SMAP_TBV_A_AMSR2_TBH_D	0.621	0.393	0.053	0.043
SMAP_TBV_A_AMSR2_TBH_D_σ40	0.604	0.393	0.051	0.041
SMAP_TBV_A_AMSR2_TBH_D_BTI	0.597	0.401	0.051	0.041
SMAP_TBV_A_AMSR2_MPDI_D	0.600	0.362	0.055	0.044
SMAP_TBV_A_AMSR2_MPDI_D_σ40	0.583	0.354	0.053	0.043
SMAP_TBV_A_AMSR2_MPDI_D_BTI	0.573	0.381	0.053	0.042

SMAP_TBV_A: Ascending Tbv in the 1.41 GHz (SMAP) band; AMSR2_TBH_D: Descending Tbh in the 10.7 GHz (AMSR2) band; AMSR2_MPDI_D: Descending MPDI in the 10.7 GHz (AMSR2) band; σ40: σ40 from ASCAT; BTI: BTI from ASCAT.

As shown in Table 3, in the 1.41 GHz (SMAP) band, Tbv has a higher spatial sensitivity to SM than Tbh, and the quality of Tbv in the ascending orbit exceeds that of Tbv in the descending orbit, with the average spatial correlation increased by 0.014 and the average temporal correlation increased by 0.087, and RMSE and MAE decreased by 0.001 and 0.002, respectively. In the 6.9, 7.3, and 10.7 GHz bands, Tbh is more sensitive to SM than is Tbv. In addition, the MPDI obtained from preprocessing in these bands also has greater spatial sensitivity to SM. Based on the AMSR2 microwave data, the 10.7 GHz band produces greater spatial and temporal correlation and lower RMSE and MAE in Qinghai Province compared with the 6.9 and 7.3 GHz bands and FY3C's 10.7 GHz band, which indicate a higher sensitivity to SM. In addition, the experiments show that Tbh and MPDI are highly similar in terms of spatial distribution in the 6.9, 7.3, and 10.7 GHz bands but differ significantly from Tbv in the 1.41 GHz band, which leads to the assumption that complementary relationships exist between them. The processed BTI data were also more sensitive to soil moisture than the original σ40 data, with an increase of 0.01 in the average spatial correlation and an increase of 0.002 in the average temporal correlation, whereas the RMSE and MAE decreased by 0.002 and 0.001, respectively. Finally, the best microwave band combination in Qinghai province was selected by joint retrieval of Tbv in the ascending orbit of 1.41Ghz (SMAP) band, Tbh and MPDI in the descending orbit of 10.7Ghz (AMSR2) band, and BTI, σ40 data of ASCAT. Figure 4 shows the raw images of these five single bands, a map of daily average SM as obtained by the NN, and a map of the temporal correlation between the NN SM and the CLDAS SM, respectively.

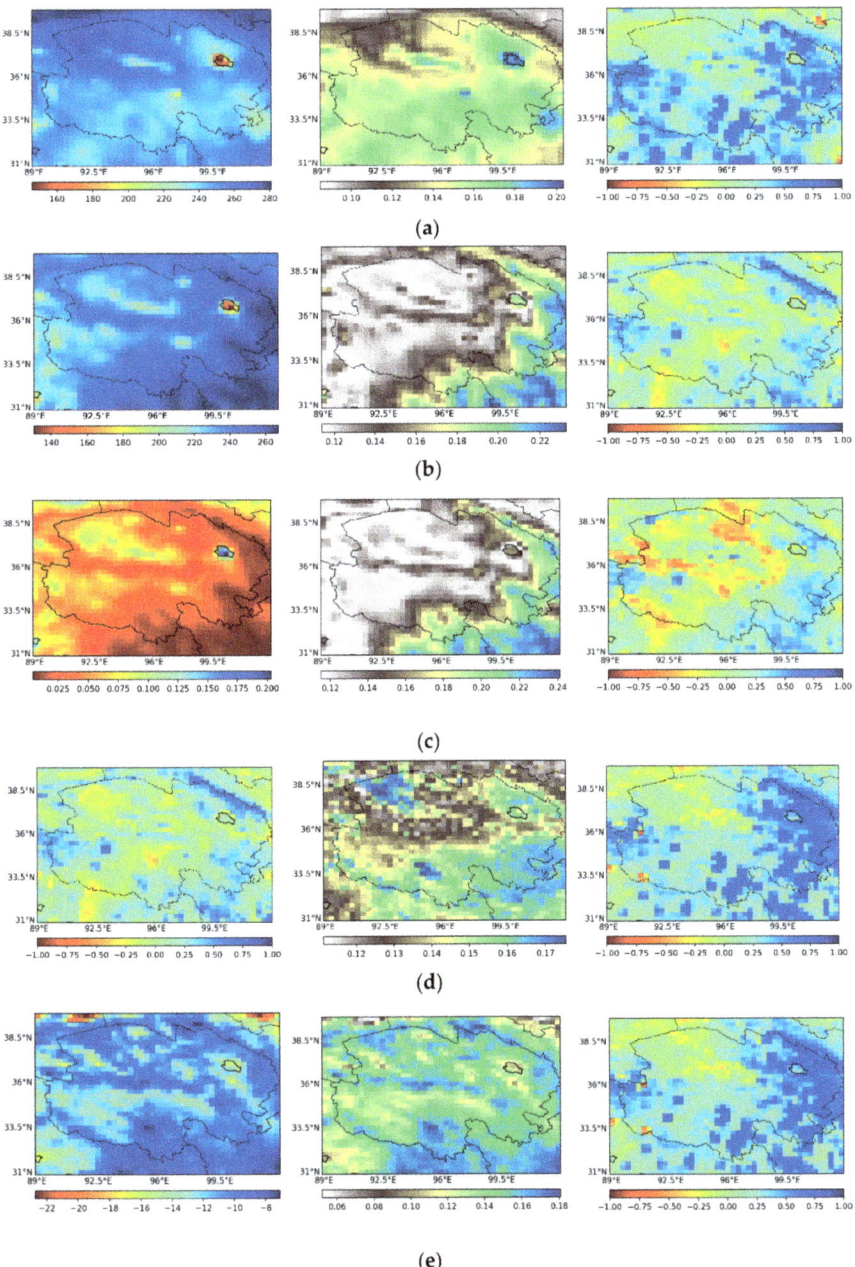

Figure 4. From top to bottom: (**a**) Ascending Tbv in 1.41 GHz band (SMAP); (**b**) Descending Tbh in 10.7 GHz band (AMSR2); (**c**) Descending MPDI in 10.7 GHz band (AMSR2); (**d**) and BTI data from ASCAT; and (**e**) σ40 data from ASCAT. From left to right: original image, map of daily average soil moisture map, as obtained by the NN, and map of temporal correlation between NN soil moisture and CLDAS soil moisture.

Figure 4 shows the SM monitoring capability of different wavebands in Qinghai Province. The first row of Figure 4 shows that the original Tbv data at 1.41 GHz indicate higher temperatures in bare land and forest areas than in grassland areas, and both bare land and forest areas have lower daily average SM, which means that more information is needed to distinguish forest areas from bare lands when retrieving SM. Meanwhile, lake areas, such as the Qinghai Lake area in the northeast, are coolest. The temporal correlation map indicates a weak negative correlation in hotter areas (higher Tbv), such as the bare lands in the northwest (Qaidam Basin) and the bare lands in the northeast corner, and a strong positive correlation in cooler areas. The original Tbh maps in the second row reveal a high sensitivity to vegetation; combining these with the maps of average daily SM shows that higher vegetation coverage and higher temperature results in greater SM. However, the poor distinction between the bare lands in the northwest and the mixture of bare lands and grassland in the southwest indicates that more information is needed to distinguish between these two areas. The observation of the MPDI image in the third row shows a high similarity in spatial distribution with the TBH in the second row. The BTI maps in the fourth row show higher BTIs in the grassland in the southeast and in the hinterland of the Qaidam Basin in the northwest, which reflects greater SM in the SM map, and the mixture of bare lands and grassland around the Qaidam Basin has a lower BTI, which reflect a lower SM. Therefore, the correlation map shows a negative correlation of BTI in the northwest corner of Qaidam Basin and a strong positive correlation for the mixture of bare lands and grassland in the southwest and the mixture of bare lands and grassland in the northeast. The observation of the $\sigma 40$ image in the fifth row shows that its spatial distribution is highly similar to that of the NN SM and CLDAS SM time correlation map of BTI in the fourth row. However, the differentiation between different areas of vegetation cover is worse in the SM map.

Figure 5 shows that the spatial distributions of SM obtained by NN retrieval of the four different combinations of data are highly similar to each other, and the overall SM increases from northwest to southeast, which clearly distinguishes bare soil areas, bare soil and grassland mixed areas, grassland areas, and forest areas, making up for the lack of detection capability of the single microwave band in Figure 4. Figure 5b,d with BTI data added at the same time do a better job distinguishing bare soil areas compared with Figure 5a,c with $\sigma 40$ added [i.e., bare soil areas in Figure 5b,d have lower SM values and are better distinguished compared with grassland areas in the same image]. The NN SM and CLDAS SM time-correlation maps show that all four images are poorly correlated with CLDAS data in the northwest bare soil region and in the southeast corner of the mixed forest-steppe region but achieve better correlation in all other regions. Figure 5a,b show greater positive correlation in the northeast region than do Figure 5c,d, which confirms that the 10.7 Ghz (AMSR2) band descending-orbit TBH is more capable of detecting SM information in Qinghai province than is the 10.7Ghz (AMSR2) band descending-orbit MPDI.

The results in Table 4 show that the combined SMAP_TBV_A and AMSR2_TBH_D produce higher-quality SM data than the combination of SMAP_TBV_A and AMSR2_MPDI_D. Taking the CLDAS soil moisture data as a benchmark, for SMAP_TBV_A and AMSR2_TBH_D the combination spatial correlation and temporal correlation reach 0.621 and 0.393, respectively, for SMAP_TBV_A and AMSR2_MPDI_D the spatial correlation and temporal correlation reach 0.600 and 0.354, respectively. Meanwhile, given the high similarity between Tbh and MPDI (see Figure 4), AMSR2_TBH_D is used as a final input variable. Furthermore, the addition of $\sigma 40$ and BTI to the above NN reduces the RMSE and MAE. Compared with the original $\sigma 40$ data, BTI data obtained after preprocessing translate into a greater temporal correlation between the SM obtained by the NN model and CLDAS data. And based on experience, active and passive microwaves have different sensitivities to SM, vegetation, and surface roughness. The 5.3 GHz observation frequency of ASCAT also differs significantly from that of SMAP (1.41 GHz) and AMSR2 (10.7 GHz). Therefore, the ASCAT dataset is considered as a potentially useful dataset that could compensate for the combination of passive microwave data in the NN. BTI is used as a final input variable.

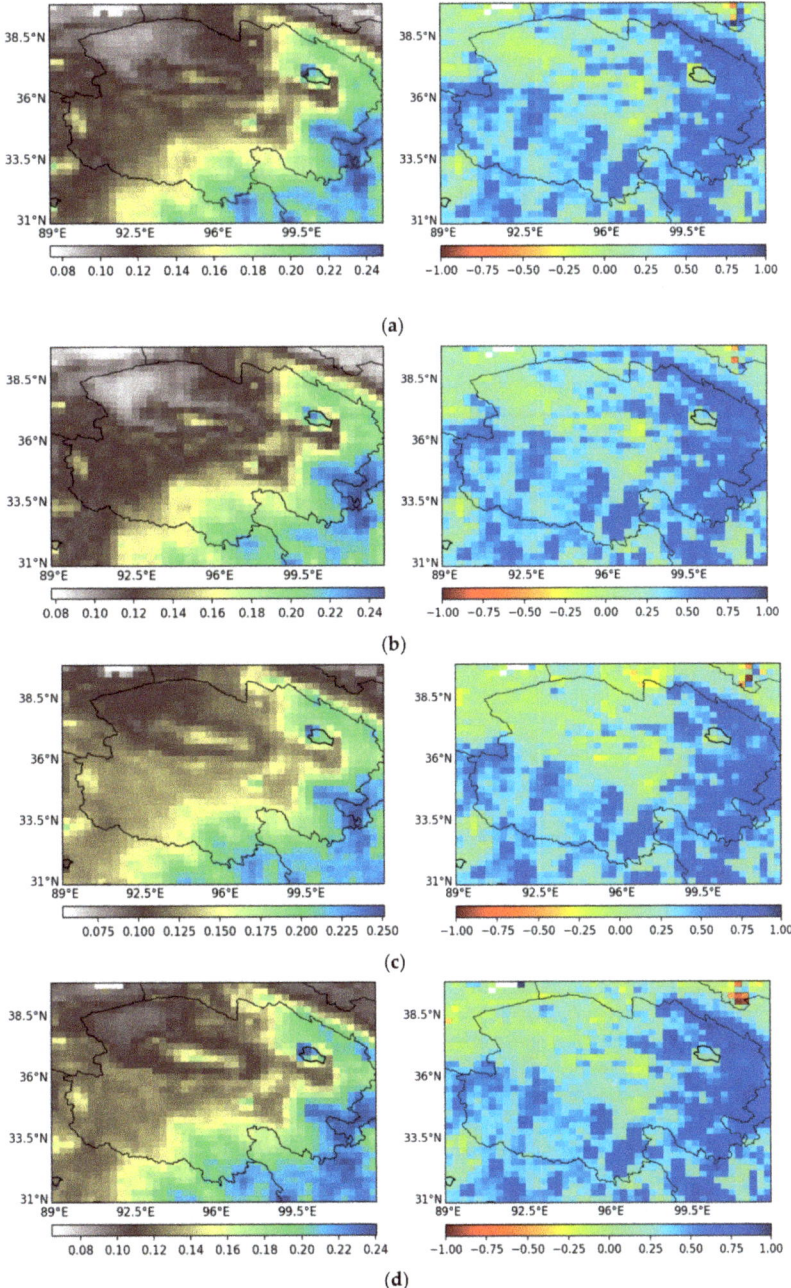

Figure 5. From left to right: Map of daily average SM retrieved by applying the NN, map of temporal correlation between NN soil moisture and CLDAS soil moisture. From top to bottom: (**a**) NN input data: SMAP_TBV_A, AMSR2_TBH_D, and σ40. (**b**) NN input data: SMAP_TBV_A, AMSR2_TBH_D, and BTI. (**c**) NN input data: SMAP_TBV_A, AMSR2_MPDI_D, and σ40. (**d**) NN input data: SMAP_TBV_A, AMSR2_MPDI_D, and BTI. SMAP_TBV_A: Ascending Tbv in the 1.41 GHz (SMAP) band; AMSR2_TBH_D: Descending Tbh in the 10.7 GHz (AMSR2) band; AMSR2_MPDI_D: Descending MPDI in the 10.7 GHz (AMSR2) band; σ40: σ40 from ASCAT; BTI: BTI from ASCAT.

Finally, the input variables are ascending Tbv in the 1.41 GHz (SMAP) band, descending Tbh in the 10.7 GHz (AMSR2) band, and BTI data from ASCAT.

3.2. Selection of Auxiliary Data

This section discusses the results of a collaborative analysis of microwave data and auxiliary input data for SM retrieval. The purpose is to determine the content and type of information that can be extracted from microwave data and auxiliary observation data and determine how to combine these data to provide maximal information for SM retrieval. Experimental trials were conducted to add and combine various auxiliary input data based on microwave data and to retrieve SM from different combinations of datasets using the NN model. These results are compared with the CLDAS data to determine the optimal combination (see detailed results in Table 5). In addition, for completeness, the SM products retrieved from all available data are compared among themselves.

Table 5. Average quality index of soil moisture calculated from auxiliary input variables compared with data from CLDAS.

Auxiliary Input Variable	ρspatial	ρtemporal	RMSE	MAE
TBV_TBH_BTI	0.597	0.401	0.051	0.041
Use of vegetation data				
TBV_TBH_BTI_NDVI	0.623	0.409	0.050	0.040
TBV_TBH_BTI_EVI	0.614	0.409	0.051	0.040
Use of terrain data				
TBV_TBH_BTI_SLOPE	0.616	0.407	0.051	0.040
TBV_TBH_BTI_DEM	0.634	0.412	0.049	0.039
TBV_TBH_BTI_SLOPE _NDVI	0.636	0.415	0.050	0.040
TBV_TBH_BTI_DEM _NDVI	0.658	0.443	0.048	0.038
TBV_TBH_BTI _DEM_SLOPE_NDVI	0.676	0.450	0.047	0.037
Use of land surface temperature data				
TBV_TBH_BTI _LST	0.604	0.441	0.049	0.039
TBV_TBH_BTI_LST _NDVI	0.617	0.448	0.049	0.039
TBV_TBH_BTI_LST _NDVI_DEM_SLOPE	0.663	0.477	0.046	0.036
Use of surface reflectance data				
TBV_TBH_BTI_NSDSI	0.608	0.410	0.051	0.040
TBV_TBH_BTI_NSDSI _NDVI	0.631	0.417	0.050	0.040
TBV_TBH_BTI_NSDSI _NDVI_DEM_SLOPE	0.684	0.453	0.047	0.037
TBV_TBH_BTI_ NSDSI _NDVI_DEM_SLOPE _LST	0.669	0.475	0.046	0.036

The brightness temperature and backscattering coefficient obtained by active and passive microwave data are all affected by the opacity of vegetation cover, which reduces the radiation from the soil surface. Therefore, information about the vegetation strongly affects SM retrieval. Table 5 also shows that adding the VI data to the microwave data improves spatial and temporal correlations and reduces MAE and RMSE. Compared with EVI, NDVI improves the spatial correlation to 0.623. Given the high correlation between NDVI and EVI, NDVI is used as a final input variable.

Terrain data such as DEM and SLOPE also play an important role in the retrieval of SM by physical models. The complex mountainous terrain reduces the quality of the microwave data retrieved. In addition, precipitation is mainly concentrated at higher altitudes in many areas of Qinghai Province, leading to relatively lush vegetation cover, which strongly affects the SM. As a result, DEM and SLOPE are also used as final input variables. Table 5 shows that adding DEM to the NN model improves the spatial correlation to 0.634 and the temporal correlation to 0.412. When NDVI, DEM, and SLOPE are all added to the NN model, the spatial correlation reaches 0.676, and the temporal correlation reaches 0.450. The surface temperature information strongly affects the soil surface emissivity, which directly affects the brightness temperature and the backscatter coefficient. Table 5 shows that when a single auxiliary input variable for the NN model, adding LST produces the greatest improvement of the temporal correlation, which attains 0.441.

Compared with CLDAS data, the SM retrieved (see Table 5) from the combination of microwave data and auxiliary inputs NSDSI, NDVI, DEM, and SLOPE produces the highest spatial correlation of 0.684, whereas the temporal correlation is only 0.453. The SM retrieved from the combination of microwave data and auxiliary inputs LST, NDVI, DEM, and SLOPE produces the highest temporal correlation of 0.477, whereas the spatial correlation is only 0.663. The SM retrieved from the combination of microwave data and auxiliary inputs NSDSI, LST, NDVI, DEM, and SLOPE produces a spatial correlation of 0.669 and a temporal correlation of 0.475, which is the most balanced combination. Therefore, we use herein the microwave data and auxiliary inputs NDVI, DEM, SLOPE, LST, NSDSI, etc. as final input variables to obtain the daily average SM map of Qinghai Province, which is shown on the left side of Figure 6. On the right side of Figure 4 is shown the correlation map between SM data retrieved from the NN and CLDAS SM data.

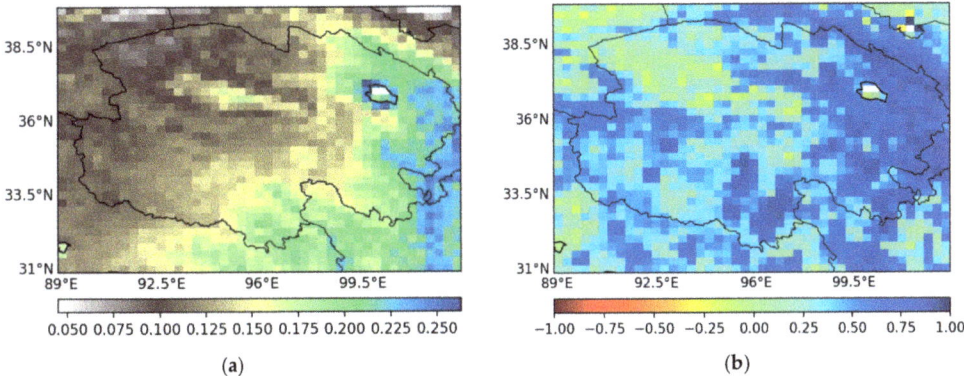

Figure 6. (**a**) Map of daily average soil moisture retrieved by applying the NN to the combination of microwave data and auxiliary inputs NDVI, LST, DEM, SLOPE, and NSDSI. (**b**) Map of temporal correlation between the NN SM data (left side) and CLDAS soil moisture data.

Figure 6 shows that the overall SM in the entire study area increases from northwest to southeast. Also, the high positive correlation in the grassland areas and poor correlations in the Qaidam Basin (northwest corner) and the mixture of forest and grassland (southeast corner) show that the above input variables do not allow us to retrieve SM from bare lands and forest areas but do allow us to retrieve SM from grassland areas.

3.3. Triple Collocation Method to Verify Soil Moisture as Determined by Neural Network

To estimate how accurately the NN model determines the SM on a large scale, we apply TC to analyze the SM from the NN. TC estimates the distribution of spatial error for each dataset by locally solving the linear relationships between the three SM datasets. One of the assumptions is that the errors in all three datasets are independent, so the FY3C SM data, which were not used to train the NN model, are combined with the GEOS-5 ground-model SM data and the SM data used to train the NN model in the form of [NN SM, FY3C SM, GEOS-5 SM] for TC. Furthermore, to ensure the accuracy of the TC results, the areas with a correlation coefficient between the three different datasets less than 0.2 are masked and are not involved in the final TC calculation. Finally, the error variance and correlation coefficient are estimated between the NN data and the actual SM data.

The spatial distribution shown in Figure 7 of the variance in TC error indicates that the variance in error between the SM retrieved from NN and FY3C and the actual SM is lowest in the Qaidam Basin in the northwest, whereas the variance in the error between the SM retrieved from GEOS-5 and the actual SM in the Qaidam Basin area is significantly greater. Combining these results with the maps of the spatial distribution of TC correlation coefficients shows that the spatial distribution of the correlation coefficient between the SM retrieved from NN and FY3C and the actual SM correlates to the error variance, meaning that the areas with greater error variance correlate more to the actual SM data.

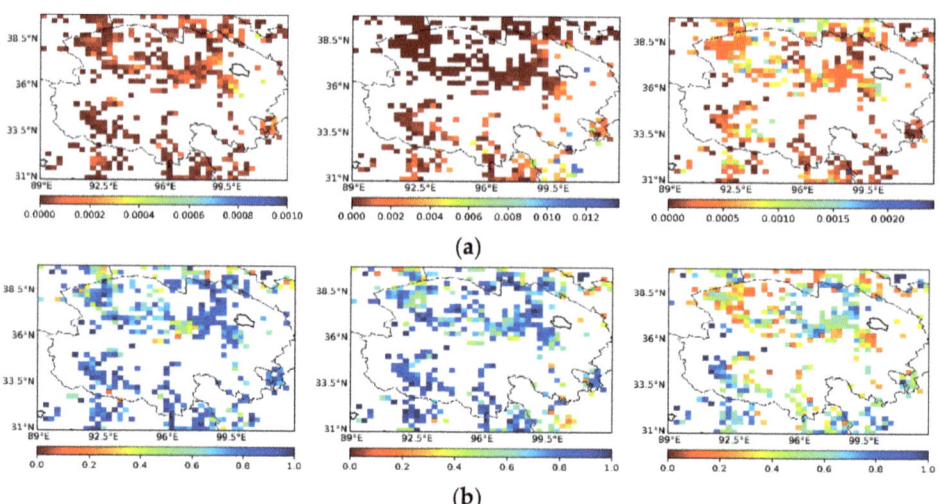

Figure 7. (**a**) Top row from left to right, Map of error variance for NN SM, FY3C SM, and GEOS-5 SM compared with actual SM data. (**b**) Bottom row from left to right, Map of correlation coefficient for NN SM, FY3C SM, and GEOS-5 SM compared with actual SM data.

Figure 8 show that the error variance between NN SM and the actual SM is much less than the error variance between (i) FY3C SM and GEOS-5 SM and (ii) the actual SM, with a median error variance of 0.0003 (NN) < 0.00017 (FY3C) < 0.00030 (GEOS-5). The correlation coefficient between (i) NN SM and FY3C SM and (ii) the actual SM is much greater than that for GEOS-5 SM, with a median correlation coefficient of 0.811 (NN) > 0.792 (FY3C) > 0.516 (GEOS-5). Among these three datasets, NN and FY3C have similar median correlation coefficients, but NN has Q1 = 0.681 and a lower-limit outlier of 0.338, which are much greater than for FY3C (Q1 = 0.594 and lower-limit outlier of 0.115). Therefore, after comprehensive analysis and comparison, the SM data retrieved by the NN model is of better quality for Qinghai Province.

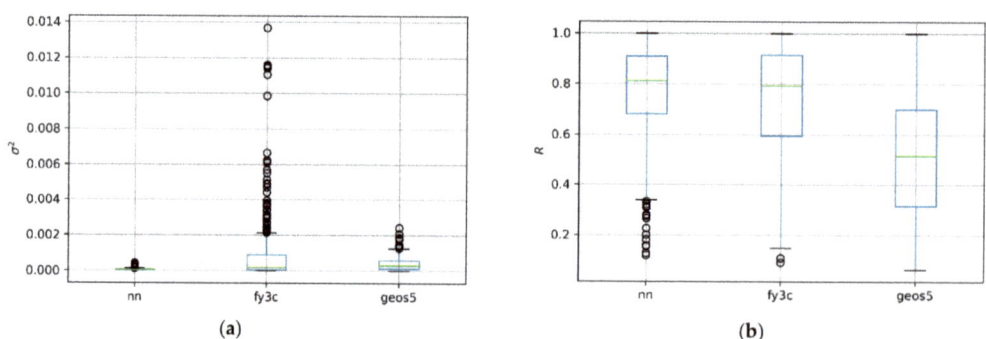

Figure 8. (**a**) Box-whisker plots of error variance for NN SM, FY3C SM, and GEOS-5 SM compared with actual SM data. (**b**) Box-whisker plots of correlation coefficients for NN SM, FY3C SM, and GEOS-5 SM compared with actual SM data.

3.4. Verification of Downscaled Soil Moisture from Neural Network

In this study, the downscaled SM dataset for Qinghai Province and the map of the daily average SM (Figure 9) were obtained by inputting MODIS data with high spatial resolution and resampled microwave data into the NN model verified by the TC. To verify the adaptability of the downscaled SM data for Qinghai Province, we apply a correlation analysis where we compare the downscaled 1 km SM data, original SMAP SM data, original AMSR2 SM data, and CLDAS SM data with the SM data collected from six ground stations in Qinghai Province. In terms of data selection, for each time series, we use data from all available at ground stations and from CLDAS, SMAP, AMSR2, and NN. In addition, each time series extends over at least 30 days to obtain good statistics. Furthermore, to determine whether the downscaled SM data capture the actual ground SM dynamics, we verify the variations over time of the downscaled SM by studying the time series of the seven ground precipitation stations (see Figure 10).

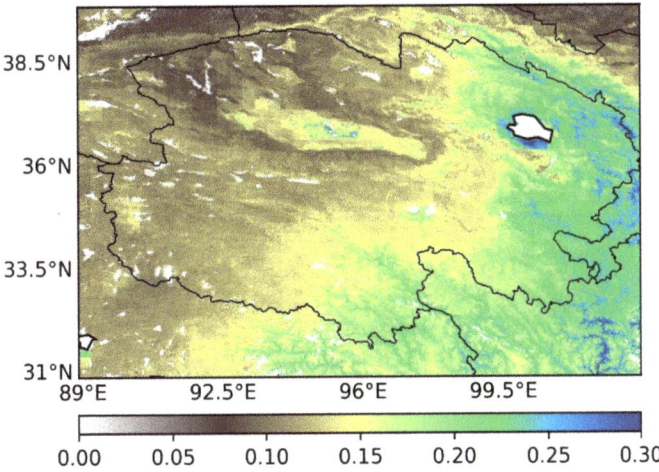

Figure 9. Map of downscaled daily average soil moisture.

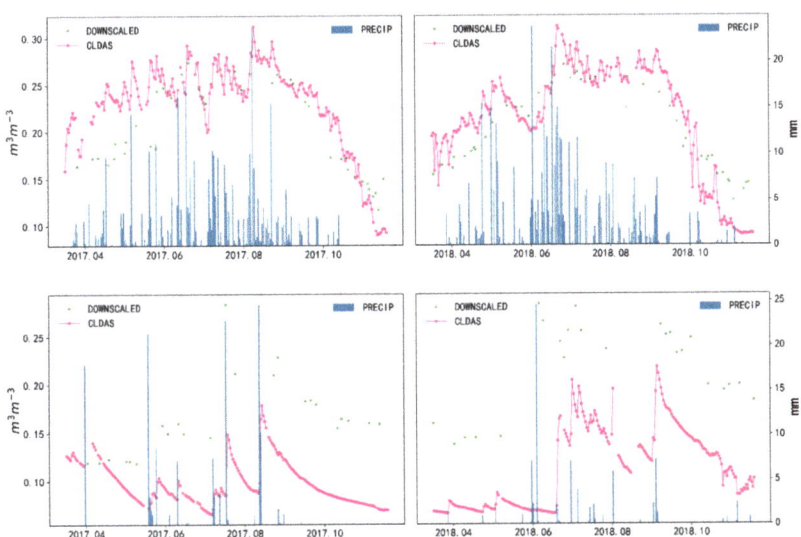

Figure 10. *Cont.*

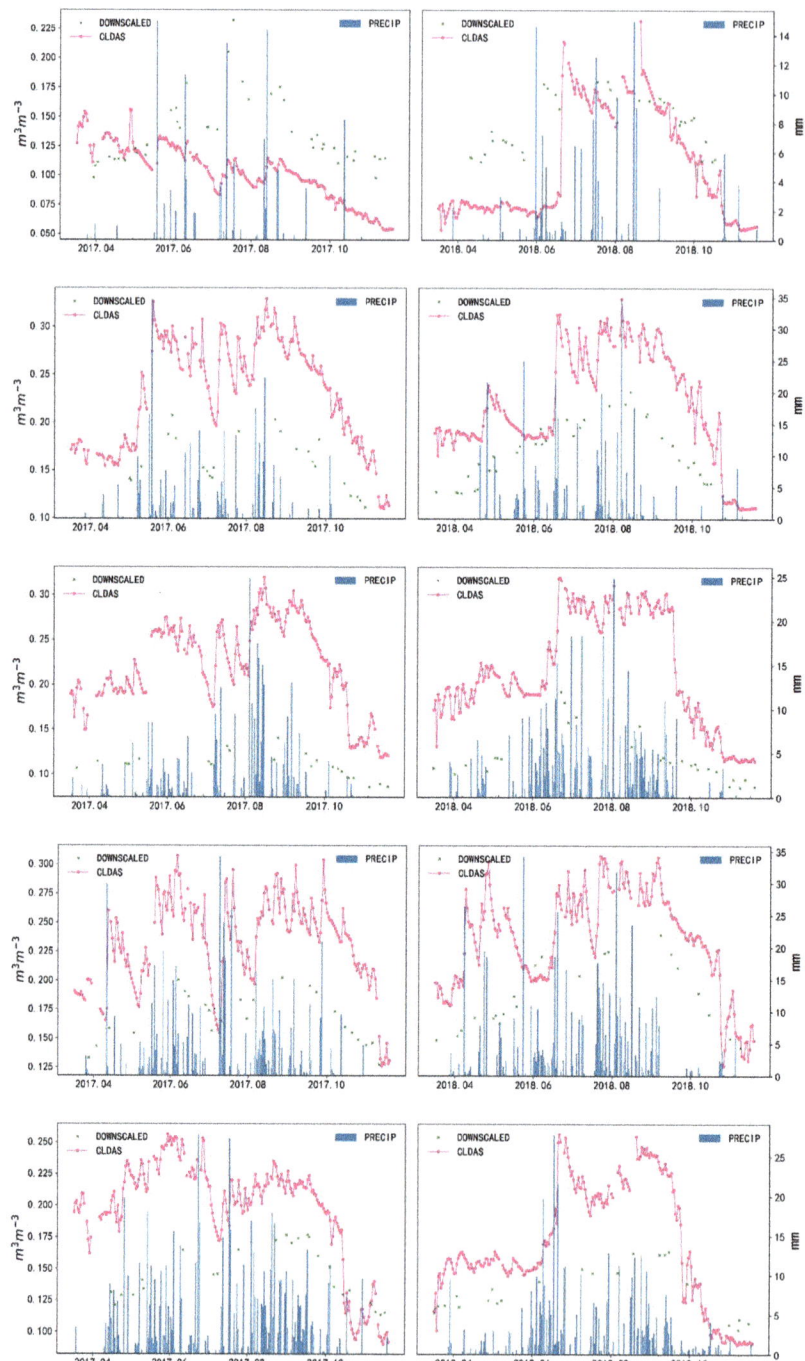

Figure 10. Daily precipitation, downscaled SM, and CLDAS SM over time for seven ground precipitation stations (Yeniugou, Xiaozaohuo, Dachaidan, Chaka, Wudaoliang, Xinghai, Qumarai).

Table 6 shows that the correlation of the downscaled SM results of the NN model at the Dulan, Tianjun, and WuUlan sites exceeds 0.6, which is a larger average than CLDAS, SMAP, and AMSR2, thereby demonstrating that the NN model properly downscales the SM. Table 6 also reveals negative correlations with CLDAS SM at both the Golmud and Nuomuhong sites, whereas SMAP, AMSR2, and NN produce negative correlations at the Golmud site but positive correlations at the Nuomuhong site. This indicates that the temporal and spatial structures based on the NN model are driven by the satellite observations rather than by the target data. Figure 7 shows a map of the daily average SM in Qinghai Province after downscaling; these results provide much more SM information than do large-scale maps of SM.

Table 6. Correlation coefficients between downscaled NN soil moisture, CLDAS soil moisture, SMAP soil moisture, AMSR2 soil moisture, and soil moisture collected from ground stations.

Station	Delingha	Dulan	Golmud	Nuomuhong	Tianjun	Wulan
CLDAS	0.427	0.759	−0.025	−0.270	0.391	0.670
SMAP	0.185	0.762	−0.587	0.193	0.548	0.776
AMSR2	0.117	0.328	−0.655	0.051	0.398	0.584
DOWNSCALED	0.212	0.768	−0.524	0.251	0.620	0.616

Figure 10 shows that the downscaled SM strongly correlates with the precipitation data because the SM increases significantly after precipitation and decreases significantly during drought. Furthermore, the downscaled SM data from Xiaozaohuo, Chaka, Wudaoliang and other sites depart significantly from the absolute value of the CLDAS data, whereas both maintain good time consistency. The results demonstrate that downscaling the SM captures better the variations in precipitation over time, which indicates that the downscaled SM better reflects the actual variations in SM over time.

4. Conclusions

This paper presents a method to retrieve soil moisture (SM) by combining multi-instrument observation data. The method is based on a neural network (NN) to retrieve SM information from passive microwave sensors SMAP and AMSR2, active microwave sensors ASCAT, as well as MODIS data (LST, NSDSI, NDVI) and topographic data (DEM, SLOPE). The greatest advantage of this method is that it can give full play to the potential of the joint retrieval of SM by each microwave sensor and also make full use of the segmentation capability of high-spatial-resolution MODIS data and topographic data.

From the microwave band selection, the best retrieval effect was achieved by the combination of Tbv in the ascending orbit for the 1.41 GHz (SMAP) band, Tbh in the descending orbit for the 10.7 GHz (AMSR2) band, and BTI data of ASCAT through the neural network method. The final NN SM dataset is obtained by combining the auxiliary data LST, NDVI, NSDSI, DEM, and SLOPE with the above three bands of microwave data. The above two models were compared with the CLDAS model SM dataset, and the result shows that the spatial correlation increases from 0.597 to 0.669, the temporal correlation increases from 0.401 to 0.475, the root mean square error decreases from 0.051 to 0.046, and the mean absolute error decreases from 0.041 to 0.036. All indicators improve, which confirms that the use of the auxiliary data improves the performance of the NN model.

The low-resolution SM products obtained from the NN retrieval in the triple collocation are higher quality than the SM products from the FY3C satellite and the ground model GEOS5 in Qinghai Province (i.e., the NN low-resolution products have the highest median correlation of 0.811, the highest correlation Q1 value of 0.681, and the lowest error variance of 0.00003).

Based on the comparison with the ground stations data, the NN SM dataset obtained on the small scale is also of better quality than the CLDAS product, and the correlation with SM at three stations, namely, Dulan (0.768), Tianjun (0.620), and Wulan (0.616), exceeds 0.6, showing strong correlation. The correlation between CLDAS SM products is greater than

0.6 only in Dulan (0.759) and Wulan (0.670). In addition, comparing with the rainfall site data shows that downscaled NN SM data also better capture the dynamic changes of SM in the study area, producing higher SM values when there is more rainfall and a decrease in SM during the long dry season. Comparing the images before and after downscaling also shows that the SM after downscaling can provide more detailed SM information. We also discuss some shortcomings in the downscaling process. The downscaled SM is susceptible to interference from clouds and rain, leading to a significant quantity of missing data, so future work will focus on data completion.

The results of this study confirm that the NN method can be used to obtain SM with high spatial resolution and can be applied to the Qinghai Province area. The data used herein can be downloaded for free from the official websites of the National Aeronautics and Space Administration (NASA), the Japan Aerospace Exploration Agency (JAXA), the European Centre for Medium-Range Weather Forecasts (ECMWF), and the China Meteorological Information Sharing Platform (CIMISS) without regional restrictions and can be used to produce sTable 1 km SM data in the Qinghai Province area.

Author Contributions: A.L. designed and conduct the study. Z.Z. performed the data analysis and wrote the manuscript. H.Z. read and edited the manuscript. All authors reviewed and approved the manuscript. All authors have read and agreed to the published version of the manuscript.

Funding: This research was funded the National Natural Science Foundation of China (41671026), the Important Science & Technology Specific Projects of Qinghai Province (2019-SF-A4-1) and Scientific Research and Promotion Projects of the Second Phase Project of Ecological Protection and Construction of the Three Rivers Source in Qinghai Province (2018-S-3).

Institutional Review Board Statement: Not applicable.

Informed Consent Statement: Not applicable.

Data Availability Statement: The data used in this study were downloaded from the National Aeronautics and Space Administration (NASA), the Japan Aerospace Exploration Agency (JAXA), the European Centre for Medium-Range Weather Forecasts (ECMWF), and the China In-tegrated Meteorological Information Service System (CIMISS).

Acknowledgments: We are very grateful to the teams at NASA, JAXA, ECMWF, and CIMISS who have made their datasets available and ready to use.

Conflicts of Interest: The authors declare no conflict of interest.

References

1. McColl, K.A.; Alemohammad, S.H.; Akbar, R.; Konings, A.G.; Yueh, S.; Entekhabi, D. The global distribution and dynamics of surface soil moisture. *Nat. Geosci.* **2017**, *10*, 100–104.
2. Peng, J.; Loew, A.; Merlin, O.; Verhoest, N.E.C. A review of spatial downscaling of satellite remotely sensed soil moisture. *Rev. Geophys.* **2017**, *55*, 341–366.
3. Keshavarz, M.R.; Vazifedoust, M.; Alizadeh, A. Drought monitoring using a Soil Wetness Deficit Index (SWDI) derived from MODIS satellite data. *Agric. Water Manag.* **2014**, *132*, 37–45.
4. Rosenzweig, C.; Tubiello, F.N.; Goldberg, R.; Mills, E.; Bloomfield, J. Increased crop damage in the US from excess precipitation under climate change. *Glob. Environ. Chang.* **2002**, *12*, 197–202.
5. Miller, G.R.; Baldocchi, D.D.; Law, B.E. Meyers An analysis of soil moisture dynamics using multi-year data from a network of micrometeorological observation sites. *Adv. Water Resour.* **2007**, *30*, 1065–1081.
6. Mladenova, I.; Lakshmi, V.; Jackson, T.J.; Walker, J.P.; Merlin, O.; Jeu, R.A.M. De Validation of AMSR-E soil moisture using L-band airborne radiometer data from National Airborne Field Experiment 2006. *Remote Sens. Environ.* **2011**, *115*, 2096–2103.
7. Zhang, P.; Yang, J.; Dong, C.; Lu, N.; Yang, Z.; Shi, J. General introduction on payloads, ground segment and data application of Fengyun 3A. *Front. Earth Sci. China* **2009**, *3*, 367–373.
8. Zabolotskikh, E.V.; Mitnik, L.M.; Chapron, B. New approach for severe marine weather study using satellite passive microwave sensing. *Geophys. Res. Lett.* **2013**, *40*, 3347–3350.
9. Kerr, Y.H.; Waldteufel, P.; Wigneron, J.P.; Delwart, S.; Cabot, F.; Boutin, J.; Escorihuela, M.J.; Font, J.; Reul, N.; Gruhier, C.; et al. The SMOS L: New tool for monitoring key elements ofthe global water cycle. *Proc. IEEE* **2010**, *98*, 666–687.
10. Medeiros, J. Design and Development of the SMAP Microwave Radiometer Electronics. In Proceedings of the Specialist Meeting on Microwave Radiometry and Remote Sensing of the Environment, Pasadena, CA, USA, 24–27 March 2014.

11. Figa-Saldaña, J.; Wilson, J.J.W.; Attema, E.; Gelsthorpe, R.; Drinkwater, M.R.; Stoffelen, A. The advanced scatterometer (ascat) on the meteorological operational (MetOp) platform: A follow on for european wind scatterometers. *Can. J. Remote Sens.* **2002**, *28*, 404–412.
12. Njoku, E.G.; Wilson, W.J.; Yueh, S.H.; Dinardo, S.J.; Bolten, J. Observations of soil moisture using a passive and active low-frequency microwave airborne sensor during SGP99. *Geosci. Remote Sens. IEEE Trans.* **2002**, *40*, 2659–2673.
13. Das, N.N.; Entekhabi, D.; Njoku, E.G. An algorithm for merging SMAP radiometer and radar data for high-resolution soil-moisture retrieval. *IEEE Trans. Geosci. Remote Sens.* **2011**, *49*, 1504–1512.
14. Zhan, X.; Houser, P.R.; Walker, J.P.; Crow, W.T. A method for retrieving high-resolution surface soil moisture from hydros L-band radiometer and radar observations. *IEEE Trans. Geosci. Remote Sens.* **2006**, *44*, 1534–1544.
15. Wilson, D.J.; Western, A.W.; Grayson, R.B. A terrain and data-based method for generating the spatial distribution of soil moisture. *Adv. Water Resour.* **2005**, *28*, 43–54.
16. Srivastava, P.K.; Han, D.; Ramirez, M.R.; Islam, T. Machine Learning Techniques for Downscaling SMOS Satellite Soil Moisture Using MODIS Land Surface Temperature for Hydrological Application. *Water Resour. Manag.* **2013**, *27*, 3127–3144.
17. Yang, K.; Zhu, L.; Chen, Y.; Zhao, L.; Qin, J.; Lu, H.; Tang, W.; Han, M.; Ding, B.; Fang, N. Land surface model calibration through microwave data assimilation for improving soil moisture simulations. *J. Hydrol.* **2016**, *533*, 266–276.
18. Chen, W.; Shen, H.; Huang, C.; Li, X. Improving soil moisture estimation with a dual ensemble Kalman smoother by jointly assimilating AMSR-E brightness temperature and MODIS LST. *Remote Sens.* **2017**, *9*, 273.
19. Chauhan, N.S.; Miller, S.; Ardanuy, P. Spaceborne soil moisture estimation at high resolution: A microwave-optical/IR synergistic approach. *Int. J. Remote Sens.* **2003**, *24*, 4599–4622.
20. Liu, Z.; Zhou, P.; Zhang, F. Spatiotemporal characteristics of dryness/wetness conditions across Qinghaiprovince, Northwest China. *Agric. For. Meteorol.* **2013**, *182–183*, 101–108.
21. Qi, Y.; Li, S.; Ran, Y.; Wang, H.; Wu, J.; Lian, X.; Luo, D. Mapping Frozen Ground in the Qilian Mountains in 2004–2019 Using Google Earth Engine Cloud Computing. *Remote Sens.* **2021**, *13*, 149.
22. Liu, S.; Zhang, Y.; Cheng, F.; Hou, X.; Zhao, S. Response of Grassland Degradation to Drought at Different Time-Scales in Qinghai Province: Spatio-Temporal Characteristics, Correlation, and Implications. *Remote Sens.* **2017**, *9*, 1329.
23. Cui, Y.; Long, D.; Hong, Y.; Zeng, C.; Zhou, J.; Han, Z.; Liu, R.; Wan, W. Validation and reconstruction of FY-3B/MWRI soil moisture using an artificial neural network based on reconstructed MODIS optical products over the Tibetan Plateau. *J. Hydrol.* **2016**, *543*, 242–254.
24. Shi, C.; Xie, Z.; Qian, H.; Liang, M.; Yang, X. China land soil moisture EnKF data assimilation based on satellite remote sensing data. *Sci. China Earth Sci.* **2011**, *54*, 1430–1440.
25. Lucchesi, R. *File Specification for GEOS-5 FP*; NASA Global Modeling and Assimilation Office (GMAO) Office Note No. 4 (Version 1.0); NASA: Washington, DC, USA, 2013.
26. Hu, F.; Wei, Z.; Zhang, W.; Dorjee, D.; Meng, L. A spatial downscaling method for SMAP soil moisture through visible and shortwave-infrared remote sensing data. *J. Hydrol.* **2020**, *590*, 125360.
27. Van der Vliet, M.; Van der Schalie, R.; Rodriguez-Fernandez, N.; Colliander, A.; de Jeu, R.; Preimesberger, W.; Scanlon, T.; Dorigo, W. Reconciling Flagging Strategies for Multi-Sensor Satellite Soil Moisture Climate Data Records. *Remote Sens.* **2020**, *12*, 3439.
28. Ghulam, A.; Qin, Q.; Teyip, T.; Li, Z.-L. Modified perpendicular drought index (MPDI): A real-time drought monitoring method. *ISPRS J. Photogramm. Remote Sens.* **2007**, *62*, 150–164.
29. Kolassa, J.; Gentine, P.; Prigent, C.; Aires, F. Soil moisture retrieval from AMSR-E and ASCAT microwave observation synergy. Part 1: Satellite data analysis. *Remote Sens. Environ.* **2016**, *173*, 1–14.
30. Scipal, K.; Holmes, T.; De Jeu, R.; Naeimi, V.; Wagner, W. A possible solution for the problem of estimating the error structure of global soil moisture data sets. *Geophys. Res. Lett.* **2008**, *35*, 2–5.
31. Gruber, A.; Su, C.-H.; Zwieback, S.; Crow, W.; Dorigo, W.; Wagner, W. Recent advances in (soil moisture) triple collocation analysis. *Int. J. Appl. Earth Obs. Geoinf.* **2016**, *45*, 200–211.
32. Chen, F.; Crow, W.T.; Bindlish, R.; Colliander, A.; Burgin, M.S.; Asanuma, J.; Aida, K. Global-scale evaluation of SMAP, SMOS and ASCAT soil moisture products using triple collocation. *Remote Sens. Environ.* **2018**, *214*, 1–13.
33. McColl, K.A.; Vogelzang, J.; Konings, A.G.; Entekhabi, D.; Piles, M.; Stoffelen, A. Extended triple collocation: Estimating errors and correlation coefficients with respect to an unknown target. *Geophys. Res. Lett.* **2014**, *41*, 6229–6236.
34. Paloscia, S.; Pettinato, S.; Santi, E.; Notarnicola, C.; Pasolli, L.; Reppucci, A. Soil moisture mapping using Sentinel-1 images: Algorithm and preliminary validation. *Remote Sens. Environ.* **2013**, *134*, 234–248.
35. Ge, L.; Hang, R.; Liu, Y.; Liu, Q. Comparing the performance of neural network and deep convolutional neural network in estimating soil moisture from satellite observations. *Remote Sens.* **2018**, *10*, 1327.
36. Kolassa, J.; Reichle, R.H.; Liu, Q.; Alemohammad, S.H.; Gentine, P.; Aida, K.; Asanuma, J.; Bircher, S.; Caldwell, T.; Colliander, A.; et al. Estimating surface soil moisture from SMAP observations using a Neural Network technique. *Remote Sens. Environ.* **2018**, *204*, 43–59. [PubMed]
37. Rodríguez-Fernández, N.J.; Aires, F.; Richaume, P.; Kerr, Y.H.; Prigent, C.; Kolassa, J.; Cabot, F.; Jiménez, C.; Mahmoodi, A.; Drusch, M. Soil moisture retrieval using neural networks: Application to SMOS. *IEEE Trans. Geosci. Remote Sens.* **2015**, *53*, 5991–6007.
38. Levenberg, K. A Method for the Solution of Certain Nonlinear Problems in Least Squares. *Q. Appl. Math.* **1944**, *2*, 164–168.

39. Marquardt, D.W. An Algorithm for Least-Squares Estimation of Nonlinear Parameters. *J. Soc. Ind. Appl. Math.* **1963**, *11*, 431–441.
40. Rumelhart, D.E.; Chauvin, Y. *Backpropagation: Theory, Architectures, and Applications*; Psychology Press: Hove, UK, 1995.
41. Piles, M.; Petropoulos, G.P.; Sánchez, N.; González-Zamora, Á.; Ireland, G. Towards improved spatio-temporal resolution soil moisture retrievals from the synergy of SMOS and MSG SEVIRI spaceborne observations. *Remote Sens. Environ.* **2016**, *180*, 403–417.
42. Piles, M.; Camps, A.; Vall-Llossera, M.; Corbella, I.; Walker, J.P. Downscaling SMOS-Derived Soil Moisture Using MODIS Visible/Infrared Data. *IEEE Trans. Geosci. Remote Sens.* **2011**, *49*, 3156–3166.

Article

Satellite-Observed Global Terrestrial Vegetation Production in Response to Water Availability

Yuan Zhang [1,2], Xiaoming Feng [1,*], Bojie Fu [1], Yongzhe Chen [1,2] and Xiaofeng Wang [3]

1. State Key Laboratory of Urban and Regional Ecology, Research Center for Eco-Environmental Sciences, Chinese Academy of Sciences, Beijing 100085, China; yuanzhang_st2018@rcees.ac.cn (Y.Z.); bfu@rcees.ac.cn (B.F.); yzchen_st@rcees.ac.cn (Y.C.)
2. University of Chinese Academy of Sciences, Beijing 100049, China
3. School of Land Engineering, Chang'an University, Xi'an 710054, China; wangxf@chd.edu.cn
* Correspondence: fengxm@rcees.ac.cn; Tel.: +86-10-6248-9102

Citation: Zhang, Y.; Feng, X.; Fu, B.; Chen, Y.; Wang, X. Satellite-Observed Global Terrestrial Vegetation Production in Response to Water Availability. *Remote Sens.* **2021**, *13*, 1289. https://doi.org/10.3390/rs13071289

Academic Editor: Jordi Cristóbal Rosselló

Received: 7 March 2021
Accepted: 25 March 2021
Published: 28 March 2021

Publisher's Note: MDPI stays neutral with regard to jurisdictional claims in published maps and institutional affiliations.

Copyright: © 2021 by the authors. Licensee MDPI, Basel, Switzerland. This article is an open access article distributed under the terms and conditions of the Creative Commons Attribution (CC BY) license (https:// creativecommons.org/licenses/by/ 4.0/).

Abstract: Water stress is one of the primary environmental factors that limits terrestrial ecosystems' productivity. Hense, the way to quantify gobal vegetation productivity's vulnerability under water stress and reveal its seasonal dynamics in response to drought is of great significance in mitigating and adapting to global changes. Here, we estimated monthly gross primary productivity (GPP) first based on light-use efficiency (LUE) models for 1982–2015. GPP's response time to water availability can be determined by correlating the monthly GPP series with the multiple timescale Standardized Precipitation Evapotranspiration Index (SPEI). Thereafter, we developed an optimal bivariate probabilistic model to derive the vegetation productivity loss probabilities under different drought scenarios using the copula method. The results showed that LUE models have a good fit and estimate GPP well (R^2 exceeded 0.7). GPP is expected to decrease in 71.91% of the global land vegetation area because of increases in radiation and temperature and decreases in soil moisture during drought periods. Largely, we found that vegetation productivity and water availability are correlated positively globally. The vegetation productivity in arid and semiarid areas depends considerably upon water availability compared to that in humid and semi-humid areas. Weak drought resistance often characterizes the land cover types that water availability influences more. In addition, under the scenario of the same level of GPP damage with different drought degrees, as droughts increase in severity, GPP loss probabilities increase as well. Further, under the same drought severity with different levels of GPP damage, drought's effect on GPP loss probabilities weaken gradually as the GPP damage level increaes. Similar patterns were observed in different seasons. Our results showed that arid and semiarid areas have higher conditional probabilities of vegetation productivity losses under different drought scenarios.

Keywords: LUE-GPP; SPEI; copula function; conditional probability

1. Introduction

The Intergovernmental Panel on Climate Change (IPCC) report shows that the world will continue to warm in the 21st century [1]. This indicates that as the future temperature rises, the occurrence and frequency of extreme weather and climate, such as heatwaves and droughts, will increase rapidly [2]. This trend and its associated adverse effects on natural and social ecosystems are expected to increase further [3]. Terrestrial ecosystems' gross primary productivity (GPP) is the largest component of global terrestrial carbon flux, and slight fluctuations in GPP have a significant influence on atmospheric CO_2 concentration [4]. On a global scale, particularly in arid and semiarid regions, water stress is the primary environmental factor that limits terrestrial ecosystems' productivity [5]. Hence, studying drought's effects on their productivity has become an important priority in global change research.

The terrestrial ecosystem's productivity characterizes its quality, it is the key to controlling the ecosystem's carbon, water, energy and nutrient cycles, and plays a decisive role in regulating its function and controlling its carbon source and sink [6]. At the same time, ecosystem productivity is also the basis of a variety of ecosystem services [7] and is the source of food, fiber, wood, livestock grains and biofuels in human society [8]. As the main outcome and important manifestation of global climate change, drought not only affects photosynthesis and ecosystem productivity directly [9], but it can also affect terrestrial ecosystem productivity in directly by changing the intensity and frequency of other forms of disturbance, such as increasing the frequency and intensity of fires, plant mortality and the occurrence of pests and diseases [10]. In recent years, many studies have investigated drought's effects on terrestrial ecosystems' productivity at different spatio-temporal scales based on long-term observations from ecological research sites, field experiments, eddy covariance measurements, large-scale satellite measurements and model simulations [11–13]. Satellites data used commonly to evaluate vegetation productivity include primarily the Moderate Resolution Imaging Spectroradiometer-Fraction of Photosynthetically Active Radiation (MODIS-FPAR) and the Global Inventory Modeling and Mapping Studies-Fraction of Photosynthetically Active Radiation (AVHRR GIMMS-FPAR), remotely-sensed data for atmosphere state, MODIS Enhanced Vegetation Index, leaf area index, spatially distributed CO_2 concentrations, and canopy information. For example, Reichstein et al. [11] found that the drought stress in Europe in summer 2003 caused the reduction of ecosystem productivity by jointing flux tower, remote sensing (including MODSI and AVHRR-GIMMS FPAR) and modelling datasets. Zhang et al. [14] emphasized that the drought in Southwestern China in spring 2010 reduced primary productivity by using primary productivity products derived from MODIS. Stocker et al. [15] used four remote sensing data-driven models and demenstrated that drought's effects on terrestrial primary production were underestimated. Previous studies have shown that different vegetation types and environmental conditions (such as climate and soil factors) largely determine biological communities' resistance and ability to recover from drought stress [16].

Although there is increasing recognition of drought's adverse effects on terrestrial vegetation productivity, few studies have focused on the seasonal dynamics in global productivity in response to drought across various climate zones and land biomes. Moreover, studies of drought's lag effect on terrestrial vegetation productivity and vegetation's responses to drought at various time-scales on a global scale are scare. The Standardized Precipitation Evapotranspiration Index (SPEI) quantifies different drought types based on the water balance, and is used widely on global and regional scales because of its multiple time scales and spatial comparability [17]. Therefore, determining the corresponding time scale when GPP and SPEI have the highest correlation based upon different time scales is of great significance in understanding the variations in different terrestrial ecosystems' productivity depending uopn drought resistance. In addition, we give greater attention here to the way to quantify the probability of varying degrees of damage to vegetation productivity under predictable drought scenarios. Copulas can be used to couple multiple variables and construct their joint probability distribution by considering their correlations [18], which can derive the vegetation productivity loss distributions conditioned on any drought scenario. For example, Madadgar et al. [19] estimated the probability of drought in agricultural production in Australia based on a copula model, while Fang et al. [20] constructed a bivariate probabilistic framework to assess vegetation vulnerability and map drought-prone ecosystems in China's Loess Plateau.

Given ecosystem productivity's key role in affecting the global carbon budget and the supply of ecosystem services, in the context of global warming, the rapid increase in atmospheric carbon dioxide concentration, and increasing human food production and resource demand, estimating ecosystem productivity accurately is important to assess drought's effects on terrestrial ecosystems precisely [21]. Three main methods are used to estimate ecosystems' productivity at the regional and global scales—remote sensing-based light-use efficiency (LUE) models [22], machine learning algorithms [23], and process-based physiological and ecological models [24]. Remote sensing observations can provide information continuously on land surface features that affect ecosystem productivity, such as ecosystem structure, vegetation phenology, biomass, soil moisture, and land cover over a large area. The LUE model relies on remote sensing observation data to estimate ecosystem productivity through the utilization rate of photosynthetically active radiation in the ecosystem [25]. We gave special attention to determining water stress and produced a data set of monthly LUE GPPs at the global scale based on various representations of water stress: Water stress based on vapor-pressure deficit, humidity deficit and root-zone soil-water content separately.

Our study had two primary objectives. The first was to determine the seasonal dynamics of global vegetation productivity's response to drought across various climate zones and land biomes, and the second was to quantify the probability of varying degrees of damage to vegetation productivity under predictable drought scenarios. First, we used a LUE model to produce a dataset of global monthly satellite-observed GPPs, which are designed primarily to determinate water stress. Second, we explored vegetation productivity's spatio-temporal evolution of dependence and response time to water availability by correlating monthly GPP series with the multiple timescale SPEIs (1- to 24-months). Finally, we develpoed an optimal bivariate probabilistic model in each pixel of the globe using the copula method, which was used to derive the vegetation productivity loss probabilities under different drought scenarios.

2. Materials and Methods

2.1. Data

2.1.1. Remote Sensing Data

The actual evapotranspiration, potential evapotranspiration and soil moisture (including surface soil and root-zone soil moisture) data were derived from the Global Land Evaporation Amsterdam Model (GLEAM), which is designed to be determined by remote sensing observations only [26]. The GLEAM_V3.3a dataset has a 0.25° spatial resolution at a monthly time step that spans the 39-year period from 1980–2018 (www.gleam.eu (accessed on 25 March 2021)). Fraction of Photosynthetically Active Radiation (FPAR)3g data were generated from AVHRR GIMMS NDVI3g using an Artificial Neural Network (ANN) derived model, with a spatial resolution of 1/12° and a 15 days time interval [27]. We used the MODIS MCD12C1 product, which contains yearly, worldwide distributions of 17 land cover types from 2001–2017 at a spatial resolution of 0.05° (https://lpdaac.usgs.gov/ (accessed on 10 December 2020)) [28]. In this study, barren land, water bodies, permanent snow, and ice were removed.

2.1.2. Meteorological data and FLUXNET data

Downward shortwave radiation, air temperature, specific humidity and air pressure were obtained from the CRU-NCEPV6.1 dataset with a spatial resolution of 0.5° and a 6-h time interval. We used gross primary productivity (GPP) data from the FLUXNET2015 dataset to compare flux site measurement of GPP and LUE GPP (https://fluxnet.org/data/fluxnet2015-dataset/ (accessed on 1 December 2020)).

2.2. Drought Index

SPEI can be calculated by fitting the difference between precipitation and potential evapotranspiration based on a log-logistic probability distribution, which combines the multi-temporal characteristic of standardized precipitation index and the sensitivity of palmer drought severity index to changes in evaporation demand. A global gridded dataset of the SPEI was calculated using monthly precipitation and potential evapotranspiration from the CRU TS3.24.01 dataset [29]. This dataset is generated at a 0.5° spatial resolution and at a monthly time step and covers the period from 1902 to 2015. We used the SPEI dataset from 1982 to 2015 in this study, which contains timescales from 1- to 24-months. A 6-month SPEI value is formulated by the cumulative water deficit or surplus from five months before to the current month. Three drought scenarios including moderate ($-1.5 < \text{SPEI} \leq 1$), severe ($-2 < \text{SPEI} \leq -1.5$), and extreme ($\text{SPEI} \leq -2$) can be differentiated using a set of SPEI thresholds [29]. The yearly aridity index (AI), which has a spatial resolution of 0.5°, was calculated with Feng and Fu's formula [30]. (Response: we cited the citation 31 in Section 2.3) According to this dataset, global land can be classified into arid ($0.05 \leq \text{AI} < 0.2$), semiarid ($0.2 \leq \text{AI} < 0.5$), semihumid ($0.5 \leq \text{AI} < 0.65$) and humid zones ($\text{AI} \geq 0.65$).

2.3. LUE GPP

The global GPP for the 1982–2015 period was calculated based on the LUE equation, as follows [31]:

$$GPP = \varepsilon_{max} \times SOL \times 0.45 \times fPAR \times f(T) \times f(W), \quad (1)$$

in which ε_{max} is vegetation type related maximum light use efficiency, SOL is downward solar shortwave radiation, and $fPAR$ represents the fraction of photosynthetically active radiation vegetation absorbs. ft and fw are temperature and water stress respectively to plant productivity. We adopted the meteorological data from the CRU-NCEPV6.1 dataset, which includes downward shortwave radiation, air temperature, specific humidity and air pressure, and so forth. We used $fPAR$ Zhu et al. [27] developed. We calculated the temperature stress with Equation (2), as follows:

$$ft = \frac{(T - T_{MIN})(T - T_{MAX})}{[(T - T_{MIN})(T - T_{MAX})] - (T - T_{opt})^2}, \quad (2)$$

in which T_{MIN} and T_{MAX} are the minimum and maximum temperature for plant's photosynthesis, with the specific parameters can be found in Zhang et al. [32].

In this study, we focused particularly on determining water stress (fw) in the LUE equation. The MODIS-GPP algorithm [33] calculates fw with the daily vapor pressure deficit (VPD) on consideration that leaf stomatal conductance is mainly restricted by cold stress and VPD, named VPDMOD expressed as in Equation (3):

$$fw = f(VPD) = \begin{cases} 1 & VPD \leq VPD_{open} \\ \frac{VPD_{close} - VPD}{VPD_{close} - VPD_{open}} & VPD_{open} < VPD < VPD_{close} \\ 0 & VPD \geq VPD_{close} \end{cases}, \quad (3)$$

in which VPD_{close} represents VPD that induces full stomatal closure, while VPD_{open} represents VPD that results in full stomatal opening, all parameters are set based on the vegetation type.

The GLO-PEM algorithm [34] considers each vegetation type's optimum growth temperature (T_{opt}). The algorithm calculates water stress with vegetation type, independently with Equation (4) named VPDGLO:

$$fw = f(VPD) = \begin{cases} 1 - 0.05\delta_q & 0 < \delta_q \leq 15 \\ 0.25 & \delta_q > 15 \end{cases}, \quad (4)$$

in which δ_q is the specific humidity deficit (g/kg), the difference between saturated and actual specific humidity.

The CASA model provides a framework that considers water stress with a multidisciplinary algorithm [35], shown in Equation (5), as follows:

$$fw = f(VPD) \times W_s$$
$$W_s = \begin{cases} 1 & VPD \text{ only model} \\ 0.5 + ETR & VPD + ETR \text{ model} \\ 0.5 + K_S & VPD + SM \text{ model} \end{cases}, \quad (5)$$

in which ETR is the ratio of actual evapotranspiration to potential evapotranspiration. According to FAO 56 [36], ETR can be equal to K_S, which is calculated based on root zone soil moisture (hereinafter SM) and soil and vegetation properties, written as Equations (6) and (7):

$$K_S = \frac{SM - WP}{(1-p) \times TAW} \quad (6)$$

$$p = p_{std} + 0.04 \times (5 - ET_0), \quad (7)$$

in which WP and TAW represent soil wilting point and total available soil water derived from the International Geosphere-Biosphere Programme Data and Information System (IGMP-DIS) dataset [37] and the World Inventory of Soil property Estimates (WISE) derived soil properties [38]. ET_0 is calculated with Hagreaves equation [39] while p_{std} is a vegetation type-specific depletion factor (allowable extent of soil water deficit before water stress) at ET_0 of 5 mm per day. Actual evapotranspiration, potential evapotranspiration, and root-zone soil moisture data were all acquired from the GLEAM dataset.

In this study, we provided a dataset of monthly plant productivity that spans the 34-year period from 1982–2015 at the global scale, composed by five LUE GPPs derived from the multidisciplinary framework in Equation (5). These five GPPs represent different considerations of water stress in the LUE equation, the VPD only model, VPD_{GLO}-ETR model, VPD_{MOD}-ETR model, VPD_{GLO}-SM model, and VPD_{MOD}-SM model. Because the LUE equation estimate GPP from multiple satellite data, for example, NDVI, FPAR, and GLEAM soil water, it is considered satellite-observed GPP.

2.4. Determining LUE GPP's Response Time to Water Availability

To screen out vegetation productivity's response time to water availability, a monthly GPP series from 1982 to 2015 was correlated with the multiple timescale SPEI (1- to 24-months). For the i-th month in a year, Spearman's correlation analysis was performed between the monthly GPP and SPEI series at different timescales as follows

$$R_j^i = corr\left(GPP^i, SPEI_j^i\right) i = 1, 2, \ldots, 12, j = 1, 2, \ldots, 24, \quad (8)$$

in which R is the correlation coefficient and a timescale j denotes the cumulative water deficit or surplus from j-1 months before to the current month. Finally, a timescale able to maximize the GPP-SPEI association, is retained as the vegetation productivity response time (VPRT) to water variations for the i-th month.

$$VPRT^i = max\left\{R_j^i\right\} j = 1, 2, \ldots, 24. \quad (9)$$

2.5. Copulas

We chose the extreme value (EV), generalized extreme value (GEV), exponential (EXP), gamma (GAM), Poisson (POISS), normal (NOR), generalized Pareto (GP) and Weibull (WBL) distribution functions to fit the monthly GPP series (Table S1). The SPEI series is a statistic distributed normally and accordingly, the normal distribution is employed as its marginal. The probability distribution functions of monthly GPP and SPEI series are defined as F_{GPP} (gpp) and F_{SPEI} (spei), respectively.

We constructed a joint distribution function in each pixel of the globe in every month based on monthly GPP and SPEI series. Considering F_{GPP} (gpp) and F_{SPEI} (spei), these two marginal distribution's joint distribution function can be defined as a copula C:

$$F_{GPP,SPEI}(gpp, spei) = C(F_{GPP}(gpp), F_{SPEI}(spei)). \quad (10)$$

In this study, we used the methods of maximum likelihood and inference functions for margins to estimate the fitted parameters in F_{GPP} (gpp), F_{SPEI} (spei) and $F_{GPP,SPEI}$ (gpp, spei) [40]. The bivariate distribution functions used commonly, including the Frank,

Clayton, Gumbel, t and Normal copulas, were constructed to couple FGPP (gpp)) and FSPEI (spei) in each pixel of the globe (as shown in Table S2).

Next, we obtained the optimal joint distribution function of F_{GPP} (gpp) and F_{SPEI} (spei) as follows by referring to the method of the goodness-of-fit test Zhang et al. [41] provided.

$$F_{GPP,SPEI}(gpp, spei) = P(GPP \leq gpp, SPEI \leq spei) = \int_{-\infty}^{gpp} \int_{-\infty}^{spei} f(gpp, spei) d_u d_v. \quad (11)$$

Ultimately, we generated a set of conditional probabilities of vegetation productivity losses for diverse drought conditions with a bivariate statistical framework. Larger values of vegetation productivity loss probabilities under the same scenario imply greater ecosystem vulnerability, and such ecosystems are categorized thereby as the drought-prone class. Given multiple drought scenarios, conditional probabilities of vegetation productivity lower than different percentiles (e.g., the 10th, 20th, 30th, 40th percentiles were considered in the study) are derived using copula-based joint distribution and the conditional distribution formulas. The conditional probabilities of different percentiles of vegetation productivity loss under moderate, severe and extreme drought scenarios can be calculated by the following formula.

$$P(GPP < gpp| \ -1.5 < SPEI \leq -1) = \frac{F_{GPP, SPEI}(gpp, -1) - F_{GPP, SPEI}(gpp, -1.5)}{F_{SPEI}(-1) - F_{SPEI}(-1.5)} \quad (12)$$

$$P(GPP < gpp|-2 < SPEI \leq -1.5) = \frac{F_{GPP, SPEI}(gpp, -1.5) - F_{GPP, SPEI}(gpp, -2)}{F_{SPEI}(-1.5) - F_{SPEI}(-2)} \quad (13)$$

$$P(GPP < gpp| \ SPEI \leq -2) = \frac{F_{GPP, SPEI}(gpp, -2)}{F_{SPEI}(-2)}. \quad (14)$$

2.6. Statistical Tests

We used the linear regression method to calculate the GPP's mean change rate over the entire temporal domain. A two-tailed t-test and the Mann-Kendall test were used to examine the significance of the trend, while the Kolmogorov–Smirnov test was used to establish the optimal marginal distribution function of GPP ($p < 0.05$). In addition, we adopted several goodness-of-fit measures to evaluate different copula models' performance, including the squared Euclidean distance (SED) between the theoretical copula and the empirical copula, the root mean squared error (RMSE) and the Akaike information criterion (AIC).

3. Results

3.1. Validation of LUE GPPs' Accuracy, Dynamics Trends and Drought's Effect on GPP

In this study, special attention was given to determining water stress (including atmosphere vapor pressure deficit, soil moisture content and the humidity deficit) in the LUE model to estimate the global GPP. By comparing with the FLUXNET GPP site data, we found that different LUE models have a good fit in estimation of LUE GPP. The fitted R^2 of VPD_{GLO}-SM, VPD_{MOD}-SM, VPD_{GLO}-ETR, and VPD_{MOD}-ETR were 0.7739, 0.7399, 0.7427, 0.7459 and 0.7628, respectively (Table 1). The average goodness of fit of multiple models reached 0.78 (Figure S2). Overall, LUE GPP had a greater response to the atmospheric vapor pressure deficit, followed by soil water content, and the humidity deficit. The LUE GPP showed greater sensitivity to soil moisture content than the vapor pressure deficit only in evergreen needleleaf forest (ENF), mixed forest (MF) (R^2 = 0.7833 and 0.8479, respectively, RMSE = 1.6399 and 1.3825, respectively). The annual average spatial distribution of GPP in multiple models is shown in Figure S1. From 1982 to 2015, the global average annual GPP of terrestrial vegetation continued to increase at a mean rate of 0.134 Pg C a $^{-1}$ ($p < 0.001$), but its growth rate declined after the mid-1990s (Figure 1a). From the spatial distribution of the mean annual GPP change trend, it can be seen that the global mean annual GPP still had a largely increasing trend, and the significant increase area accounted for 36.83% of the global terrestrial vegetation area, located primarily in the mid-high latitudes of the northern hemisphere, India, Southeastern China, Southeastern South America, Northern and Southern Africa, and Eastern Australia. 11.62% of the regional GPP showed a significant downward trend, largely in Eastern Brazil and Central and Southern Africa (Figure 1b).

Taking SPEI-12 as an example, a drought episode with an SPEI value less than -1 and no less than 3 consecutive months was considered a drought event. We counted the total frequency of droughts from 1982 to 2015 in each grid around the world. It can be seen that the global drought-prone areas are concentrated primarily in Central Russia, Southeast Asia, the Mediterranean coast and most parts of Africa, and the frequency of droughts has increased more than 10 times (Figure 2a). We divided the entire time period from 1982 to 2015 into a dry period and a non-dry period, and compared the changes in LUE GPP during the dry period relative to the non-dry period. It can be seen that in the high latitudes of the Northern hemisphere, Eastern Asia, the Amazon basin and Central Africa, vegetation productivity under drought conditions increased (accounting for 28.09% of the total global land vegetation area), while in 71.91% of the total land vegetation area in the world, the occurrence of drought led to a decrease in GPP (Figure 2b). To reveal the cause of this phenomenon further, we considered the effect of climate factors (including temperature, solar radiation and soil moisture) on vegetation productivity. As shown in Figure 3, temperature and solar radiation play important roles in the changes in GPP in the high latitudes of the Northern hemisphere, Eastern Asia, the Amazon basin and Central Africa during drought periods, while in other regions, soil moisture limits GPP primarily. In arid and semiarid areas, GPP is correlated negatively with temperature and solar radiation, while in humid and semi-humid areas, the converse is true. GPP is correlated positively with soil moisture globally.

Table 1. Regression statistics for different light-use efficiency models between each land-use type's measured FLUXNET GPP and LUE GPPs.

Type	Statistics	Light-Use Efficiency Models				
		VPD only	VPD$_{GLO}$-SM	VPD$_{MOD}$-SM	VPD$_{GLO}$-ETR	VPD$_{MOD}$-ETR
DBF (N: 1693)	R^2	**0.7822**	0.7467	0.7610	0.7275	0.7454
	RMSE	**2.1887**	2.3839	2.3068	2.4894	2.3932
EBF (N: 857)	R^2	0.6388	0.3798	0.3663	0.5599	**0.6662**
	RMSE	1.9865	2.8458	2.8692	2.5501	**1.9500**
ENF (N: 2968)	R^2	0.7706	**0.7833**	0.7733	0.7671	0.7590
	RMSE	1.6990	**1.6399**	1.6919	1.7278	1.7697
MF (N: 863)	R^2	0.8480	**0.8479**	0.8484	0.8486	0.8496
	RMSE	1.3900	**1.3825**	1.3877	1.3873	1.3904
WET (N: 541)	R^2	**0.6860**	0.5871	0.6297	0.5953	0.6406
	RMSE	**2.0777**	2.5013	2.3196	2.4670	2.2742
CSH/OSH (N: 317)	R^2	0.6563	0.4718	0.4369	**0.6829**	0.6609
	RMSE	1.3422	1.7781	1.8739	**1.3185**	1.3611
WSA (N: 545)	R^2	**0.8103**	0.7806	0.7947	0.7784	0.8001
	RMSE	**1.1185**	1.3406	1.2351	1.3736	1.2354
SAV (N: 427)	R^2	**0.6776**	0.6511	0.6517	0.6749	0.6746
	RMSE	**1.3769**	1.6252	1.6233	1.4901	1.4919
GRA (N: 1767)	R^2	0.7667	0.7727	0.7730	0.7754	**0.7763**
	RMSE	1.9214	1.9175	1.9088	1.9137	**1.9012**
CRO (N: 1392)	R^2	**0.5290**	0.4686	0.5047	0.4564	0.4953
	RMSE	**3.5701**			/	
All sites except cropland (N: 9978)	R^2	**0.7739**	0.7399	0.7427	0.7459	0.7628
	RMSE	**1.8089**	1.9817	1.9679	1.9739	1.8855

Note: The sites are deciduous broadleaf forest (DBF), evergreen broadleaf forest (EBF), evergreen needleleaf forest (ENF), mixed forest (MF), wetlands (WET), closed shrublands/open shrublands (CSH/OSH), woody savanna (WSA), savanna (SAV), grassland (GRA) and cropland (CRO). VPD only, VPD$_{GLO}$-SM, VPD$_{MOD}$-SM, VPD$_{GLO}$-ETR and VPD$_{MOD}$-ETR stand for the moisture stress algorithms. N represents the total number of data points; R^2 is the R^2e values of linear regressions with intercept. RMSE represents root mean square error. For each land-use type, the corresponding "best algorithm" is highlighted in bold.

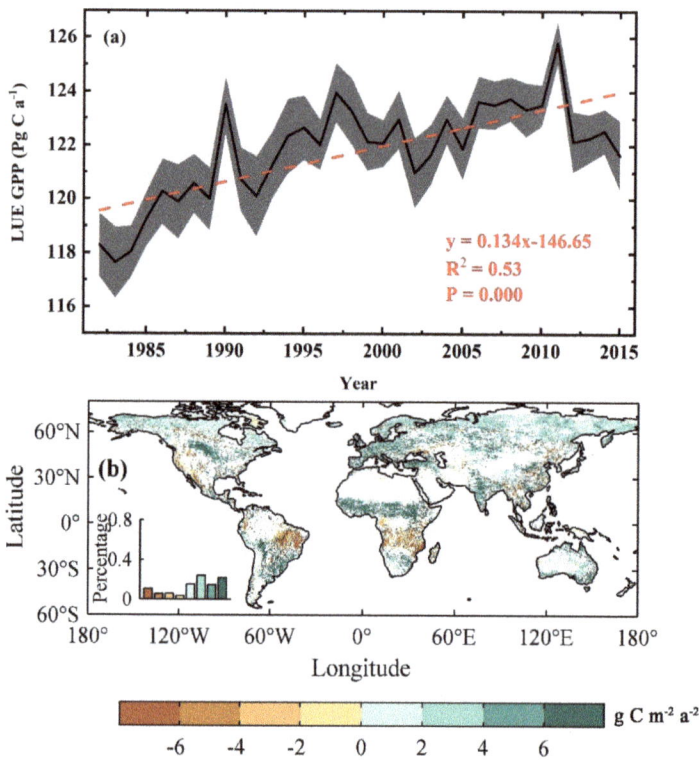

Figure 1. The dynamics trend and spatial distribution of the annual ensemble mean light-use efficiency (LUE) model gross primary productivities (GPPs) from 1982 to 2015. (**a**) Linear trend of annual ensemble mean LUE model GPPs. The shaded areas denote one standard deviation of each corresponding ensemble. (**b**) Spatial distribution of the linear change rate of GPP during 1982–2015. Only statistically significant trends ($p < 0.05$) are shown in this figure. The inset shows the corresponding values' frequency distribution.

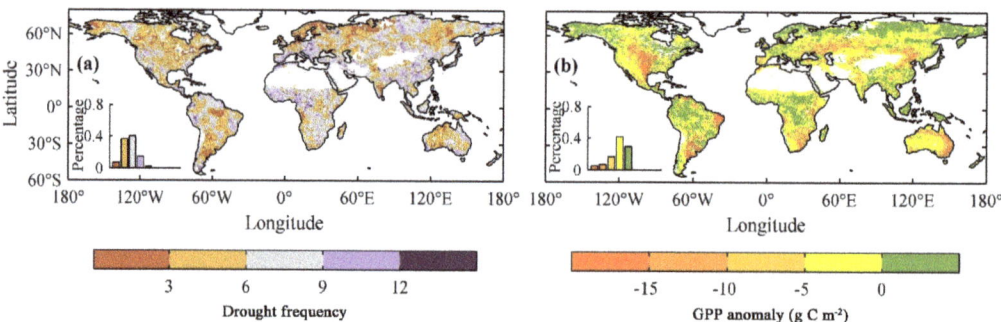

Figure 2. Drought frequency and LUE GPP changes' spatial distribution during the drought period. (**a**) Geographical pattern of drought frequency from 1982–2015 (based on Standardized Precipitation Evapotranspiration Index (SPEI)–12 month). (**b**) GPP anomalies during drought period relative to non-drought period.

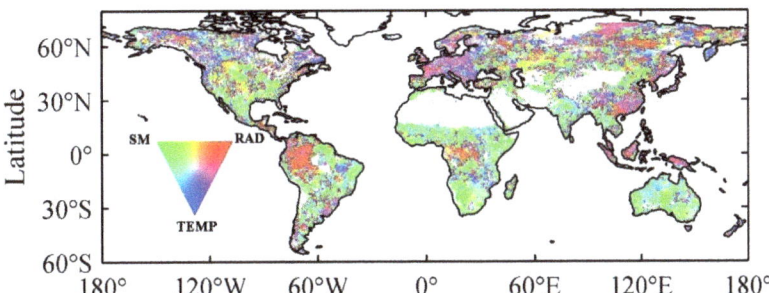

Figure 3. Geographic distributions of potential climate factors (temperature, radiation and soil moisture) to GPP from 1982 to 2015.

3.2. Spatio-Temporal Dynamics of Vegetation Productivity's Dependence on Water Availability

The monthly LUE GPP series was correlated with the SPEI of multiple time scales (1–24 months) to determine the extent to which water resources determine global vegetation productivity changes. As shown in Figure 4a, vegetation productivity and water availability are largely correlated positively globally, indicating that vegetation productivity tends to increase (decrease) as water supply increases (decreases). In 47.30% of the global terrestrial ecosystem, the correlation coefficient between GPP and SPEI can exceed 0.6. Particularly in the Southwest United States, Southeastern Argentina, South Africa, Central Asia, and Australia, the correlation coefficient between GPP and SPEI can exceed 0.8, indicating that vegetation productivity in these areas depends more upon water supply. Seasonal changes also affect the correlation between GPP and SPEI. As Figure S3a–d shows, the regions with a high correlation between GPP and SPEI in the United States and Central Asia were located primarily in the Southwest of these two regions in March–May, and from June to August, the scope and intensity of GPP that SPEI affected in these two regions reached the maximum (the correlation coefficient was 0.6 or higher). As the rainy season in the Northern hemisphere fades, the correlation between GPP and SPEI in the United States and Central Asia weakens gradually, and from December to February of the following year, the correlation coefficient between GPP and SPEI in this region fell below 0.4. Vegetation productivity and water availability's dependence in the high latitudes of the Northern hemisphere is relatively stable, and remains largely between 0.2 and 0.4, while the correlation between GPP and SPEI in Australia is strong throughout the year.

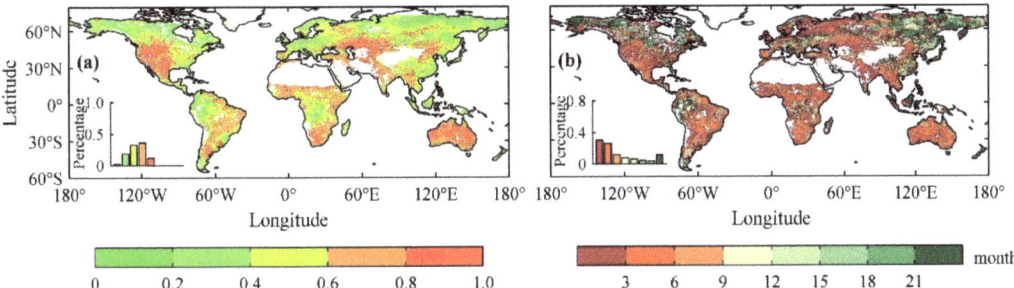

Figure 4. Geographical patterns of vegetation productivity's dependence upon, and response to, water availability. (**a**) Spatial distribution of the maximum correlation coefficients recorded between LUE GPP and SPEI in each pixel during 1982–2015. (**b**) SPEI timescales at which the maximum correlation coefficient between LUE GPP and SPEI was found over the period 1982–2015.

As shown in Figure 5a, an analysis of the changes in the correlation between GPP and SPEI in various climate zones indicated that as the climatic conditions become gradually humid, the dependence between vegetation productivity and water availability continues to decrease (the correlation coefficient between GPP and SPEI declined from 0.76 to 0.47). Different seasons showed consistent characteristics, indicating that the vegetation productivity in arid and semiarid areas is more sensitive to changes in water availability than in humid and semi-humid areas. By comparing the maximum correlation coefficients of GPP and SPEI under different land cover types (Figure 6a), it can be seen that the productivity of GRA, SAV and DBF has a greater correlation with water availability (the correlation coefficients are 0.69, 0.64 and 0.61 respectively), followed by CRO, MF, DNF, ENF, OS and EBF (correlation coefficients of 0.57, 0.55, 0.53, 0.51, 0.50, and 0.50 in order). WSA and WET's productivity had the weakest correlation with water resources. Our results showed that various land cover types have different adaptation strategies to the increase and loss of water resources.

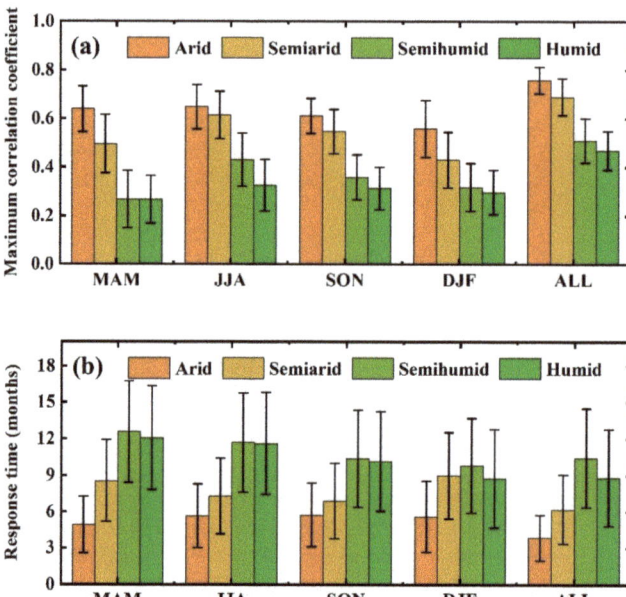

Figure 5. The correlation coefficient (**a**) and response time (**b**) of vegetation productivity to water availability in various climatic regions and different seasons. MAM, JJA, SON, DJF, and ALL represent March–May, June–August, September–November, December–February and the entire year, respectively. The error bars indicate ±1 standard error.

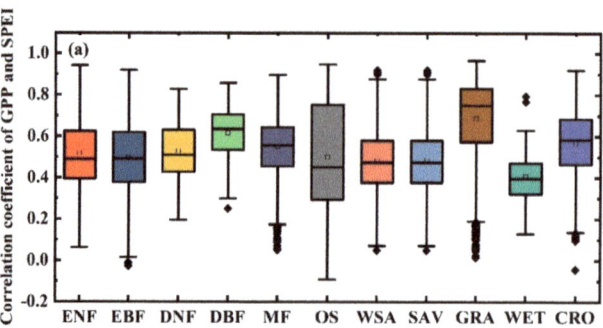

Figure 6. *Cont.*

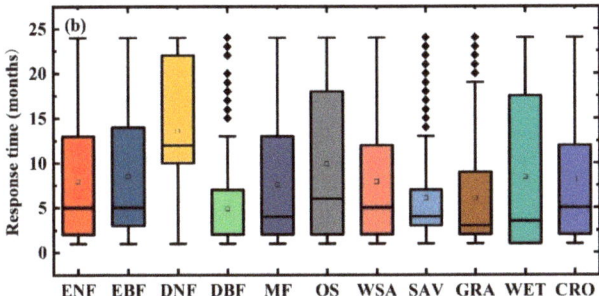

Figure 6. The correlation coefficient (**a**) and response time (**b**) of vegetation productivity to water availability in different land cover types. ENF, EBF, DNF, DBF, MF, OS, WSA, SAV, GRA, WET and CRO represent evergreen needleleaf forest, evergreen broadleaf forest, deciduous needleleaf forest, deciduous broadleaf forest, mixed forest, open shrublands, woody savannas, savannas, grasslands, wetlands and croplands, respectively. Boxes represent interquartile ranges of the values of 25th and 75th percentiles (Q_{25}, Q_{75}), whiskers cover $Q_{25} - 1.5 \times (Q_{75} - Q_{25})$ to Q_{75} $1.5 \times (Q_{75} - Q_{25})$, horizontal lines represent the median, black empty squares represent the mean, and black solid diamonds indicate outliers.

3.3. Spatio-Temporal Dynamics of Vegetation Productivity's Response Time to Water Availability

We analyzed the temporal and spatial dynamics of vegetation productivity's response time of to water availability changes further. Productivity's response time to water availability is defined as timescales over which the maximum GPP-SPEI correlations are recorded. It can be seen from the spatial distribution of productivity's response time to water availability (Figure 4b) that the response time of 56.8% of the global terrestrial ecosystems to water resources is based primarily on short-term and medium-term time scales. In Canada, Northwestern Colombia, Eastern Russia, and Northeastern and Southeastern China, the response time of vegetation productivity to SPEI is relatively long, typically more than 12 months. The longer the response time to water availability changes, the stronger the ability of vegetation productivity in these areas to withstand long-term water shortages. Seasonal changes also affect the length of time in which GPP responds to SPEI in different regions. It can be seen from Figure S3e–h that 34.92% of the global terrestrial ecosystem GPP requires a long time to respond to SPEI during the period from March to May and is concentrated largely in the mid-high latitudes of the Northern hemisphere. With the passage of seasons, the proportion of GPPs in these high latitude regions with a longer response time to SPEI decrease gradually from June to November. From December to November of the following year in Australia, GPP's response time to SPEI changes from a short-term time scale (less than 6 months) to a medium- and long-term time scale (6–9 months).

As shown in Figure 5b, an analysis of the changes GPP's in the response time to SPEI in various climate zones indicates that as the climatic conditions become gradually humid, the response time of vegetation productivity to changes in water availability increased (GPP's mean response time to SPEI extended from 3.9 months to 8.9 months). The characteristics were consistent in different seasons, indicating that vegetation's productivity capacity to withstand long-term water shortages is weaker in arid and semiarid areas than in humid and semi-humid areas. By comparing GPP's response time to SPEI under different land cover types (Figure 6b), it can be seen that DNF has the strongest ability to resist long-term water shortages (the mean response time reached 13.6 months), followed by OS, EBF, WET, CRO, ENF, WSA and MF (mean response time of 9.9, 8.5, 8.4, 8.1, 7.9, 7.8 and 7.5 months, respectively). SAV, GRA, and DBF have poor ability to resist and mitigate drought (mean response time of 6.1, 6, and 4.9 months, respectively). It is worth noting that the land cover types that are more relevant to water availability are often accompanied by weak drought resistance.

3.4. Vegetation Productivity Loss Probability under Different Drought Scenarios

We first fitted GPP's optimal marginal distribution during different months based on the Kolmogorov–Smirnov test. According to the criterion that the smaller the SED, RMSE, and AIC values, the better the copula function's fit effect, we fitted the optimal copula function with the GPP and SPEI in each grid. We set GPP's damage degree to four levels, the 10th, 20th, 30th, and 40th percentile values of GPP in different seasons. Figure 7 shows the spatial distribution of the LUE GPP values at the four percentiles over the globe in March–May, and those in June–August, September–November and December–February are provided in Figures S4–S6. It can be seen that there are obvious differences in the level of global GPP damage during different seasons. At the same time, three drought scenarios were set based on SPEI, moderate ($-1.5 <$ SPEI ≤ 1), severe ($-2 <$ SPEI ≤ -1.5) and extreme (SPEI ≤ -2). In this way, the temporal and spatial distribution characteristics of the probability of different degrees of damage to vegetation productivity under different drought scenarios were studied to determine areas around the globe susceptible to drought during different seasons.

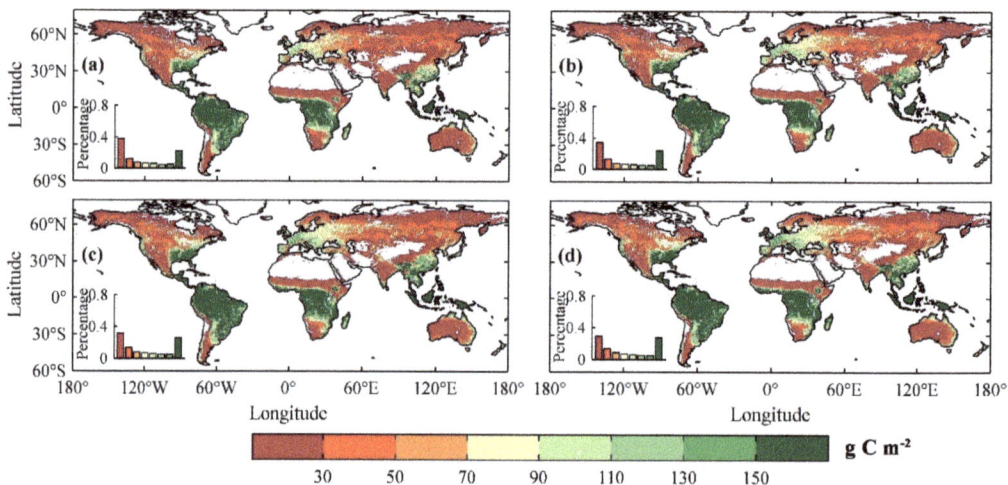

Figure 7. Spatial distribution of global LUE GPP values at 10th (**a**), 20th (**b**), 30th (**c**) and 40th percentiles (**d**) during March–May.

As Figure 8d,h,i shows, the proportion of high-probability areas in the global terrestrial ecosystem increases with the increase in drought intensity when the GPP is lower than the 40th percentile in March–May. Under moderate drought conditions, the area with a probability of occurrence higher than 75% covers only 5.04% of the global terrestrial vegetation area, while under extreme drought conditions, the proportion increases to 27.78%. Further, when the GPP is lower than the 10th, 20th, and 30th percentiles, it still shows the same characteristics of change (the area with a probability of occurrence higher than 75% increases from 0.005% to 2.13%, from 0.12% to 9.24%, and from 1.65% to 18.71%). In other seasons, as the degree of drought increases, the conditional probabilities that GPP will be damaged at different levels increase as well (Figures S7–S9). Under the same drought severity with different levels of GPP damage, drought's influence on GPP loss probabilities weaken gradually as the GPP damage level increases (Figure 8a–d). We explored the seasonal evolution in vegetation productivity's loss probability further under the same drought scenario and the same level of damage of GPP. Using Figure 8h, Figures S7h, S8h and S9h as examples, it can be seen that under severe drought conditions, the area with a conditional probability higher than 75% was 17.48, 21.89, 17.37, and 14.25% in March–May, June–August, September–November, and December–February respectively, in which vegetation productivity was more sensitive to drought in June–August. The primary reason for this phenomenon is drought's effect on GPP in mid- and high-latitude regions of the Northern hemisphere, as this period is the growing season of vegetation in those regions. The occurrence of drought during the growing season reduced the carbon sequestration capacity of vegetation greatly, and increased the GPP loss probability thereby.

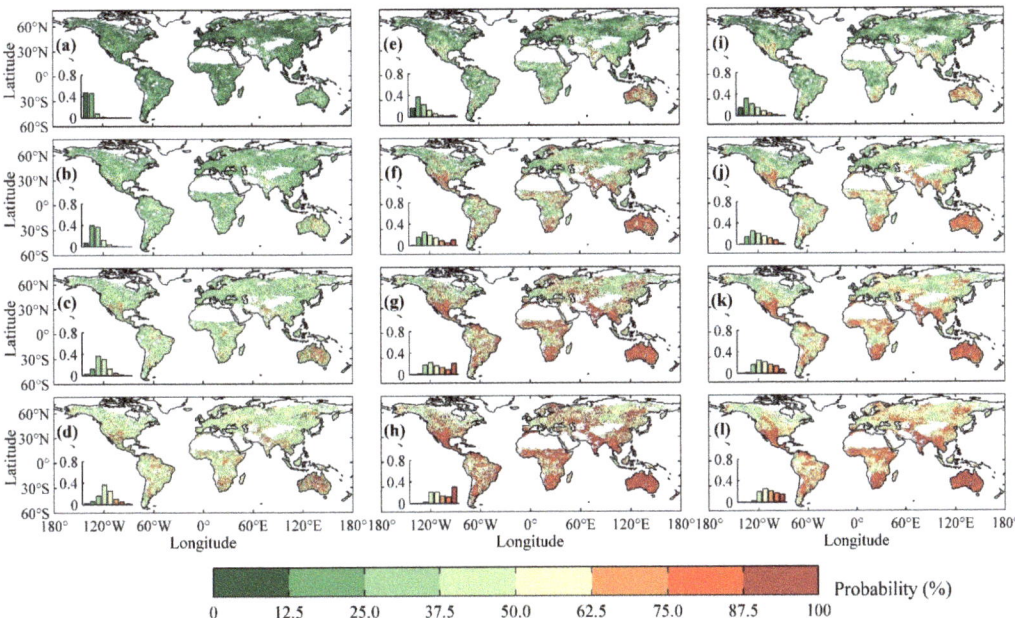

Figure 8. Conditional probabilities of vegetation productivity losses under different drought scenarios in March–May. (a–d) represent the conditional probabilities when the GPP ≤ 10th, ≤ 20th, ≤ 30th, and ≤ 40th percentiles in the moderate drought scenario, respectively. (e–h) represent the conditional probabilities when the GPP ≤ 10th, ≤ 20th, ≤ 30th, and ≤ 40th percentiles in the severe drought scenario, respectively. (i–l) represent the conditional probabilities when the GPP ≤ 10th, ≤ 20th, ≤ 30th, and ≤ 40th percentiles in the extreme drought scenario, respectively.

Further, we analyzed the conditional probabilities of vegetation productivity losses under different drought scenarios in various climatic regions. Take the conditional probability when GPP is less than the 40th percentile from March to May as an example (as shown in Figure 9). Under the moderate drought scenario, the conditional probabilities of GPP loss are 0.54, 0.49, 0.45, and 0.44 in arid, semi-arid, semi-humid, and humid regions, respectively. (They are 0.74, 0.67, 0.56, and 0.56 under the severe drought scenario, and are 0.81, 0.72, 0.59, and 0.58 under the extreme drought scenario). Our results showed that arid and semiarid areas are more likely to suffer different levels of damage than are humid and semi-humid areas under different drought scenarios. As the degree of drought increases, the conditional probabilities of GPP loss also increase in different climate zones (similar results were also evident in other seasons (Figures S10–S12). By comparing the probability of vegetation productivity losses under different drought scenarios in different land cover types when the GPP ≤ 40th percentile (Figures 10 and S13–S15), it can be seen that as the degree of drought increases, the productivity loss probability of EBF, DNF, OS, SAV, and GRA show an increasing trend in different seasons. From March to August, for other land cover types, such as ENF, compared with extreme drought, severe drought has the greatest effect on the degree of damage to vegetation productivity, indicating that different land types respond differently to drought during different seasons.

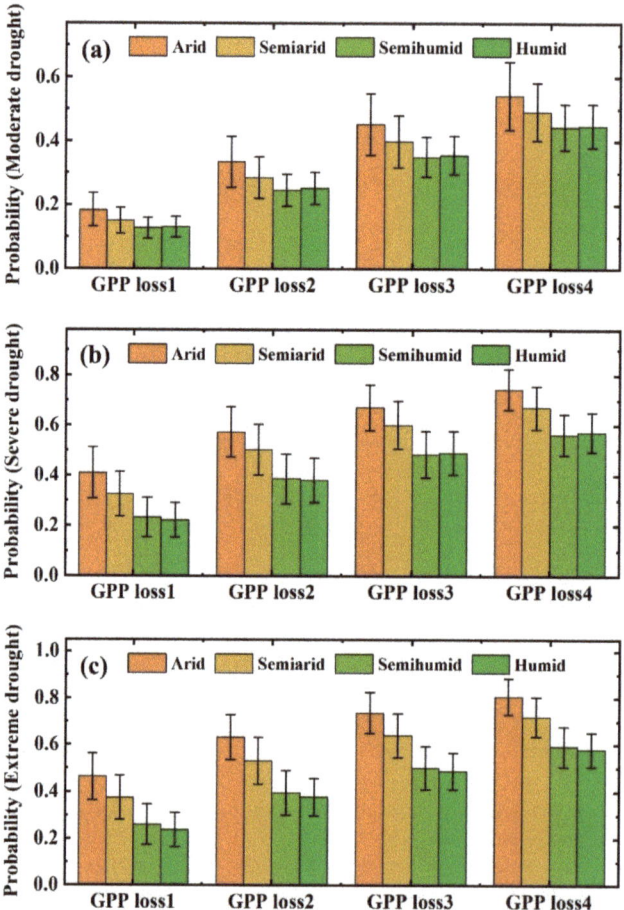

Figure 9. Conditional probabilities of vegetation productivity losses under different scenarios in various climatic regions in March–May. (**a**–**c**) represent conditional probabilities of vegetation productivity losses under moderate, severe and extreme drought, respectively. GPP loss1, GPP loss2, GPP loss3 and GPP loss4 represent GPP $\leq$ 10th, $\leq$ 20th, $\leq$ 30th, and $\leq$ 40th percentiles, respectively. The error bars indicate ±1 standard error.

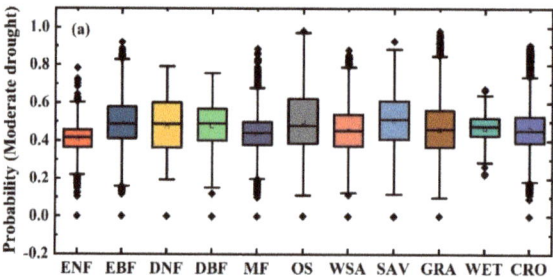

Figure 10. *Cont.*

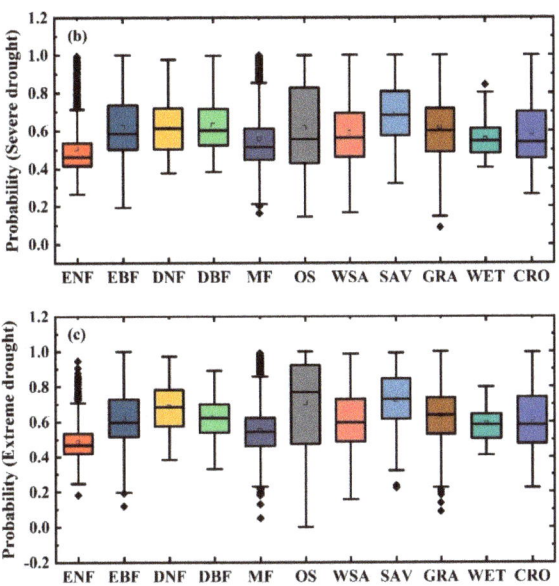

Figure 10. Conditional probabilities of vegetation productivity losses under different drought scenarios in different land cover types when the GPP ≤ 40th percentile in March–May. (**a**–**c**) represent conditional probabilities of vegetation productivity losses under moderate drought, severe drought and extreme drought, respectively. Boxes represent interquartile ranges of the values of 25th and 75th percentiles (Q_{25}, Q_{75}), whiskers cover $Q_{25} - 1.5 \times (Q_{75} - Q_{25})$ to $Q_{75} + 1.5 \times (Q_{75} - Q_{25})$, horizontal lines represent the median, black empty squares represent the mean, and black solid diamonds indicate outliers.

4. Discussion

4.1. Estimating GPP and Drought's Effects on GPP

In this study, special attention was given to determining water stress (including atmosphere vapor pressure deficit, soil moisture content and the humidity deficit) in the LUE model to estimate the global GPP. We found that different LUE models have a good fit-in estimating the LUE GPP (R^2 exceed 0.7). Cai et al. [42] showed that the range of global GPP estimated by different light energy utilization efficiency models was 95–140 Pg C yr^{-1}, which is consistent with this study's estimation results. Generally, because of the lack of water during drought periods, vegetation's stomatal conductance tends to decrease to minimize water loss and prevent hydraulic conductivity loss [43,44], which reduces GPP. Many studies have shown that drought reduces terrestrial ecosystem carbon sinks significantly, and may even transform them into carbon sources [13,45]. However, our results showed that approximately 30% of the global GPP values increased after being disturbed by drought (Figure 2b), a phenomenon attributable to climatic factors' influence. Our results showed that GPP is correlated negatively with temperature and solar radiation in arid and semiarid areas, while the converse is true in humid and semi-humid areas. Further, GPP is correlated positively with soil moisture globally. When a drought episode occurs, because of lower radiation and temperature, and the replenishment of soil moisture in the high latitudes of the Northern Hemisphere, Eastern Asia, the Amazon River basin and Central Africa, it causes an increase in GPP in these regions relative to those in non-dry periods (as shown in Figure 11). This result can be confirmed by Zhao and Runing [46]. In contrast, in most parts of the world, GPP decrease during drought periods because of the increase in radiation and temperature and the decrease in soil moisture.

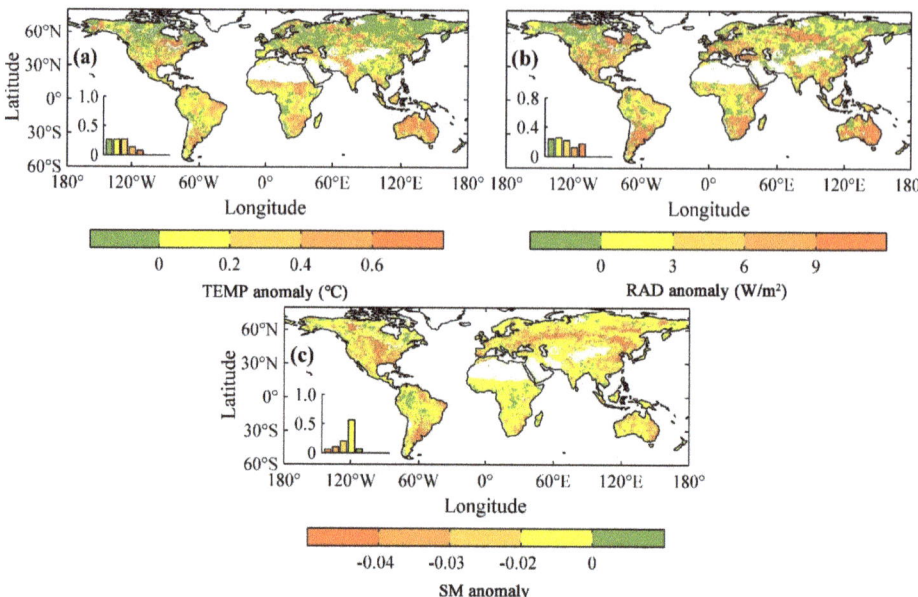

Figure 11. Changes in climate factors during the drought period. (**a–c**) show the anomalies of TEMP (Temperature), RAD (Radiation), and SM (soil moisture), respectively, during drought periods compared to those in non-dry periods.

4.2. Terrestrial Ecosystems' Drought Resistance

Terrestrial ecosystems' drought resistance can be characterized as: a water deficit must continue for a period of time before negative anomalies occur in ecosystem variables [47]. In this study, response time of productivity to water availability was defined according to timescales over which the maximum GPP-SPEI correlations were recorded. To some extent, the response time of vegetation productivity to drought can characterize terrestrial ecosystems' drought resistance. Our research results showed that 56.8% of the global terrestrial ecosystems' response time to water availability is based largely on short- and medium-term time scales. Because different hydrological backgrounds and vegetation composition affect terrestrial ecosystems, their drought resistance demonstrates strong spatial heterogeneity. In Canada, Northwestern Colombia, Eastern Russia, and Northeastern and Southeastern regions China, the response time of vegetation productivity to SPEI is relatively long, usually greater than 12 months. The longer the response time to water availability changes in these regions, the stronger the vegetation productivity's capacity to withstand long-term water shortages. Seasonal changes also affect the resistance of the ecosystem to drought in the same region, particularly in mid-to-high latitude regions (Figure S3). Taking the mid-to-high latitude regions of the Northern hemisphere as an example, it can be seen that the proportion of GPPs in the high latitude with a longer response time to SPEI decreases gradually with the passage of seasons. This result indicates that vegetation has a strong resistance to drought in the early stage of growth, and as subsequent vegetation growth requires considerable water, the occurrence of drought at this stage is more likely to cause vegetation productivity to decline.

Climatic conditions and different types of land cover often affect terrestrial ecosystems' drought resistance [48,49]. Our results showed that as the climatic conditions become gradually humid, the response time of vegetation productivity to changes in water availability increases. This indicates that vegetation productivity in arid and semiarid areas is more vulnerable to drought than in humid and semi-humid areas. Vicente-Serrano et al. [49] also found that arid biomes respond to drought at short time-scales. This may be related to the adaptation strategies of vegetation to water availability in different climate zones.

When a water deficit occurs, arid ecosystems adapt quickly to water shortages by reducing water loss and respiratory costs and increasing growth rates. Fang et al. [20] demonstrated that, as there is a long-term water surplus in humid areas, a short-term drought can not cause changes in vegetation productivity easily. Further, our results revealed different land biomes' role in drought resistance. By comparing GPP's response time to SPEI under different land cover types, we found that DNF, OS, EBF, and ENF have a greater ability to adapt to drought, while WSA, MF, SAV, and GRA have a weak ability to resist and mitigate drought. Forests generally have deep root systems, and thus, under severe drought conditions, they can use the water stored in the deep soil to achieve strong resistance to drought. Whether drought disturbs a forest ecosystem is related closely to its intensity and duration. The water use strategy of shrubs is generally to increase the use of surface soil water during non-arid periods and absorb deep soil water during dry periods to maintain their own production needs [50]. However, herbaceous plants' xylem system has low water and carbon storage capacity compared with that of woodland and shrubs, so it is less resistant to drought. Ivits et al. [51] found that steppic ecosystems also showed weak resistance against drought. Our results are more consistent with the research above.

4.3. The Significance for Ecosystem Management of Estimating the GPP Loss Probability

We developed an optimal bivariate probabilistic model to derive the vegetation productivity loss probabilities under different drought scenarios using copula method. The vegetation productivity loss probability drought causes can also reflect the ability of terrestrial ecosystems to resist interference. Under the same drought conditions, the higher the probability of vegetation loss, the weaker the resistance to drought. Our results showed that arid and semiarid areas have higher conditional probabilities of vegetation productivity losses under different drought scenarios. The productivity loss probability of EBF, DNF, OS, SAV, and GRA showed an increasing trend during different seasons. Quantifying the probability of varying degrees of damage to vegetation productivity under predictable drought scenarios has great significance in mitigating and adapting to global changes. However, there is none of the comparisons either between vegetation are types or the prediction is self are not statistically significant. Huang et al. [52] found that the persistence of a water deficit (11 months) with an intensity of -1.64 (SPEI) led to negligible growth of conifer species. Berdugo et al. [53] showed that aridification can lead to systemic and abrupt changes in multiple ecosystem attributes. Therefore, in future studies, to determine how much drought can cause significant changes in ecosystem productivity and to estimate the corresponding GPP loss probability can better reflect the differences between arid to humid areas.

5. Conclusions

In this study, we explored satellite-observed global terrestrial vegetation production in response to water avaliability by determining gobal vegetation productivity's seasonal dynamics in response to drought in various climate zones and land biomes and quantifying its vulnerability under predictable drought scenarios. Our primary conclusions are as follows:

1. Different LUE models have a good fit effect in estimating GPP. The fitting R^2 of VPD_{GLO}-SM, VPD_{MOD}-SM, VPD_{GLO}-ETR and VPD_{MOD}-ETR were 0.7739, 0.7399, 0.7427, 0.7459 and 0.7628, respectively. From 1982 to 2015, the global mean annual GPP of terrestrial vegetation continued to increase at an average rate of 0.134 Pg C a $^{-1}$ ($p < 0.001$), but its growth rate declined after the mid-1990s. GPP is expected to decrease in 71.91% of the global land vegetation area because of increases in radiation and temperature and decreases in soil moisture during drought periods.
2. Vegetation productivity and water availability are largely correlated positively globally. Further, seasonal changes also affect vegetation productivity's dependence upon water availability. The correlation coefficient between GPP and SPEI declined from 0.76 to 0.47 as the climatic conditions became gradually humid, indicating that the

vegetation productivity in arid and semiarid areas depends more heavily on water availability than that in humid and semi-humid areas. Various land cover types have different adaptation strategies to the increase and loss of water resources, and the productivity of GRA, SAV, and DBF has a higher correlation with water availability.
3. 56.8% of the global terrestrial ecosystems' response time to water resources is based primarily on short and medium-term time scales (3–6 months). The GPP's mean response time to SPEI increased from 3.9 to 8.9 months as the climatic conditions became gradually humid, which indicates that the capacity of productivity of vegetation in arid and semiarid areas to withstand long-term water shortages is weaker than that in humid and semi-humid areas. The land cover types that are more relevant to water availability are often accompanied by weak drought resistance, while DNF, OS, EBF and WET have a stronger ability to resist long-term water deficits.
4. Under the scenario of the same level of GPP damage with different drought degrees, as droughts increase in severity, GPP loss probabilities increase as well. Further, under the same drought severity with different levels of GPP damage, drought's effect on GPP loss probabilities weakens gradually as the GPP damage level increases. Similar patterns were observed in different seasons. Our results showed that arid and semiarid areas have higher conditional probabilities of vegetation productivity losses under different drought scenarios. The productivity loss probability of EBF, DNF, OS, SAV, and GRA show an increasing trend in different seasons, and different land types have different responses to drought in different seasons.

Supplementary Materials: The following are available online at https://www.mdpi.com/article/10.3390/rs13071289/s1, Figure S1: Geographical patterns of the annual ensemble mean LUE model GPPs from 1982 to 2015, Figure S2: The comparison between monthly flux site measurement of GPP and monthly LUE GPP, Figure S3: Geographical patterns of the dependence and response time of vegetation productivity to water availability in different seasons, Figure S4: Spatial distribution of global LUE GPP values at 10th (a), 20th (b), 30th (c) and 40th percentiles (d) during June–August, Figure S5: Spatial distribution of global LUE GPP values at 10th (a), 20th (b), 30th (c) and 40th percentiles (d) during September–November, Figure S6: Spatial distribution of global LUE GPP values at 10th (a), 20th (b), 30th (c) and 40th percentiles (d) during December–February, Figure S7: Conditional probabilities of vegetation productivity losses under different drought scenarios in June–August, Figure S8: Conditional probabilities of vegetation productivity losses under different drought scenarios in September–November, Figure S9: Conditional probabilities of vegetation productivity losses under different drought scenarios in December–February, Figure S10: Conditional probabilities of vegetation productivity losses under different scenarios in various climatic regions in June–August, Figure S11: Conditional probabilities of vegetation productivity losses under different scenarios in various climatic regions in September–November, Figure S12: Conditional probabilities of vegetation productivity losses under different scenarios in various climatic regions in December–February, Figure S13: Conditional probabilities of vegetation productivity losses under different scenarios in different land cover types when the GPP $\leq$ 40th percentile in June–August, Figure S14: Conditional probabilities of vegetation productivity losses under different scenarios in different land cover types when the GPP $\leq$ 40th percentile in September–November, Figure S15: Conditional probabilities of vegetation productivity losses under different scenarios in different land cover types when the GPP $\leq$ 40th percentile in December–February, Table S1: Univariate margin distribution functions, Table S2: Common two-dimensional copula function families.

Author Contributions: Data curation, Y.C.; Supervision, B.F.; Validation, X.W.; Writing—original draft, Y.Z.; Writing—review & editing, X.F. and B.F. All authors have read and agreed to the published version of the manuscript.

Funding: This research was funded by the National Key Research and Development Program of China (2017YFA0604700), National Natural Science Foundation of China (41991233) and Chinese Academy of Sciences (QYZDY-SSW-DQC025 and 2019DC0027).

Institutional Review Board Statement: Not applicable.

Informed Consent Statement: Not applicable.

Data Availability Statement: The data used in the study can be downloaded through the corresponding link provided in Section 2.1.

Conflicts of Interest: The authors declare no conflict of interest.

References

1. Field, C.B. *Climate Change 2014—Impacts, Adaptation and Vulnerability: Regional Aspects*; Cambridge University Press: Cambridge, UK, 2014.
2. Cai, W.; Borlace, S.; Lengaigne, M.; Van Rensch, P.; Collins, M.; Vecchi, G.; Timmermann, A.; Santoso, A.; McPhaden, M.J.; Wu, L. Increasing frequency of extreme El Niño events due to greenhouse warming. *Nat. Clim. Chang.* **2014**, *4*, 111–116. [CrossRef]
3. Diffenbaugh, N.S.; Giorgi, F. Climate change hotspots in the CMIP5 global climate model ensemble. *Clim. Chang.* **2012**, *114*, 813–822. [CrossRef] [PubMed]
4. Beer, C.; Reichstein, M.; Tomelleri, E.; Ciais, P.; Jung, M.; Carvalhais, N.; Rödenbeck, C.; Arain, M.A.; Baldocchi, D.; Bonan, G.B. Terrestrial gross carbon dioxide uptake: Global distribution and covariation with climate. *Science* **2010**, *329*, 834–838. [CrossRef] [PubMed]
5. Boyer, J.S. Plant Productivity and Environment. *Science* **1982**, *218*, 443–448. [CrossRef]
6. Anav, A.; Friedlingstein, P.; Beer, C.; Ciais, P.; Harper, A.; Jones, C.; Murray-Tortarolo, G.; Papale, D.; Parazoo, N.C.; Peylin, P. Spatiotemporal patterns of terrestrial gross primary production: A review. *Rev. Geophys.* **2015**, *53*, 785–818. [CrossRef]
7. Sannigrahi, S.; Zhang, Q.; Joshi, P.K.; Sutton, P.C.; Sen, S. Examining effects of climate change and land use dynamic on biophysical and economic values of ecosystem services of a natural reserve region. *J. Clean. Prod.* **2020**, *257*, 120424. [CrossRef]
8. Ryu, Y.; Berry, J.A.; Baldocchi, D.D. What is global photosynthesis? History, uncertainties and opportunities. *Remote Sens. Environ.* **2019**, *223*, 95–114. [CrossRef]
9. Zargar, S.M.; Gupta, N.; Nazir, M.; Mahajan, R.; Malik, F.A.; Sofi, N.R.; Shikari, A.B.; Salgotra, R. Impact of drought on photosynthesis: Molecular perspective. *Plant Gene* **2017**, *11*, 154–159. [CrossRef]
10. Van Nieuwstadt, M.G.; Sheil, D. Drought, fire and tree survival in a Borneo rain forest, East Kalimantan, Indonesia. *J. Ecol.* **2005**, *93*, 191–201. [CrossRef]
11. Reichstein, M.; Ciais, P.; Papale, D.; Valentini, R.; Running, S.; Viovy, N.; Cramer, W.; Granier, A.; OGÉE, J.; Allard, V. Reduction of ecosystem productivity and respiration during the European summer 2003 climate anomaly: A joint flux tower, remote sensing and modelling analysis. *Glob. Chang. Biol.* **2010**, *13*, 634–651. [CrossRef]
12. Jentsch, A.; Kreyling, J.; Elmer, M.; Gellesch, E.; Glaser, B.; Grant, K.; Hein, R.; Lara, M.; Mirzae, H.; Nadler, S.E. Climate extremes initiate ecosystem-regulating functions while maintaining productivity. *J. Ecol.* **2011**, *99*, 689–702. [CrossRef]
13. Schwalm, C.R.; Williams, C.A.; Schaefer, K.; Baldocchi, D.; Black, T.A.; Goldstein, A.H.; Law, B.E.; Oechel, W.C.; Paw, U.K.T.; Scott, R.L. Reduction in carbon uptake during turn of the century drought in western North America. *Nat. Geosci.* **2012**, *5*, 551–556. [CrossRef]
14. Zhang, L.; Xiao, J.; Li, J.; Wang, K.; Lei, L.; Guo, H. The 2010 spring drought reduced primary productivity in southwestern China. *Environ. Res. Lett.* **2012**, *7*, 045706. [CrossRef]
15. Stocker, B.D.; Zscheischler, J.; Keenan, T.F.; Prentice, I.C.; Seneviratne, S.I.; Peñuelas, J. Drought impacts on terrestrial primary production underestimated by satellite monitoring. *Nat. Geosci.* **2019**, *12*, 264–270. [CrossRef]
16. Craine, J.M.; Ocheltree, T.W.; Nippert, J.B.; Towne, E.G.; Skibbe, A.M.; Kembel, S.W.; Fargione, J.E. Global diversity of drought tolerance and grassland climate-change resilience. *Nat. Clim. Chang.* **2013**, *3*, 63–67. [CrossRef]
17. Beguería, S.; Vicente-Serrano, S.M.; Reig, F.; Latorre, B. Standardized precipitation evapotranspiration index (SPEI) revisited: Parameter fitting, evapotranspiration models, tools, datasets and drought monitoring. *Int. J. Climatol.* **2014**, *34*, 3001–3023. [CrossRef]
18. Bárdossy, A. Copula-based geostatistical models for groundwater quality parameters. *Water Resour. Res.* **2006**, *42*. [CrossRef]
19. Madadgar, S.; AghaKouchak, A.; Farahmand, A.; Davis, S.J. Probabilistic estimates of drought impacts on agricultural production. *Geophys. Res. Lett.* **2017**, *44*, 7799–7807. [CrossRef]
20. Fang, W.; Huang, S.; Huang, Q.; Huang, G.; Wang, H.; Leng, G.; Wang, L.; Guo, Y. Probabilistic assessment of remote sensing-based terrestrial vegetation vulnerability to drought stress of the Loess Plateau in China. *Remote Sens. Environ.* **2019**, *232*, 111290. [CrossRef]
21. Winkler, A.J.; Myneni, R.B.; Alexandrov, G.A.; Brovkin, V. Earth system models underestimate carbon fixation by plants in the high latitudes. *Nat. Commun.* **2019**, *10*, 885. [CrossRef]
22. Yuan, W.; Liu, S.; Zhou, G.; Zhou, G.; Tieszen, L.L.; Baldocchi, D.; Bernhofer, C.; Gholz, H.; Goldstein, A.H.; Goulden, M.L. Deriving a light use efficiency model from eddy covariance flux data for predicting daily gross primary production across biomes. *Agric. For. Meteorol.* **2007**, *143*, 189–207. [CrossRef]
23. Bodesheim, P.; Jung, M.; Gans, F.; Mahecha, M.D.; Reichstein, M. Upscaled diurnal cycles of land-atmosphere fluxes: A new global half-hourly data product. *Earth Syst. Sci. Data Discuss.* **2018**, *10*, 1327–1365. [CrossRef]
24. Melillo, J.M.; Mcguire, A.D.; Kicklighter, D.W.; Moore, B.; Vorosmarty, C.J.; Schloss, A.L. Global climate change and terrestrial net primary production. *Nature* **1993**, *363*, 234–240. [CrossRef]
25. Running, S.W.; Nemani, R.R.; Ann, H.F.; Maosheng, Z.; Matt, R.; Hirofumi, H. A Continuous Satellite-Derived Measure of Global Terrestrial Primary Production. *Bioscience* **2004**, *54*, 547–560. [CrossRef]

26. Brecht, M.; Miralles, D.G.; Hans, L.; Robin, V.D.S.; De, J.R.A.M.; Diego, F.-P.; Beck, H.E.; Dorigo, W.A.; Verhoest, N.E.C. GLEAM v3: Satellite-based land evaporation and root-zone soil moisture. *Geosci. Model Dev.* **2017**, *10*, 1903–1925.
27. Zhu, Z.; Bi, J.; Pan, Y.; Ganguly, S.; Anav, A.; Xu, L.; Samanta, A.; Piao, S.; Nemani, R.R.; Myneni, R.B. Global data sets of vegetation leaf area index (LAI) 3g and fraction of photosynthetically active radiation (FPAR) 3g derived from global inventory modeling and mapping studies (GIMMS) normalized difference vegetation index (NDVI3g) for the period 1981 to 2011. *Remote Sens.* **2013**, *5*, 927–948.
28. Friedl, M.A.; Sulla-Menashe, D.; Tan, B.; Schneider, A.; Ramankutty, N.; Sibley, A.; Huang, X. MODIS Collection 5 global land cover: Algorithm refinements and characterization of new datasets. *Remote Sens. Environ.* **2010**, *114*, 168–182. [CrossRef]
29. Vicente-Serrano, S.M.; Beguería, S.; López-Moreno, J.I. A multiscalar drought index sensitive to global warming: The standardized precipitation evapotranspiration index. *J. Clim.* **2010**, *23*, 1696–1718. [CrossRef]
30. Feng, S.; Fu, Q. Expansion of global drylands under a warming climate. *Atmos. Chem. Phys.* **2013**, *13*, 10081–10094. [CrossRef]
31. Running, S.W.; Thornton, P.E.; Nemani, R.; Glassy, J.M. Global terrestrial gross and net primary productivity from the Earth Observing System. *Methods Ecosyst. Sci.* **2000**, *3*, 44–45.
32. Zhang, Y.; Xiao, X.; Wu, X.; Zhou, S.; Zhang, G.; Qin, Y.; Dong, J. A global moderate resolution dataset of gross primary production of vegetation for 2000–2016. *Sci. Data* **2017**, *4*, 170165. [CrossRef] [PubMed]
33. Mu, Q.; Heinsch, F.A.; Zhao, M.; Running, S.W. Development of a global evapotranspiration algorithm based on MODIS and global meteorology data. *Remote Sens. Environ.* **2007**, *111*, 519–536. [CrossRef]
34. Cramer, W.; Kicklighter, D.; Bondeau, A.; Iii, B.M.; Churkina, G.; Nemry, B.; Ruimy, A.; Schloss, A. Comparing global models of terrestrial net primary productivity (NPP): Overview and key results. *Glob. Chang. Biol.* **1999**, *5*, iii–iv. [CrossRef]
35. Potter, C.S.; Randerson, J.T.; Field, C.B.; Matson, P.A.; Klooster, S.A. Terrestrial Ecosystem Production: A Process Model Based on Global Satellite and Surface Data. *Glob. Biogeochem. Cycles* **1993**, *7*, 811–841. [CrossRef]
36. Allen, R.; Pereira, L.; Raes, D.; Smith, M.; Allen, R.G.; Pereira, L.S.; Martin, S. *Crop Evapotranspiration: Guidelines for Computing Crop Water Requirements*; FAO Irrigation and Drainage Paper 56; FAO: Rome, Italy, 1998; p. 56.
37. Task, G.S.D. Global Gridded Surfaces of Selected Soil Characteristics (IGBP-DIS). Ornl Daac. 2000. Available online: https://daac.ornl.gov/cgi-bin/dsviewer.pl?ds_id=569 (accessed on 1 December 2020).
38. Batjes, N.H. Harmonized soil property values for broad-scale modelling (WISE30sec) with estimates of global soil carbon stocks. *Geoderma* **2016**, *269*, 61–68. [CrossRef]
39. Hargreaves, G.H.; Samani, Z.A. Estimating Potential Evapotranspiration. *J. Irri. Drain. Div.* **1982**, *108*, 225–230. [CrossRef]
40. Lee, T.; Modarres, R.; Ouarda, T.B. Data-based analysis of bivariate copula tail dependence for drought duration and severity. *Hydrol. Process.* **2013**, *27*, 1454–1463. [CrossRef]
41. Zhang, Y.; Feng, X.; Wang, X.; Fu, B. Characterizing drought in terms of changes in the precipitation–runoff relationship: A case study of the Loess Plateau, China. *Hydrol. Earth Syst. Sci.* **2018**, *22*, 1749–1766. [CrossRef]
42. Wenwen, C.; Wenping, Y.; Shunlin, L.; Shuguang, L.; Wenjie, D.; Yang, C.; Dan, L.; Haicheng, Z. Large Differences in Terrestrial Vegetation Production Derived from Satellite-Based Light Use Efficiency Models. *Remote Sens.* **2014**, *6*, 8945–8965.
43. Zhou, S.; Zhang, Y.; Williams, A.P.; Gentine, P. Projected increases in intensity, frequency, and terrestrial carbon costs of compound drought and aridity events. *Sci. Adv.* **2019**, *5*, eaau5740. [CrossRef]
44. Liu, L.; Gudmundsson, L.; Hauser, M.; Qin, D.; Li, S.; Seneviratne, S.I. Soil moisture dominates dryness stress on ecosystem production globally. *Nat. Commun.* **2020**, *11*, 4892. [CrossRef] [PubMed]
45. Piao, S.; Zhang, X.; Chen, A.; Liu, Q.; Wu, X. The impacts of climate extremes on the terrestrial carbon cycle: A review. *Sci. China Earth Sci.* **2019**, *62*, 1551–1563. [CrossRef]
46. Zhao, M.; Running, S.W. Drought-Induced Reduction in Global Terrestrial Net Primary Production from 2000 through 2009. *Science* **2010**, *329*, 940–943. [CrossRef] [PubMed]
47. Xu, H.-J.; Wang, X.-P.; Zhao, C.-Y.; Yang, X.-M. Diverse responses of vegetation growth to meteorological drought across climate zones and land biomes in northern China from 1981 to 2014. *Agric. For. Meteorol.* **2018**, *262*, 1–13. [CrossRef]
48. Grossiord, C.; Granier, A.; Ratcliffe, S.; Bouriaud, O.; Bruelheide, H.; Cheko, E.; Forrester, D.I.; Dawud, S.M.; Finér, L.; Pollastrini, M. Tree diversity does not always improve resistance of forest ecosystems to drought. *Proc. Natl. Acad. Sci. USA* **2014**, *111*, 14812–14815. [CrossRef] [PubMed]
49. Vicente-Serrano, S.M.; Gouveia, C.; Camarero, J.J.; Beguería, S.; Sanchez-Lorenzo, A. Response of vegetation to drought time-scales across global land biomes. *Proc. Natl. Acad. Sci. USA* **2012**, *110*, 52–57. [CrossRef]
50. West, A.G.; Dawson, T.E.; February, E.C.; Midgley, G.F.; Bond, W.J.; Aston, T.L. Diverse functional responses to drought in a Mediterranean-type shrubland in South Africa. *New Phytol.* **2012**, *195*, 396–407. [CrossRef]
51. Ivits, E.; Horion, S.; Erhard, M.; Fensholt, R. Assessing European ecosystem stability to drought in the vegetation growing season: Ecosystem stability to drought. *Glob. Ecol. Biogeogr.* **2016**, *25*, 1131–1143. [CrossRef]
52. Huang, K.; Yi, C.; Wu, D.; Zhou, T.; Zhao, X.; Blanford, W.J.; Wei, S.; Wu, H.; Ling, D.; Li, Z. Tipping point of a conifer forest ecosystem under severe drought. *Environ. Res. Lett.* **2015**, *10*, 024011. [CrossRef]
53. Berdugo, M.; Delgado-Baquerizo, M.; Soliveres, S.; Hernández-Clemente, R.; Zhao, Y.; Gaitán, J.J.; Gross, N.; Saiz, H.; Maire, V.; Lehman, A. Global ecosystem thresholds driven by aridity. *Science* **2020**, *367*, 787–790. [CrossRef]

Article

Analysis of the Spatiotemporal Changes in Watershed Landscape Pattern and Its Influencing Factors in Rapidly Urbanizing Areas Using Satellite Data

Zhenjie Zhu [1], Bingjun Liu [2,3,*], Hailong Wang [2,3] and Maochuan Hu [2,3]

1. School of Geography and Planning, Sun Yat-sen University, Guangzhou 510275, China; zhuzhj7@mail2.sysu.edu.cn
2. School of Civil Engineering, Sun Yat-sen University, Guangzhou 510275, China; wanghlong3@mail.sysu.edu.cn (H.W.); humch3@mail.sysu.edu.cn (M.H.)
3. Guangdong Engineering Technology Research Center of Water Security Regulation and Control for Southern China, Sun Yat-sen University, Guangzhou 510275, China
* Correspondence: liubj@mail.sysu.edu.cn; Tel.: +86-020-84114575

Citation: Zhu, Z.; Liu, B.; Wang, H.; Hu, M. Analysis of the Spatiotemporal Changes in Watershed Landscape Pattern and Its Influencing Factors in Rapidly Urbanizing Areas Using Satellite Data. *Remote Sens.* **2021**, *13*, 1168. https://doi.org/10.3390/rs13061168

Academic Editor: Francesca Cigna

Received: 14 January 2021
Accepted: 16 March 2021
Published: 18 March 2021

Publisher's Note: MDPI stays neutral with regard to jurisdictional claims in published maps and institutional affiliations.

Copyright: © 2021 by the authors. Licensee MDPI, Basel, Switzerland. This article is an open access article distributed under the terms and conditions of the Creative Commons Attribution (CC BY) license (https://creativecommons.org/licenses/by/4.0/).

Abstract: Analyzing the spatiotemporal characteristics and causes of landscape pattern changes in watersheds around big cities is essential for understanding the ecological consequence of urbanization and provides a basic reference for the watershed management. This study used a land-use transition matrix and landscape indices to explore the spatiotemporal change of land use and landscape pattern over Liuxihe River basin of Guangzhou in the southeast of China from 1980 to 2015 with multitemporal Landsat satellite data in response to the rapid urbanization process. Primary temporal and spatial influencing factors were first quantitatively identified through grey relation analysis (calculating correlation degree between land use changes and influencing factors) and Geodetector (detecting landscape spatial heterogeneity and its driving factors), respectively. Considerable spatial and temporal differences in land use and landscape pattern changes were observed herein, thus determining the influencing factors of these differences in the Liuxihe River basin. These changes were characterized by a large increase in construction land converted from cropland, particularly in the middle and lower reaches of the basin from 2000 to 2010, causing dramatic fragmentation and homogenization of the landscape pattern there. Meanwhile, the landscape pattern gradually transitioned from an agricultural land use dominant landscape to a construction land use dominant landscape in these regions. Furthermore, the rapid growth of a nonagricultural population and the transformation of industry primarily caused the temporal changes of landscape pattern, and the landscape spatial heterogeneity was mainly caused by the interaction of complicated geomorphology and anthropogenic activities in different spatial locations, particularly after 2000. This study not only provides an improved approach to quantifying the main spatiotemporal influencing factors of landscape pattern changes during different time periods, but also offers a reference for decision-makers to formulate optimal strategies on ecological protection and urban sustainable development of different regions in this study area.

Keywords: landscape pattern; spatiotemporal changes; influencing factors; watershed; China SE; satellite data

1. Introduction

The increasing expansion of big cities has been a common social and economic phenomenon taking place all around the world, especially in the developing countries since the 21st century [1,2]. This process, with no sign of slowing down, may be the most critical anthropogenic force that has brought about dramatic changes in land use and landscape pattern at local, regional, and global scales [3–5]. Numerous studies have found that these immense changes can not only contribute to various environmental issues [6,7], but also affect the structure, function, and health of the ecosystem [8,9], and further threaten the sustainable development of big cities [10,11]. Therein, watersheds around rapidly urbanizing areas are more sensitive to these changes due to its richer and fragile natural ecosystems [12]. Moreover, a watershed is a complete natural and unnatural circulation unit, which is more conducive to conduct the ecological protection and restoration. The problem related to landscape pattern changes in these kinds of watersheds has been receiving more and more attention from international scholars in recent decades [13,14]. Therefore, gaining a deep understanding of the processes and causes of landscape pattern changes is crucial for protection, management, and sustainable planning of these areas under rapid urbanization [15,16].

Previous studies have illustrated that the analysis of land use changes is usually regarded as the basis for studying the landscape patterns change [17], because the landscape pattern is usually defined as the spatial arrangement of various landscape patches of different types, sizes, and shapes, which are classified by different land use types [18]. Changes in the landscape pattern were proved to be the results of changes in various land use types [19]. Most scholars choose a land use transition matrix to reflect the mutual transformation characteristics between any two different land use types [20,21], and use landscape metrics to detect the characteristics of spatial-structural composition and configuration in different landscape patches [22,23]. Therein, the former emphasizes changes of land surface properties in different periods [24], and the latter stresses the changes of potential ecological pattern [25]. When studying the changing characteristics of landscape pattern, it is necessary to analyze land use changes first, and emphasize both the temporal and spatial changes of them.

Currently, previous studies on the change of watershed landscape pattern in rapidly urbanizing areas seldom quantified spatiotemporal processes and causes of landscape pattern changes comprehensively. For example, Su et al. [26] analyzed the land use and landscape pattern in a different period to reflect its spatiotemporal changes characteristics and its causes, but ignored the overall spatial heterogeneity of landscape pattern. Zhang et al. [27] and Shi et al. [28] systematically analyzed the spatiotemporal changes processes of land use and landscape pattern in watersheds. The former study only quantified the temporal influencing factors of land use changes, and the latter study described the temporal and spatial influencing factors respectively, but not quantitatively. In addition, when analyzing the influencing factors of landscape pattern changes quantitatively, many studies failed to solve the problem of insufficient multi-temporal land use data [29–31], nor did they consider the interactive effects of different factors on the spatial landscape heterogeneity [32,33]. Some studies even analyzed the transition driving forces of different land use types in different locations of different period to meet the requirements of large quantities data for commonly used analysis methods [34]. But the causes of the spatial characteristics of landscape pattern were ignored in this case. Considering the fact that the analysis of both spatial and temporal causes were all important for guiding the management and protection of the natural ecosystem in watershed [35]. There is no doubt that researchers should carry out a systematic analysis on the process and cause of watershed landscape pattern spatiotemporal changes quantitatively.

With the continuous improvement of the remote sensing technology, more and more remote sensing data with different sensors, time periods, spectrum and spatial resolution can be acquired [36]. It provides the most stable and accurate multi-temporal data source for land use analysis, thus the land use and landscape pattern changes under the rapid urbanization can be monitored and analyzed spatiotemporally [37]. There existed many studies that using these kinds of data analyzed the land use change of different regions in Guangzhou city [38,39], and compared the relationships between these changes and urban expansion [40,41]. Liu et al. [31] also discussed the land use change and its causes based on the Landsat satellite data with remote sensing technology. These researches all illustrated the feasibility of using land use data provided by remote sensing technology to analyze the changes of landscape patterns in Guangzhou city. Besides, along with the rapidly urban expansion of Guangzhou city, the intensive interaction between natural and human elements in the LXH has brought about the large transition of construction land from lots of forest and cropland [42]. The water quality and natural environment of the LXH were degraded, especially in the downstream [43,44]. The Liuxihe River basin (LXH), as the final ecological barrier on the northwest of Guangzhou city, is an important water resource conservation area. Therefore, compared with areas divided by the administrative units in Guangzhou city, analyzing the landscape pattern changes in the LXH has greater ecological significance and protection value for Guangzhou.

This study mainly analyzes the spatiotemporal differences of process and causes on landscape pattern changes under rapid urbanization of the Guangzhou city. The specific objectives are: (1) To analyze the changes of different land use types in time and space; (2) to characterize the spatial configuration of the landscape pattern in time and space; (3) to establish the relationship between changes in different land use types and different influencing factors temporally; and (4) to determine the impact of different factors on the landscape heterogeneity. Thus, these differences in the spatiotemporal changes and causes of landscape pattern in the LXH under the rapid urbanization since 1980 are revealed. The decision-makers will be more clear about how to formulate an appropriate strategy for planning and management.

2. Materials and Methods

2.1. Study Area

The Liuxihe River basin is in the rapidly developing and urbanizing city of Guangzhou, southeast of China (Figure 1). The river, about 171 km in length with an area of 2300 km^2, flows through Guangzhou, and eventually empties into the Beijiang River, a tributary of the Pearl River. The annual precipitation rate over the LXH is 1750 mm and more than 80% of precipitation occurs from April to September. Its daily mean air temperature is about 20 °C and annual rate of evaporation is about 1200 mm [45]. The elevation of the LXH falls gradually from northeast to southwest, characterized by mountains in the upstream, hills in the midstream, and plains in the downstream, thus making the upstream more difficult to develop than the middle and lower watershed. At present, the distribution of land use in the LXH is characterized by forests in the upstream, croplands and orchards in the midstream, and construction lands in the downstream. In addition, the speed of the urbanization process in the midstream and downstream of the LXH is faster and stronger than in the upstream [46]. The special location conditions, the unbalancing effect of urbanization, the difference in natural topography, and the land uses/land covers all make the LXH an appropriate case for the research processes and influencing factors for spatiotemporal changes of watershed landscape pattern in rapidly urbanizing areas.

2.2. Data and Data Processing

Satellite images taken in 1980, 1990, 1995, 2000, 2005, 2008, 2010, and 2015 (Landsat MSS/TM/ETM, Landsat 8) with a spatial resolution of 30 m were provided by Data Center for Resources and Environmental Sciences, Chinese Academy of Sciences (http://www.resdc.cn, accessed on 30 December 2017). The method of visual interpretation was used to derive thematic land use maps based on the land resource classification system of Chinese academy of sciences [47]. Meanwhile, the accuracy of interpretation was improved through reference data, such as geomorphic maps, vegetation maps, ground truth data at different sample points, and local resident interview data. The calculated results of Kappa coefficient were larger than 0.80, which verified the accuracy and reliability of these land use maps [48]. This paper reclassified these land use maps into nine types including cropland, forest, orchard, grassland, shrub, water, floodplain, construction land, and unused land (Appendix A Table A1).

Besides, data from 1980 to 2015 about demographic factors, socioeconomic factors, and urbanized activities of the LXH were collected from the Guangzhou Statistical Yearbook (http://tjj.gz.gov.cn/, accessed on 1 April 2018) and the Outline of Guangzhou Urban

Construction Overall Strategic Concept Plan (http://ghzyj.gz.gov.cn/, accessed on 30 October 2018), which is provided by the Chinese government. The detailed factors are total population (TP), proportion of non-agricultural population (PNAP), gross domestic product (GDP), proportion of primary industry (PPI), proportion of secondary industry (PSI), proportion of tertiary industry (PTI), annual per capital income (APCI), and total investment in real estate development (IRE).

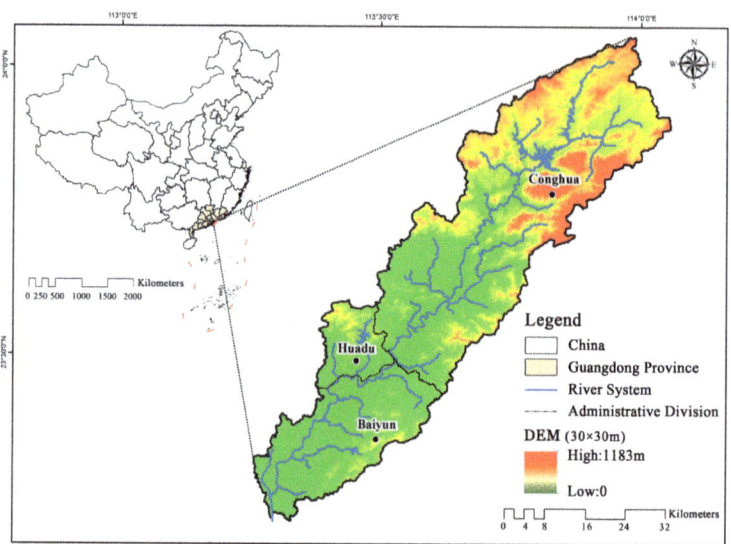

Figure 1. The location and digital elevation map (DEM) of the Liuxihe River basin (LXH).

Spatial data such as topographical elements (digital elevation map (DEM) and SLOPE) were obtained from NASA's Earth Observing System Data and Information System (https://search.earthdata.nasa.gov/search, accessed on 30 December 2018). And other spatial data were obtained from Data Center for Resources and Environmental Sciences, Chinese Academy of Sciences (Source: China, Satellite images, http://www.resdc.cn, accessed on 30 December 2018), including population density (TP), socioeconomic level (GDP), urbanized activities (NLD), and land use intensity (LUIN). These main parameters of the LXH with a spatial resolution of 1000 m (except for the DEM and SLOPE data with a spatial resolution of 30 m) were all cut out by the ArcGIS software.

2.3. Methods

2.3.1. Selection of Landscape Metrics and Influencing Factors

- Landscape Metrics

The factor analysis method [49] was taken to screen the notable landscape metrics from 23 frequently used metrics (Appendix B Figure A1). First, if the absolute value of the correlation coefficient between two indices was more than 0.9, only one was used; second, indices representing different aspects of landscape characteristics were selected to reduce the information redundancy among them [50]. Five representative indices were used in the final analysis, including the patch density (PD), aggregation index (AI), largest patch index (LPI), area-weighted mean patch fractal dimension (AWMPFD), and Shannon's diversity index (SHDI). These landscape metrics were calculated for 1980, 1990, 2000, 2010, and 2015 using the public domain software FRAGSTATS 4.2 at both the class-level and landscape-level in 14 different granularities (30 m, 50 m, 100 m, 200 m, 300 m, 400 m, 500 m, 600 m, 700 m, 800 m, 900 m, 1000 m, 1100 m, 1200 m). FRAGSTATS is a computer software program designed to compute a wide variety of landscape metrics for categorical map patterns at different levels (the detailed introduction of this software can be found at https://www.umass.edu/landeco/research/fragstats/fragstats.html, accessed on 1 February 2018). Thus, the characteristic scale interval which is appropriate for spatial analysis of the Liuxihe River basin was determined (Appendix B Figure A2). The calculation formula and ecological significance of these indices are given in Table 1.

Table 1. Landscape metrics used in this study, their formula and ecological interpretation.

Index	Definition	Equation	Ecological Significance	Scale Level
Patch Density (PD)	Number of patches per unit area.	$PD = N_i/A$	Representing the degree of landscape fragmentation and heterogeneity.	Land use class/landscape
Aggregation Index (AI)	By calculating the adjacent matrix between different types of patches, AI is used to describe the aggregation degree of different patches.	$AI = 2\ln(m) + \sum_{i=1}^{m}\sum_{j=1}^{n} P_{ij} \ln(P_{ij})$	Representing the degree of landscape connectivity and fragmentation.	Land use class/landscape
Largest Patch Index (LPI)	Quantify the percentage of the largest patches in the total landscape area.	$LPI = \frac{Max(a_{ij})}{A}(100)$	Representing the degree of landscape dominance.	Land use class/landscape
Area-weighted Mean Patch Fractal Dimension (AWMPFD)	Fractal dimension theory is used to measure the shape and structure complexity of patches and landscape (ranging from 1 to 2).	$AWMPFD = \sum_{i=1}^{m}\sum_{j=1}^{n}[\frac{2\ln(0.25p_{ij})}{\ln(a_{ij})}]/N$	Representing the interference degree of human activities to some extent.	Land use class/landscape
Shannon's Diversity Index (SHDI)	An index based on the relative area proportion of each landscape type and the total number of types. It is somewhat more sensitive to rare patch types than Simpson's diversity index.	$SHDI = -\sum_{i=1}^{m}(P_i \cdot \ln P_i)$	Representing the degree of landscape heterogeneity and diversity.	Landscape

Note: i = 1 ... m patch types (classes); j = 1 ... n patches; A = total area of each landscape type (m^2); a$_{ij}$ = area (m^2) of patch ij; P$_{ij}$ = perimeter (m) of patch ij; N$_i$ = number of patches in the landscape of patch type (class) i; m = number of patch types (classes) present in the landscape, excluding the landscape border if present; P$_i$ = proportion of the landscape occupied by patch type (class) i.

- Land Use Transition Matrix

The changes in landscapes were detected by calculating each land use type transition matrix of any two adjacent periods from 1980 to 2015. The following equation (Equation (1)) was applied to calculate the matrix:

$$p = \begin{bmatrix} P_{11} & P_{12} & \cdots & P_{1j} \\ P_{21} & P_{22} & \cdots & P_{2j} \\ \vdots & \vdots & \vdots & \vdots \\ P_{i1} & P_{i2} & \cdots & P_{ij} \end{bmatrix} \quad (1)$$

where p_{ij} indicates the area in transition from landscape i to j. Each element of the transition matrix meets two standards: (1) p_{ij} is non-negative, and (2) $\sum_{j=1}^{n} p_{ij} = 1$.

For a better characterization of landscape changes, the transition matrix between any two adjacent periods was displayed in a two-dimensional table by many researchers [51]. Therein, the diagonal entries of the table reflect the total size of persistent land use types whereas the off-diagonal entries show the transition size of one land use type to another. Besides, the gross gain and gross loss of each land use type were also displayed in the table, which could be used to calculate the total exchange (sum of the gross gain and the gross loss) and net change (the gross gain minus the gross loss).

- Selection of Landscape Change Influencing Factors

The influencing factors on the development of landscape pattern vary at different spatial and temporal scales [52], which were categorized into natural and anthropogenic factors in most of the related researches [53,54]. This study focused on the small-scale watershed around rapidly urbanization city in a short research period. In these areas, anthropogenic factors such as population growth, socioeconomic activities, urbanization activities, and related policies usually play a major role [29]. In terms of natural factors, they are relatively stable and unchanged in short term compared with anthropogenic factors [55], thus their impact on landscape changes can be ignored in this case [56–58]. Finally, only anthropogenic factors were selected to reflect the temporal influencing factors in this study, including the total population (TP), the proportion of the non-agricultural population (PNAP), gross domestic product (GDP), the proportion of the primary industry (PPI), the proportion of the secondary industry (PSI), the proportion of the tertiary industry (PTI), annual per capita income (APCI), and total investment in real estate development (IRE). These factors are about the demographic, socioeconomic, and urbanized activities of a region. Besides these, the policies related to the rapid urbanization were also considered.

In addition, researchers found that climate conditions had no significant influence on the spatial distribution of landscape in a small-scale catchment [33], and topographical and anthropogenic factors were commonly used to interpret the landscape spatial heterogeneity in this case [28,59]. As we all know, the spatial difference of soil and hydrological conditions always depend on topographical factors. Therefore, this study mainly analyzed the spatial difference of topographical and anthropogenic factors on landscape spatial heterogeneity. Herein, the spatially distributed data of GDP and TP were selected to represent the spatial difference between socioeconomic level and population density. The nighttime light (NLD) spatial data were used to represent the urbanization degree. Moreover, the spatial distribution of the LUIN was calculated using the land use comprehensive degree index, representing the urbanization activities related to land use changes. Additional details on the calculation of land use comprehensive degree index can be found in Yu et al. [60]. Besides, the topographical variables (DEM and SLOPE) were also considered.

2.3.2. Quantifying the Influence on the Change of Landscape Pattern

Considering the spatiotemporal characteristics of land use and landscape pattern changes and their interactions, this study mainly analyzed the factors that influence the temporal change of land use types and spatial differences of landscape patterns (i.e., the spatial heterogeneity of landscape). Since the data for each year could not be obtained, it made the commonly used quantitative statistical analysis impractical. This study applied grey correlation analysis to explore the impact of different influencing factors on land use changes in an attempt to effectively solve the problem of insufficient sample data. Meanwhile, spatial correlation analysis using the Geodetector was carried out to assess the influence degree and interaction of each influencing factor toward different spatial characteristics of landscape index.

- Grey Correlation Analysis

Grey correlation analysis is an impacting factor measurement method in the grey system theory proposed by Deng [61] in 1982, which analyzes the uncertain relationship between a main factor and all other influencing factors in a given system. It can complement the defects of statistical analysis methods and can work with small amounts of irregular data; it also negates the inconsistency between quantitative and qualitative results. The analysis method mainly compares the time series of each influencing factor to determine which one is dominant. That is, when the trend of changes between an independent variable and a dependent variable is consistent or the degree of synchronization change is high, a strong correlation results [62]. The relationship is often expressed by grey correlation degree (Equation (2)). The greater the degree of grey correlation is, the more the influence degree of the factor will be and vice versa.

$$\gamma_{ij} = \frac{1}{n}\sum_{k=1}^{n} \frac{\min_i \min_k \Delta_i(k) + \xi \max_i \max_k \Delta_i(k)}{\Delta_i(k) + \xi \max_i \max_k \Delta_i(k)}, \Delta_i(k) = |x_j(k) - x_i(k)| \quad (2)$$

where x_i and x_j are the independent variable series and the dependent variable series, respectively; γ_{ij} is the grey relational degree between independent variables x_i and dependent variables x_j; ξ is the resolution coefficient, $\xi \in (0,1)$, and usually ξ takes a value of 0.5; $k = 1, 2, \cdots, n$ is the time series.

- Spatial Correlation Analysis

Spatial autocorrelation can be defined as the coincidence of value similarity with location similarity, and is used to detect patterns of spatial association [63]. In this study, the global Moran's I index [64,65] (Equation (3)) was adopted to analyze the spatial autocorrelation of each landscape metric in its characteristic scale interval (500~1200 m), providing appropriate spatial scales for launching the bivariate spatial correlation analysis between each landscape metric and the influencing factors (Appendix A Table A2).

$$I = \frac{n}{\sum_i \sum_j w_{ij}} \times \frac{\sum_i \sum_j w_{ij}(x_i - \bar{x})}{\sum_i (x_i - \bar{x})^2} \tag{3}$$

where w_{ij} is the spatial weight matrix between observation unit i and its neighboring units j: i and j are established by diving the study area into uniform grids based on its appropriate scale; x_i and x_j are the observed values of adjacent research area i and j, respectively; n is the number of spatial units of the research area, and $\bar{x}$ is the average value of all observed values in the sample. Index I ranges from -1 to 1, and as the absolute value of I increases, the spatial correlation gets stronger. $I = 0$ indicates a random spatial distribution.

Further, bivariate Moran's I_{xy} [66] (Equation (4)), which is based on the principle of univariate spatial correlation, has been adopted on the specific scale of 1000×1000 m. On this spatial scale, the analysis scale between each landscape metric and each influencing factor can be unified. The spatial autocorrelation analysis of each landscape metric at this scale was extremely significant. Through the spatial autocorrelation analysis, the relationship between the landscape metric and spatial influencing factors was captured and the strength of the association between the two variables was measured over the study area.

$$I_{xy} = \frac{n}{\sum_i \sum_j w_{ij}} \times \frac{\sum_i \sum_j w_{ij}(x_i - \bar{x})(y_j - \bar{y})}{\sqrt{\sum_i (x_i - \bar{x})^2} \sqrt{\sum_j (y_j - \bar{y})^2}} \tag{4}$$

where I_{xy} also ranges from -1 to 1, x_i is the attribute values of adjacent research areas i and x, while x_j is the attribute values of adjacent research areas j and y; $\bar{x}$ and $\bar{y}$ are the average attribute values of x and y in the sample, respectively; and n is the number of the spatial units of the research area.

- Geographical Detector Model

The geographical detector model (Geodetector), which is based on the theory of spatial stratified heterogeneity [67], was used to analyze the interaction between landscape metrics and various influencing factors in this study. First, we obtained the spatial data of independent variables through discrete classification for various influencing factors using the geometrical interval method [68–70], and then analyzed the influence of these variables on each landscape metric of the same spatial scales via Geodetector.

Specifically, the factor force q in Equation (5), ranging from 0 to 1, quantified the effect of different influencing factors on the spatial distribution of landscape metrics [32], and reflected the degree of spatial stratified heterogeneity of the metrics. The larger the q became, the more heterogeneous the landscape pattern.

$$q = 1 - \frac{1}{N\sigma^2} \sum_{i=1}^{L} N_i \sigma_i^2 \tag{5}$$

where N and σ^2 stand for the number of units and the variance of the dependent variable, respectively; $i = 1 \cdots L$ is the stratification of the dependent or independent variable; N_i and σ_i^2 stand for the number of units and the variance of the dependent variable in stratification layer i, respectively.

3. Results

3.1. Spatiotemporal Variations of Land Use Types

As shown in Figure 2, the overall landscape of the LXH is dominated by forests and croplands, while other seven land use types (including forest, shrub, orchard, grassland, water, floodplain and unused land) occupy a relatively small portion, accounting for less than a quarter of the total basin. Clearly, it is noticed that the proportion change in land use types were not large in general, but their temporal and spatial differences were obvious. Therein, the proportion changes of cropland and construction land were more prominent than any other land use types, particularly in the middle and lower watershed during 2000 and 2010. Temporally, the decreases of cropland and increases of construction land in this decade were more than 50% of that in total 35 years (See subfigures b in Figure 2). Spatially, it can be seen from the results of (c) in Figure 2 that the cropland decreased by 16.96% in the lower watershed and 2.70% in the middle watershed, respectively. The construction land increased by 17.78% in the lower watershed and 3.33% in the middle watershed, respectively. Their changes were all less than 1% in the upper watershed.

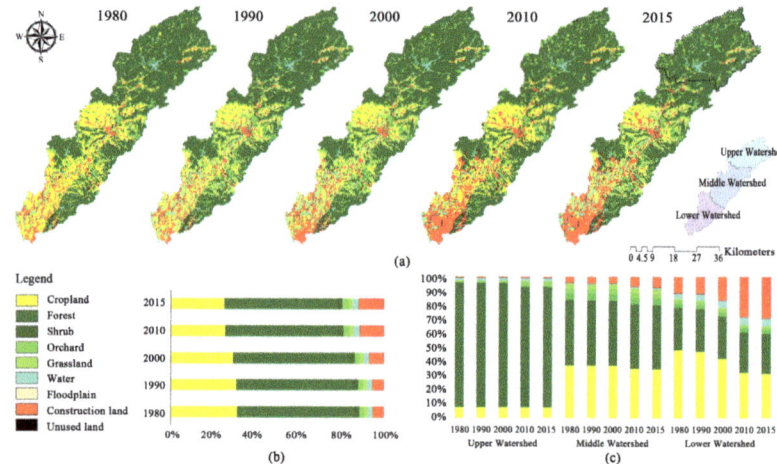

Figure 2. The types of land use in the LXH (1980–2015) (**a**) Spatial distribution of land use; (**b**) Percent coverage of the land use types; and (**c**) percent coverage of the land use types in the upper, middle and lower watershed of the river basin.

More importantly, land use changes, characterized by the transition from one type to another, were extremely prominent. From the land use conversion matrix between 1980 and 2015 in Table 2, it can be calculated that the total exchange area is 578.93 km², or 24.71% of the total catchment area. Specifically, the conversions among cropland, forest, orchard, water, and construction land comprised 94.78% of the total exchange. Among them, construction land has increased by 158.67 km² or 127.47%. Cropland and forest have decreased by 146.33 km² or 20.00% and 39.24 km² or 3.08%, respectively. Other changes were all less than 20.00 km². It can be observed that cropland and forests primarily contributed to land use exchanges and were the major land use types encroached on by urbanization. Of the 158.67 km² increases in construction land, 89.17% resulted from conversion of cropland and 10.35% resulted from conversion of forests.

Table 2. Land use types conversion matrix between 1980 and 2015 (km2).

1980	2015									2015 Total	Gain
	Cropland	Forest	Shrub	Orchard	Grassland	Water	Floodplain	Construction Land	Unused Land		
Cropland	555.34	12.47	2.96	2.14	0.91	1.68	0.16	9.65	0.20	585.51	30.17
Forest	13.37	1213.02	1.13	2.09	2.52	2.11	0.13	1.79		1236.17	23.15
Shrub	2.64	1.66	57.76	0.07	0.09	0.09		0.28	0.01	62.60	4.83
Orchard	2.30	22.86	0.42	37.14	0.08	0.19	0.07	0.60		63.65	26.51
Grassland	1.11	4.32	0.13	0.73	36.11	0.08		0.06		42.55	6.44
Water	15.52	4.55	0.22	0.20	0.25	41.60	2.66	1.73		66.73	25.13
Floodplain	0.08	0.11		0.13		0.04	1.79			2.16	0.36
Construction land	141.48	16.43	3.59	4.84	2.57	3.57	0.15	110.35	0.16	283.14	172.78
Unused land			0.08			0.01			0.38	0.48	0.09
1980 Total	731.84	1275.40	66.30	47.34	42.53	49.37	4.96	124.47	0.76		
Loss	176.50	62.38	8.53	10.20	6.42	7.77	3.17	14.12	0.38		

Besides, there also exist great temporal and spatial differences in land use exchanges. Temporally, the Sankey diagram in Figure 3 visualizes exchanges of each land use type over different time periods. Therein, exchanges in almost all land use types during 2000–2010 are the most significant. Notably, the increase of construction land in this decade accounted for more than 50% of the total increase in the entire period, and 78.11% of the construction land increase came from conversion of cropland and 10.10% from conversion of forest. Spatially, Figure 4 shows that the conversions mainly occurred in the middle and lower reaches of the basin where a large amount of cropland was converted into construction land. Particularly, the lower reaches experienced the most drastic changes in the river basin regarding croplands and construction areas.

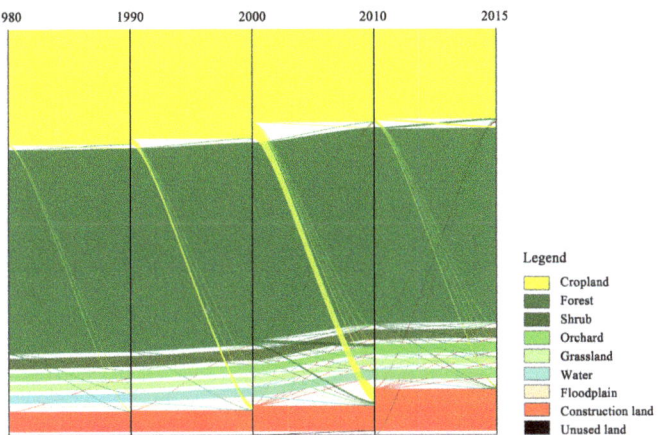

Figure 3. Comparison of exchanges of land use types in four time periods.

3.2. Spatiotemporal Variations of Landscape Patterns

The different landscape pattern indices were calculated at both landscape and land use class levels on the specific scale of 1000 × 1000 m. The former represents the overall spatial arrangement characteristics of each landscape patch, and the latter reflects the spatial arrangement characteristics of each landscape patch in different types. That is, the landscape pattern of each landscape patches at the class level can determine it at the landscape level, but not vice versa.

Figure 5 shows that changes of those indices are all obvious at the landscape level. For example, the landscape fragmentation index PD has an increasing trend and increased by 8.03%, while AI shows a decreasing trend and decreased by 2.25%. Meanwhile, the interference index AWMPFD and dominance index LPI decreased by 0.71% and 4.86%, respectively. The diversity index SHDI shows an increasing trend and increased by 5.25%. Temporally, these indices altogether indicated an increasing fragmentation and homogenization in landscapes, intensified human interference, and a weakening dominance of the once-dominant landscape (forest and cropland) in the early years. Similar to the changes in land use types, the decade from 2000–2010 experienced the most significant changes in landscape pattern.

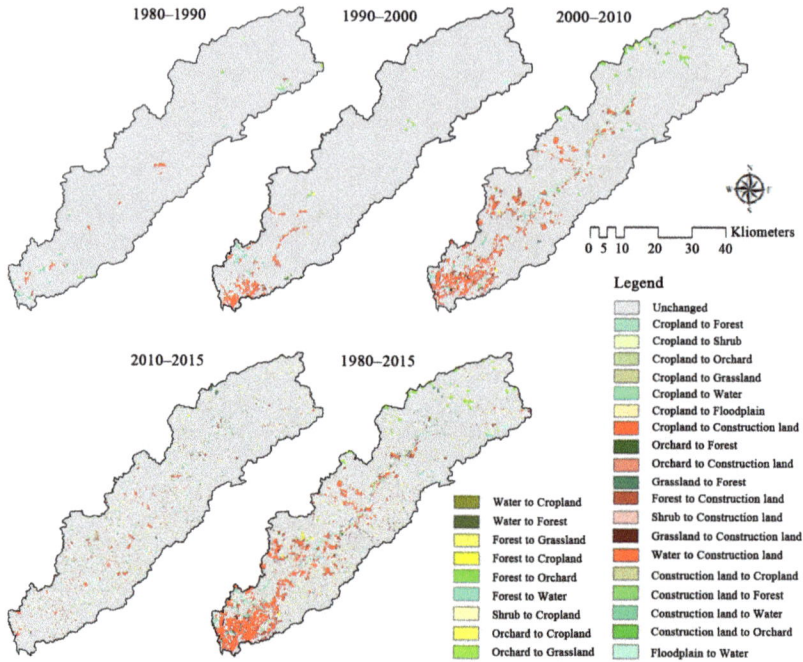

Figure 4. Spatial variations of land use type transition in different time intervals (1980–1990, 1990–2000, 2000–2010, 2010–2015 and 1980–2015).

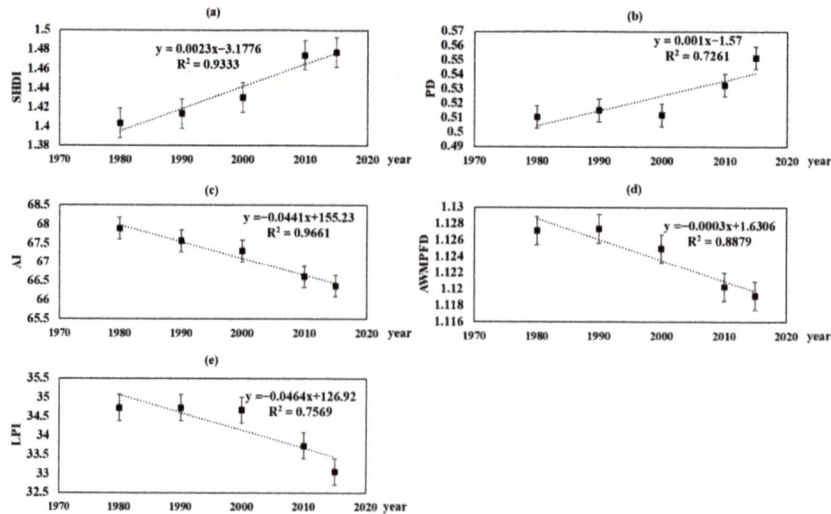

Figure 5. Changes of landscape metrics (**a**) Shannon's diversity index (SHDI), (**b**) patch density (PD), (**c**) aggregation index (AI), (**d**) area-weighted mean patch fractal dimension (AWMPFD), and (**e**) largest patch index (LPI) at the landscape level from 1980 to 2015.

Moreover, similar to land use changes, landscape pattern also showed distinct spatial characteristics, reflecting the landscape heterogeneity in the LXH. As shown in Figure 6, the PD and SHDI are relatively small while AI, AWMPFD, and LPI are relatively large in the upstream area. However, this pattern reverses in the middle and lower watershed. This indicated that the landscape in the middle and lower watershed was more fragmented and heterogeneous than the landscape in the upper watershed, with a higher interference and lower dominance. Moreover, by comparing the values of these landscape metrics in 1980 and 2015 shown in subfigures c in Figure 6, we find that they are almost unchanged in most regions, except for part of the middle and lower watershed where cropland had been largely converted to construction land.

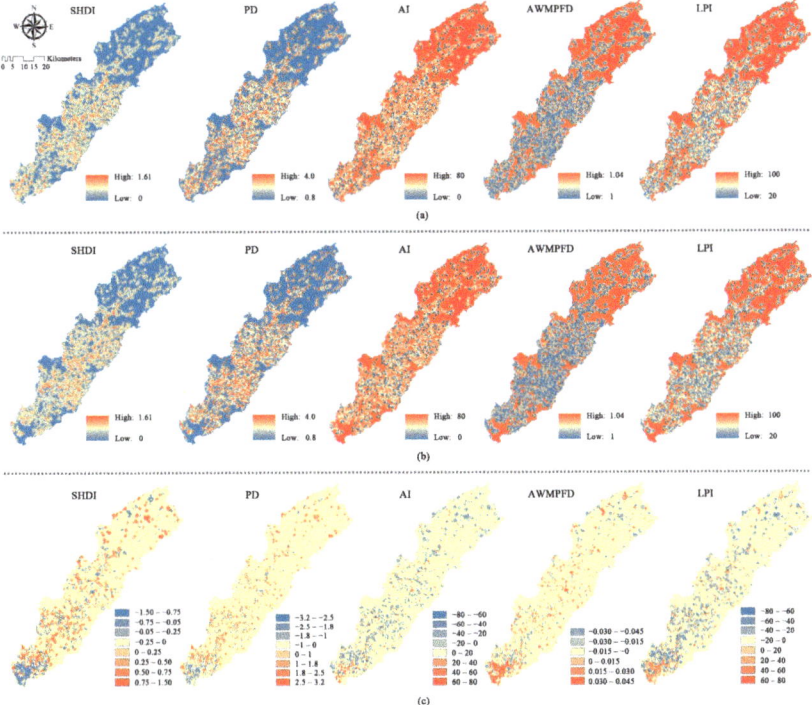

Figure 6. Spatial distribution of the changes in landscape metrics between (**a**) 1980 and (**b**) 2015, and (**c**) showed the results of subtraction of the indices in 1980 from 2015.

At the land use class-level, it can be found that the PD and AI value of cropland and forest changed more significantly than the other types during the total study period. The PD of cropland and forest increased by 42.26% and 41.59%, and the AI decreased by 1.68% and 0.21%, respectively. Meanwhile, the LPI values of cropland and forest decreased by 67.45% and 3.07%, respectively. The LPI of construction land in 2015 was 12 times that of the LPI in 1980, second only to the LPI of cropland. Among all land use types, the AWMPFD of cropland decreased most significantly, by 4.39% in total, while the net change of the AWMPFD of other types was less than 1.00%. That is, the degree of fragmentation, homogenization, and human alteration of cropland in this river basin was the most significant of all terrains. At the same time, the degree of landscape dominance of cropland was greatly reduced, while that of construction land was greatly improved. This indicated that the fragmentation and homogenization of cropland was mainly contributed by occupation of the construction land.

3.3. Impacts of the Anthropogenic Factors on Temporal Landscape Changes

Since the effect of natural factors on landscape changes was minor compared with anthropogenic factors in a short period of time, this paper selected eight anthropogenic factors involving demographic factors, socioeconomic factors, and urbanized activities to reflect their impacts on landscape changes temporally (refer to Figure 7 for detailed indices and their change trends). Meanwhile, due to the difficulty of quantitatively expressing the related policies, they were not included in the following quantitative analysis.

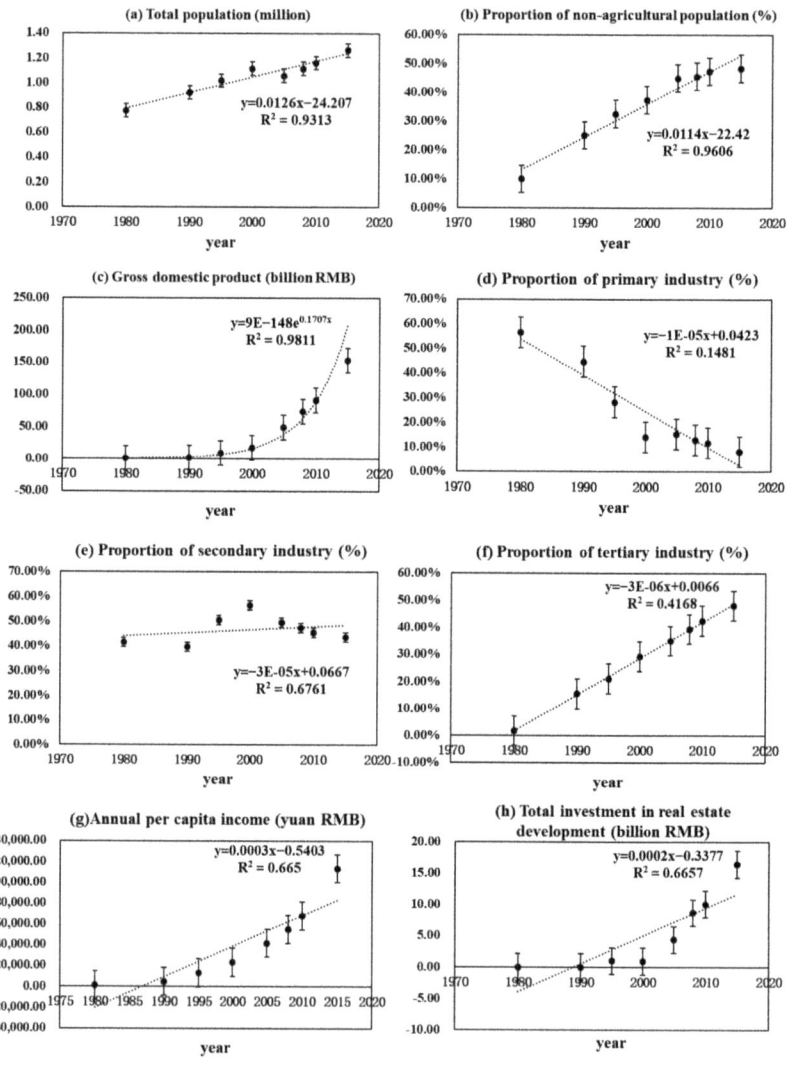

Figure 7. Temporal trends of influencing factors (a) Total population, (b) Proportion of non-agricultural population, (c) Gross domestic product, (d) Proportion of primary industry, (e) Proportion of secondary industry, (f) Proportion of tertiary industry, (g) Annual per capita income and (h) Total investment in real estate development during 1980 to 2015.

First, from the Figures 7 and 8, it can be found that the temporal changes of some influencing factors and land use types are non-linear; we chose the method of grey correlation analysis in this context. According to the results from the relational analysis in Table 3, all the grey correlation coefficients are greater than 0.55, indicating that the changes of these eight anthropogenic factors are all significantly correlated with the changes of various land use types in the LXH temporally. More specifically, the four most correlated factors on each land use type were the TP, the PNAP, the PSI, and the PTI, which were demographic and socioeconomic factors and their correlations were all above 0.73. Among them, for cropland and each natural land (forest, shrub, grassland, water, floodplain, and unused land), the most correlated factors were PSI and TP; while for construction land, the most correlated factors were PNAP and PTI. From the Figure 7, it is clear that the TP, PNAP, and PTI all increase dramatically from 1980 to 2015 in the LXH. The PSI shows a trend of first increasing and then decreasing, but it also increases on the whole. Therefore, combined with the characteristics of the proportion changes in various land use types in Figure 8, it can be concluded that the increase of construction land was mainly correlated with the increase of non-agricultural population and the continuous development of the tertiary industry in the LXH. The decreasing of cropland and each natural land was mainly correlated with the increase of TP and the changes in secondary industry. This might be attributed to the growing population (especially the growth of urban population) and the transformation of industry (especially the growth of tertiary industry), which has accelerated the encroachment on cropland and natural lands to meet the demands for more construction land in the LXH [71,72]. Alternately, the other four influencing factors were also crucial to the changes of different land use types, but they had less of an effect compared with these main influencing factors.

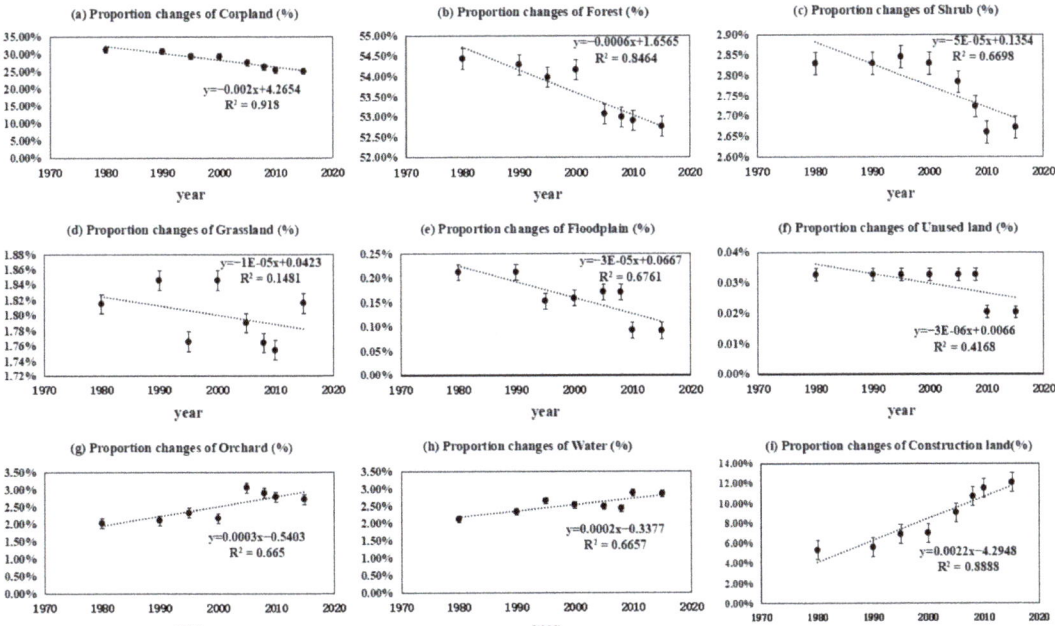

Figure 8. Temporal trends of each land use type (**a**) Proportion changes of Cropland, (**b**) Proportion changes of Forest, (**c**) Proportion changes of Shrub, (**d**) Proportion changes of Grassland, (**e**) Proportion changes of Floodplain, (**f**) Proportion changes of Unused land, (**g**) Proportion changes of Orchard, (**h**) Proportion changes of Water and (**i**) Proportion changes of Construction land during 1980 to 2015.

Table 3. Grey correlation coefficients of influencing factors on land use changes.

Grey Correlation Coefficients	TP		PNAP		GDP		PPI		PSI		PTI		APCI		IRE	
	Degree	Rank	Degree	Rank	Degree	Rank	Degree	Rank	Degree	Rank	Degree	Rank	Degree	Rank	Degree	Rank
Cropland	0.89	2	0.79	3	0.60	7	0.71	5	0.92	1	0.73	4	0.63	6	0.57	8
Forest	0.91	2	0.80	3	0.60	7	0.67	5	0.93	1	0.74	4	0.64	6	0.57	8
Shrub	0.90	2	0.80	3	0.60	7	0.68	5	0.93	1	0.74	4	0.64	6	0.57	8
Orchard	0.93	1	0.87	3	0.62	7	0.63	6	0.89	2	0.79	4	0.67	5	0.58	8
Grassland	0.91	2	0.81	3	0.60	7	0.67	5	0.92	1	0.74	4	0.63	6	0.56	8
Water	0.95	1	0.83	3	0.61	7	0.65	5	0.92	2	0.76	4	0.64	6	0.57	8
Floodplain	0.84	2	0.77	3	0.60	7	0.76	4	0.84	1	0.73	5	0.64	6	0.57	8
Construction land	0.83	3	0.89	1	0.64	6	0.55	8	0.77	4	0.84	2	0.70	5	0.59	7
Unused land	0.87	2	0.80	3	0.61	7	0.74	5	0.91	1	0.75	4	0.65	6	0.58	8

3.4. Impacts of Anthropogenic and Natural Factors on Spatial Landscape Changes

Considering the fact that landscape pattern experienced the most significant changes from 2000 to 2010, we took this period as an example to study the influencing factors of spatial heterogeneity toward landscape pattern in this river basin. Table 4 manifests the significant spatial correlation between each landscape metric and each investigated influencing factor in 2000 and 2010 respectively. Among them, topographic elements DEM and SLOPE were all spatially negatively correlated with PD and SHDI, and positively correlated with AI, AWMPFD, and LPI. This indicated that in areas with low elevation and gentle slopes, the degree of landscape fragmentation, landscape interference, and landscape homogenization was stronger, and that the landscape dominance was weak. Table 4 also shows that all anthropogenic influencing factors (GDP, TP, and NLD) are positively correlated with PD and SHDI, and negatively correlated with AI, AWMPFD, and LPI spatially. This illustrates that the degree of landscape fragmentation, landscape interference, and landscape homogenization is relatively strong in more developed regions. In terms of the impact of land use changes brought by rapid urbanization on landscape patterns, the LUIN was positively correlated with PD and SHDI spatially and negatively correlated with AI, AWMPFD, and LPI. This reflected that areas with high LUIN were usually accompanied with a relatively stronger degree of landscape fragmentation, landscape interference, and landscape homogenization.

Table 4. Bivariate Moran's I correlation analysis between landscape metrics and influencing factors in the spatial dimension in 2000 and 2010.

Moran's I	PD		AI		AWMPFD		LPI		SHDI	
	2000	2010	2000	2010	2000	2010	2000	2010	2000	2010
DEM	−0.46	−0.39	0.38	0.32	0.51	0.42	0.46	0.39	−0.52	−0.44
Slope	−0.30	−0.26	0.25	0.21	0.22	0.28	0.30	0.26	−0.33	−0.29
GDP	0.18	0.14	−0.14	−0.11	−0.21	−0.15	−0.18	−0.14	0.19	0.14
TP	0.15	0.11	−0.11	−0.08	−0.17	−0.12	−0.14	−0.11	0.15	0.11
NLD	0.30	0.23	−0.23	−0.19	−0.34	−0.25	−0.30	−0.23	0.33	0.26
LUIN	0.49	0.49	−0.40	−0.23	−0.53	−0.31	−0.49	−0.29	0.55	0.31

Note: Permutation test was used to test in this study, and the P value of each group of variables was equal to 0.001, indicating that the spatial correlation was significant under 99.9% confidence.

Geographic detector analysis showed that the interpretation of DEM on the spatial distribution of each landscape metric was the largest among all influencing factors in 2000 or 2010, with the highest average q value being 0.28 or 0.23, followed by GDP, TP, and NLD, while Slope had the lowest q value (Figure 9). Results indicated that the spatial distribution of elevation was the key factor that induced the spatial heterogeneity of landscape pattern, and the spatial distribution of socioeconomic level, population density, and urbanized activities also played an important part. The analysis of interactions between the influencing factors and landscape metrics (Figure 10) showed that the interpretation of spatial distribution characteristics of landscape pattern by any two influencing factors was greater than that of any single influencing factor, indicating that the formation of spatial heterogeneity was the result of interactions between various influencing factors. Specifically, from the subfigures a in Figure 10, the interaction between DEM and LUIN is the strongest among other factors. The interaction between LUIN and other four influencing factors on each landscape metric was almost stronger than that between any other two influencing factors in 2000. This indicated that the spatial differences of DEM and LUIN jointly resulted in the spatial heterogeneity of landscape pattern in 2000, stronger than the interactions between DEM and GDP, TP, NLD, Slope comparatively. However, compared with the results from 2000, since the interaction between DEM and TP, GDP, and NLD was strengthened, the interaction between DEM and LUIN was no longer the strongest among the other factors in 2010 (see subfigures b in Figure 10). These indicated that the spatial differences of the topographic elements and other influencing factors also jointly contributed to the spatial heterogeneity of landscape pattern in 2010.

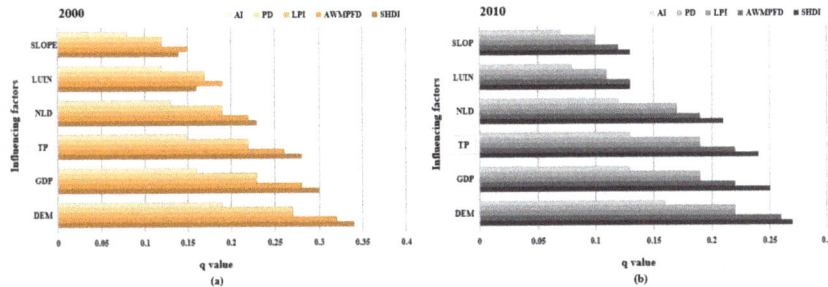

Figure 9. The force q among each influencing factor on each landscape pattern metric in (a) 2000 and (b) 2010.

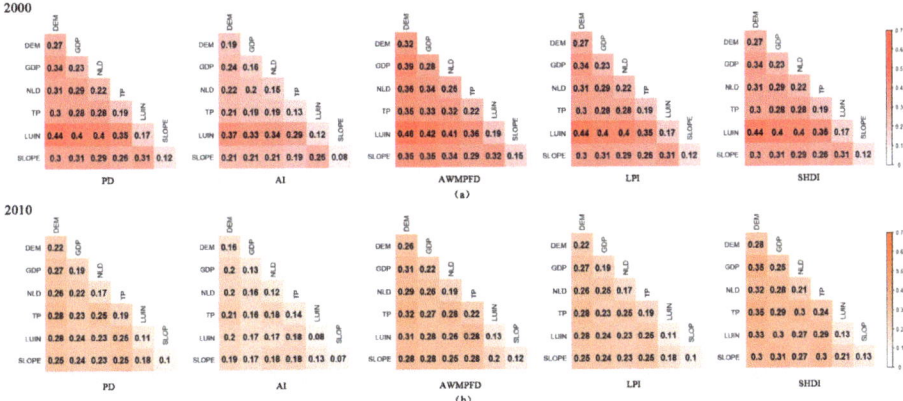

Figure 10. The force q among any two influencing factors on each landscape pattern metric interactively in (a) 2000 and (b) 2010.

4. Discussion

4.1. Spatiotemporal Changes of Land Use and Landscape Pattern

The results of this study showed that in the LXH, there exist large spatial and temporal differences in land use changes and landscape pattern changes. These changes appeared to be more prominent in the middle and lower watershed, and their changing rates were fastest during 2000 to 2010. Specifically, the land use change was featured by the increasing transition of cropland and forest to construction land, and the fragmentation and homogenization of landscape pattern was contributed to the encroachment of construction land on forest and cropland. That is, the decrease of cropland and forest was accompanied with the decreased degree of the cropland and forest landscape dominance and the increased degree of the cropland and forest landscape fragmentation and homogenization. These further proved the synchronization characteristics and interaction relationship between land use changes and landscape pattern changes proposed by scholars [17,19]. Thus, it is possible for researchers to use the temporal change of a land use type to reflect the temporal change pattern of a certain landscape type in a period of time.

Besides, our findings of the land use and landscape pattern changes are consistent with the previous research in the whole Guangzhou city. For example, Zhang et al. [73] and Gong et al. [74] also found that the increase of construction land in new urban areas of Guangzhou city mainly came from cropland, forest, and other ecological land, especially after 2000. Gong et al. [41] also confirmed that the fragmentation and homogenization of cropland in Guangzhou was mainly contributed by the expansion of construction land. However, their research paid more attention to the urbanization expansion pattern of Guangzhou by comparing the changing differences of certain land use types and landscape patterns in different jurisdictions, instead of focusing on these spatiotemporal changes brought by urbanization [38,75]. Researches on the analysis of land use and landscape pattern changes in a watershed under the urbanization expansion pattern also exist, which provide theoretical basis and method reference for this study. But watersheds selected in their studies are relatively large, spanning multiple cities [9,21]. As a case study in this paper, the LXH was relatively small and in the range of Guangzhou city. Its middle and lower watershed is adjacent to the central urban area of Guangzhou, while the upstream area is far away from the central urban. This pushed the gradual widening of the difference between the northern and southern parts of the watershed influenced by urbanization. Our results also found that changes of land use and landscape pattern were different between the northern and southern parts.

Moreover, analysis results above reflected that the time period from 2000 to 2010 and the southern parts of the LXH with the most prominent changes should be taken seriously by relevant stakeholders. First, changes of land use and landscape pattern in southern parts of the LXH should be slow down and controlled, and the northern parts should be protected timely under the rapidly urbanizing trends. Then, the special time period from 2000 to 2010 needs to pay much attention to in related researches about the LXH. It means that this study gives not only a supplement to previous studies in these regions, but also is of great value for managers, planners, and scholars to make appropriate strategies.

4.2. The Temporal and Spatial Influencing Factors

In terms of the influencing factors of the changes of land use types and landscape patterns, previous studies mainly discussed the reasons for land use type conversion at different locations [32,76] and in different time periods [21,29], but few analyzed the factors responsible for the spatial heterogeneity of landscape patterns in river basins, nor did they comprehensively quantify the factors that contributed to the spatiotemporal change of land use types and landscape patterns. In this study, considering the fact that various land use type changes emphasized the transition of different landscape patches in different time periods, and that changes of landscape patterns reflected the difference of spatial configuration characteristics in different landscape patches, we analyzed the influencing factors on the temporal change of land use types and spatial heterogeneity of landscape pattern, respectively. We found that there was a greater difference in the spatiotemporal influencing factors of land use and landscape pattern changes in the LXH. Thus, it is very important to propose a targeted protection and development strategy, which can meet the current needs of the different regions in the LXH. Temporally, we found that the demographic factors, socioeconomic factors, and urbanized activities were important in shaping the temporal variations of land use types in this river basin, and changes of major land use types were more sensitive to the increasing of non-agricultural population and transformation of industry than any other factors. This was consistent with the findings in other studies on the influencing factors of land use types changes in other river basins [15,28]; they also found that the growth of urban population and changes of industries contributed to the increase of construction land [4,77]. Moreover, similar studies elsewhere underlined that government policy also played an important role in the change of land use types during different periods [78–80]. Although there was no appropriate method for analyzing the impact of the related policies on land use changes quantitatively, we found that various land use types in the LXH have undergone significant changes in the last three decades after 1980, and these changes were particularly dramatic after 2000. Zhang et al. [73] found that the implementation of China's reform and development policy in 1978 was an important driving force for economic development and population migration of Guangzhou, pushing the continuous expansion of construction land to gradually occupy the cropland, forest, shrub, and grassland in the suburbs. Thus, the observed expansion of the construction land in the LXH after the early 1980s may be attributed to the implementation of this policy. In addition, the overall urban development master plan of Guangzhou in 2000 had put forward the strategy of expanding the urban area to the north and built Guangzhou into an international metropolis by 2010, which could further accelerate the expansion of construction land if practices continue. Correspondingly, the land use types in the LXH changed significantly after 2000 compared with the pre-2000 practices, accounting for more than 50% of the total variation in 35 years. Moreover, Baiyun, Huadu, and Conghua districts in the north had successively merged into the jurisdiction of Guangzhou in the year 2000, 2010, and 2015, respectively. The different speed of urbanization in different regions altered the variation characteristics of land use types. We also found that the lower watershed that contains the Baiyun and Huadu districts had the largest proportion of cropland conversion to construction land, which was five times higher than that of the upper and middle watershed. Therefore, apart from the demographic factors, socioeconomic factors, and urbanized activities, the relevant government policy, which is difficult to quantify, also significantly affected the variations of land use types.

Focusing on the impact of various influencing factors on landscape pattern changes in spatial dimension will be very useful in identifying and controlling the major driving forces, guiding the watershed protective management and sustainable planning. However, most of the current studies adopted the classification method to describe the spatial characteristics of landscape pattern and their influencing factors of different regions, and seldom analyzed the spatial relationships among variables quantitatively [9,33]. In the research of Ju et al. [70], the applicability of the geographic detector model in analyzing the driving force of construction land expansion was proved, providing a quantitative method for the analysis of the interaction among various spatial factors. But their research did not conduct a comparative analysis of the spatial driving relationships in different periods. Here, using the model of geographic detector by Wang et al. [69], this study compared the impact of different influencing factors on spatial landscape heterogeneity during 2000 and 2010, when the most dramatic land use changes happened. It can be found that the spatial distribution of LUIN and elevation were the two critical factors for the formation of landscape heterogeneity in 2000 compared with other factors, while the interaction between elevation and other human factors was strengthened in 2010; this illustrated that elevation was always a basic factor that directly determined the spatial distribution of landscape pattern. Liu et al. [61] also proposed that it was difficult for people to break through existing natural obstacles in the hilly regions of southern China, and this difficulty had largely restricted human activities. From Figure 11, it is clear that great differences in the terrain conditions exist in this study. Because areas with high elevation or greater slopes were difficult to develop and not suitable for urban construction, they were seldom disturbed by human activities [33], so that the degree of landscape fragmentation and homogenization in the upper reaches was low and the degree of landscape dominance was high. Therefore, elevation was a prerequisite for the impact of anthropogenic factors to occur. On the other hand, based on the difference of elevation in different regions of LXH, the human influencing factors, such as population and socioeconomic and urbanized activities, played an increasing role in the formation of the heterogeneous characteristics of the landscape pattern after 2000. This may be due to the fact that the spatial difference of human influence factors increased significantly after 2000, as it shown in subfigure c in Figure 11.

Figure 11. Spatial distributions of the selected influencing factors selected in this study in (**a**) 2000 and (**b**) 2010, and (**c**) shows the results of the subtraction of each influencing factor between 2000 and 2010.

The above discussion illustrates that it is necessary to control the increasing trends of non-agricultural population and the continuous development of secondary and tertiary industry in the future, thus the demand for more construction land will be decreased. Meanwhile, relevant policies should try to meet these demands. We also should pay much attention to the southern parts of the LXH, and strengthen its adjustment ability to deal with the intensive population density, higher GDP, and greater urban construction. For example, the urban occupancy rate in these areas can be increased, artificial green land can be increased, the native forest and grassland must be strictly protected, etc. This means that establishing the spatiotemporal change trends and causes of land use and landscape pattern in a rapidly urbanizing watershed is very important for guiding the diagnosing of urbanization problems, clarifying the main protection areas and main control factors.

4.3. The Limitations and Potential Outlooks

This study produces a quantitative estimate of the spatiotemporal variations in land use types and landscape patterns and analyzes the dominant influencing factors leading to these changes in LXH quantitatively, which provides a systematic integration and deepening of previous studies. The main land use maps used in this study interpreted by the common method of visual interpretation, and their errors mainly came from the personal subjective judgment of the interpreters and the similarity of the tones and textures of the satellite image. Although these errors in the interpretation process were considered and improved through some reference maps (including topographic maps, vegetation maps, ground truth data at different sample points, and local resident interview data), there also existed uncertainties in the data measurement and description [36]. These errors will also affect the accuracy of the research results to some degree. Therefore, it is necessary to compare any two measurement methods to improve the analysis accuracy of land use maps and express them on different scales as much as possible in the future. Besides, due to the scarcity of historical landscape information, e.g., land use maps from 1980 to 2015,

Topography data and Nighttime Light Image Data in 2015, we were unable to accurately establish the relationship between each land use type and different influencing factors, nor could we compare the driving factors of spatial heterogeneity of landscape patterns in each period. In addition, there are some factors that cannot be quantified, such as policy factors, which make the analysis of influencing factors still not comprehensive enough. In the future, it may be possible to construct a comprehensive model combining qualitative and quantitative analysis on all possible influencing factors of land use changes.

Resolution of available spatial data set of spatial influencing factors was also a limitation. We conducted the driving analysis between the influencing factors and landscape indices at a 1 km grid. Although on this scale the influencing factors also had good spatial correlation, the resolution limitation of influencing factors might affect the accuracy of the analysis results to some degree. Therefore, spatial data with appropriate accuracy at a higher resolution and longer periods could substantially improve the accuracy when analyzing the change of landscape patterns and their driving forces. In that sense, models of land use change would be a good alternative for such studies by simulating the land use in different years to increase the range and length of land use data [81], and hence guide the urbanization development of this region by analyzing the change of land use types and landscape patterns in the future. Moreover, the extensive establishment of the real-time monitoring data platform of different spatial influencing factors such as social economic activities, population density, and urbanization activities in the future with higher resolution will improve the accuracy of spatial analysis, thereby realizing its dynamic analysis.

Moreover, based on our comprehensive analysis of the spatiotemporal changes and causes of landscape pattern in the LXH and its ecological and hydrological effects in related researches [43], it is more urgent to establish specific strategies to guide the sustainable development of LXH in the future. The Hellwig classification and measurement method introduced by Hellwig in 1968 provides a decision-making method for formulating sustainable development strategies based on the evaluating of urban development [82]. This method was first applied in the sustainable decision-making process of urban green space biodiversity management in Lublin, eastern of Poland, thus the main ecological areas that should be protected can be established [83]. Then, other scholars used and extended this method at different scales in European Union to formulate sustainable development strategies based on the different goals. These application of the Hellwig method in different researches prove its effectiveness in evaluating the level of development of different regions in different fields at different scales, which provide a new direction for the future research of establishing the sustainable development strategy in the LXH based on the analysis of its changes about land use, landscape pattern, hydrological and ecological conditions under the rapid urbanization. After that, the specific areas of the LXH under the rapid urbanization process, in which its land use transition and landscape pattern fragmentation should be extremely controlled can be found.

5. Conclusions

In this study, we analyzed the spatiotemporal changes of land use types and landscape pattern of the LXH from 1980 to 2015 under the rapid urbanization of Guangzhou city, as well as quantified the major influencing factors temporally and spatially. The main conclusion can be concluded as one sentence that there exist great spatiotemporal differences in land use and landscape pattern changes and its causes in the LXH during the past 35 years. Specifically, it can be drawn as follows:

- The most obvious land use change was characterized as the large transition from cropland to construction land, bringing about the fragmentation of cropland that was encroached on by the construction land. The landscape pattern showed an increasing trend of landscape fragmentation, homogenization, and landscape interference, and a decreasing trend in landscape dominance. These changes mainly occurred in the lower watershed, particularly between 2000 and 2010. Therein, these changes were more than 50% in this decade compared with total 35 years.

- Many influencing factors affected the temporal variations in landscapes, including population growth, economic and industrial development, urbanized activities, and relevant policies. Among them, changes of major land use types were more sensitive to the increase of a non-agricultural population and transformation of industries than other factors. In addition, the spatial distribution of land use types and elevation were found to be the two key factors for the formation of landscape heterogeneity in 2000, while the spatial distribution of the other three human factors and elevation gradually became the same important factors after 2000.
- Our research shows that the temporal and spatial difference of changes in land use and landscape pattern at a watershed with unbalance urbanization degree in different regions was great. This is not only affected by the difference of the degree in socioeconomic level, population growth rate, and urbanizing expansion in different time and space, but also determined by the related policies. Besides, the topographical factors were also the basis of the formation on landscape pattern. When developing, we need to consider both the geographical conditions and the urbanizing degree of the watershed, thus a sustainable development strategy could be formulated and the goals of protecting and restoring the watershed ecosystem can be achieved.

The findings are of great significance for review and outlook of the ecological protection and sustainable development of the watershed around the rapidly urbanizing areas. It can not only allow decision-makers to clarify their main problems, but also guide them to clarify the key protection areas and control indicators. However, the analysis of the landscape patterns above was limited to the period from 1980 to 2015, and the comparison of influencing factors on spatial landscape configurations focused only on 2000 and 2010. Nonetheless, results in this study are insightful, although they could be more generalized with the analysis over a longer period.

Author Contributions: Conceptualization, Z.Z. and B.L.; methodology, Z.Z.; software, Z.Z.; validation, Z.Z., H.W., and M.H.; formal analysis, Z.Z. H.W., and M.H.; investigation, Z.Z. and B.L.; resources, B.L.; data curation, B.L.; writing—original draft preparation, Z.Z.; writing—review and editing, Z.Z., B.L., H.W., and M.H.; visualization, Z.Z.; supervision, B.L.; project administration, B.L.; funding acquisition, B.L. All authors have read and agreed to the published version of the manuscript.

Funding: This research was funded by the National Natural Science Foundation of China (Grant Nos. 51879289), the Guangdong Basic and Applied Basic Research Foundation (2019B1515120052) and the Innovation Fund of Guangzhou City water science and technology (GZSWKJ-2020-2).

Institutional Review Board Statement: Not applicable.

Informed Consent Statement: Not applicable.

Data Availability Statement: The data presented in this study are available on request from the corresponding website.

Acknowledgments: We would like to thank the editor Madalina Buzatu and three anonymous reviewers for their constructive comments.

Conflicts of Interest: The authors declare no conflict of interest.

Appendix A

Table A1. Description of land use types in the LXH.

Land Use Types	Description
Cropland	Arable agricultural land, including paddy fields and dry land
Forest	Natural and semi-natural manmade woodland
Shrub	Dwarf woodland (height < 2 m) and shrubbery
Orchard	Intensively managed orchards (fruit orchards, mulberry orchards, tea orchards) and plant nursery
Grassland	Natural and artificial grassland
Water	Rivers, creeks, canals, ponds, lakes, reservoirs, and bays
Floodplain	Permanent and seasonal floodplains
Construction land	Mainly urban and rural settlements, mining land, transportation land, and other special construction land
Unused land	Mainly land without vegetation cover and difficult to use, including bare soil, sandy land, desert, saline, and landfills

Table A1 gives a detailed description of the content about each land use type in this study, which can better display the classification standard of land use types.

Table A2. Analysis results of spatial autocorrelation of each landscape metric in different scales in 2000 and 2015 of LXH

Scales/(m × m)	PD		AI		AWMPFD		LPI		SHDI	
	2000	2015	2000	2015	2000	2015	2000	2015	2000	2015
500 × 500 m	0.47	0.47	0.45	0.45	0.46	0.47	0.47	0.46	0.47	0.47
600 × 600 m	0.47	0.47	0.45	0.45	0.46	0.47	0.47	0.46	0.47	0.47
700 × 700 m	0.46	0.46	0.45	0.45	0.46	0.46	0.46	0.46	0.46	0.46
800 × 800 m	0.46	0.46	0.43	0.43	0.45	0.46	0.46	0.45	0.46	0.46
900 × 900 m	0.46	0.46	0.43	0.43	0.45	0.46	0.46	0.45	0.46	0.46
1000 × 1000 m	0.44	0.44	0.42	0.42	0.43	0.44	0.44	0.43	0.44	0.44
1100 × 1100 m	0.44	0.44	0.41	0.41	0.43	0.44	0.44	0.43	0.44	0.44
1200 × 1200 m	0.44	0.44	0.44	0.44	0.44	0.44	0.44	0.44	0.44	0.44

Note: Permutation test was used to test in this study, and the P value of each landscape metrics in different scales was equal to 0, indicating that the spatial correlation was significant under 99.9% confidence. The Z value of them were all >1.96, reflecting that there exists extremely significant spatial autocorrelation among these landscape metrics in different spatial scales.

Table A2 reflects the degree of spatial autocorrelation about each landscape metric selected in this paper. It confirms that the spatial autocorrelation of these landscape metrics is extremely significant in different spatial scales ranging from 500 to 1200 m. Therefore, it can be proved that these spatial scales are all appropriate for analyzing the bivariate spatial correlation between each landscape metric and each influencing factor.

Appendix B

This Figure A1 is provided to screen the notable landscape metrics from 23 frequently used metrics. It demonstrates the correlation between any two kinds of landscape metrics among these 23 metrics above. Clearly, most of them were highly correlated with each other. Thus, when the absolute value of correlation coefficient between two indices is more than 0.9, only one is used; second, indices representing different aspects of landscape characteristics were selected to reduce the information redundancy among them. Finally, five representative indices were selected in this paper, including patch density (PD), aggregation index (AI), largest patch index (LPI), area-weighted mean patch fractal dimension (AWMPFD), and Shannon's diversity index (SHDI), which represent different aspects of landscape characteristics.

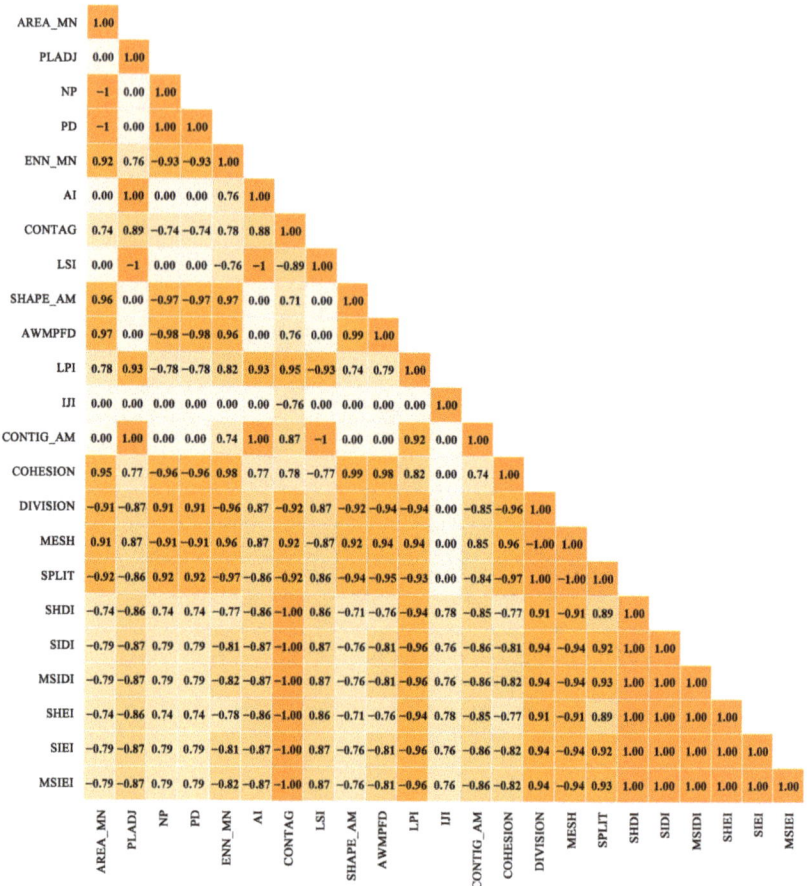

Figure A1. Results of the factor analysis among 23 common metrics.

Figure A2 presents the changing value of five selected landscape metrics at landscape-level in 14 different granularities (including 30 m, 50 m, 100 m, 200 m, 300 m, 400 m, 500 m, 600 m, 700 m, 800 m, 900 m, 1000 m, 1100 m, 1200 m). It proves that 500~1200 m was the common characteristics interval of these landscape metrics. Thus, the spatial scale for analyzing the spatial autocorrelation of each landscape metric has been established.

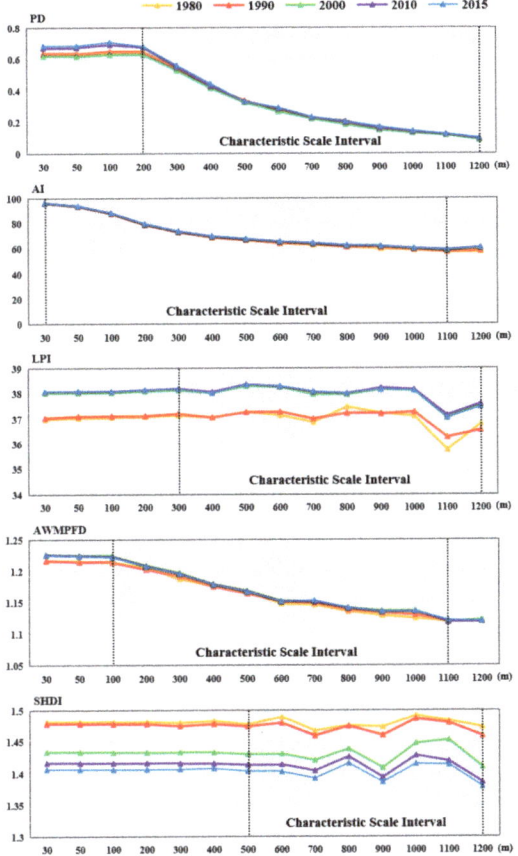

Figure A2. Analysis results of the characteristic scale interval in 14 different granularities.

References

1. Deng, J.S.; Wang, K.; Hong, Y.; Qi, J.G. Spatio-temporal dynamics and evolution of land use change and landscape pattern in response to rapid urbanization. *Landsc. Urban Plan.* **2009**, *92*, 187–198. [CrossRef]
2. Sui, D.Z.; Zeng, H. Modeling the dynamics of landscape structure in Asia's emerging desakota regions: A case study in Shenzhen. *Landsc. Urban Plan.* **2002**, *53*, 37–52. [CrossRef]
3. Liu, Y.; Yao, C.; Wang, G.; Bao, S. An integrated sustainable development approach to modeling the eco-environmental effects from urbanization. *Ecol. Indic.* **2011**, *11*, 1599–1608. [CrossRef]
4. Liu, Y.; Luo, T.; Liu, Z.; Kong, X.; Li, J.; Tan, R. A comparative analysis of urban and rural construction land use change and driving forces: Implications for urban-rural coordination development in Wuhan, Central China. *Habitat Int.* **2015**, *47*, 113–125. [CrossRef]
5. Meyfroidt, P.; Lambin, E.F.; Erb, K.H.; Hertel, T.W. Globalization of land use: Distant drivers of land change and geographic displacement of land use. *Curr. Opin. Environ. Sustain.* **2013**, *5*, 438–444. [CrossRef]
6. Luo, P.; Mu, D.; Xue, H.; Ngo-Duc, T.; Dang-Dinh, K.; Takara, K.; Nover, D.; Schladow, G. Flood inundation assessment for the Hanoi Central Area, Vietnam under historical and extreme rainfall conditions. *Sci. Rep.* **2018**, *8*, 12623. [CrossRef]
7. Zhu, Y.; Luo, P.; Zhang, S.; Sun, B. Spatiotemporal analysis of hydrological variations and their impacts on vegetation in semiarid areas from multiple satellite data. *Remote Sens.* **2020**, *12*, 4177. [CrossRef]
8. Edge, C.B.; Fortin, M.J.; Jackson, D.A.; Lawrie, D.; Stanfield, L.; Shrestha, N. Habitat alteration and habitat fragmentation differentially affect beta diversity of stream fish communities. *Landsc. Ecol.* **2017**, *32*, 647–662. [CrossRef]
9. Tang, J.; Li, Y.; Cui, S.; Xu, L.; Ding, S.; Nie, W. Linking land-use change, landscape patterns, and ecosystem services in a coastal watershed of southeastern China. *Glob. Ecol. Conserv.* **2020**, *23*, e01177. [CrossRef]

10. Xiao, J.; Shen, Y.; Ge, J.; Tateishi, R.; Tang, C.; Liang, Y.; Huang, Z. Evaluating urban expansion and land use change in Shijiazhuang, China, by using GIS and remote sensing. *Landsc. Urban Plan.* **2006**, *75*, 69–80. [CrossRef]
11. Li, X.; Yeh, A.G. Analyzing spatial restructuring of land use patterns in a fast growing region using remote sensing and GIS. *Landsc. Urban Plan.* **2004**, *69*, 335–354. [CrossRef]
12. Li, C.; Zhang, Y.; Kharel, G.; Zou, C.B. Impact of Climate Variability and Landscape Patterns on Water Budget and Nutrient Loads in a Peri-urban Watershed: A Coupled Analysis Using Process-based Hydrological Model and Landscape Indices. *Environ. Manag.* **2018**, *61*, 954–967. [CrossRef]
13. Liu, J.; Shen, Z.; Chen, L. Assessing how spatial variations of land use pattern affect water quality across a typical urbanized watershed in Beijing, China. *Landsc. Urban Plan.* **2018**. [CrossRef]
14. Shen, Z.; Hou, X.; Li, W.; Aini, G.; Chen, L.; Gong, Y. Impact of landscape pattern at multiple spatial scales on water quality: A case study in a typical urbanised watershed in China. *Ecol. Indic.* **2015**. [CrossRef]
15. Zhao, R.; Chen, Y.; Shi, P.; Zhang, L.; Pan, J.; Zhao, H. Land use and land cover change and driving mechanism in the arid inland river basin: A case study of Tarim River, Xinjiang, China. *Environ. Earth Sci.* **2013**, *68*, 591–604. [CrossRef]
16. Mallinis, G.; Koutsias, N.; Arianoutsou, M. Monitoring land use/land cover transformations from 1945 to 2007 in two peri-urban mountainous areas of Athens metropolitan area, Greece. *Sci. Total Environ.* **2014**, *490*, 262–278. [CrossRef]
17. Křováková, K.; Semerádová, S.; Mudrochová, M.; Skaloš, J. Landscape functions and their change—A review on methodological approaches. *Ecol. Eng.* **2015**, *75*, 378–383. [CrossRef]
18. Forman, R.T.T. Some general principles of landscape and regional ecology. *Landsc. Ecol.* **1995**. [CrossRef]
19. Li, X.; Lu, L.; Cheng, G.; Xiao, H. Quantifying landscape structure of the Heihe River Basin, north-west China using FRAGSTATS. *J. Arid Environ.* **2001**. [CrossRef]
20. Teixeira, Z.; Teixeira, H.; Marques, J.C. Systematic processes of land use/land cover change to identify relevant driving forces: Implications on water quality. *Sci. Total Environ.* **2014**, *470–471*, 1320–1335. [CrossRef]
21. Gao, C.; Zhou, P.; Jia, P.; Liu, Z.; Wei, L.; Tian, H. Spatial driving forces of dominant land use/land cover transformations in the Dongjiang River watershed, Southern China. *Environ. Monit. Assess.* **2016**, *188*. [CrossRef]
22. Dadashpoor, H.; Azizi, P.; Moghadasi, M. Land use change, urbanization, and change in landscape pattern in a metropolitan area. *Sci. Total Environ.* **2019**, *655*, 707–719. [CrossRef]
23. Pan, D.; Domon, G.; Marceau, D.; Bouchard, A. Spatial pattern of coniferous and deciduous forest patches in an Eastern North America agricultural landscape: The influence of land use and physical attributes. *Landsc. Ecol.* **2001**. [CrossRef]
24. Song, X.P.; Hansen, M.C.; Stehman, S.V.; Potapov, P.V.; Tyukavina, A.; Vermote, E.F.; Townshend, J.R. Global land change from 1982 to 2016. *Nature* **2018**. [CrossRef]
25. Da Silva, A.M.; Huang, C.H.; Francesconi, W.; Saintil, T.; Villegas, J. Using landscape metrics to analyze micro-scale soil erosion processes. *Ecol. Indic.* **2015**, *56*, 184–193. [CrossRef]
26. Su, S.; Hu, Y.; Luo, F.; Mai, G.; Wang, Y. Farmland fragmentation due to anthropogenic activity in rapidly developing region. *Agric. Syst.* **2014**, *131*, 87–93. [CrossRef]
27. Zhang, W.; Yu, N.; Liu, M.; Hu, Y.M. Landscape pattern and driving forces in the upper reaches of Minjiang River, China. In Proceedings of the 2010 3rd International Congress on Image and Signal Processing, Yantai, China, 16–18 October 2010; Volume 5, pp. 2189–2193. [CrossRef]
28. Shi, Y.; Xiao, J.; Shen, Y.; Yamaguchi, Y. Quantifying the spatial differences of landscape change in the Hai River Basin, China, in the 1990s. *Int. J. Remote Sens.* **2012**, *33*, 4482–4501. [CrossRef]
29. Su, S.; Wang, Y.; Luo, F.; Mai, G.; Pu, J. Peri-urban vegetated landscape pattern changes in relation to socioeconomic development. *Ecol. Indic.* **2014**, *46*, 477–486. [CrossRef]
30. Zhang, F.; Kung, H.T.; Johnson, V.C. Assessment of land-cover/land-use change and landscape patterns in the two national nature reserves of Ebinur Lake Watershed, Xinjiang, China. *Sustainability* **2017**, *9*, 724. [CrossRef]
31. Liu, S.; Yu, Q.; Wei, C. Spatial-Temporal Dynamic Analysis of Land Use and Landscape Pattern in Guangzhou, China: Exploring the Driving Forces from an Urban Sustainability Perspective. *Sustainability* **2019**, *11*, 6675. [CrossRef]
32. Gong, Y.; Li, J.; Li, Y. Spatiotemporal characteristics and driving mechanisms of arable land in the Beijing-Tianjin-Hebei region during 1990–2015. *Socioecon. Plann. Sci.* **2019**. [CrossRef]
33. Wang, L.J.; Wu, L.; Hou, X.Y.; Zheng, B.H.; Li, H.; Norra, S. Role of reservoir construction in regional land use change in Pengxi River basin upstream of the Three Gorges Reservoir in China. *Environ. Earth Sci.* **2016**, *75*. [CrossRef]
34. López-Barrera, F.; Manson, R.H.; Landgrave, R. Identifying deforestation attractors and patterns of fragmentation for seasonally dry tropical forest in central Veracruz, Mexico. *Land Use Policy* **2014**, *41*, 274–283. [CrossRef]
35. McSherry, L.; Steiner, F.; Ozkeresteci, I.; Panickera, S. From knowledge to action: Lessons and planning strategies from studies of the upper San Pedro basin. *Landsc. Urban Plan.* **2006**, *74*, 81–101. [CrossRef]
36. Piovan, S.E. Remote Sensing. In *The Geohistorical Approach*; Springer Geography; Springer: Cham, Switzerland, 2020; pp. 171–197. [CrossRef]
37. Southworth, J.; Nagendra, H.; Tucker, C. Fragmentation of a landscape: Incorporating landscape metrics into satellite analyses of land-cover change. *Landsc. Res.* **2002**. [CrossRef]
38. Yu, X.; Ng, C. An integrated evaluation of landscape change using remote sensing and landscape metrics: A case study of Panyu, Guangzhou. *Int. J. Remote Sens.* **2006**. [CrossRef]

39. He, Y.; Wang, W.; Chen, Y.; Yan, H. Assessing spatio-temporal patterns and driving force of ecosystem service value in the main urban area of Guangzhou. *Sci. Rep.* **2021**. [CrossRef]
40. Guo, L.; Xia, B.; Liu, W.; Jiang, X. Spatio-temporal change and gradient differentiation of landscape pattern in Guangzhou City during its urbanization. *Chin. J. Appl. Ecol.* **2006**, *17*, 1671–1676.
41. Gong, J.; Hu, Z.; Chen, W.; Liu, Y.; Wang, J. Urban expansion dynamics and modes in metropolitan Guangzhou, China. *Land Use Policy* **2018**. [CrossRef]
42. Zhang, Y.; Wang, T.; Cai, C.; Li, C.; Liu, Y.; Bao, Y.; Guan, W. Landscape pattern and transition under natural and anthropogenic disturbance in an arid region of northwestern China. *Int. J. Appl. Earth Obs. Geoinf.* **2016**, *44*, 1–10. [CrossRef]
43. Zhao, P.; Xia, B.; Hu, Y.; Yang, Y. A spatial multi-criteria planning scheme for evaluating riparian buffer restoration priorities. *Ecol. Eng.* **2013**, *54*, 155–164. [CrossRef]
44. Li, Q.; Xu, X.L.; Huang, J.R. Length-weight relationships of 16 fish species from the Liuxihe national aquatic germplasm resources conservation area, Guangdong, China. *J. Appl. Ichthyol.* **2014**, *30*, 434–435. [CrossRef]
45. Zhang, C.; Xia, B.; Lin, J. A basin-scale estimation of carbon stocks of a forest ecosystem characterized by spatial distribution and contributive features in the Liuxihe River basin of pearl river delta. *Forests* **2016**, *7*, 299. [CrossRef]
46. Yu, X.J.; Ng, C.N. Spatial and temporal dynamics of urban sprawl along two urban-rural transects: A case study of Guangzhou, China. *Landsc. Urban. Plan.* **2007**, *79*, 96–109. [CrossRef]
47. Jiyuan, L.; Mingliang, L.; Xiangzheng, D.; Dafang, Z.; Zengxiang, Z.; Di, L. The land use and land cover change database and its relative studies in China. *J. Geogr. Sci.* **2002**, *12*, 275–282. [CrossRef]
48. Liu, J.; Liu, M.; Tian, H.; Zhuang, D.; Zhang, Z.; Zhang, W.; Tang, X.; Deng, X. Spatial and temporal patterns of China's cropland during 1990–2000: An analysis based on Landsat TM data. *Remote Sens. Environ.* **2005**, *98*, 442–456. [CrossRef]
49. Riitters, K.H.; O'Neill, R.V.; Hunsaker, C.T.; Wickham, J.D.; Yankee, D.H.; Timmins, S.P.; Jones, K.B.; Jackson, B.L. A factor analysis of landscape pattern and structure metrics. *Landsc. Ecol.* **1995**. [CrossRef]
50. Cushman, S.A.; McGarigal, K.; Neel, M.C. Parsimony in landscape metrics: Strength, universality, and consistency. *Ecol. Indic.* **2008**, *8*, 691–703. [CrossRef]
51. Pontius, R.G.; Shusas, E.; McEachern, M. Detecting important categorical land changes while accounting for persistence. *Agric. Ecosyst. Environ.* **2004**. [CrossRef]
52. Schneeberger, N.; Bürgi, M.; Hersperger, A.M.; Ewald, K.C. Driving forces and rates of landscape change as a promising combination for landscape change research-An application on the northern fringe of the Swiss Alps. *Land Use Policy* **2007**. [CrossRef]
53. Su, C.; Fu, B.; Lu, Y.; Lu, N.; Zeng, Y.; He, A.; Halina, L. Land use change and anthropogenic driving forces: A case study in Yanhe River Basin. *Chin. Geogr. Sci.* **2011**, *21*, 587–599. [CrossRef]
54. Kavian, A.; Jafarian Jeloudar, Z. Land use/cover change and driving force analyses in parts of northern Iran using RS and GIS techniques. *Arab. J. Geosci.* **2011**, *4*, 401–411. [CrossRef]
55. Fang, G.; Zhang, Y.; Yang, J. Evolution of urban landscape pattern in Suzhou City during 1987–2009. *Appl. Mech. Mater.* **2012**, *178–181*, 332–336. [CrossRef]
56. Wang, Z.; Zhang, Y.; Zhang, B.; Song, K.; Guo, Z.; Liu, D.; Li, F. Landscape dynamics and driving factors in Da'an County of Jilin Province in Northeast China During 1956–2000. *Chin. Geogr. Sci.* **2008**. [CrossRef]
57. Bürgi, M.; Straub, A.; Gimmi, U.; Salzmann, D. The recent landscape history of Limpach valley, Switzerland: Considering three empirical hypotheses on driving forces of landscape change. *Landsc. Ecol.* **2010**, *25*, 287–297. [CrossRef]
58. Hersperger, A.M.; Bürgi, M. Going beyond landscape change description: Quantifying the importance of driving forces of landscape change in a Central Europe case study. *Land Use Policy* **2009**, *26*, 640–648. [CrossRef]
59. Liu, X.; Li, Y.; Shen, J.; Fu, X.; Xiao, R.; Wu, J. Landscape pattern changes at a catchment scale: A case study in the upper Jinjing river catchment in subtropical central China from 1933 to 2005. *Landsc. Ecol. Eng.* **2014**, *10*, 263–276. [CrossRef]
60. Yu, G.; Li, M.; Tu, Z.; Yu, Q.; Jie, Y.; Xu, L.; Dang, Y.; Chen, X. Conjugated evolution of regional social-ecological system driven by land use and land cover change. *Ecol. Indic.* **2018**, *89*, 213–226. [CrossRef]
61. Ju-Long, D. Control problems of grey systems. *Syst. Control. Lett.* **1982**. [CrossRef]
62. Chen, M.; Lu, Y.; Ling, L.; Wan, Y.; Luo, Z.; Huang, H. Drivers of changes in ecosystem service values in Ganjiang upstream watershed. *Land Use Policy* **2015**, *47*, 247–252. [CrossRef]
63. Tobler, W.R. A Computer Movie Simulating Urban Growth in the Detroit Region. *Econ. Geogr.* **1970**. [CrossRef]
64. Moran, P.A. Notes on continuous stochastic phenomena. *Biometrika* **1950**. [CrossRef]
65. Cui, C.; Wang, J.; Wu, Z.; Ni, J.; Qian, T. The socio-spatial distribution of leisure venues: A case study of karaoke bars in Nanjing, China. *ISPRS Int. J. Geo-Inf.* **2016**, *5*, 150. [CrossRef]
66. Wartenberg, D. Multivariate Spatial Correlation: A Method for Exploratory Geographical Analysis. *Geogr. Anal.* **1985**, *17*, 263–283. [CrossRef]
67. Wang, J.F.; Li, X.H.; Christakos, G.; Liao, Y.L.; Zhang, T.; Gu, X.; Zheng, X.Y. Geographical detectors-based health risk assessment and its application in the neural tube defects study of the Heshun Region, China. *Int. J. Geogr. Inf. Sci.* **2010**. [CrossRef]
68. Ju, H.; Zhang, Z.; Zuo, L.; Wang, J.; Zhang, S.; Wang, X.; Zhao, X. Driving forces and their interactions of built-up land expansion based on the geographical detector—A case study of Beijing, China. *Int. J. Geogr. Inf. Sci.* **2016**, *30*, 2188–2207. [CrossRef]

69. Kromroy, K.; Ward, K.; Castillo, P.; Juzwik, J. Relationships between urbanization and the oak resource of the Minneapolis/St. Paul Metropolitan area from 1991 to 1998. *Landsc. Urban. Plan.* **2007**, *80*, 375–385. [CrossRef]
70. Gong, C.; Yu, S.; Joesting, H.; Chen, J. Determining socioeconomic drivers of urban forest fragmentation with historical remote sensing images. *Landsc. Urban. Plan.* **2013**, *117*, 57–65. [CrossRef]
71. Zhang, H.; Ning, X.; Shao, Z.; Wang, H. Spatiotemporal pattern analysis of China's cities based on high-resolution imagery from 2000 to 2015. *ISPRS Int. J. Geo-Inf.* **2019**, *8*, 241. [CrossRef]
72. Gong, J.; Jiang, C.; Chen, W.; Chen, X.; Liu, Y. Spatiotemporal dynamics in the cultivated and built-up land of Guangzhou: Insights from zoning. *Habitat Int.* **2018**. [CrossRef]
73. Sun, Y.; Zhang, X.; Zhao, Y.; Xin, Q. Monitoring annual urbanization activities in Guangzhou using Landsat images (1987–2015). *Int. J. Remote Sens.* **2017**, *38*, 1258–1276. [CrossRef]
74. Garcia, A.S.; Ballester, M.V.R. Land cover and land use changes in a Brazilian Cerrado landscape: Drivers, processes, and patterns. *J. Land Use Sci.* **2016**, *11*, 538–559. [CrossRef]
75. Yin, K.; Li, X.; Zhang, G.; Xiao, L. Analysis of socio-economic driving forces on built-up area expansion in Xiamen. *Int. J. Sustain. Dev. World Ecol.* **2010**. [CrossRef]
76. Wang, S.Y.; Liu, J.S.; Ma, T.B. Dynamics and changes in spatial patterns of land use in Yellow River Basin, China. *Land Use Policy* **2010**. [CrossRef]
77. Biazin, B.; Sterk, G. Drought vulnerability drives land-use and land cover changes in the Rift Valley dry lands of Ethiopia. *Agric. Ecosyst. Environ.* **2013**. [CrossRef]
78. Gebremicael, T.G.; Mohamed, Y.A.; van der Zaag, P.; Hagos, E.Y. Quantifying longitudinal land use change from land degradation to rehabilitation in the headwaters of Tekeze-Atbara Basin, Ethiopia. *Sci. Total Environ.* **2018**, *622–623*, 1581–1589. [CrossRef]
79. Zubair, O.A.; Ji, W.; Weilert, T.E. Modeling the impact of urban landscape change on urban wetlands using similarityweighted instance-based machine learning and Markov model. *Sustainability* **2017**, *9*, 2223. [CrossRef]
80. Pomianek, I.; Chrzanowska, M. A spatial comparison of semi-urban and rural gminas in Poland in terms of their level of socio-economic development using Hellwig's method. *Bull. Geogr.* **2016**, *33*, 103–117. [CrossRef]
81. Łopucki, R.; Kiersztyn, A. Urban green space conservation and management based on biodiversity of terrestrial fauna—A decision support tool. *Urban. For. Urban. Green.* **2015**, *14*, 508–518. [CrossRef]
82. Reiff, M.; Surmanová, K.; Balcerzak, A.P.; Pietrzak, M.B. Multiple criteria analysis of European union agriculture. *J. Int. Stud.* **2016**, *9*, 62–74. [CrossRef]
83. Roszkowska, E.; Filipowicz-Chomko, M. Measuring Sustainable Development Using an Extended Hellwig Method: A Case Study of Education. *Soc. Indic. Res.* **2021**, *153*, 299–322. [CrossRef]

Article

Assimilation of Multi-Source Precipitation Data over Southeast China Using a Nonparametric Framework

Yuanyuan Zhou [1,2], Nianxiu Qin [3], Qiuhong Tang [4,5], Huabin Shi [1] and Liang Gao [1,2,*]

1. State Key Laboratory of Internet of Things for Smart City and Department of Civil and Environmental Engineering, University of Macau, Macao 999078, China; yb87424@um.edu.mo (Y.Z.); huabinshi@um.edu.mo (H.S.)
2. Center for Ocean Research in Hong Kong and Macau (CORE), Hong Kong 999077, China
3. Key Laboratory of Beibu Gulf Environment Change and Resources Use, Ministry of Education, Nanning Normal University, Nanning 530001, China; qinnianxiu@nnnu.edu.cn
4. Key Laboratory of Water Cycle and Related Land Surface Processes, Institute of Geographical Sciences and Natural Resources Research, Chinese Academy of Sciences, Beijing 100101, China; tangqh@igsnrr.ac.cn
5. Institute of Geographic Sciences and Natural Resources Research, University of Chinese Academy of Sciences, Beijing 100049, China
* Correspondence: gaoliang@um.edu.mo; Tel.: +853-8822-9092

Citation: Zhou, Y.; Qin, N.; Tang, Q.; Shi, H.; Gao, L. Assimilation of Multi-Source Precipitation Data over Southeast China Using a Nonparametric Framework. *Remote Sens.* 2021, 13, 1057. https://doi.org/10.3390/rs13061057

Academic Editor: Shreedhar Maskey

Received: 10 January 2021
Accepted: 2 March 2021
Published: 11 March 2021

Publisher's Note: MDPI stays neutral with regard to jurisdictional claims in published maps and institutional affiliations.

Copyright: © 2021 by the authors. Licensee MDPI, Basel, Switzerland. This article is an open access article distributed under the terms and conditions of the Creative Commons Attribution (CC BY) license (https://creativecommons.org/licenses/by/4.0/).

Abstract: The accuracy of the rain distribution could be enhanced by assimilating the remotely sensed and gauge-based precipitation data. In this study, a new nonparametric general regression (NGR) framework was proposed to assimilate satellite- and gauge-based rainfall data over southeast China (SEC). The assimilated rainfall data in Meiyu and Typhoon seasons, in different months, as well as during rainfall events with various rainfall intensities were evaluated to assess the performance of this proposed framework. In rainy season (Meiyu and Typhoon seasons), the proposed method obtained the estimates with smaller total absolute deviations than those of the other satellite products (i.e., 3B42RT and 3B42V7). In general, the NGR framework outperformed the original satellites generally on root-mean-square error (RMSE) and mean absolute error (MAE), especially on Nash-Sutcliffe coefficient of efficiency (NSE). At monthly scale, the performance of assimilated data by NGR was better than those of satellite-based products in most months, by exhibiting larger correlation coefficients (CC) in 6 months, smaller RMSE and MAE in at least 9 months and larger NSE in 9 months, respectively. Moreover, the estimates from NGR have been proven to perform better than the two satellite-based products with respect to the simulation of the gauge observations under different rainfall scenarios (i.e., light rain, moderate rain and heavy rain).

Keywords: precipitation; assimilation; nonparametric modeling; multi-source

1. Introduction

As a key component within the water and energy cycle system, precipitation plays a crucial role in the fields of hydrology, meteorology and water resources management [1–7]. Accurate precipitation is an essential model input to predict the hydrological responses of the selected watershed and the potential rain-induced hazards [8–11]. Therefore, attention is drawn to estimating the precipitation distribution using different methods. The ground rain gauge is a common approach for measuring precipitation at specific locations during a prescribed period, which is of high credibility after calibration. In many cases, however, the sparsely distributed rain gauges could not provide sufficient precipitation data which can represent its spatial variability in detail [1,12]. Alternatively, remote sensing techniques can supply precipitation data on a global scale [13], which is exempt from the topographic restriction.

During the past two decades, on the merits of satellite sensors and signal-processing algorithms, rainfall products are emerging, such as the Precipitation Estimation from Remotely Sensed Information Using Artificial Neural Networks (PERSIANN) (as listed in Table 1) [14], the precipitation dataset based on the Climate Prediction Center (CPC) Morphing (CMORPH) technique [15] derived using the motion vectors and morphed method, the Integrated Multi-satellite Retrievals for Global Precipitation Measurement (IMERG) dataset [16] and the Tropical Rainfall Measuring Mission (TRMM) Multi-satellite Precipitation Analysis (TMPA) [17]. Particularly, the TRMM satellite started to serve on 27 November 1997 and was decommissioned in 2015, nevertheless the corresponding blended rainfall data is still provided to the public until the transition (from TRMM to IMERG) is completed. The TMPA precipitation dataset including post-real time product (3B42V7) and near-real-time product (3B42RT) has been widely used over China [18–21].

Table 1. The information of rainfall datasets employed in this study.

Products	Spatial/Temporal Resolution	Time Period Available	Coverage	Source of Data
3B42V7	0.25°/3 h	January 1998 to January 2020	50° S to 50° N	Goddard Space Flight Center (GSFC)
3B42RT	0.25°/3 h	February 2000 to January 2020	60° S to 60° N	GSFC
PERSIANN	0.25°/3 h	March 2000 to present	60° S to 60° N	Center for Hydrometeorology and Remote Sensing (CHRS)
Rain gauge observation	Point/Daily	1951 to present	China	China Meteorological Data Service Center (CMDC)

The satellite-based rainfall products with fine spatio-temporal resolution is desirable, but the uncertainty and error originated from indirect measurements of precipitations inferred from micro-wave and infrared radar measurements are non-negligible [22]. Moreover, according to the evaluation of the satellite-based precipitation products over China [20,23,24], the performance of these products varies with different spatial and temporal scales. For instance, 3B42RT and PERSIANN significantly overestimate rainfall amounts across the Tibetan Plateau [25], but 3B42RT can detect the most flood warning events compared to IMERG [23]. Guo et al. [20] reported that 3B42V7 performed relatively better in northwestern China, but overestimated rain rates in southern China. Therefore, to obtain more accurate estimates which incorporate the merits of satellite-based and ground-based rainfall, multi-source precipitation datasets need to be assimilated. The satellite-based rainfall, ground-based gauge/radar rainfall data and some reanalysis precipitation datasets are typically selected as one of the assimilated sources [22,26,27]. Moreover, meteorological and land surface data, such as temperature, elevation and soil moisture, could also be adopted to estimate precipitation [22,28,29]. Introducing the meteorological and land surface data, however, might cause uncertainties due to the relatively low correlation between precipitation and the corresponding factors at a daily scale [30]. Moreover, the meteorological factors may involve lag effects or/and spatial variance, which should be investigated and discussed ahead. In addition, the accuracy of the precipitation products as part of the source datasets could be evaluated specifically with comparison to the gauge data under a certain assimilation framework. Therefore, two groups of TMPA dataset, namely the real-time product of 3B42RT and the post-real-time product of 3B42V7, were employed as source data in this study.

In general, the methodologies for assimilating multi-source precipitation datasets can be categorized into two major types, i.e., parametric and nonparametric methods [31,32]. In terms of parametric algorithms, a functional form with finite number of parameters must be specified by users, and the unknown parameters can be determined by evaluating the attributes of input–output data [33]. Nonparametric methods, alternatively, can reduce the complexity of determining the unknown parameters, which can construct the input–output relationship without prior knowledge of specifying functional form [34]. Moreover, nonparametric methods can be exempt from limitation of data types, such as spatial non-stationary rainfall data [24], and modeling of the relationships among independent and dependent variables. That is, nonparametric methods employ relatively weaker assumptions of data than traditional parametric approaches and model the nonadditive effects without explicit functional form.

In light of its advantages, some nonparametric algorithms have been developed recently and applied to assimilate the rainfall data. Bhuiyan et al. [35] combined multiple precipitation datasets using quantile regression forests (QRF) and evaluated the results from the perspective of stream simulations on the Iberian Peninsula. Ma et al. [36] derived the merged rainfall data over the Tibet Plateau by adopting the dynamic Bayesian model averaging scheme, and also evaluated the assimilated precipitation data in four seasons and at different elevations over Tibet. The artificial neural networks (ANNs) have also been used to assimilate multi-source precipitation data including satellite-based, gauge-based and radar datasets in different regions [37–39]. There are also other nonparametric methods, such as the general regression neural network (GRNN) [40] and Bayesian nonparametric general regression [41]. The performance of these nonparametric models in assimilating the rainfall data has not been tested. Nevertheless, studies to evaluate the application scenarios, such as the rainfall events with different intensities on different time scales, are still insufficient. The applicability of a certain fusion algorithm needs to be assessed for rainfall in Meiyu and Typhoon seasons, in different months, as well as rainfall events with various rainfall intensities.

In this study, a framework based on a nonparametric general regression (NGR) is proposed for assimilating gauge- and satellite-based precipitation data, and then it is applied to southeast China (SEC). Besides, this study yields more insights into evaluations of assimilated data on multiple scales. The study area and precipitation data resources are introduced first. Then, the proposed framework is depicted. Thereafter, the performance of the nonparametric framework is analyzed and the comparisons of assimilated results using NGR and multiple linear regression (MLR), as well as PERSIANN products, are conducted. In the end, some major conclusions are drawn.

2. Materials and Methods

2.1. Study Area

The southeast China (SEC) was selected as the study area, ranging from (15° N, 105° E) to (35° N, 125° E). Figure 1 shows the location of the study area, and the distribution of rain gauges. In this area, East Asian monsoon dominates. Influenced by the summer monsoon, the majority of rainfall occurs in summer, which accounts for 60–85% of the annual total precipitation in SEC [42]. The precipitation is characterized by the trend of increasing from northeast to southeast, which shares a similar pattern with that of temperature over this region [43]. It is in general warm and humid in summer, while mild in winter [44]. The complex topography and climate features of SEC result in prominent spatiotemporal variability of precipitation [45]. Due to the increasing extreme precipitation events, SEC is becoming more and more prone to floods, landslides and other natural disasters [46].

2.2. Data Sources

Under the influence of the super El Nino, 20 severe rainstorms occurred in 2016 over SEC [47]. As a result, deadly floods and landslides were triggered, leading to serious damages [48]. Furthermore, 88% of severe rainstorms occurred from June to September. Therefore, the daily rainfall data at 330 rain stations (as shown in Figure 1) across SEC, covering a period from 01 January to 31 December in 2016, were adopted in this study. The gauge dataset was provided by China Meteorological Data Service Center (CMDC), which has been examined by extreme values check, internal consistency check and spatial consistency check [36].

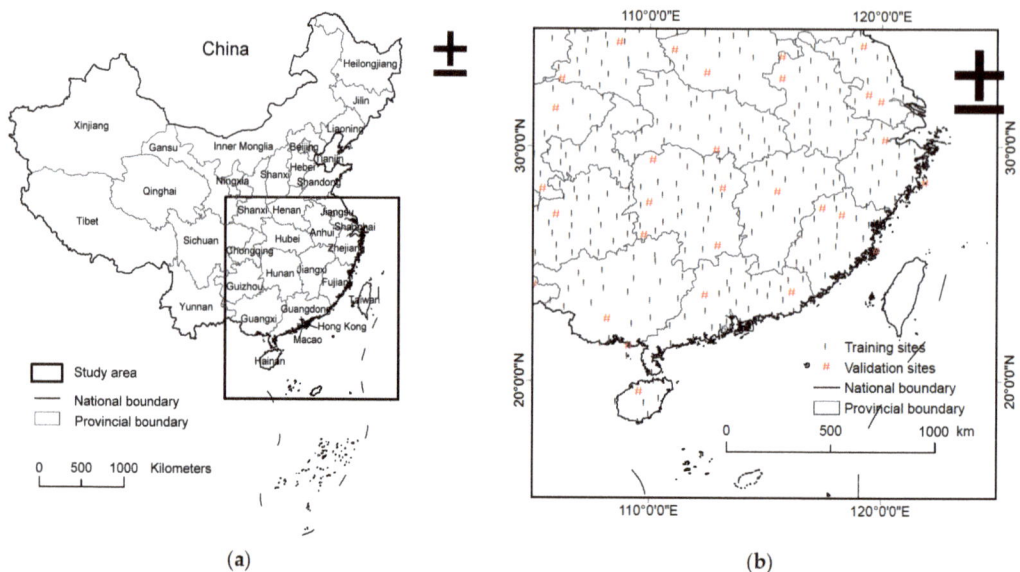

Figure 1. (**a**) The location of study area, and (**b**) rain gauge stations.

The latest Version-7 TRMM TMPA near-real-time (3B42RT) and post-real-time (3B42V7) products were adopted in this study. The National Aeronautics and Space Administration (NASA) Goddard Space Flight Center (GSFC) developed 3B42V7 and 3B42RT datasets with the spatial resolution of 0.25° × 0.25° and the temporal resolution of 3 h, respectively [49]. In order to match the temporal resolution between gauge and satellite-based data, the 3-hourly satellite-based products were adjusted to daily accumulated datasets in Beijing time. To keep consistent with the format of gauge data, the rainfall value at the corresponding location was derived from the satellite product (in grid format) using the inverse distance weighting (IDW) method [50]. The information of rainfall data employed in this study is listed in Table 1.

2.3. Methods

2.3.1. The Framework Based on Nonparametric General Regression

In this study, a new framework based on nonparametric general regression is proposed. This method is composed of the general regression network and the parameter identifying model. The nonparametric general regression network is designed as follows. Let $\mathbf{T} = [\mathbf{T}_1, \mathbf{T}_2, \ldots, \mathbf{T}_N] \in \mathbb{R}^{2 \times N}$ be the satellite-derived datasets, namely 3B42V7 and 3B42RT datasets, and $\mathbf{G} = [G_1, G_2, \ldots, G_N] \in \mathbb{R}^N$ denotes the gauge-based data in this study. N is the number of samples, i.e., $N = N_{days} * N_{stations}$, where N_{days} and $N_{stations}$ are the number of days and stations, respectively. There is the following relationship between $\mathbf{T}$ and $\mathbf{G}$:

$$\mathbf{G} = F(\mathbf{T}), \tag{1}$$

Then, use θ to represent the unknown parameter vector in the nonparametric general regression network. The conditional probability density function (PDF) of $\mathbf{G}$ based on $\mathbf{T}$ and θ can be expressed by Equation (2), which is also called the likelihood in a frequentist framework.

$$p(\mathbf{G}|\theta, \mathbf{T}) = p(G_1, G_2, \ldots, G_N|\theta, \mathbf{T}) = \prod_{m=1}^{N} p(G_m|G_1, \ldots, G_{m-1}, \theta, \mathbf{T}), \tag{2}$$

The conditional PDFs in the right hand of Equation (2) can be given by:

$$p(G_m|G_1,\ldots,G_{m-1},\boldsymbol{\theta},\mathbf{T}) = (2\pi\sigma_{2,m}^2)^{-1/2}\exp\left\{-\frac{[G_m - \hat{G}_{m|m-1}(\mathbf{T}_m)]^2}{2\sigma_{2,m}^2}\right\}, \quad (3)$$

where $\hat{G}_{m|m-1}(\mathbf{T}_m)$ is

$$\hat{G}_{m|m-1}(\mathbf{T}_m) = \frac{\sum_{n=1}^{m-1} G_n \exp[-(\mathbf{T}_m - \mathbf{T}_n)^2/(2\sigma_{1,m}^2)]}{\sum_{n=1}^{m-1} \exp[-(\mathbf{T}_m - \mathbf{T}_n)^2/(2\sigma_{1,m}^2)]}, \quad (4)$$

where $\sigma_{1,m}^2$ and $\sigma_{2,m}^2$ are the smooth parameters and the prediction-error variances respectively, $m = 1, 2, \ldots, N$. $\hat{G}_{m|m-1}(\mathbf{T}_m)$ is one estimate of $\mathbf{G}$. $\sigma_{1,m}^2$ and $\sigma_{2,m}^2$ are computed using the following forms:

$$\sigma_{1,m}^2 = \frac{v_1}{m-1}\sum_{n=1}^{m-1}(\mathbf{T}_m - \mathbf{T}_n)^2, \quad (5)$$

$$\sigma_{2,m}^2 = \frac{v_2}{\sum_{n=1}^{m-1}\exp[-2(\mathbf{T}_m - \mathbf{T}_n)^2]}, \quad (6)$$

where v_1 and v_2 are two unknowns: $\boldsymbol{\theta} = [v_1, v_2]^\mathrm{T}$.

Based on the general regression network, there are now two unknown parameters to be determined. Note that we can rewrite the likelihood in Equation (2) in terms of the unknown parameters as:

$$p(\mathbf{G}|v_1, v_2, \mathbf{T}) \propto (v_2)^{(-\frac{N}{2})}\exp[-\frac{1}{2v_2}\sum_{m=1}^{N}\Omega_m(G_m - \hat{G}_{m|m-1,v_1}(\mathbf{T}_m))^2], \quad (7)$$

where Ω_m can be given by:

$$\Omega_m = \sum_{n=1}^{m-1}\exp[-2(\mathbf{T}_m - \mathbf{T}_n)^2], \quad (8)$$

Particularly, if v_1 is given, $\hat{v}_2$ ($\hat{v}_2$ is the estimation of v_2) can be expressed by Equation (9) by solving $\frac{\partial p(\mathbf{G}|v_1,v_2,\mathbf{T})}{\partial v_2} = 0$, which means that only one parameter needs to be calculated.

$$\hat{v}_2(v_1) = \frac{1}{N}\sum_{m=1}^{N}\Omega_m(G_m - \hat{G}_{m|m-1,v_1}(\mathbf{T}_m))^2, \quad (9)$$

$\hat{v}_1$ (estimation of v_1) can be obtained by maximizing the function of v_1: $f(v_1) = p(\mathbf{G}|v_1, \hat{v}_2(v_1), \mathbf{T})$, which is usually realized by standard optimization algorithms, such as genetic algorithm (GA) herein. Thereafter, $\hat{v}_1$, $\hat{v}_2$ and $\hat{\mathbf{G}}$ can be obtained.

2.3.2. Data Processing for the Framework Validation

To comprehensively assess the NGR framework, k-fold cross-validation was performed. In this study, k was set to 11. In the 11-fold cross-validation, the data derived from the 330 stations is divided into 11 mutually exclusive subsets, one of which is employed as a validation dataset, while the other 10 are used as the training datasets. This process needs to be repeated 11 times. When k equals to 1, k-fold cross-validation is a special case, which is also termed as hold-out validation. The hold-out validation method is mainly conducted in this study as suggested by previous studies [28,36,51]. The data is divided into two non-overlapping sets. One is referred to as training dataset, which is adopted to train the framework, and the other is referred to as validation dataset, which is used to compare with the assimilated rainfall to assess the performance of NGR. The flowchart of training and validating the framework for assimilating multi-source rainfall datasets based on hold-out validation is shown in Figure 2. Under the framework, the 330 sites over SEC were assigned into training and validation sites from which the training and validation data were extracted, respectively. With reference to previous studies, the ratio of the training data to the validation data was set to be 10:1 [36,38,52]. That is, 30 out of 330 sites were selected randomly as validation sites, and the remaining 300 sites were set as training sites (in Figures 1b and 2). Note that the satellite-based data was derived from the original gridded satellite-based data using the inverse distance weighting (IDW) method. In the training process, the proposed nonparametric framework was trained using the satellite-based training data extracted at 300 training sites as inputs and the gauge-based training data recorded at 300 training sites. After that, the gauge-based validation data recorded by 30 validation sites was adopted to validate the performance of the NGR framework.

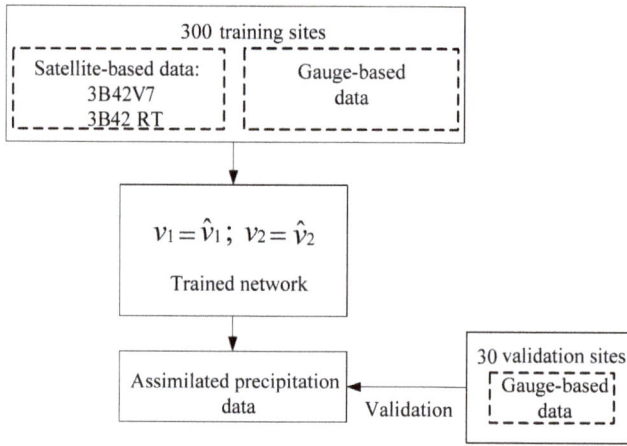

Figure 2. The flowchart of the framework based on NGR for assimilating multiple-source rainfall data based on hold-out cross-validation.

2.3.3. Statistical Metrics for Evaluating the Performance of the NGR Framework

In order to compare the outputs (assimilated rainfall data) of the trained NGR framework from different perspectives, four statistical metrics, i.e., Pearson correlation coefficient (CC), root mean square error (RMSE), mean absolute error (MAE) and Nash-Sutcliffe coefficient of efficiency (NSE), were adopted in this study. CC denotes the linear agreement between the assimilated data and the validation gauge observations. RMSE and MAE are the measures of errors between the estimated and the gauge data. NSE, whose best value is 1, is used to assess the fit of two data pairs. The mentioned statistical indices are calculated by the following formulas:

$$CC = \frac{\sum_{i=1}^{k}(\hat{y}_i - \bar{\hat{y}})\sum_{i=1}^{k}(y_i - \bar{y})}{\sqrt{\sum_{i=1}^{k}(\hat{y}_i - \bar{\hat{y}})^2}\sqrt{\sum_{i=1}^{k}(y_i - \bar{y})^2}}, \quad (10)$$

$$RMSE = \sqrt{\frac{1}{k}\sum_{i=1}^{k}(\hat{y}_i - y_i)^2}, \quad (11)$$

$$MAE = \frac{\sum_{i=1}^{k}|\hat{y}_i - y_i|}{k}, \quad (12)$$

$$NSE = 1 - \frac{\sum_{i=1}^{k}(\hat{y}_i - y_i)^2}{\sum_{i=1}^{k}(y_i - \bar{y}_i)^2}, \quad (13)$$

where k is the number of samples, y_i is the ith data of the validation rainfall dataset y, $\hat{y}_i$ is the assimilated rainfall data, and $\bar{\hat{y}}$ and $\bar{y}$ are the mean values of the assimilated and gauge-based validation data, respectively.

In addition, the Kling-Gupta efficiency (KGE) [53], as a statistical metric combining with correlation coefficient, standard deviation and simulation mean, is increasingly employed to evaluate models. It can be expressed as:

$$\text{KGE} = 1 - \sqrt{(CC-1)^2 + \left(\frac{\sigma_{estimates}}{\sigma_{observations}} - 1\right)^2 + \left(\frac{\mu_{estimates}}{\mu_{observations}} - 1\right)^2}, \quad (14)$$

where $\sigma_{estimates}$ and $\mu_{estimates}$ are the standard deviation and mean of estimates respectively, and $\sigma_{observations}$ and $\mu_{observations}$ stand for the standard deviation and mean of gauge-based observations. According to these studies [54–56], although KGE = 1 indicates perfect agreement between the estimates and observations, various KGE values should be set as the index of good agreement in order to ensure more accurate evaluation of different models. Therefore, negative KGE values are considered as bad agreement between estimates and observations in this study.

2.3.4. Multiple Linear Regression Method

The MLR method [57] is usually adopted to model the linear relationship between dependent and independent variables, which is described by the following general form:

$$\mathbf{Y} = a_0 + a_1 \times \mathbf{X}_1 + \ldots + a_M \times \mathbf{X}_M + \varepsilon, \quad (15)$$

where $\mathbf{Y}$ is the dependent variable, $\mathbf{X}_1, \mathbf{X}_2, \ldots, \mathbf{X}_M$ are the independent variables, $a_0, a_1, a_2, \ldots, a_M$ are the coefficients for independent variables, M is the number of independent variables and ε is the model's error term. In this study, the independent variables and dependent variables denote two satellite-based datasets and assimilated rainfall data, respectively. According to the form of MLR, it is obvious that the mapping relationship between independent and dependent variables has been set to be linear in advance, whereas it is unnecessary to prescribe the mapping function when using NGR. Based on the mentioned characteristics of MLR and NGR, comparison was performed to evaluate the blended results calculated from the two schemes.

3. Results

The mean values of daily statistical metric of rainfall estimates originated from the eleven-fold cross-validation are listed in Table 2. The proposed scheme in general performed better on RMSE, MAE and NSE, while a little worse on CC. Although all the KGE values are positive, 3B42V7 obtained the largest KGE.

Table 2. The mean values of daily statistical metric of rainfall estimates originated from the eleven-fold cross-validation.

Products	CC	RMSE (mm)	MAE (mm)	NSE	KGE
Estimates	0.68	**9.76**	**3.61**	**0.45**	0.58
3B42V7	**0.70**	9.98	3.78	0.43	**0.70**
3B42RT	0.67	11.38	4.19	0.25	0.63

Note: The numbers in bold indicate the optimum values for the indices.

To evaluate the applicability of the framework, the rainfall in Meiyu (June and July) and Typhoon (July, August and September) seasons, in different months and rainfall events with different rainfall intensities, were included. According to the China Meteorological Association (http://www.cma.gov.cn, accessed on 11 January 2020), the severity of rain events in China can be categorized in terms of the 24 h accumulated rainfall, which are light rain (0.1–10 mm/day), moderate rain (10–25 mm/day), heavy rain (25–50 mm/day), rainstorm (50–100 mm/day), heavy rainstorm (100–250 mm/day) and severe rainstorm

(>250 mm/day). In this study, the rainfall events with rainfall intensity > 50 mm/day were considered as a rainstorm. Since there were quite a few heavy and severe rainstorms in SEC during 2016, only four rainfall intensities (i.e., light rain, moderate rain, heavy rain and rainstorm) were discussed in this study. Note that in order to show the performance of rainfall estimates spatially, the assimilated rainfall data at the 30 validation sites (Figure 1b) from the hold-out validation was evaluated at different scales in the following sections.

3.1. Assimilated Precipitation Data at Meiyu Seasons

Figure 3 shows the bias from 3B42V7, 3B42RT and NGR at 30 selected validation sites (Figure 1b) during Meiyu season, which was the absolute deviation between the mean daily estimates and gauge-based observations at each validation site. A bounding circle in Figure 3 indicates that the estimates yield the smallest absolute deviation at this validation site compared to those from the other two products at the same location. Table 3 summarizes the numbers of stations corresponding to the best performance of estimates on CC, RMSE, MAE and smallest absolute deviation in Meiyu and Typhoon seasons. The absolute deviation from NGR exhibited the smallest value at 18 validation sites, followed by 3B42V7 (11 validation sites) and 3B42RT (1 validation site), respectively. Specifically, the large deviations from 3B42RT data (in Figure 3b) corresponded to the sites in the south of Guangxi, Hunan province, and coastal areas, where smaller errors were obtained by 3B42V7 and NGR. Regarding to the spatial distribution of errors, NGR and 3B42V7 tended to exhibit lager bias in inland areas, while the major errors from 3B42RT were discovered across the middle and south of the study area. From the perspective of error values, 3B42RT yielded the largest bias with value of 8.40 mm at the site located at the south of Guangxi province, while 3B42V7 and NGR obtained relatively smaller bias values of 4.61 and 5.21 mm, respectively. The minimum bias with value of 0.05 mm was from NGR, followed by 0.07 mm from 3B42V7 and 0.41 mm from 3B42RT. The mean value of the total absolute deviation at the 30 validation sites from 3B42RT was the largest with value of 2.97 mm, followed by 3B42V7 with value of 1.30 mm and NGR with value of 1.17 mm.

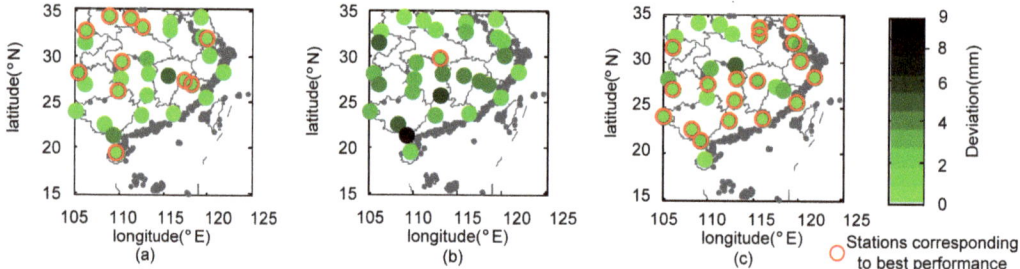

Figure 3. Absolute deviation of mean daily rainfall estimates against gauge observations at each validation site in Meiyu season from (**a**) 3B42V7, (**b**) 3B42RT and (**c**) NGR in 2016.

Figure 4 presents the distribution of the statistical metrics between estimates (assimilated NGR data and satellite products) and gauge observations at each validation site in Meiyu season. In general, the spatial variations of CCs from the three products are of high spatial consistency, especially those between 3B42V7 and NGR. Moreover, NGR exhibited the largest CC values at 40% of validation sites, but 3B42V7 and 3B42RT data corresponding to 36% and 24% of validation sites were highly correlated with gauge observations (Table 3). As for RMSE, the indicator from 3B42RT corresponding to the majority of validation sites was larger than those from NGR and 3B42V7. Meanwhile, there were 19 out of 30 stations having smaller RMSEs from estimated datasets compared to satellite products. The largest MAE was originated from 3B42RT and located south of Sichuan province, where the MAE values from 3B42V7 and NGR were relatively smaller. MAE from NGR at 16 validation sites were smaller than those from 3B42V7 (14 validation sites) and 3B42RT (none of the validation sites), as shown in Table 3. According to the definition of NSE, the closer the value is to 1, the better fit between the two models. Therefore, as for NSE values, the estimated rainfall data at 12 sites from 3B42RT, 9 sites from 3B42V7 and 2 sites from NGR did not match the gauge observations well (i.e., NSE was smaller than 0). The proposed nonparametric framework yielded the largest NSE values at the majority of validation sites, which were mainly located at in inland areas of the study area.

Table 3. The number of validation stations corresponding to the best performance in Meiyu and Typhoon seasons.

Statistical Metrics	Products	Number of Stations (Meiyu)	Number of Stations (Typhoon)
CC	3B42V7	11	12
	3B42RT	7	6
	Estimates	**12**	**12**
RMSE	3B42V7	11	8
	3B42RT	0	2
	Estimates	**19**	**20**
MAE	3B42V7	14	10
	3B42RT	0	3
	Estimates	**16**	**17**
NSE	3B42V7	11	8
	3B42RT	0	2
	Estimates	**19**	**20**
Deviation	3B42V7	11	8
	3B42RT	1	4
	Estimates	**18**	**18**

Note: The numbers in bold indicate the maximum number of stations.

Figure 5 shows the box plots for statistical metrics of daily precipitation at the 30 validation sites. In terms of CC, the performance of three datasets was in general in the same level, whereas the median value from 3B42V7 was the largest. Besides, the values of CC at the 25th and 75th percentile corresponding to NGR were both higher than those from the other two products. NGR yielded the lowest median values for RMSE and MAE (in Figure 5b,c). As for NSE in Figure 5d, the outliers based on the assimilated NGR dataset were closer to the median line. In contrast to satellite-based products, NGR yielded larger NSE values at the 25th and 75th percentile, as well as a smaller range between these two quartiles, indicating that the assimilated rainfall data using NGR agreed better with gauge data overall.

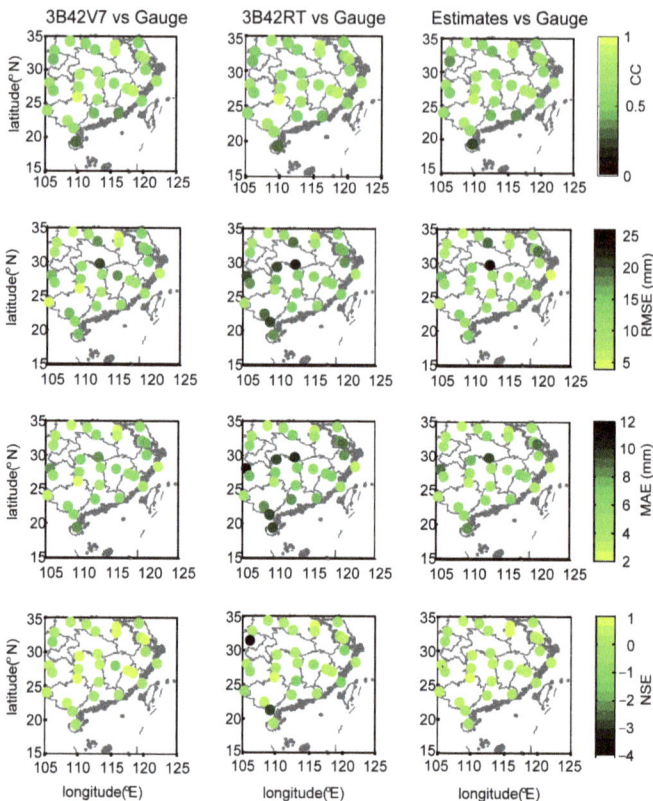

Figure 4. Spatial distribution of statistical metrics of daily precipitation at each validation site during Meiyu season in 2016.

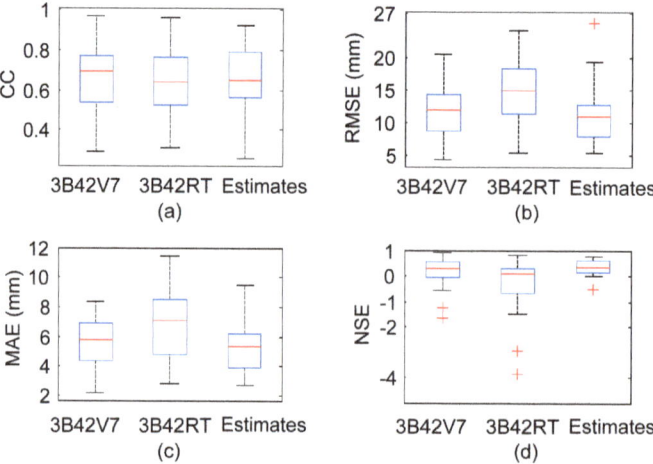

Figure 5. Box plots depicting statistical metrics including (**a**) CC, (**b**) RMSE, (**c**) MAE and (**d**) NSE for daily precipitation at each validation site during Meiyu season in 2016. The line in the box stands for the median value.

3.2. Assimilated Precipitation Data at Typhoon Seasons

The blended precipitation in Typhoon season, as another rainy period in SEC, was also evaluated. Figure 6 shows the spatial distributions of absolute deviation of the mean merged precipitation products against mean gauge data at each validation site in the Typhoon season of 2016. Neither satellite-based datasets can accurately estimate the rainfall amounts on the seashores of Guangxi, Jiangxi, Fujian and Zhejiang provinces, as shown in Figure 6a,b. Moreover, the largest errors from 3B42RT were marked at the sites in Sichuan and Guangxi provinces, where the estimates (Figure 6c) attained comparatively smaller errors. The total errors from NGR were substantially smaller than those generated by 3B42RT and 3B42V7 in the Typhoon season. From Table 3, estimates based on the NGR framework obtained the smallest absolute deviations at 18 sites, while the 3B42V7 and 3BN42RT yielded the smallest errors at 8 sites and 4 sites, respectively. NGR tended to obtain the estimates with the smallest deviations along coastal lines. In general, the proposed approach was capable of effectively diminishing more absolute errors compared to the two satellite-based products in the Typhoon season of 2016 across SEC.

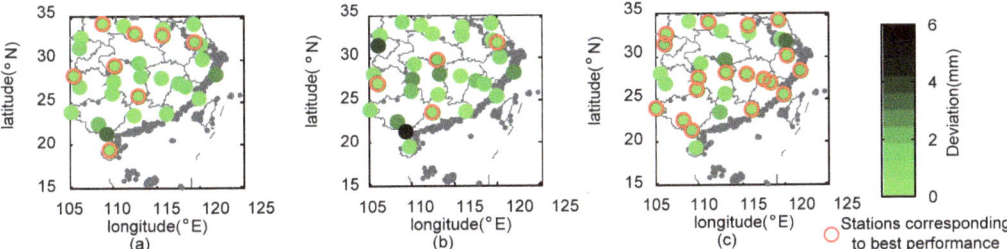

Figure 6. Absolute deviation of mean daily rainfall estimates against gauge observations at each validation site in Typhoon season from (**a**) 3B42V7, (**b**) 3B42RT and (**c**) NGR in 2016.

Figure 7 demonstrates the spatial patterns of daily metrics at each validation site in the Typhoon season over SEC. There were no significant spatial variances among NGR-, 3B42V7- and 3B42RT-Gauge CCs, but obvious spatial variances for RMSE, MAE and NSE. Specifically, the assimilated rainfall and satellite products exhibited highly different RMSE across SEC, with the range between 0 and 23 mm. The larger RMSE from 3B42V7 and 3B42RT was found in Hainan province, while a lower value was observed from NGR in this area. Moreover, 3B42RT tended to obtain larger RMSE values than the other two products over SEC. In terms of MAE, all the maximum values of MAE from the three approaches appeared in the south of the study area, where NGR exhibited the best performance, followed by 3B42V7 and 3B42RT. There were more stations with smaller RMSE (20 out of 30 sites) and MAE (17 out of 30 sites) yielded by NGR than those from 3B42V7 and 3B42RT. For NSE (in Figure 7), there were more NSE values from satellite-based datasets far smaller than 1. In other words, the estimates from the nonparametric framework at each validation site matched the gauge precipitation better than those by 3B42V7 and 3B42RT in the Typhoon season.

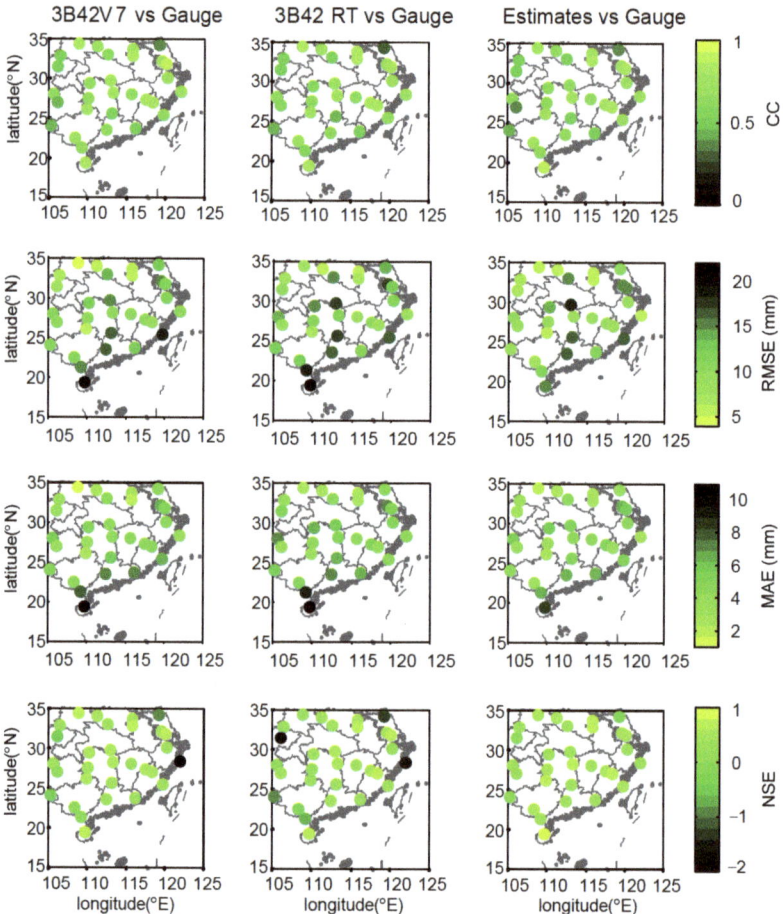

Figure 7. Spatial distribution of statistical metrics for precipitation at daily scale from 3B42V7 data, 3B42RT data and estimated rainfall data at 30 validation sites during the Typhoon season in 2016 over SEC.

Figure 8 depicts the box plots of metrics of the indices in the Typhoon season. The maximum CC value was obtained by NGR while the minimum was attained by 3B42RT, whereas the median lines from the three products were almost at the same level. Although 3B42V7 exhibited the smallest range between upper quartile and lower quartile in terms of RMSE (in Figure 8b) and MAE (in Figure 8c), the median values from NGR were the smallest. Figure 8d shows that the 25th/75th percentile and the upper/lower end of outliers from NGR were much closer to 1 compared to the corresponding values from satellite-based data, indicating that the estimates obtained by the proposed scheme better captured the gauge observations at each validation site in the Typhoon season.

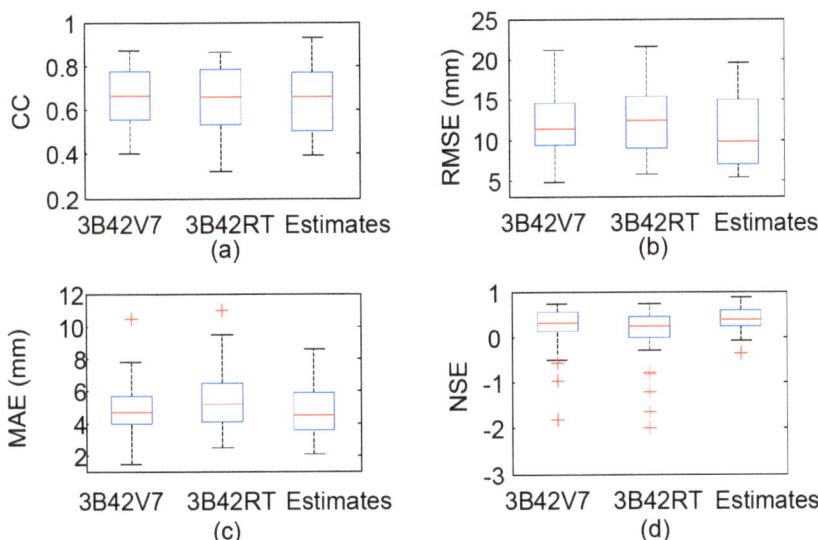

Figure 8. Box plots depicting the difference of statistical metrics including (**a**) CC, (**b**) RMSE, (**c**) MAE and (**d**) NSE for daily assimilated and satellite-based rainfall datasets at each validation site in the Typhoon season in 2016. The line in the box stands for the median value. NGR performed the best on the median values of RMSE, MAE and NSE.

3.3. Assimilated Daily Precipitation at Monthly Scale

Due to the climatic features in SEC, precipitation amounts vary significantly at different time scales. Therefore, in order to capture the accurate temporal patterns of rainfall, the accuracy of precipitation at monthly scale needs to be evaluated. Figure 9 demonstrates the statistical metrics of blended and original satellite-based daily rainfall data from 30 validation sites in 12 months over SEC. All three datasets had similar trends of RMSE and MAE, which decreased from January to February, increased from February to June and then decreased from July to December. CCs dominated by values larger than 0.5 and varied slightly in each month, whereas RMSEs, MAEs and NSE changed significantly from month to month. According to the three datasets, 3B42RT performed worst, as indicated by the smallest CCs and NSE, largest RMSEs and MAEs in almost all of months except for October. Moreover, compared to satellite-based data, the NGR-based rainfall data exhibited larger CC values in 6 months, smaller RMSE in 9 months and smaller MAE in 10 months, as well as larger NSE in 9 months. CCs from 3B42V7 in February, March, May, July and November were higher than those from NGR, whereas NGR performed better on RMSE, MAE and NSE in two of the five months. Overall, compared to these two satellite-based schemes, the estimates based on the proposed NGR framework exhibited the best performance with respect to the four statistical metrics in April, June, August and September of 2016 over SEC.

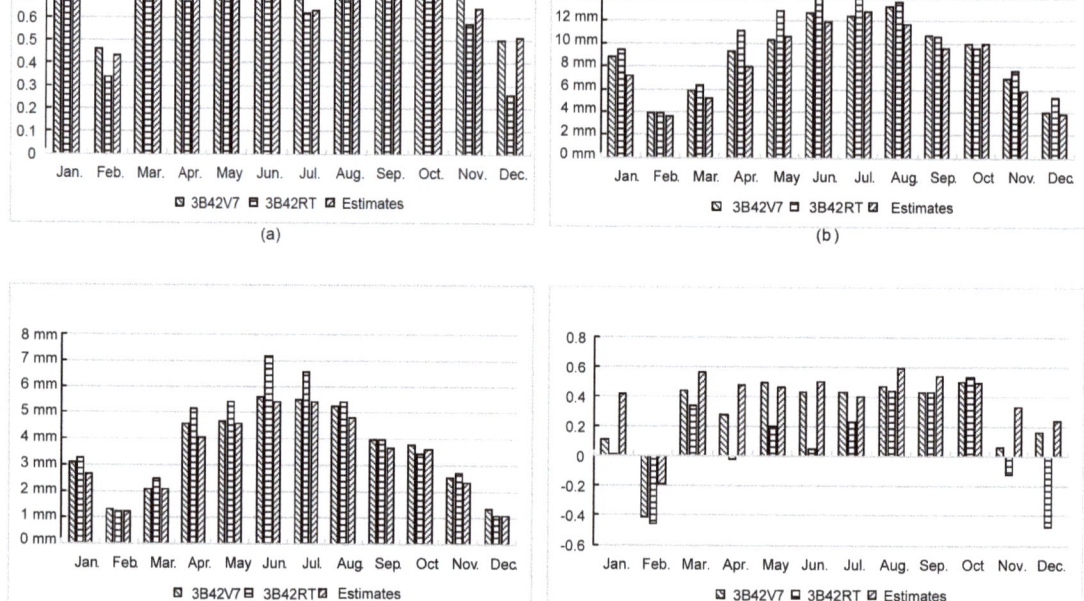

Figure 9. Bar graphs of daily metrics of (**a**) CC, (**b**) RMSE, (**c**) MAE and (**d**) NSE from the estimated and satellite-based precipitation datasets at validation sites in different months of 2016.

3.4. Assimilated Rainfall with Different Intensities

The metrics from 3B42V7, 3B42RT and NGR precipitation datasets with different rainfall intensities during 2016 are listed in Table 4. All the CCs were relatively small and mainly ranged from 0.124 to 0.295, except for those corresponding to rainstorm events, whereas the CC from NGR was the largest in each category. In terms of errors, both RMSE and MAE increased with the rainfall intensities, indicating that as the rainfall amounts increased, the inaccuracy of estimated rainfall datasets was enlarged, even though, when the rainfall intensity is light rain, moderate rain, as well as heavy rain, NGR yielded estimates with smaller RMSE and MAE than the other two satellite products. As for NSE, all the values were negative, but compared to those from 3B42V7 and 3B42RT, the NSE values from NGR were the largest with rainfall intensities of light rain, moderate rain and heavy rain, indicating that the estimated data can simulate the gauge observations better when rainfall intensity was less than 50 mm/day. The metrics, especially RMSE and MAE from rainstorm events, were quite large, and the root relative mean squared errors (RRMSE) from 3B42V7, 3B42RT and NGR rainfall datasets were more than 50%. According to Chen and Li [58], the monthly satellite-based datasets were unreliable if the RRMSE was more than 50%. Thus, all three products cannot precisely estimate the large precipitation amounts, especially under the circumstances that the rainfall is more than 50 mm.

Table 4. Statistical metrics for daily precipitation with various rainfall intensities at 30 validation sites in 2016.

Classification of Rainfall Intensities	Products	CC	RMSE (mm)	MAE (mm)	NSE
Light rain	3B42V7	0.284	8.75	4.61	−9.86
	3B42RT	0.263	9.99	5.01	−13.16
	Estimates	**0.295**	**6.77**	**4.01**	**−5.45**
Moderate rain	3B42V7	0.161	17.01	13.00	−14.41
	3B42RT	0.124	20.27	14.45	−20.90
	Estimates	**0.163**	**12.63**	**10.28**	**−7.49**
Heavy rain	3B42V7	0.148	24.09	19.79	−11.95
	3B42RT	0.150	27.42	22.23	−15.79
	Estimates	**0.152**	**21.47**	**18.77**	**−9.29**
Rainstorm	3B42V7	0.541	**44.88**	**34.89**	**−0.33**
	3B42RT	0.501	47.05	37.38	−0.46
	Estimates	**0.600**	53.11	43.88	−0.86

Note: The numbers in bold indicate the optimum values for the indices.

4. Discussion

4.1. Comparison with the Blended Rainfall Data Obtained by MLR and ANN

The assimilated precipitation data from the multiple linear regression (MLR) method and PERSIANN product was adopted for comparison to the proposed approach. Table 5 summarizes the daily statistical metrics in the rainy season (from June to September) of assimilated precipitation computed by the NGR, MLR and ANN approaches at 30 validation sites of SEC in 2016. For the daily statistical metrics in the rainy season, compared with those of satellite-based and MLR methods, as well as PERSIANN rainfall data, the performance of NGR was better in terms of CC, RMSE and NSE, with values of 0.715, 11.54 and 0.51 mm respectively, and marginally larger MAE (4.83 versus 4.76 mm from the MLR method). MLR estimates are better than the PERSIANN products, as indicated by the indicators in Table 5. The daily KGE of rainfall estimates from four methods against gauge-based observations in the rainy season are shown in Figure 10. Positive KGE values can be observed from MLR and NGR, indicating that MLR and NGR rainfall data in the rainy season can simulate the gauge-based rainfall well. Furthermore, KGE values from NGR at 18 validation sites were larger than those from MLR at the same sites, which means that NGR can achieve better results at these stations compared to MLR. However, negative KGE values (one from 3B42V7 and three from 3B43RT) and fluctuant variation of KGE of the two satellite products were observed, indicating worse consistency compared to the estimates from the proposed NGR framework. Figure 11 shows the spatial distribution of absolute deviations of mean daily rainfall estimates from MLR and NGR against gauge-based observations. Obviously, in comparison to NGR, MLR tended to underestimate or overestimate the mean rainfall amount in the rainy season at some validation sites, especially at Sichuan and Hainan provinces. In addition, the mean value (0.91 mm) of the total absolute deviation from MLR was larger than that (0.80 mm) of NGR, indicating that NGR can reduce errors more effectively than the MLR method in the rainy season.

Table 5. Daily metrics of assimilated precipitation obtained by the proposed framework, MLR and ANN at validation sites in rainy season over SEC.

Products	CC	RMSE (mm)	MAE (mm)	NSE
Estimates	**0.715**	**11.54**	4.83	**0.51**
MLR	0.701	11.79	**4.76**	0.49
3B42V7	0.700	12.31	5.08	0.44
3B42RT	0.673	13.94	5.78	0.29
PERSIANN	0.571	13.73	5.49	0.31

Note: The bold text stands for the optimum values for the indices.

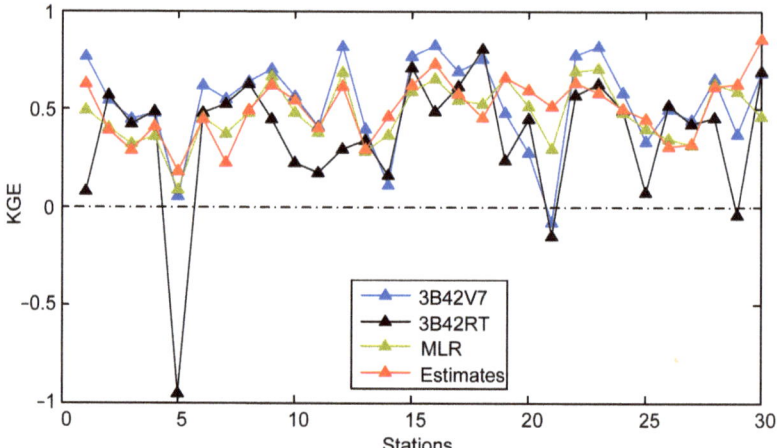

Figure 10. Daily KGE of rainfall estimates against gauge-based rainfall in the rainy season at each validation site in 2016.

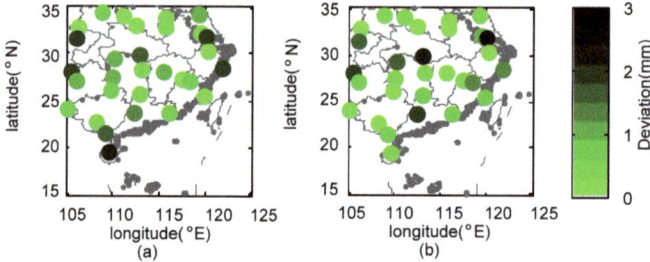

Figure 11. Deviation of mean daily rainfall amounts in the rainy season from (**a**) MLR data and (**b**) NGR data against gauge observations at each validation site in 2016.

4.2. Uncertainties, Strengths and Weaknesses

Uncertainty, as a factor that disturbs the accuracy of evaluation, should be considered. The uncertainty may be from several aspects. In this study, gauge data was used as a reference to verify the assimilated rainfall data. Nevertheless, gauge precipitation data also suffers from errors. Ye et al. [59] reported that the annual rainfall amount recorded by gauges over China was increased by 8 to 740 mm after bias corrections by considering wind-induced under-catch, wetting loss and light rain. Hence, these error-induced factors should be considered and eliminated as much as possible. Moreover, the scale discrepancy also introduces uncertainty. In order to transform the gridded satellite rainfall data into point-based data, the IDW method was employed during the training and validation process, which is likely to induce errors.

The modeling errors between the estimates and gauge-based rainfall data are assumed to follow a Gaussian distribution, which is suggested by the previous study [21]. For each data point, the obtained $\sigma_{2,m}^2$ in Equation (3) represents the variance of the modeling error. Then, the confidence interval (CI) of the estimated value of a data point can be directly acquired with the assumption of Gaussian residuals. Figure 12 shows the percentages of gauge-based data falling in different confidence intervals of estimates based on the nonparametric framework under light rain, moderate rain, heavy rain and rainstorm scenarios. The proposed model can provide accurate quantifications of the uncertainties for the large confidence intervals (CI) under the light, moderate and heavy rain scenarios. Specifically, the percentage corresponding to 95% CI is the largest one (in Figure 12a) among the three, indicating that most of the gauge-based rainfall data falls within 95% CI during light rain scenarios. That is, estimates from NGR during light rain are the most accurate, followed by the ones during moderate rain, heavy rain and rainstorms.

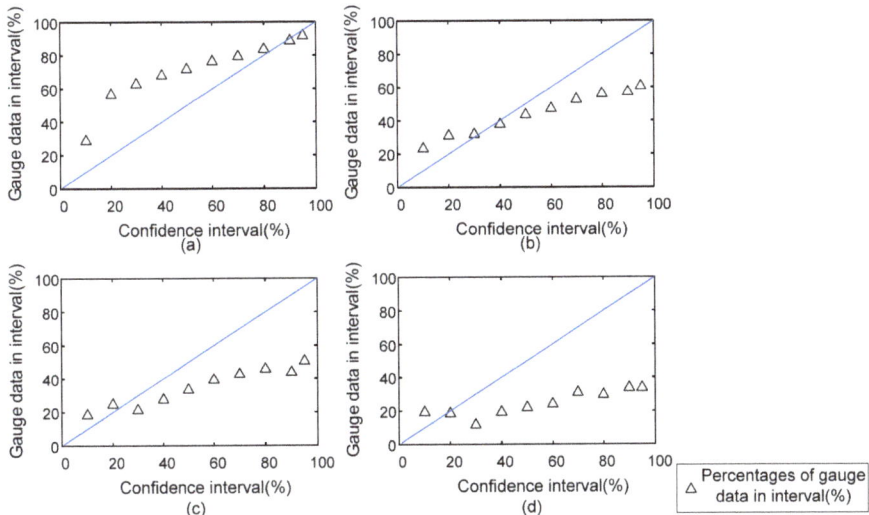

Figure 12. The percentages of gauge-based data falling in different confidence intervals of estimates during (**a**) light rain, (**b**) moderate rain, (**c**) heavy rain and (**d**) rainstorm events.

Although uncertainties were inevitable, the estimated NGR rainfall data were substantially improved upon almost all of the statistical indicators, except for the similar daily CCs in Meiyu and Typhoon seasons (in Figures 5 and 8). According to the aforementioned comparisons, the 3B42V7 data, in general, performed better than 3B42RT data at 30 validation sites across SEC in 2016. Figure 13 plots daily assimilated and satellite-based rainfall data in Meiyu and Typhoon seasons at 30 validation sites. The CCs between the estimates and the satellite-based data were calculated and marked in the sub-figures. The CC between NGR and 3B42V7 daily rainfall data was larger than that between NGR and 3B42RT daily rainfall data, indicating that the 3B42V7 dataset, as one of the data sources, contributed more to the NGR rainfall data than those from 3B42RT. In addition, because of the relatively worse performance of 3B42RT on statistical indexes, less information from the 3B42RT dataset and more details from the 3B42V7 dataset were retained by NGR during the process of framework construction. Thus, although similar CC values were observed between the NGR and 3B42V7 rainfall data in Meiyu and Typhoon seasons, the NGR framework is capable of automatically selecting the original satellite-based dataset with better performance. Moreover, this proposed NGR framework can not only be used in SEC, but also in other places where the derived satellite-based rainfall data is available. Nevertheless, the performance of this proposed framework applied in other regions, especially the data-gap areas, still needs to be evaluated.

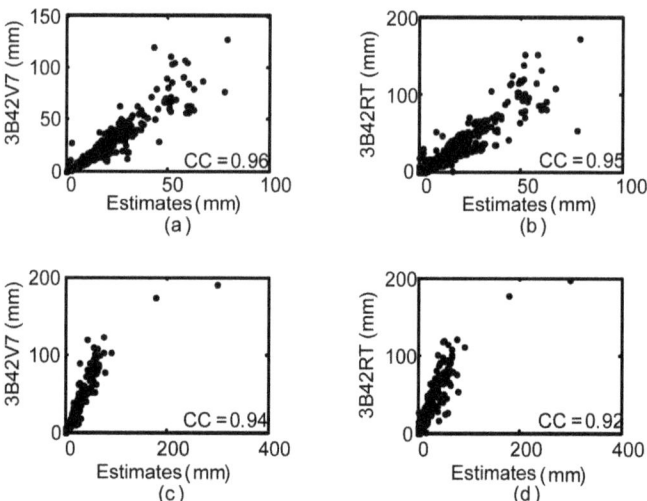

Figure 13. Scatterplots of daily precipitation from (**a**) the estimates versus 3B42V7, (**b**) the estimates versus 3B42RT in Meiyu season, (**c**) the estimates versus 3B42V7 and (**d**) the estimates versus 3B42RT in Typhoon season at 30 validation sites in 2016.

The proposed framework also has its limitations. As listed in Table 4, the statistical indictors of RMSE and MAE became more and more fluctuant as the rainfall intensity increased, especially for rainstorm events. NGR cannot precisely estimate the large precipitation amounts based solely on two satellite-based rainfall data as merged sources, as indicated by Figure 12. The uncertainty of assimilated precipitation data using NGR originated from the satellite-based datasets, i.e., 3B42V7 and 3B42RT, whose RRMSEs were both more than 50% during rainstorm events. Thus, to improve the performance of merged data during rainstorm events, higher quality of remote sensing rainfall data needs to be utilized as the blended sources.

5. Conclusions

In this study, a new framework was proposed to assimilate multi-source precipitation datasets in regions of SEC based on nonparametric general regression. The daily training datasets, including 3B42V7, 3B42RT and gauge-based data, corresponding to 300 training sites in 2016, were adopted to train the NGR framework. The gauge-based data at 30 validation sites was used to assess the trained framework. To evaluate the applicability of the framework, the rainfall in Meiyu and Typhoon seasons, in different months and rainfall events with different rainfall intensities, were included. Based on the study, the major findings were summarized as follows:

(1) During Meiyu season, the proposed framework in general outperformed 3B42V7 and 3B42RT on the mean value of the total absolute deviation, with a value of 1.17 mm. NGR exhibited the largest CC values at 40% of validation sites and the minimum RMSE at 19 out of 30 validation sites. For NSE, the estimates from NGR can match the gauge observations much better at 28 validation sites.

(2) During Typhoon season, the total absolute deviation from NGR was smaller than those from satellite-based schemes. Except for similar CC over SEC, NGR exhibited smaller RMSE and MAE, as well as larger NSE at most of the validation sites.

(3) At a monthly scale, NGR performed better on CC in 6 months, RMSE in 9 months and MAE in 10 months, as well as NSE in 9 months. Compared with 3B42V7 and 3B42RT, NGR yielded estimates with larger CC, smaller RMSE and MAE, as well as larger NSE, when the rainfall intensity was less than 50 mm/day.

(4) The 3B42V7 data, in general, performed better than 3B42RT data at 30 validation sites across SEC in 2016, which contributed more to the assimilated rainfall data than those from 3B42RT. The NGR framework is capable of automatically selecting the original satellite-based dataset with better performance.

Author Contributions: Conceptualization, L.G.; methodology, Y.Z.; formal analysis, H.S. and N.Q.; data curation, Q.T. and Y.Z.; writing—original draft preparation, Y.Z.; writing—review and editing, L.G., H.S., Q.T., and N.Q.; supervision, L.G.; project administration, L.G. All authors have read and agreed to the published version of the manuscript.

Funding: This paper is financially supported by the Science and Technology Development Fund, Macau SAR (File no.: SKL-IOTSC-2021-2023, 0030/2020/A1, and 0021/2020/ASC), UM Research Grant (File no.: SRG2019-00193-IOTSC, SRG2020-00020-IOTSC, and MYRG2020-00072-IOTSC), Guangdong–Hong Kong-Macau Joint Laboratory Program (Project No.: 2020B1212030009), National Natural Science Foundation of China (41730645), and CORE (EF017/IOTSC-GL/2020/HKUST). CORE is a joint research center for ocean research between QNLM and HKUST.

Institutional Review Board Statement: Not applicable.

Informed Consent Statement: Not applicable.

Data Availability Statement: The data in this study are subject to third party restrictions. The data that support the findings of this study are available from the National Climate Centre in Beijing, China. Restrictions apply to the availability of these data, which were used under license for this study. Data are available at https://data.cma.cn/en, accessed on 11 January 2020, with the permission of the National Climate Centre in Beijing, China.

Acknowledgments: The authors would like to thank the National Climate Centre in Beijing, China, for providing climate data described in this paper.

Conflicts of Interest: The authors declare no conflict of interest.

References

1. Wu, Y.; Chen, J. Analyzing the Water Budget and Hydrological Characteristics and Responses to Land Use in a Monsoonal Climate River Basin in South China. *Environ. Manag.* **2013**, *51*, 1174–1186. [CrossRef] [PubMed]
2. Ma, Y.; Tang, G.; Long, D.; Yong, B.; Zhong, L.; Wan, W.; Hong, Y. Similarity and Error Intercomparison of the GPM and Its Predecessor-TRMM Multisatellite Precipitation Analysis Using the Best Available Hourly Gauge Network over the Tibetan Plateau. *Remote. Sens.* **2016**, *8*, 569. [CrossRef]
3. Gao, L.; Zhang, L.; Lu, M. Characterizing the spatial variations and correlations of large rainstorms for landslide study. *Hydrol. Earth Syst. Sci.* **2017**, *21*, 4573–4589. [CrossRef]
4. Gao, L.; Zhang, L.M.; Cheung, R.W.M. Relationships between natural terrain landslide magnitudes and triggering rainfall based on a large landslide inventory in Hong Kong. *Landslides* **2018**, *15*, 727–740. [CrossRef]
5. Gao, L.; Zhang, L.; Li, X.; Zhou, S. Evaluating Metropolitan Flood Coping Capabilities under Heavy Storms. *J. Hydrol. Eng.* **2019**, *24*, 05019011. [CrossRef]
6. Luo, P.; Sun, Y.; Wang, S.; Wang, S.; Lyu, J.; Zhou, M.; Nakagami, K.; Takara, K.; Nover, D. Historical assessment and future sustainability challenges of Egyptian water resources management. *J. Clean. Prod.* **2020**, *263*, 121154. [CrossRef]
7. Zhu, Y.; Luo, P.; Zhang, S.; Sun, B. Spatiotemporal Analysis of Hydrological Variations and Their Impacts on Vegetation in Semiarid Areas from Multiple Satellite Data. *Remote. Sens.* **2020**, *12*, 4177. [CrossRef]
8. Su, F.; Hong, Y.; Lettenmaier, D.P. Evaluation of TRMM Multisatellite Precipitation Analysis (TMPA) and Its Utility in Hydrologic Prediction in the La Plata Basin. *J. Hydrometeorol.* **2008**, *9*, 622–640. [CrossRef]
9. Lee, T.; Ouarda, T.B.M.J. Long-term prediction of precipitation and hydrologic extremes with nonstationary oscillation processes. *J. Geophys. Res. Athmos.* **2010**, *115*, D13. [CrossRef]
10. Yong, B.; Hong, Y.; Ren, L.-L.; Gourley, J.J.; Huffman, G.J.; Chen, X.; Wang, W.; Khan, S.I. Assessment of evolving TRMM-based multisatellite real-time precipitation estimation methods and their impacts on hydrologic prediction in a high latitude basin. *J. Geophys. Res. Atmos.* **2012**, *117*, D9. [CrossRef]
11. Mu, D.; Luo, P.; Lyu, J.; Zhou, M.; Huo, A.; Duan, W.; Nover, D.; He, B.; Zhao, X. Impact of temporal rainfall patterns on flash floods in Hue City, Vietnam. *J. Flood Risk Manag.* **2020**, e12668. [CrossRef]
12. Zhu, H.; Li, Y.; Huang, Y.; Li, Y.; Hou, C.; Shi, X. Evaluation and hydrological application of satellite-based precipitation datasets in driving hydrological models over the Huifa river basin in Northeast China. *Atmos. Res.* **2018**, *207*, 28–41. [CrossRef]
13. Trinh-Tuan, L.; Matsumoto, J.; Ngo-Duc, T.; Nodzu, M.I.; Inoue, T. Evaluation of satellite precipitation products over Central Vietnam. *Prog. Earth Planet. Sci.* **2019**, *6*, 54. [CrossRef]

14. Ashouri, H.; Hsu, K.-L.; Sorooshian, S.; Braithwaite, D.K.; Knapp, K.R.; Cecil, L.D.; Nelson, B.R.; Prat, O.P. PERSIANN-CDR: Daily Precipitation Climate Data Record from Multisatellite Observations for Hydrological and Climate Studies. *Bull. Am. Meteorol. Soc.* **2015**, *96*, 69–83. [CrossRef]
15. Joyce, R.J.; Janowiak, J.E.; Arkin, P.A.; Xie, P. CMORPH: A method that produces global precipitation estimates from passive microwave and infrared data at high spatial and temporal resolution. *J. Hydrometeorol.* **2004**, *5*, 487–503. [CrossRef]
16. Hou, A.Y.; Kakar, R.K.; Neeck, S.; Azarbarzin, A.A.; Kummerow, C.D.; Kojima, M.; Oki, R.; Nakamura, K.; Iguchi, T. The Global Precipitation Measurement Mission. *Bull. Am. Meteorol. Soc.* **2014**, *95*, 701–722. [CrossRef]
17. Huffman, G.J.; Adler, R.F.; Bolvin, D.T.; Nelkin, E.J. *The TRMM Multi-Satellite Precipitation Analysis (TMPA) in Satellite Rainfall Applications for Surface Hydrology*; Springer: Dordrecht, The Netherlands, 2010; pp. 3–22. ISBN 978-90-481-2914-0.
18. Wu, L.; Xu, Y.; Wang, S. Comparison of TMPA-3B42RT Legacy Product and the Equivalent IMERG Products over Mainland China. *Remote Sens.* **2018**, *10*, 1778. [CrossRef]
19. Cao, Y.; Zhang, W.; Wang, W. Evaluation of TRMM 3B43 data over the Yangtze River Delta of China. *Sci. Rep.* **2018**, *8*, 1–12. [CrossRef]
20. Guo, H.; Chen, S.; Bao, A.; Behrangi, A.; Hong, Y.; Ndayisaba, F.; Hu, J.; Stepanian, P.M. Early assessment of Integrated Multi-satellite Retrievals for Global Precipitation Measurement over China. *Atmos. Res.* **2016**, *176*, 121–133. [CrossRef]
21. Wang, Y.; Chen, J.; Yang, D. Bayesian Assimilation of Multiscale Precipitation Data and Sparse Ground Gauge Observations in Mountainous Areas. *J. Hydrometeorol.* **2019**, *20*, 1473–1494. [CrossRef]
22. Bhuiyan, M.A.E.; Yang, F.; Biswas, N.K.; Rahat, S.H.; Neelam, T.J. Machine Learning-Based Error Modeling to Improve GPM IMERG Precipitation Product over the Brahmaputra River Basin. *Forecast* **2020**, *2*, 14. [CrossRef]
23. Tang, G.; Zeng, Z.; Ma, M.; Liu, R.; Wen, Y.; Hong, Y. Can Near-Real-Time Satellite Precipitation Products Capture Rainstorms and Guide Flood Warning for the 2016 Summer in South China? *IEEE Geosci. Remote Sens. Lett.* **2017**, *14*, 1208–1212. [CrossRef]
24. Chao, L.; Zhang, K.; Li, Z.; Zhu, Y.; Wang, J.; Yu, Z. Geographically weighted regression based methods for merging satellite and gauge precipitation. *J. Hydrol.* **2018**, *558*, 275–289. [CrossRef]
25. Tong, K.; Su, F.; Yang, D.; Hao, Z. Evaluation of satellite precipitation retrievals and their potential utilities in hydrologic modeling over the Tibetan Plateau. *J. Hydrol.* **2014**, *519*, 423–437. [CrossRef]
26. Zhang, L.; Li, X.; Zheng, D.; Zhang, K.; Ma, Q.; Zhao, Y.; Ge, Y. Merging multiple satellite-based precipitation products and gauge observations using a novel double machine learning approach. *J. Hydrol.* **2021**, *594*, 125969. [CrossRef]
27. Ma, Y.; Sun, X.; Chen, H.; Hong, Y.; Zhang, Y. A two-stage blending approach for merging multiple satellite precipitation estimates and rain gauge observations: An experiment in the northeastern Tibetan Plateau. *Hydrol. Earth Syst. Sci.* **2021**, *25*, 359–374. [CrossRef]
28. Bhuiyan, M.A.E.; Nikolopoulos, E.I.; Anagnostou, E.N. Machine Learning—Based Blending of Satellite and Reanalysis Precipitation Datasets: A Multiregional Tropical Complex Terrain Evaluation. *J. Hydrometeorol.* **2019**, *20*, 2147–2161. [CrossRef]
29. Yin, J.; Guo, S.; Gu, L.; Zeng, Z.; Liu, D.; Chen, J.; Shen, Y.; Xu, C.-Y. Blending multi-satellite, atmospheric reanalysis and gauge precipitation products to facilitate hydrological modelling. *J. Hydrol.* **2021**, *593*, 125878. [CrossRef]
30. Chen, S.; Xiong, L.; Ma, Q.; Kim, J.-S.; Chen, J.; Xu, C.-Y. Improving daily spatial precipitation estimates by merging gauge observation with multiple satellite-based precipitation products based on the geographically weighted ridge regression method. *J. Hydrol.* **2020**, *589*, 125156. [CrossRef]
31. Metered, H.; Bonello, P.; Oyadiji, S. Nonparametric Identification Modeling of Magnetorheological Damper Using Chebyshev Polynomials Fits. *SAE Int. J. Passeng. Cars Mech. Syst.* **2009**, *2*, 1125–1135. [CrossRef]
32. Kuok, S.C.; Yuen, K.V. Broad learning for nonparametric spatial modeling with application to seismic attenuation. *Comput. Aided Civ. Infrastruct. Eng.* **2020**, *35*, 203–218. [CrossRef]
33. Fan, J.; Huang, L.-S. Goodness-of-Fit Tests for Parametric Regression Models. *J. Am. Stat. Assoc.* **2001**, *96*, 640–652. [CrossRef]
34. Hill, J.L. Bayesian Nonparametric Modeling for Causal Inference. *J. Comput. Graph. Stat.* **2011**, *20*, 217–240. [CrossRef]
35. Bhuiyan, M.A.E.; Nikolopoulos, E.I.; Anagnostou, E.N.; Quintana-Seguí, P.; Barella-Ortiz, A. A nonparametric statistical technique for combining global precipitation datasets: Development and hydrological evaluation over the Iberian Peninsula. *Hydrol. Earth Syst. Sci.* **2018**, *22*, 1371–1389. [CrossRef]
36. Ma, Y.; Hong, Y.; Chen, Y.; Yang, Y.; Tang, G.; Yao, Y.; Long, D.; Li, C.; Han, Z.; Liu, R. Performance of Optimally Merged Multisatellite Precipitation Products Using the Dynamic Bayesian Model Averaging Scheme Over the Tibetan Plateau. *J. Geophys. Res. Atmos.* **2018**, *123*, 814–834. [CrossRef]
37. Matsoukas, C.; Islam, S.; Kothari, R. Fusion of radar and rain gage measurements for an accurate estimation of rainfall. *J. Geophys. Res. Atmos.* **1999**, *104*, 31437–31450. [CrossRef]
38. Xu, G.; Wang, Z.; Xia, T. Mapping Areal Precipitation with Fusion Data by ANN Machine Learning in Sparse Gauged Region. *Appl. Sci.* **2019**, *9*, 2294. [CrossRef]
39. Wehbe, Y.; Temimi, M.; Adler, R.F. Enhancing Precipitation Estimates Through the Fusion of Weather Radar, Satellite Retrievals, and Surface Parameters. *Remote. Sens.* **2020**, *12*, 1342. [CrossRef]
40. Specht, D.F. A general regression neural network. *IEEE Trans. Neural Netw.* **1991**, *2*, 568–576. [CrossRef] [PubMed]
41. Yuen, K.-V.; Ortiz, G.A. Bayesian Nonparametric General Regression. *Int. J. Uncertain. Quantif.* **2016**, *6*, 195–213. [CrossRef]
42. Chen, W.; Jiang, Z.; Li, L.; Yiou, P. Simulation of regional climate change under the IPCC A2 scenario in southeast China. *Clim. Dyn.* **2011**, *36*, 491–507. [CrossRef]

43. Gao, X.; Shi, Y.; Song, R.; Giorgi, F.; Wang, Y.; Zhang, D. Reduction of future monsoon precipitation over China: Comparison between a high resolution RCM simulation and the driving GCM. *Meteorol. Atmos. Phys.* **2008**, *100*, 73–86. [CrossRef]
44. Zheng, J.; Han, W.; Jiang, B.; Ma, W.; Zhang, Y. Infectious Diseases and Tropical Cyclones in Southeast China. *Int. J. Environ. Res. Public Health* **2017**, *14*, 494. [CrossRef]
45. Wu, Y.; Chen, J. Investigating the effects of point source and nonpoint source pollution on the water quality of the East River (Dongjiang) in South China. *Ecol. Indic.* **2013**, *32*, 294–304. [CrossRef]
46. Yang, L.; Scheffran, J.; Qin, H.; You, Q. Climate-related flood risks and urban responses in the Pearl River Delta, China. *Reg. Environ. Chang.* **2015**, *15*, 379–391. [CrossRef]
47. Zhao, X.; Niu, R. Similarities and differences of summer persistent heavy rainfall and atmospheric circulation characteristics in the middle and lower reaches of the Yangtze River between 2016 and 1998. *Torrential Rain Disasters* **2019**, *38*, 615–623. [CrossRef]
48. Luo, P.; Mu, D.; Xue, H.; Ngo-Duc, T.; Dang-Dinh, K.; Takara, K.; Nover, D.; Schladow, G. Flood inundation assessment for the Hanoi Central Area, Vietnam under historical and extreme rainfall conditions. *Sci. Rep.* **2018**, *8*, 1–11. [CrossRef]
49. Huffman, G.J.; Bolvin, D.T.; Nelkin, E.J.; Wolff, D.B.; Adler, R.F.; Gu, G.; Hong, Y.; Bowman, K.P.; Stocker, E.F. The TRMM Multisatellite Precipitation Analysis (TMPA): Quasi-Global, Multiyear, Combined-Sensor Precipitation Estimates at Fine Scales. *J. Hydrometeorol.* **2007**, *8*, 38–55. [CrossRef]
50. Fotheringham Stewart, A.; Brunsdon, C.; Charlton, M. *GWR and Spatial Autocorrelation in Geographically Weighted Regression: The Analysis of Spatially Varying Relationships*; John Wiley: New York, NY, USA, 2002; pp. 103–124. ISBN 0-471-49616-2.
51. Giarno, G.; Hadi, M.P.; Suprayogi, S.; Murti, S.H. Suitable Proportion Sample of Holdout Validation for Spatial Rainfall Interpolation in Surrounding the Makassar Strait. *Forum Geogr.* **2020**, *33*, 219–232. [CrossRef]
52. Raftery, A.E.; Gneiting, T.; Balabdaoui, F.; Polakowski, M. Using Bayesian Model Averaging to Calibrate Forecast Ensembles. *Mon. Weather Rev.* **2005**, *133*, 1155–1174. [CrossRef]
53. Gupta, H.V.; Kling, H.; Yilmaz, K.K.; Martinez, G.F. Decomposition of the mean squared error and NSE performance criteria: Implications for improving hydrological modelling. *J. Hydrol.* **2009**, *377*, 80–91. [CrossRef]
54. Knoben, W.J.M.; Freer, J.E.; Woods, R.A. Technical note: Inherent benchmark or not? Comparing Nash-Sutcliffe and Kling-Gupta efficiency scores. *Hydrol. Earth Syst. Sci.* **2019**, *23*, 4323–4331. [CrossRef]
55. Castaneda-Gonzalez, M.; Poulin, A.; Romero-Lopez, R.; Arsenault, R.; Brissette, F.; Chaumont, D.; Paquin, D. Impacts of Regional Climate Model Spatial Resolution on Summer Flood Simulation. *EPiC Ser. Eng.* **2018**, *3*, 372–380. [CrossRef]
56. Andersson, J.C.; Arheimer, B.; Traoré, F.; Gustafsson, D.; Ali, A. Process refinements improve a hydrological model concept applied to the Niger River basin. *Hydrol. Process.* **2017**, *31*, 4540–4554. [CrossRef]
57. Weisberg, S. *Simple Linear Regression in Applied Linear Regression*; John Wiley & Sons: Hoboken, NJ, USA, 2005; pp. 19–33. ISBN 0-471-66379-4.
58. Chen, F.; Li, X. Evaluation of IMERG and TRMM 3B43 Monthly Precipitation Products over Mainland China. *Remote Sens.* **2016**, *8*, 472. [CrossRef]
59. Ye, B.; Yang, D.; Ding, Y.; Han, T.; Koike, T. A Bias-Corrected Precipitation Climatology for China. *J. Hydrometeorol.* **2004**, *5*, 1147–1160. [CrossRef]

Article

Comparison of the Hydrological Dynamics of Poyang Lake in the Wet and Dry Seasons

Fangdi Sun [1], Ronghua Ma [2], Caixia Liu [3,*] and Bin He [4,5]

1. School of Geography and Remote Sensing, Guangzhou University, Guangzhou 510006, China; sfd_geo@gzhu.edu.cn
2. Key Laboratory of Watershed Geographic Sciences, Nanjing Institute of Geography and Limnology, Chinese Academy of Sciences, Nanjing 210008, China; rhma@niglas.ac.cn
3. State Key Laboratory of Remote Sensing Science, Aerospace Information Research Institute, Chinese Academy of Sciences, Beijing 100101, China
4. Guangdong Key Laboratory of Integrated Agro-Environmental Pollution Control and Management, Institute of Eco-Environmental and Soil Sciences, Guangdong Academy of Sciences, Guangzhou 510650, China; bhe@soil.gd.cn
5. National-Regional Joint Engineering Research Center for Soil Pollution Control and Remediation in South China, Guangzhou 510650, China
* Correspondence: liucx@radi.ac.cn

Citation: Sun, F.; Ma, R.; Liu, C.; He, B. Comparison of the Hydrological Dynamics of Poyang Lake in the Wet and Dry Seasons. *Remote Sens.* 2021, *13*, 985. https://doi.org/10.3390/rs13050985

Academic Editor: Igor Ogashawara

Received: 31 January 2021
Accepted: 26 February 2021
Published: 5 March 2021

Publisher's Note: MDPI stays neutral with regard to jurisdictional claims in published maps and institutional affiliations.

Copyright: © 2021 by the authors. Licensee MDPI, Basel, Switzerland. This article is an open access article distributed under the terms and conditions of the Creative Commons Attribution (CC BY) license (https://creativecommons.org/licenses/by/4.0/).

Abstract: Poyang Lake is the largest freshwater lake connecting the Yangtze River in China. It undergoes dramatic dynamics from the wet to the dry seasons. A comparison of the hydrological changes between the wet and dry seasons may be useful for understanding the water flows between Poyang Lake and Yangtze River or the river system in the watershed. Gauged measurements and remote sensing datasets were combined to reveal lake area, level and volume changes during 2000–2020, and water exchanges between Poyang Lake and Yangtze River were presented based on the water balance equation. The results showed that in the wet seasons, the lake was usually around 1301.85–3840.24 km^2, with an average value of 2800.79 km^2. In the dry seasons, the area was around 618.82–2498.70 km^2, with an average value of 1242.03 km^2. The inundations in the wet seasons were approximately quadruple those in the dry seasons. In summer months, the lake surface tended to be flat, while in winter months, it was inclined, with the angles at around 10″–16″. The mean water levels of the wet and dry seasons were separately 13.51 m and 9.06 m, with respective deviations of around 0–2.38 m and 0.38–2.15 m. Monthly lake volume changes were about 7.5–22.64 km^3 and 1–5.80 km^3 in the wet and dry seasons, respectively. In the wet seasons, the overall contributions of ground runoff, precipitation on the lake surface and lake evaporation were less than the volume flowing into Yangtze River. In the dry seasons, the three contributions decreased by 50%, 50% and 65.75%, respectively. Therefore, lake storages presented a decrease (−7.42 km^3/yr) in the wet seasons and an increase (9.39 km^3/yr) in the dry seasons. The monthly exchanges between Poyang Lake and Yangtze River were at around −14.22–32.86 km^3. Water all flowed from the lake to the river in the wet seasons, and the chance of water flowing from Yangtze River in the dry seasons was only 5.26%.

Keywords: Poyang Lake; Yangtze River; hydrological changes; water balance

1. Introduction

As the largest freshwater lake in China [1], Poyang Lake has drawn more and more attention [2–5], especially after the implementation of the Three Gorges Dam (TGD), which is located upstream of Yangtze River and began to impound water in 2003 [6–8]. The inundation extent of Poyang Lake showed a declining trend of around 30.2 km^2/yr during 2000–2010 [9]. Some research pointed out that the discharge flowed from Poyang Lake into Yangtze River increased by 7.86 km^3 after the implementation of TGD [10]. Nearly 1/3 of the Nanjishan Wetland National Nature Reserve has transformed from water into vegetation area even in the wet seasons during 2000–2010 [11]. With the water level decreased, the western part of the lake region was changed to emerged land, and many kinds of vegetation began to grow. In 2016, a dam was proposed, which would be built on the northern end of the lake to keep Poyang Lake in a sustainable state by managing the river–lake water flow, and this proposal was finally rejected from the view of ecology. The deteriorating hydrological conditions of Poyang Lake will finally lead to a negative impact on the diversity of the aquatic vegetation and marsh wildlife. Revealing the hydrologic changes of Poyang Lake is very important to understand the water flows between Poyang Lake and Yangtze River or the river system in the watershed. Though there are several hydrological stations around Poyang Lake, there are some restrictions in terms of hydrological data sharing, especially in recent observations. In addition, hydrological stations tend to be decentralized and punctate and thus may not reflect the comprehensive and objective dynamics of the whole lake.

Remote sensing can be used to monitor lake hydrologic conditions and their changes [12–14]. Altimeter data have been widely used to continuously monitor the water level changes of large rivers, lakes, and flood plains [15,16]. Since the 1990s, 25 years of altimeter data have been collected, which cover the globe with the highest frequency of 10 days, such as the Topex/Poseidon (T/P), Jason-1, and Jason-2 datasets. At present, there are four kinds of water level databases for large rivers, lakes, and reservoirs derived from altimeter data in the world: the Database for Hydrological Time Series of Inland Waters (DAHITI) [17], Global Reservoir and Lake Monitor (GRLM) [18], River Lake Hydrology product (RLH) [19], and Hydroweb [20]. T/P data during 1992–2002 were used in the six largest lakes in China, and the derived water level changes, with the precipitation and south oscillation, were analyzed [21]. Zheng et al. (2016) used T/P and ENVIromental Satellite (ENVISat) data during 1992–2010 to monitor the water level changes of Hulun Lake in Northeast China and found that the lake presented a decreasing trend, with the rate of -0.36 m/yr, and climate warming was the main cause [22]. Ice, Cloud, and Land Elevation Satellite (ICESat) data during 2003–2009 were applied to 56 lakes in China, which showed that the surface level of the lakes in Inner Mongolia and Xinjiang presented a decreasing appearance, while the lakes in the eastern plain fluctuated [23]. In addition, T/P data during 1992–1999 were applied to rivers with a width of more than 1 km in the Amazon watershed [16]. Chipman and Lillesand (2007) revealed the shrinkage of the Toshak lakes in Southern Egypt based on the ICESat data [24]. Additionally, remotely sensed images are able to capture lake area fluctuations occurring in a short period and to reveal long-term changes. Feng et al. (2012) used MODerate-resolution Imaging Spectro-radiometer (MODIS) images to monitor dynamic changes of Poyang Lake during 2000–2010 and found that the lake was 714.1 km^2 in the dry season and 3162.9 km^2 in the wet season [9]. Sun et al. (2014) used MODIS images to study the inundation changes of more than 600 large lakes in China during 2000–2010 [25]. Multisource remote sensing images were employed to delineate the monthly spatial distribution of global land surface water bodies during 1993–2004 [26,27].

In order to quantify the water storage of a water body, bottom topography is necessary. The traditional method for obtaining a bathymetry map was to survey the depth of water using sonar sensors. However, this method consumed a lot of time, labor, and money [28]. The Airborn Lidar was also used to detect underwater topography near the ocean up to a depth of 40 m; although this technique was sensitive to water turbidity, surface waves, and sun glint, its maximum detectable depth was only 2–3 times of the Secchi depth [14,29]. Some researchers have studied volume changes of large rivers and lakes based on altimeter data and remote sensing images. Water mass changes of the Negro River basin were revealed by Synthetic Aperture Radar (SAR), T/P, and in situ water level observations [30]. The ICESat data and Landsat images were used to construct area–level curves for 30 lakes on the Tibetan Plateau in order to study their volume changes, and the result showed an increase of 92.43 km^3 in volume for the 30 lakes from the 1970s to 2011 [31]. Cai et al. (2016) constructed area–volume models for 128 lakes and 108 reservoirs in the Yangtze River watershed, according to gauged measurements and MODIS images. The research found that 53.91% of lakes were shrinking at a rate of 14×10^6 m^3/m, while reservoirs were expanding at a rate of 177×10^6 m^3/m [10]. Crétaux et al. (2005) used bottom topography and water levels derived from T/P altimeter data to construct the water volume changes of the Aral Sea [32]. Medina et al. (2010) applied gauged water level measurements, ENVISat and Advanced Synthetic Aperture Radar (ASAR) images to estimate the storage changes of Lake Izabal [33].

Based on the above researches, it is practical to describe the detailed hydrological changes of Poyang Lake. The aim of this research was to obtain the variations of hydrological aspects of Poyang Lake during 2000–2020. An accurate and automatic method of extracting water-land boundaries was used to accomplish high frequency mapping of Poyang Lake. Water level records were obtained based on gauged observations and DAHITI. Then variations of lake storages were calculated by combining the surface area and water level data. The water flows between the lake and Yangtze River were derived from the view of water balance. Finally, driving forces were analyzed to illustrate the quantitative contributions of inflow (ground runoff, precipitation on the lake surface) and outflow (lake evaporation and exchanges with Yangtze River).

2. Study Area

Poyang Lake is located in the south of the Yangtze River and it is the largest lake directly connected to the Yangtze River. Poyang Lake absorbs water from five tributaries (Ganjiang river, Fu river, Xinjiang river, Rao river, and Xiu river) and flows into the Yangtze River at Hukou connection in most of time. The geographical range of Poyang Lake is 28°11′N–29°51′N and 115°49′E–116°46′E. The lake spans around 173.0 km from north to south, and the average west–east width is around 16.9 km. The width of northern part of the lake is only 5–8 km due to the restriction of the neighboring mountains, while the southern part of the lake tends to be an open surface, with a width of up to 60 km, as shown in Figure 1. The watershed of Poyang Lake has an area of about 162068.68 km^2, which is nearly 9% of Yangtze River basin and 97% of Jiangxi Province [1].

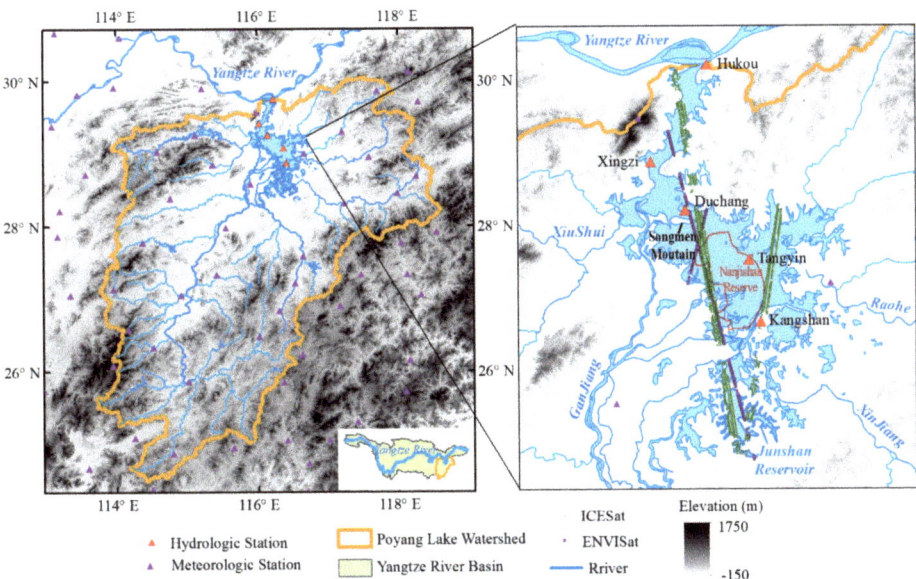

Figure 1. Spatial distribution of Poyang Lake and the feeding rivers in the watershed, and the paths of altimeter data on the lake.

The local climate is a subtropical monsoon climate. The local precipitation shows an obvious intra-annual variety and the annual average is around 1570 mm. Precipitation mainly occurs during April–June, accounting for about 45–50% of the annual rainfall. The annual average temperate is 16.5–17.8 °C. In summer, the temperature can reach 28.4–30.0 °C, while in winter, it is around 4.2–7.2 °C [1].

In the wet season (Jun–Sep), the lake surface usually presents a flat state, with the maximum inundation of more than 3000 km^2. Conversely, in the dry season (Nov–next Feb), with less rainfall and water flows from the south to Yangtze River, the corresponding water extent can shrink to less than 1000 km^2, showing a narrow and inclined state. The drop of the water level at Hukou Station can reach 3 m from summer to winter. The average water flow from Poyang Lake to Yangtze River is 1436.0 × 10^8 m^3 each year, accounting for about 15.5% of the annual Yangtze River discharge [11].

The seasonal changes of water level and inundation were favorable for Poyang Lake to create habitats for rich biodiversity/diversity of life. The famous Nanjishan Reserve is located in the main body of the lake. In hot summers, subtropical vegetation prospers and in cool and wet winters, temperate vegetation is productive [34]. In addition, over 98% of the population of the endangered Siberian crane, *Leucogeranus*, gathers in this reserve in winter [35].

3. Data and Methods
3.1. Data
3.1.1. Hydrological Data

Daily measurements of the flow rate of the five feeding rivers during 2001–2006 were obtained to calculate the total ground runoff flowing into Poyang Lake. The daily gauged water level at five hydrologic stations during 2001–2013 were used to present the fluctuation of the lake.

3.1.2. DAHITI

DAHITI is a database, presenting the time-series water level of 457 global lakes/reservoirs and rivers. DAHITI combines T/P, Jason-1, Jason-2, European Remote Sensing Satellites (ERS)-2, ENVISat, and Satellite with ARgos and ALtika (SARAL) altimetry data [17]. Compared with gauged measurements, the accuracies for lakes/reservoirs and rivers are 4–36 cm and 8–114 cm, respectively. DAHITI was used to analyze the fluctuation of Poyang Lake during 2000–2017.

3.1.3. MODIS Images

The 8-day level-3 composited product, MOD09A, with a 500 m resolution, available in the Earth Observing System (https://reverb.echo.nasa.gov/reverb/, accessed on 5 March 2021), was able to capture short-term and rapid fluctuations of inundations. MODIS images in the wet and dry seasons for each year from 2000–2020 were selected. Some images showed that Poyang Lake was covered by thick clouds, especially in the rainy seasons, and the lake could not be recognized correctly. In order to accurately depict the changes, these kinds of images were discarded, and 349 scenes were finally used in this research.

3.1.4. Meteorological Data

The daily gauged precipitation, evaporation, and temperature data of the stations from 2000–2010 were obtained from the China Meteorological Data Sharing Service System (http://cdc.cma.gov.cn/, accessed on 5 March 2021). The precipitations of the whole watershed were estimated by Kriging interpolation of the measured data, based on 66 meteorological stations, shown in Figure 1. The research assessed Kriging interpolated results of rainfall data in both wet and dry months in Lijiang River basin, and obtained that the average accuracy was 94.74% [36]. The land evaporation in the basin was calculated based on gauged observations, and the lake surface evaporation was estimated from the nearest in situ stations, according to the Penman–Monteith equation [37]. Penman–Monteith model was evaluated by gauged data in Taihu Lake and the accuracy was 93.50% [38].

3.2. Method

3.2.1. Inundation Extraction

An accurate water–land discrimination method was applied to delineate lake surface dynamics between 2000 and 2020. The method used the automatic selection of training data and Support Vector Machine (SVM) classifier. First, the classification system, including the water body, bare soil (including urban area), vegetation, ice, snow, and clouds was determined. Second, the training data of each image were collected based on six rules, considering the spectral characteristics of each class. Then, k-means and automated water extraction index (AWEI) were integrated and iterated to remove noise from the training samples. Finally, the filtered training data and SVM were combined to extract water bodies. The procedure can be implemented on long series of images automatically. The details of this method are illustrated in the literature [39]. This method has been used for the surface extraction of several major lakes on the Tibetan Plateau and Aral Sea, and the omission errors and commission errors were 0.9–1.5% and 2.94–4.23%, respectively [40,41].

Compared with several water indexes, such as the normalized differenced water index (NDWI), modified NDWI (MNDWI), and AWEI, this method has a high robustness. Water indexes need suitable thresholds, which depend on imaging environments, such as aerosol interference and viewing geometry. However, the proposed method was solely based on the spectrum presentations of the pixels of each image. The automatic selection of training data and the filtration of noise through iterated clustering can help obtain a high-accuracy water extraction, without manual intervention.

3.2.2. Water Level

The water level of Poyang Lake during 2000–2020 consisted of three kinds of data sources. The first part is gauged observations of five hydrological stations, which were taken daily from 2001 to 2013. The second part is DAHITI records from 2001 to 2017, and the third part is the left lake levels derived from the level–area relationship to match the length of the lake area data.

The five hydrological stations located around Poyang Lake were Hukou, Xingzi, Duchang, Tangyin, and Kangshan, from north to south. The available observed data from the five stations were for the following periods: 2001–2009, 2001–2013, 2001–2013, 2001–2007, and 2001–2007 respectively. As the inter- and intra-annual variabilities of Poyang Lake, the water level fluctuated greatly. In general, in the dry seasons, the 5 gauged water levels had large standard deviations, and the southern level was higher, suggesting that the lake surface was in an inclined state, supplying Yangtze River. In the wet seasons, the observations were high, and the corresponding deviations were small, indicating that there was little difference among them, and the surface tended to be flat.

To supplement data and create long-term records on the water level, DAHITI results were used in this research. DAHITI results start in 2002 and are missing for 2011–2012. To maintain consistency with the DAHITI data, the gauged heights of the lake surface relative to Wusong were converted to WGS-84. The DAHITI results showed a similar fluctuation with the average in-situ measurements, while they showed higher values. The DAHITI results were usually 3.39–5.02 m and 3.58–8.51 m higher than the gauged records in the wet and dry seasons, respectively. To assess the accuracy of DAHITI, comparisons of DAHITI results and the gauged records were executed separately for the wet and dry seasons, as shown in Figure 2.

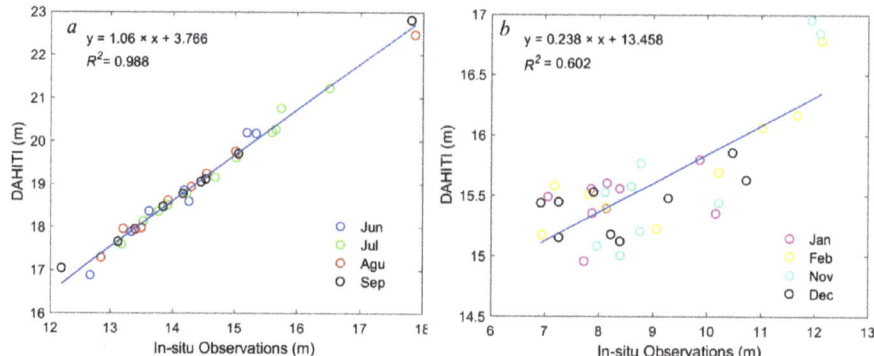

Figure 2. Comparisons of the gauged measurements and DAHITI records in the wet (**a**) and dry (**b**) seasons.

The same dates during 2001–2007 for the two datasets were selected. There were 36 pairs of data in the wet seasons and the R^2 value of their relation reached 0.99 (Figure 2a), indicating that DAHITI results were able to capture level changes in the wet seasons. There were also 36 pairs of data in the dry seasons, while the R^2 value was only 0.60 (Figure 2b), indicating that DAHITI results had large errors. In winters, the gauged data had standard deviations of around 0.96–2.17 m, while the deviations of DAHITI were around 0.33 m. The winter results of DAHITI were not able to present surface fluctuations, as in dry seasons, when Poyang Lake shrunk to a small lake of less than 1000 km^2, altimeter footprints may fall on the lakeside and the returned signals involved the wetland or vegetation, showing a low accuracy. The footprints of two kinds of altimeter data ICESat and ENVISat were shown in Figure 1. In the wet seasons, the lake was large, and the footprints could fully fall on the water surface. Thus, DAHITI could correctly delineate the lake level changes. Based on the linear relation shown in Figure 2a, 29 DAHITI results in the wet seasons during 2002–2017 were transformed into the gauged measurement standard.

Based on gauged observations and converted DAHITI results, there were 169 water level results, including 140 observed records, and each record was the mean value of the five observations. However, there were 349 records in the lake area data. To match the length between the level and area, the 180 missing water level data were derived according to the level–area relation, shown in Figure 3, which was constructed from the available level and area datasets.

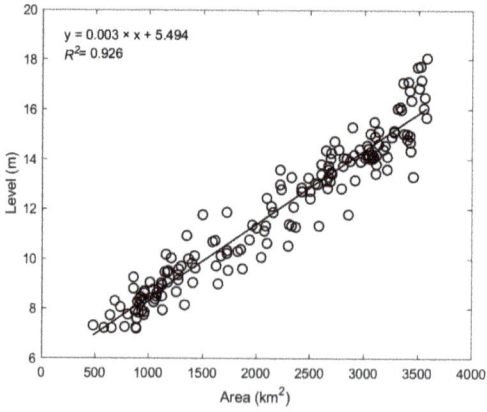

Figure 3. The relationship between the area and water level of Poyang Lake during 2000–2020.

Finally, the 349 water levels of Poyang Lake in the wet and dry seasons between 2000 and 2020 were integrated based on 29 DAHITI results, 140 in-situ measurements, and 180 area–level relation-derived data.

3.2.3. Lake Storage Changes

In this research, we assumed the lake to be a conical frustum [42,43], and the variation of the lake volume from one state to another was deduced by the following Formula (1). Volume changes of Poyang Lake were computed with the aid of the 349 pairs of level and area data acquired on the most proximate dates.

$$\Delta V = \frac{1}{3}(H_2 - H_1) \times \left(A_1 + A_2 + \sqrt{A_1 \times A_2}\right) \quad (1)$$

where ΔV means the changed lake storage from one state with level H_1 and area A_1 to another state with level H_2 and area A_2.

3.2.4. Water Balance of the Watershed

The water balance equation of Poyang Lake, considering precipitation, ground runoff, evaporation, and water exchange with the Yangtze River, was established based on climate data and gauged measurements. The main replenishments of Poyang Lake were rainfall and the five feeding rivers in the basin. The outflows were lake surface evaporation and water flowing to Yangtze River. The equation is as follows:

$$A_t \times P + R - E - W + V_t = V_{t+1} \quad (2)$$

where A_t is the area of Poyang Lake at time t, and P is the corresponding precipitation on the lake surface. R is the accumulated runoff, which is the total discharge from the five feeding rivers, and E is the evaporation of the lake. W indicates the water flowing from Poyang Lake to Yangtze River. When W is less than 0, this indicates that the water flows from Yangtze River to Poyang Lake. V_t and V_{t+1} are the water storages of Poyang Lake at two consecutive moments. In addition, according to the research [44,45], the infiltration of the lake was very stable and accounted for only 1.30% of the whole water resources in this region. Therefore, the infiltration was neglected in this research.

As the lake volume change data were on a monthly scale, daily observation data on precipitation and evaporation were accumulated on a monthly basis, so the monthly measurements were interpolated in the study area to calculate the land precipitation and land evaporation of the watershed. The evaporation of Poyang Lake was estimated from three nearest in-situ stations, according to the Penman-Monteith equation.

The gauged flowing data of the five feeding rivers from 2001–2006 on a monthly scale were obtained from Jiangxi Hydrologic station. Figure 4 shows that the total discharge of the five tributaries was highly related to net basin precipitation, which was the effect of precipitation on the land of the watershed. The net basin precipitation was the result of land precipitation minus land evaporation. Therefore, the total discharge from the five rivers of the rest years from 2000–2020 was deduced based on this linear relationship and the net basin precipitation.

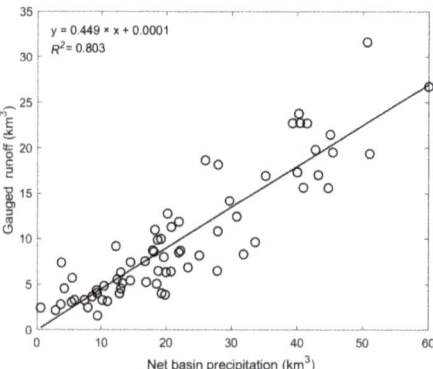

Figure 4. The relationship between the total discharge of the five rivers and net basin precipitation during 2001–2006.

Based on the above variables and lake volume changes, the water exchange W between Poyang Lake and Yangtze River was derived from the water balance equation.

4. Results

4.1. Comparison of the Water Surface in the Wet and Dry Seasons

The inundation areas in the wet and dry seasons during 2000–2020 were calculated, and the fluctuations of the surface extents are shown in Figure 5.

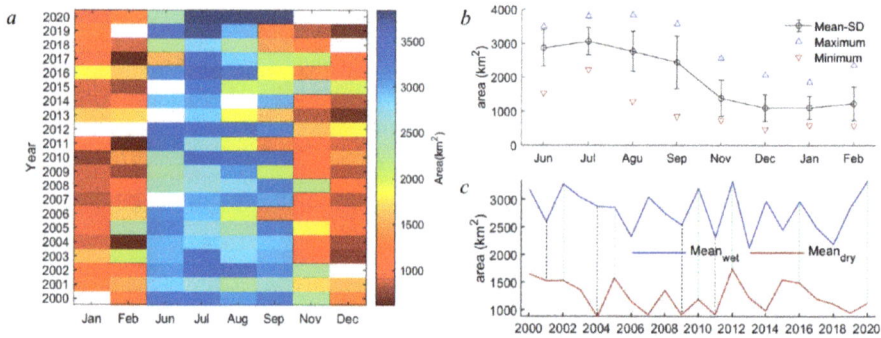

Figure 5. Inundation dynamics of Poyang Lake in the wet and dry seasons from 2000–2020 (**a**), and its seasonal (**b**) and yearly (**c**) variations. A white rectangle in (**a**) indicates that the data is not available for this month. The green and black dotted lines in (**c**) separately indicate wet and dry years, as the values in the wet and dry seasons were both local maximums (wet years) or local minimums (dry years).

In the wet seasons, the lake was usually around 1301.85–3840.24 km^2, with an average value of 2800.79 km^2. The maximum extent occurred in August 2020. In the dry seasons, the area was around 618.82–2498.70 km^2, with an average value of 1242.03 km^2, and Poyang Lake shrank and separated into several small water bodies. The smallest surface area occurred in February 2004. The lake underwent dramatic fluctuations, and the area in the wet seasons was usually 4 times of that in the dry seasons.

Poyang Lake usually begins to increase from May and then shrink in September. In the wet seasons, the lake usually had the highest extents in July, at around 3071.56 ± 399.00 km^2, and tended to be in a small state in each September, with an area of 2445.02 ± 778.41 km^2, as shown in Figure 5b. In the dry seasons, the lake presented a medium state, at around

1385.67 ± 530.56 km², and reached its minimum in December, with an area of 1104.39 ± 395.26 km².

In the years 2006, 2011, 2013 and 2018, the lake had small areas in the wet seasons, at around 2118.13–2326.42 km², and it even shrank to 1900 km² in August in the years 2006, 2011 and 2013. In the years 2000, 2002, 2010, 2012 and 2020, the lake presented large extents in the wet seasons with an area of around 3000–3349.82 km².

In the years 2004, 2007, 2009, 2011, 2014 and 2019, the lake was frequently in a small state in the dry seasons, with an average value of less than 1000 km². In the years 2000, 2005, 2012, 2015, and 2002, the surface in the dry seasons showed relatively ample states, with an area of around 1527.01–1744.58 km².

From Figure 5c, the average areas in the wet and dry seasons were both local maximums in the years 2002, 2005, 2010, 2012, 2016 and 2020, indicating these years were wet years and Poyang Lake was in an ample state. However, the values were both minimums in the wet and dry seasons in the years 2001, 2004, 2009 and 2011, implying these years were very dry. This result coincided with the drought and flood events in researches and news reports [46–49]. In the years 2003, 2006, 2013, 2017 and 2018, though the areas in the wet and dry seasons were not all local minimums, the lake tended to be in droughts. In a word, for Poyang Lake, the number of wet years were less than dry years during the studied period. Some researches indicated that the drought frequency and intensity in the Poyang Lake region increased after TGD began to impound water in 2003 [5,7,33,48].

In each extracted result of lake surface, water pixels were with the value "1" and no water pixels were with the value "0". To obtain a clear picture of the spatial fluctuations of Poyang surface extents, 179 results in the wet seasons and 170 results in the dry seasons were separately overlaid and added to reflect the inundation frequency of each part. For each pixel on the summed images of the wet and dry seasons, the value ranged from 1 to 180, and this value indicated the inundated times, as shown in Figure 6.

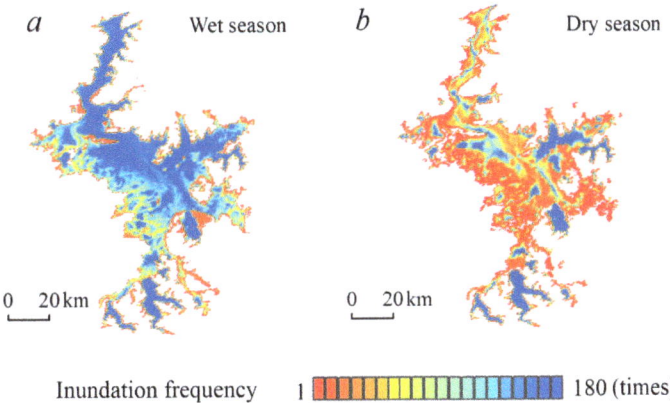

Figure 6. The inundation frequency of Poyang Lake in the wet (**a**) and dry (**b**) seasons during 2000–2020.

In the wet seasons, most regions were frequently inundated. In the dry seasons, the most frequently inundated regions were the central channel and several low-lying lakes, including Junshan Lake. In the south of Poyang Lake, Junshan Lake maintained a stable coverage in both the wet and dry seasons. In fact, Junshan Lake has been a reservoir since the 1950s, when the floodgates were constructed to separate it from the main lake. Thus, it was lightly influenced by the water flow between Poyang Lake and Yangtze River. In the dry seasons, the edge region of the lake had large dynamics, with the water and wetland replacing each other and the wetland vegetation period appearing longer year-by-year. In the central part of Poyang Lake, near Songmen mountain, the wetland vegetation area was becoming more abundant and prospering. Some research showed that in this area, the mudflats of the Nanjishan Wetland National Nature Reserve presented a shrinking trend, with a rate of $-12.1 \text{km}^2/\text{yr}$, during the last three decades [11].

In addition, in the wet seasons, the lake, with an inundated frequency greater than 150, 120, 90, 60, and 30 during the studied period, had areas of 1687.65 km^2, 2470.51 km^2, 2926.66 km^2, 3311.97 km^2, and 3640.61 km^2, respectively. In the dry seasons, these results changed to 504.66 km^2, 741.00 km^2, 1007.18 km^2, 1378.11 km^2, and 2055.79 km^2, respectively. The differences between these several states revealed the drastic dynamics of Poyang Lake.

4.2. The Inclination of the Lake Surface

Figure 7 presents the daily fluctuations of gauged water level of the five hydrologic stations. As two stations had data during 2001–2007, one station had data during 2001–2009 and the rest two stations in the south had data during 2001–2013, curves in Figure 7 shows low variability after 2007. The gauged measurements were comparatively high in the wet seasons, with little difference (0–0.07 m), implying that the surface was flat. However, the water levels varied a lot in the dry seasons, with a standard deviation of 2–3 m. The minimum standard deviation of the five observations was 0.0027 m in June, 2011. The maximum standard deviation was 2.57 m in February 2002. The difference of the water level between the upper south and lower north could reach 7 m, which occurred in January–March, and only reached around 0.15 m during summer. When the mean water level was greater than 15.18 m, which usually occurred in summer, the standard deviation of the five measurements was less than 0.17 m. When the mean water level was less than 13.72 m, which generally occurred in winter, the standard deviation was usually greater than 1.05 m, showing fluctuations of the lake.

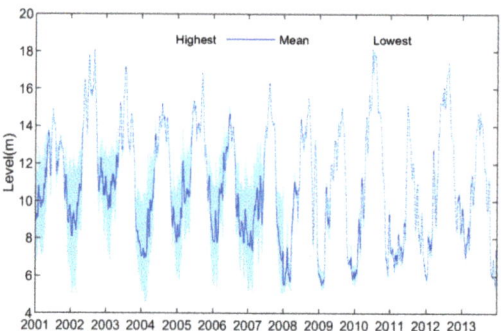

Figure 7. Daily in situ water level observations from the five stations during 2001–2013. The highest, mean and lowest lines indicate the maximum, mean and minimum of the five observations. Because only two observations were available after 2009, the three lines overlap with each other.

In the dry seasons, the lake levels had their lowest values from Dec–next Feb, and the five observations were all lower than 13.00 m. The gauged levels increased from north to south, with great deviations, by around 1.52–2.59 m. The northernmost station Hukou had large fluctuations during 2001–2009 of around 4.71–8.75 m, with a mean level of 6.35 m. The lake levels in the southernmost station of Kangshan were around 10.09–12.67 m, with the minimum occurring in April 2004. In the wet seasons, the gauged levels had high values greater than 15.2 m, with small deviations of around 0.01–1.36 m. Hukou station varied from 8.32 m to 16.51 m, and the mean value was 13.69 m, while Kangshan station fluctuated from 11.60 m to 16.53m, with a mean value of 14.20 m.

The research indicated that there was an obvious linear relationship between the latitudes and observed water levels of the stations in winter [9]. The correlations between the latitudes and daily water levels of the five stations were evaluated in this research. Nearly 50% of the relationships had R^2 values of more than 0.90, especially in the dry seasons from November to February, as shown in Figure 8. In the dry seasons, R^2 had high values and little variance. If the lake surface was supposed to be a plane, then the corresponding inclined angles could be derived by the gradient of the linear relation. Based on this supposition, the inclined angles were calculated and they were usually greater than 10″ in the winter months. Conversely, in the summer months, the R^2 values showed fluctuations, and sometimes they were less than 0.3, indicating that there were no strong relationships between the latitudes and water levels. In these cases, the corresponding derived angels were nearly 0″, especially in July. In addition, the negative values for the angles meant that the surfaces declined from south to north.

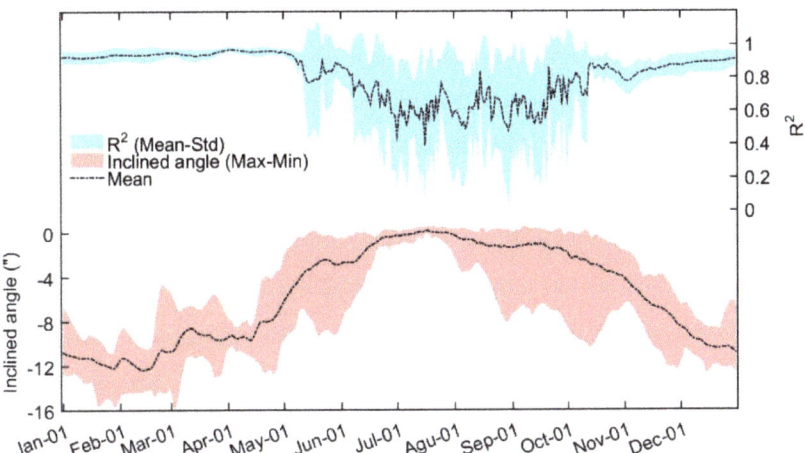

Figure 8. Daily changes of R^2 and inclined angles based on in situ measurements. The black dashed lines are the daily mean values of R^2 and inclined angles. The area in shallow blue shows the ranges of standard deviation, relative to the means. The area in pink shows the daily max and min ranges of the inclined angles, which were determined based on the linear relationship of the latitudes and daily measurements of the gauged stations.

4.3. Variations of the Lake Level and Volume

In the wet seasons, the water level had relatively low values of around 11.94–12.87 m in the years 2001, 2006, 2011, 2013, 2015, and 2017, while it had high values of around 15–16.34 m in the years 2002, 2007, 2010, 2012, 2016, and 2020, as shown in Figure 9. The deviation of the water level in the wet seasons was around 0–2.38 m. In the dry seasons, the lake had low mean levels of around 8.02–8.32 m in 2004, 2007, 2009, 2011, 2014, and 2019 and high values between 10.00 m and 10.70 m in the years of 2000, 2012, 2015, and 2016. The deviation of the water level in the dry seasons was around 0.38–2.15 m. High levels were usually accompanied by large deviations of around 1.50–2.15 m in 2000, 2008, 2015, and 2017. Low levels with small deviations of less than 0.50 m occurred in 2006, 2009, 2011, and 2014. As for monthly fluctuations, the largest variation reached 2.30 m, which occurred in September, followed by 1.85 m in August. Several months in the dry seasons had low variations of around 1.00 m.

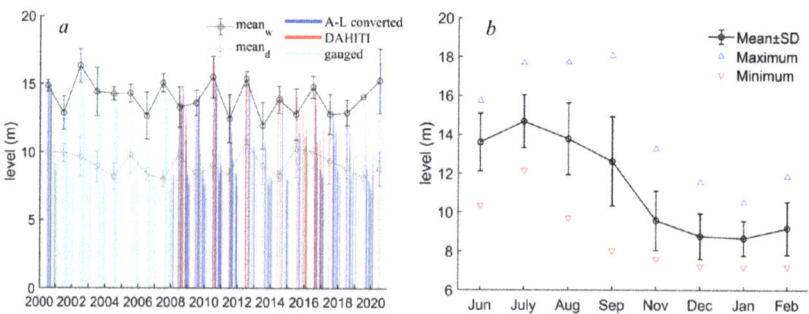

Figure 9. The fluctuations of the lake level from the gauged observations, results converted from DAHITI records, and results converted from the area–level relationship (**a**) and its seasonal variations (**b**).

From Figure 10, in the wet seasons, the monthly volume changes were usually greater than 20 km^3 in 2002, 2010, and 2020, mainly from Jul–Aug. The maximum was 22.64 km^3, which occurred in August 2002. The volume changes were low in the wet seasons of 2006, 2011, and 2013, with a monthly mean value of around 7.5 km^3. In the dry seasons, the volume changes had low values of less than 1 km^3 in the years 2003, 2007, 2013, and 2014, while high values of between 4.36–5.80 km^3 were found in the years 2002, 2012, and 2015. The lake volumes from November to December were usually 1.61 km^3 higher than those from January to February. Considering monthly variations, the largest monthly variation reached 5.28 km^3, which occurred in September, followed by 4.72 km^3 in August. Several months in the dry seasons had low variations of around 1.19–2.54 km^3.

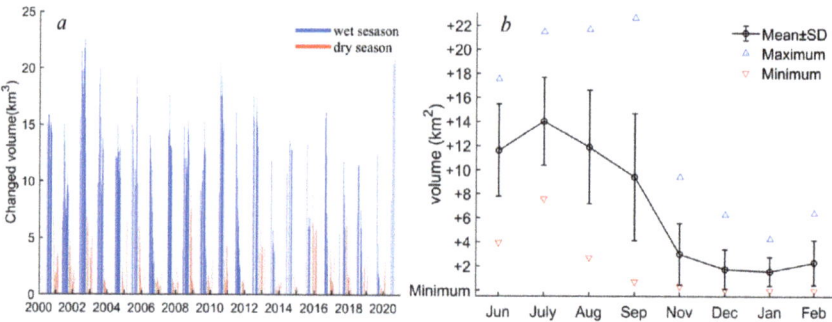

Figure 10. Lake storage changes in wet and dry seasons on a monthly scale (**a**) and their seasonal variations (**b**). The red bars in (**a**) indicate the changed volumes in the dry seasons, and the blue bars indicate those in the wet seasons. Because the volume under the smallest inundated area during 2000–2020 was unknown, the unknown volume was considered as the minimum in (**b**).

4.4. Water Flowing into Yangtze River

As the data were only available on a monthly scale, the water exchange between Poyang Lake and Yangtze River can only be derived according to the volume changes between two adjacent months. The wet and dry seasons both consist of four consecutive months; therefore, the water exchanges over six months (Jun, Jul, Aug, Nov, Dec, and Jan) for each year were calculated. In total, there were 110 values indicating the monthly water exchanges, as shown in Figure 11. They ranged from −14.22 km^3 to 32.86 km^3, with 53 values in the wet seasons and 57 values in the dry seasons. Positive values imply that Poyang Lake supplied Yangtze River, while negative values mean that water flowed from the river to the lake, which occurred occasionally.

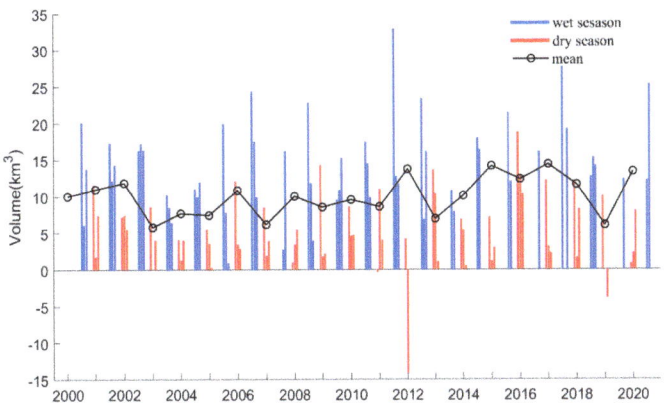

Figure 11. Water exchange between Poyang Lake and Yangtze River on a monthly scale during 2000–2020. The red bars indicate the water flows in the dry seasons, the blue bars indicate those in the wet seasons, and the circle-line symbol indicates mean value of the wet and dry seasons.

Water all flowed from the lake to the river in the wet seasons, with a value of around 0.94–32.86 km^3/m. The values in June were usually higher than those in July and August. In the last two decades, the average volume that flowed to the Yangtze River in June was 18.49 km^3, followed by 12.66 km^3 in July and 12.04 km^3 in August. Some studies have pointed out that the summer monsoon was in the south of Yangtze River during May–Jun, causing increased precipitation in the watershed of Poyang Lake. Therefore, the discharge from the five tributaries increased in June, and more water flowed to Yangtze River. However, the summer monsoon moved to the north of Yangtze River during Jul–Aug, resulting in more rainfall in the upstream of the river. Thus, the increased discharge of Yangtze River flowed backward to the supply from Poyang Lake. The annual mean flow discharge from Poyang Lake to the river in the whole wet seasons was 14.36 km^3 from 2000–2020, with a maximum of 23.44 km^3 occurring in 2017. The exchange was low in the years 2003 and 2013, with values of 8.38 km^3 and 9.36 km^3, respectively.

The exchanges in the dry seasons were around −14.22–18.75 km^3/m. In total, there were three negative values, indicating that the chance of water flowing from Yangtze River in the dry seasons was only 5.26%. The maximum value of the water flow from the river was −14.22 km^3 in December 2011. The mean exchanges in January, November, and December were 3.90 km^3, 8.35 km^3, and 3.50 km^3, respectively. During 2000–2020, the mean water flowing from the lake to the river in the dry seasons was 4.96 km^3/yr, with a maximum of 13.06 km^3 in 2015 and minimum of −2.50 km^3 in 2011.

In 2002, 2012, 2015, 2017, and 2020, the exchanges were higher than those in the adjacent years, with values of between 10.56 km^3/m and 11.32 km^3/m. The exchanges in 2003, 2007, 2013, and 2019 were low, at around 5.77–6.91 km^3/m.

5. Discussion

5.1. Driving Forces

Based on the water balance equation, including ground runoff (R), lake surface precipitation (P), lake surface evaporation (E), and water exchange (W), the driving forces of lake storage changes (ΔV) were analyzed. The monthly contributions of these factors are listed in Table 1.

Table 1. Contributions of the factors to lake storage changes.

Month	R (km^3)	P (km^3)	E (km^3)	W (km^3)	ΔV (km^3)
June	21.03	1.40	1.25	18.49	2.69
July	11.21	0.91	1.70	12.66	−2.25
August	10.60	0.72	1.59	12.04	−2.30
September	5.98	0.39	1.29	10.63	−5.56
Wet season	48.82	3.43	5.84	53.83	−7.42
November	6.94	0.44	0.65	8.35	−1.61
December	4.12	0.26	0.47	3.50	0.42
January	5.19	0.38	0.41	3.90	1.26
February	6.68	0.49	0.47	−2.63	9.33
Dry season	22.93	1.57	2.00	13.11	9.39

In the wet seasons, the monthly ground runoff was around 5.98–21.03 km^3/m, with a mean value of 12.21 km^3/m. The maximum value was 21.03 km^3/m in June. In the dry seasons, the monthly ground runoff was between 4.12 km^3/m and 6.94 km^3/m, with a mean value of 5.73 km^3/m. The average total ground runoff had values of 48.82 km^3 and 22.93 km^3 in the wet and dry seasons, respectively.

The total lake surface evaporation in the wet seasons was 5.84 km^3/yr, which is about 2.92 times that in the dry seasons. The evaporation reached a maximum of 1.70 km^3/m in July. The monthly mean value of evaporation in the wet seasons was 1.46 km^3/m, and that in the dry seasons was 0.50 km^3/m.

The monthly precipitation on the lake surface was 0.86 km^3/m, with a maximum of 1.40 km^3/m in June. The monthly mean values of lake surface precipitation in the wet and dry seasons were 0.86 km^3/m and 0.39 km^3/m, respectively

The lake evaporations were higher than the precipitations on the lake surface in both the wet and dry seasons, and they occupied 11.96% and 8.72% of the supply from the ground runoff in the wet and dry seasons, respectively.

The ground runoff and precipitation on the lake surface gradually decreased as the rainfall usually reduced from around 500 mm in June to less than 100 mm in September in the watershed. As the lake evaporation remained stable in the wet seasons, the water flowing to Yangtze River decreased from 18.49 km^3 in June to 10.63 km^3 in September. In the wet seasons, the overall contributions of runoff, precipitation, and evaporation were less than the volume supplying Yangtze River. Therefore, the lake storages presented a decrease, at a rate of −7.42 km^3/yr.

It is worth mentioning that as the rainfall decreased to around 10–15 mm in September in the years 2001 and 2019, the ground runoff had relatively low values of 1.69 km^3 and 1.85 km^3, respectively. Therefore, the monthly ground runoff in September was lower than that in February and November.

In the dry seasons, the three factors, ground runoff, precipitation on the lake surface, and lake evaporation, occupied 50%, 50%, and 34.25% of those in the wet seasons, respectively. The average volume of water supplying Yangtze River was 13.11 km^3, occupying 58.27% of the whole input of the lake. Therefore, Poyang lake showed an increase, at a rate of 9.39 km^3/yr.

The monthly basin precipitation and lake storage changes showed a similar pattern on annual and seasonal scales as shown in Figure 12a. On average, the monthly basin precipitation and lake volume were correlated in the research period, although several discrepancies existed in some detailed changes. In 2002, 2010, 2012, and 2020, the rainfall was higher than in the other years, and the corresponding lake storages also increased. However, during 2006–2007, the precipitation and lake volume in Poyang showed opposite performances. Poyang Lake was at the local minimum in 2006, whereas the precipitation appeared to be normal. The lake storage had a low value in 2006 and got better in 2007, while the precipitation in 2006 was higher than that in 2007. The precipitation in the basin increased in 2019, while the corresponding storage had no obvious changes. These discrepancies may be because the precipitation needs to convert to ground runoff in order to feed the lake, and there may be a delay of the effect from rainfall. Moreover, besides the basin precipitation, the constant discharge flowing into Yangtze River also had an effect on the lake storage changes. The temperatures of the three nearby stations presented stable states and had no relation with the lake storage changes (Figure 12b), indicating that the lake evaporation induced by temperature was not the main driving factor. On the whole, basin precipitation was the most important driving force.

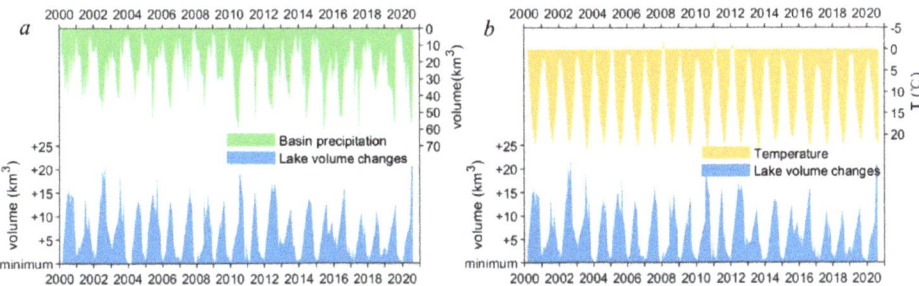

Figure 12. Comparisons of the lake storage changes and the equivalent precipitation (**a**) and temperatures (**b**) in the basin on a monthly scale. Because the volume of the smallest inundated area during 2000–2020 was unknown, the unknown volume was considered as the minimum.

5.2. Accuracy Assessment

Two 30 m interpretation results based on Landsat images in the years of 2009 and 2016 were collected to check the accuracy of the inundation results of this research. The interpretation results showed a higher accuracy (96%) [39].

To ensure that the lake states were consistent, the MODIS results on the nearest dates to the 30 m Landsat results were selected. The two pairs of water boundaries were presented in Figure 13. Evaluations were carried out spatially, and the outcome showed that the omission errors were 11.56% and 2.56%, and the commission errors were 10.94% and 5.47% for the MODIS results in the years 2009 and 2016, respectively. The inundation area of the lakes was higher in the 30 m results. The area differences were 9.31% and 12.76% for the selected inundation results in the wet and dry seasons, respectively. The boundaries of some small tributaries were not correctly depicted in the MODIS images due to its coarse resolution. Nevertheless, the overall accuracy of the MODIS results was greater than 85%, and they indicated that the results were convincible to study lake inundation changes.

Hukou station is located at the intersection of Poyang Lake and Yangtze River. The gauged monthly average flow velocities at Hukou station were available during 2000–2008 and 2013–2014. The fluctuations of the exchanged water coincided with the dynamics of the flow velocity at Hukou station, as shown in Figure 14. The observed velocities were all greater than zero, and the simultaneous water exchanges were all positive values, indicating the water flowing into Yangtze River. The high velocities usually matched large exchanges, and the low velocities corresponded to a small water flow at Hukou station. The flow velocities tended to be high in the wet seasons and had low values in the dry seasons. The maximum was 12,600 m^3/s, occurring in June 2006, when the exchange also reached the peak of the adjacent years. The minimum velocity was 895 m^3/s in February 2004. In addition, June and July usually had higher velocities than August and September, and this phenomenon was consistent with the results of this study, which found that the water exchange in June was higher than in other months. The 25 pairs of velocity and volume data occurred in the same months had the R^2 value of 0.72. Therefore, a similar pattern between the fluctuations of the flow rate and water exchange changes showed the credibility of this research.

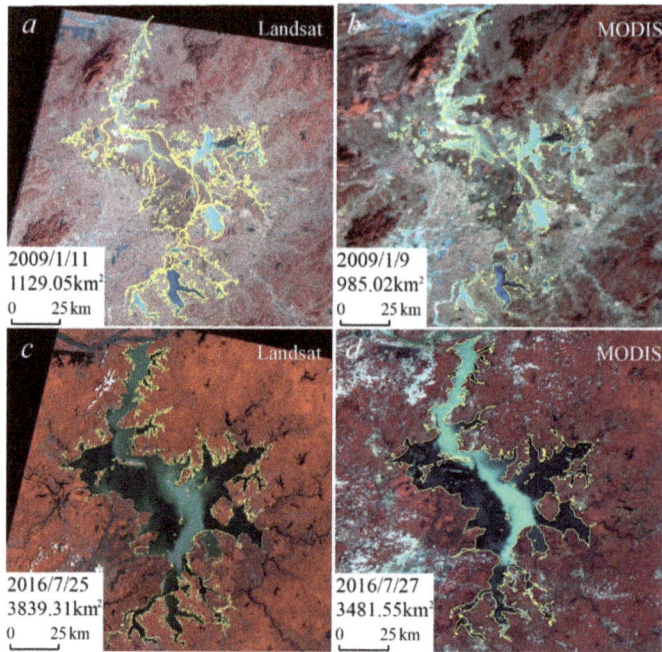

Figure 13. Comparison of 30 m interpretation products and the extracted extents based on MODIS images. (**a,b**) indicate water boundaries extracted from Landsat and MODIS images in 2009. (**c,d**) separately indicate Landsat and MODIS results in 2016.

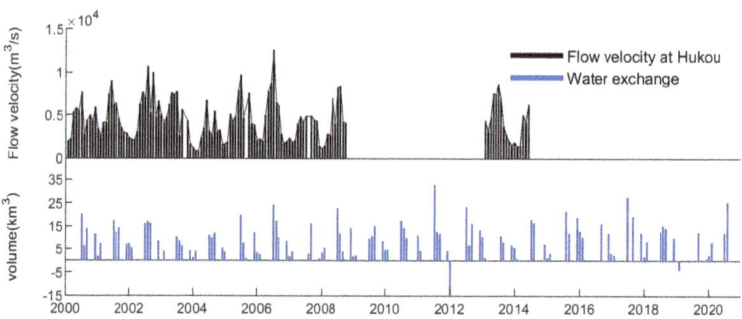

Figure 14. Gauged observations of the flow velocities (2000–2008, 2013–2014) at Hukou station and water exchanges between Poyang Lake and Yangtze River (2000–2020).

6. Conclusions

In this research, gauged observations, altimeter database, and MODIS images were combined to depict the changes of several hydrologic variables. The water extents were delineated with a high accuracy when evaluated using the 30 m interpretation results. The surface extents of Poyang Lake expressed great dynamics and seasonality. The five hydrologic stations around Poyang Lake showed disagreement in most of the years, suggesting that the lake was not flat, and the water was flowing. The lake surface inclined from south to north, with an angle of around $0''$–$16''$, and it was usually greater than $10''$ in the winter seasons. According to the appearance of the water flowing into Yangtze River, it can be concluded that, in the wet seasons, water all flowed from south to north, and there was a chance of only 5.26% in the dry seasons that it flowed backward. Precipitation was the main source of the ground runoff flowing into the lake. Thus, rainfall can be regarded as the primary influencing factor of Poyang Lake. However, there were some discrepancies between precipitation and water storage changes, as the state of Poyang Lake was also affected by the water quantity of Yangtze River.

There were several uncertainties in this research. The bathymetry of Poyang Lake during 2000–2020 was assumed to be unchanged when calculating the storage changes. Though several dredging activities have been reported in the past, they mainly occurred in the tributaries of Poyang Lake. Therefore, the changes in the lake bottom topography can be ignored, considering its large span. The ground runoff of the five tributaries flowing into Poyang Lake was estimated according to the relation between the basin precipitations and gauged discharges. It was inevitable that there were some errors in this estimation. However, the similar pattern and high correlation between the observed Hukou flow velocities and water exchanges proved the practicability of this method. In the respective of lake volume changes, the Formula (1) which treated the lake as a conical frustum definitely caused uncertainties. Though the real bathymetry of Poyang Lake has been surveyed by sonar devices, the bathymetry map was not available due to restrictions on data sharing. Considering the formula has been widely applied in some researches [22,41–43] and the accuracy assessment on water exchanges, the studied results can reveal the volume changes of Poyang Lake to a certain extent.

This paper analyzed the driving factors in the water balance equation. The effect of human activities was not determined. As for human actions, 9603 dams have been built on the five feeding rivers, compounding around 27.9 billion m^3 water until 2001 [50], and this may affect the natural flowability of water in the basin. The TGD resulted in a decrease of the water inflow to the downstream Yangtze River and caused more water to flow from Poyang Lake to the Yangtze River, especially during late autumn and winter [5]. Some researchers have pointed out that lake precipitation decreased and the evaporation increased during the post-TGD periods, compared with those during the pre-TGD periods [51]. In addition, the construction of dikes for fish ponds may affect the variation of the local flow [52].

On the whole, this research compared variations of Poyang Lake between the wet and dry seasons, quantified contribution factors of volume changes, and derived exchanges between the lake and Yangtze River. The results can serve as important information to better understand the water cycle of the watershed, and the studied datasets may also be used in hydrologic modeling and wetland studies.

Author Contributions: Writing—original draft, F.S.; supervision, C.L. and B.H.; validation, R.M. All authors have read and agreed to the published version of the manuscript.

Funding: This work was supported by the National Natural Science Foundation of China (grant No. 42001353), the Scientific Program of Guangzhou University (grant Nos. YG2020019 and SJ201911), the Open Research Fund Program of Guangdong Key Laboratory of Ocean Remote Sensing (South China Sea Institute of Oceanology Chinese Academy of Sciences) (grant No. 2017B030301005-LORS2006), the Project of Science and Technology Development of Guangdong Academy of Sciences (grant No. 2020GDASYL-20200102013) and the Scientific Program of Guangzhou Bureau of Education (grant No.1201430672).

Institutional Review Board Statement: Not applicable.

Informed Consent Statement: Not applicable.

Data Availability Statement: Publicly available datasets were analyzed in this study. These include MODIS images from https://reverb.echo.nasa.gov/reverb/, meteorological Data from http://cdc.cma.gov.cn/, and DAHITI database from https://dahiti.dgfi.tum.de/en/ (accessed on 5 March 2021).

Acknowledgments: We sincerely appreciate all valuable comments and suggestions from three anonymous reviewers.

Conflicts of Interest: The authors declare that they have no known competing financial interests or personal relationships that could have appeared to influence the work reported in this paper.

References

1. Wang, S.; Dou, H. *China Lake Catalogue*; Science Press: Beijing, China, 1998.
2. Gao, J.H.; Jia, J.; Kettner, A.J.; Xing, F.; Wang, Y.P.; Xu, X.N.; Yang, Y.; Zou, X.Q.; Gao, S.; Qi, S.; et al. Changes in water and sediment exchange between the Changjiang River and Poyang Lake under natural and anthropogenic conditions. *China Sci. Total Environ.* **2014**, *481*, 542–553. [CrossRef]
3. Ye, X.; Li, Y.; Li, X.; Zhang, Q. Factors influencing water level changes in China's largest freshwater lake, Poyang Lake, in the past 50 years. *Water Int.* **2014**, *39*, 983–999. [CrossRef]
4. Zhang, D.; Liao, Q.; Zhang, L.; Wang, D.; Luo, L.; Chen, Y.; Zhong, J.; Liu, J. Occurrence and spatial distributions of microcystins in Poyang Lake, the largest freshwater lake in China. *Ecotoxicology* **2014**, *24*, 19–28. [CrossRef]
5. Guo, H.; Hu, Q.; Zhang, Q.; Feng, S. Effects of the Three Gorges Dam on Yangtze River flow and river interaction with Poyang Lake, China: 2003–2008. *J. Hydrol.* **2012**, *416–417*, 19–27. [CrossRef]
6. Dai, Z.; Liu, J.T. Impacts of large dams on downstream fluvial sedimentation: An example of the Three Gorges Dam (TGD) on the Changjiang (Yangtze River). *J. Hydrol.* **2013**, *480*, 10–18. [CrossRef]
7. Feng, L.; Hu, C.; Chen, X.; Zhao, X. Dramatic Inundation Changes of China's Two Largest Freshwater Lakes Linked to the Three Gorges Dam. *Env. Sci. Technol.* **2013**, *47*, 9628–9634. [CrossRef]
8. Dai, Z.; Liu, J.T.; Chen, J. Detection of the Three Gorges Dam influence on the Changjiang (Yangtze River) submerged delta. *Sci. Rep.* **2014**, *4*, 6600. [CrossRef] [PubMed]
9. Feng, L.; Hu, C.; Chen, X.; Cai, X.; Tian, L.; Gan, W. Assessment of inundation changes of Poyang Lake using MODIS observations between 2000 and 2010. *Remote Sens. Environ.* **2012**, *121*, 80–92. [CrossRef]
10. Cai, X.; Feng, L.; Hou, X.; Chen, X. Remote Sensing of the Water Storage Dynamics of Large Lakes and Reservoirs in the Yangtze River Basin from 2000 to 2014. *Sci. Rep.* **2016**, *6*, 36405. [CrossRef] [PubMed]
11. Han, X.; Chen, X.; Feng, L. Four decades of winter wetland changes in Poyang Lake based on Landsat observations between 1973 and 2013. *Remote Sens. Environ.* **2015**, *156*, 426–437. [CrossRef]
12. Smith, L.C. Satellite remote sensing of river inundation area, stage and processes: A review. *Hydrol. Process.* **1997**, *11*, 1427–1439. [CrossRef]
13. Cazenave, A.; Milly, P.C.D.; Douville, H.; Benveniste, J.; Kosuth, P.; Lettenmaier, D.P. Space techniques used to measure change in terrestrial waters. *Eos Trans. Am. Geophys. Union* **2004**, *85*, 59–60.
14. Alsdorf, D.E.; Rodríguez, E.; Lettenmaier, D.P. Measuring surface water from space. *Rev. Geophys.* **2007**, *45*, RG2002. [CrossRef]
15. Birkett, C.M. The contribution of TOPEX/POSEIDON to the global monitoring of climatically sensitive lakes. *J. Geophys. Res. Ocean.* **1995**, *100*, 25179–25204. [CrossRef]
16. Birkett, C.M.; Mertes, L.A.K.; Dunne, T.; Costa, M.H.; Jasinski, M.J. Surface water dynamics in the Amazon Basin: Application of satellite radar altimetry. *J. Geophys. Res.* **2002**, *107*, D20. [CrossRef]
17. Schwatke, C.; Dettmering, D.; Bosch, W.; Seitz, F. DAHITI-an innovative approach for estimating water level time series over inland waters using multi-mission satellite altimetry. *Hydrol. Earth Syst. Sci.* **2015**, *19*, 4345–4364. [CrossRef]
18. Birkett, C.M.; Reynolds, C.; Beckley, B.; Doorn, B. From research to operations: The USDA global reservoir and lake monitor. In *Coastal Altimetry*; Vignudelli, S., Kostianoy, A., Cipollini, P., Benveniste, J., Eds.; Springer: Berlin/Heidelberg, Germany, 2011; pp. 19–50.
19. Berry, P.A.M.; Wheeler, J.L. JASON2-ENVISAT Exploitation—Development of Algorithms for the Exploitation of JASON2-ENVISAT Altimetry for the Generation of a River and Lake Product. In *Product Handbook v3.5*; De Montfort University: Leicester, UK, 2009.
20. Crétaux, J.F.; Jelinski, W.; Calmant, S.; Kouraev, A.; Vuglinski, V.; Berge-Nguyen, M.; Gennero, M.-C.; Nino, F.; Abarca Del Rio, R.; Cazenave, A.; et al. A lake database to monitor in the Near Real Time water level and storage variations from remote sensing data. *Adv. Space Res.* **2011**, *4*, 1497–1507.
21. Hwang, C.; Peng, M.; Ning, J.; Luo, J.; Sui, C. Lake level variations in China from TOPEX/Poseidon altimetry: Data quality assessment and links to precipitation and ENSO. *Geophys. J. Int.* **2005**, *161*, 1–11. [CrossRef]
22. Zheng, J.; Ke, C.; Shao, Z.; Li, F. Monitoring changes in the water volume of Hulun Lake by integrating satellite altimetry data and Landsat images between 1992 and 2010. *J. Appl. Remote Sens.* **2016**, *10*, 16029. [CrossRef]
23. Wang, X.; Gong, P.; Zhao, Y.; Xu, Y.; Cheng, X.; Niu, Z.; Luo, Z.; Huang, H.; Sun, F.; Li, X. Water-level changes in China's large lakes determined from ICESat/GLAS data. *Remote Sens. Environ.* **2013**, *132*, 131–144. [CrossRef]
24. Chipman, J.W.; Lillesand, T.M. Satellite-based assessment of the dynamics of new lakes in southern Egypt. *Int. J. Remote Sens.* **2007**, *28*, 4365–4379. [CrossRef]
25. Sun, F.D.; Zhao, Y.Y.; Gong, P.; Ma, R.H.; Dai, Y.J. Monitoring dynamic changes of global land cover types: Fluctuations of major lakes in China every 8 days 2000–2010. *Chin. Sci. Bull.* **2014**, *59*, 171–189. [CrossRef]

26. Prigent, C.; Papa, F.; Aires, F.; Rossow, W.B.; Matthews, E. Global inundation dynamics inferred from multiple satellite observations, 1993–2000. *J. Geophys. Res.* **2007**, *112*, D12107. [CrossRef]
27. Papa, F.; Prigent, C.; Aires, F.; Jimenez, C.; Rossow, W.B.; Matthews, E. Interannual variability of surface water extent at the global scale, 1993–2004. *J. Geophys. Res.* **2010**, *115*. [CrossRef]
28. Peng, D.Z.; Guo, S.L.; Liu, P.; Liu, T. Reservoir storage curve estimation based on remote sensing data. *J. Hydrol. Eng.* **2006**, *11*, 165–172. [CrossRef]
29. Davis, P.A. *Review of Results and Recommendations from the GCMRC 2000–2003 Remote-Sensing Initiative for Monitoring Environmental Resources Within the Colorado River Ecosystem*; U.S. Geological Survey: Reston, VA, USA, 2004; p. 1206.
30. Frappart, F.; Seylerb, F.; Martinezc, J.M.; Leónb, J.G.; Cazenavea, A. Floodplain water storage in the Negro River basin estimated from microwave remote sensing of inundation area and water levels. *Remote Sens. Environ.* **2005**, *99*, 387–399. [CrossRef]
31. Song, C.; Huang, B.; Ke, L. Modeling and analysis of lake water storage changes on the Tibetan Plateau using multi-mission satellite data. *Remote Sens. Environ.* **2013**, *135*, 25–35. [CrossRef]
32. Crétaux, J.F.; Kouraev, A.K.; Papa, F.; Bergé-Nguyen, M.; Cazenave, A.; Aladin, N.V.; Plotnikov, I.S. Water balance of the Big Aral Sea from satellite remote sensing and in situ observations. *Great Lakes Res.* **2005**, *31*, 520–534. [CrossRef]
33. Mei, X.; Dai, Z.; Du, J.; Chen, J. Linkage between Three Gorges Dam impacts and the dramatic recessions in China's largest freshwater lake, Poyang Lake. *Sci. Rep.* **2015**, *5*, 18197. [CrossRef]
34. Zheng, Y. Prediction of the Distribution of C3 and C4 Plant Species from a GIS-Based Model: A Case Study in Poyang Lake, China. Master's Thesis, ITC, Enschede, The Netherlands, 2009.
35. Harris, J.; Zhuang, H. *An Ecosystem Approach to Resolving Conflicts among Ecological and Economic Priorities for Poyang Lake Wetlands*; International Crane Foundation/IUCN: Gland, Switzerland, 2010.
36. Fan, Y.J.; Yu, X.X.; Zhang, H.X.; Song, M.H. Comparison between Kriging interpolation and Inverse Weighting Tension for precipitation data analysis: Taking Lijing river basin as a study case. *J. China Hydrol.* **2014**, *34*, 61–66. (In Chinese)
37. Allen, R.G.; Pereira, L.S.; Raes, D.; Smith, M. *Crop Evapotranspiration: Guidelines for Computing Crop Water Requirements-FAO Irrigation and Drainage Paper*; FAO: Rome, Italy, 1998.
38. Gao, Y.Q.; Wang, Y.W.; Hu, C.; Wang, W.; Liu, S. Variability of evaporation from Lake Taihu in 2012 and evaluation of a range of evaporation models. *Clim. Environ. Res.* **2016**, *21*, 393–404. (In Chinese)
39. Sun, F.D.; Ma, R. Hydrologic changes of Poyang Lake based on radar altimeter and optical sensor. *J. Geogr. Sci.* **2020**, *75*, 544–557.
40. Sun, F.D.; Ma, R. Hydrologic changes of Aral Sea: A reveal by the combination of radar altimeter data and optical images. *Ann. Gis* **2019**, *25*, 247–261. [CrossRef]
41. Sun, F.D.; Ma, R.; He, B.; Zhao, X.; Zeng, Y.; Zhang, S.; Tang, S. Changing Patterns of Lakes on The Southern Tibetan Plateau Based on Multi-Source Satellite Data. *Remote Sens.* **2020**, *12*, 3450. [CrossRef]
42. Liu, Y.; Yue, H. Estimating the fluctuation of Lake Hulun, China, during 1975–2015 from satellite altimetry data. *Environ. Monit. Assess.* **2017**, *189*, 630. [CrossRef] [PubMed]
43. Zhang, G.; Chen, W.; Xie, H. Tibetan Plateau's lake level and volume changes from NASA's ICESat/ICESat-2 and Landsat Missions. *Geophys. Res. Lett.* **2019**, *46*, 13107–13118. [CrossRef]
44. Wan, X.; Xu, X. Analysis of supply and demand balance o of water resources around Poyang Lake. *Yangtze River* **2010**, *41*, 43–47.
45. Feng, L.; Hu, C.; Chen, X.; Li, R. Satellite observations make it possible to estimate Poyang lake's water budget. *Environ. Res. Lett.* **2011**, *6*, 44023. [CrossRef]
46. Liu, Y.; Wu, G.; Zhao, X. Recent declines in China's largest freshwater lake: Trend or regime shift? *Env. Res. Lett.* **2013**, *8*, 14010. [CrossRef]
47. Lai, X.; Shankman, D.; Huber, C.; Yesou, H.; Huang, Q.; Jiang, J. Sand mining and increasing Poyang Lake's discharge ability: A reassessment of causes for lake decline in China. *J. Hydrol.* **2014**, *519*, 1698–1706. [CrossRef]
48. Zhang, Q.; Ye, X.-C.; Werner, A.D.; Li, Y.-L.; Yao, J.; Li, X.-H.; Xu, C.-Y. An investigation of enhanced recessions in Poyang Lake: Comparison of Yangtze River and local catchment impacts. *J. Hydrol.* **2014**, *517*, 425–434. [CrossRef]
49. Xinhua Net. China's Largest Freshwater Lake Sees Record Water Rise. 12 July 2020. Available online: http://www.xinhuanet.com/english/2020-07/12/c_139207311.htm (accessed on 16 July 2020).
50. Liu, J.; Zhang, Q.; Xu, C.; Zhang, Z. Characteristics of run off variation of Poyang Lake watershed in the past 50 years. *Trop Geogr.* **2009**, *29*, 213–218.
51. Zhang, Z.; Huang, Y.; Xu, C.Y.; Chen, X.; Moss, E.M.; Jin, Q.; Bailey, A.M. Analysis of Poyang Lake water balance and its indication of river–lake interaction. *SpringerPlus* **2016**, *5*, 1555. [CrossRef] [PubMed]
52. de Leeuw, J.; Shankman, D.; Wu, G.; de Boer, W.F.; Burnham, J.; He, Q.; Yesou, H.; Xiao, J. Strategic assessment of the magnitude and impacts of sand mining in Poyang Lake, China. *Reg. Environ. Chang.* **2010**, *10*, 95–102. [CrossRef]

Article

Evaluating the Latest IMERG Products in a Subtropical Climate: The Case of Paraná State, Brazil

Jéssica G. Nascimento [1,*], Daniel Althoff [2], Helizani C. Bazame [1], Christopher M. U. Neale [3], Sergio N. Duarte [1], Anderson L. Ruhoff [4] and Ivo Z. Gonçalves [1]

1 Biosystems Engineering Department, College of Agriculture "Luiz de Queiroz"—University of São Paulo (ESALQ/USP), Av. Pádua Dias, 11, Piracicaba 13418-900, Brazil; helizanicouto@usp.br (H.C.B.); snduarte@usp.br (S.N.D.); zution@usp.br (I.Z.G.)
2 Agricultural Engineering Department, Federal University of Viçosa (UFV), Av. Peter Henry Rolfs, s.n., Viçosa 36570-900, Brazil; daniel.althoff@ufv.br
3 Daugherty Water for Food Global Institute, University of Nebraska, Nebraska Innovation Campus, 2021 Transformation Dr. Street, 3220, Lincoln, NE 68588, USA; cneale@nebraska.edu
4 Hydraulic Research Institute, Federal University of Rio Grande do Sul, Porto Alegre 91501-970, Brazil; anderson.ruhoff@ufrgs.br
* Correspondence: jessicagarcia@usp.br

Citation: G. Nascimento, J.; Althoff, D.; C. Bazame, H.; M. U. Neale, C.; N. Duarte, S.; L. Ruhoff, A.; Z. Gonçalves, I. Evaluating the Latest IMERG Products in a Subtropical Climate: The Case of Paraná State, Brazil. *Remote Sens.* 2021, *13*, 906. https://doi.org/10.3390/rs13050906

Academic Editors: Magaly Koch and Weili Duan

Received: 27 January 2021
Accepted: 23 February 2021
Published: 28 February 2021

Publisher's Note: MDPI stays neutral with regard to jurisdictional claims in published maps and institutional affiliations.

Copyright: © 2021 by the authors. Licensee MDPI, Basel, Switzerland. This article is an open access article distributed under the terms and conditions of the Creative Commons Attribution (CC BY) license (https://creativecommons.org/licenses/by/4.0/).

Abstract: The lack of measurement of precipitation in large areas using fine-resolution data is a limitation in water management, particularly in developing countries. However, Version 6 of the Integrated Multi-satellitE Retrievals for GPM (IMERG) has provided a new source of precipitation information with high spatial and temporal resolution. In this study, the performance of the GPM products (Final run) in the state of Paraná, located in the southern region of Brazil, from June 2000 to December 2018 was evaluated. The daily and monthly products of IMERG were compared to the gauge data spatially distributed across the study area. Quantitative and qualitative metrics were used to analyze the performance of IMERG products to detect precipitation events and anomalies. In general, the products performed positively in the estimation of monthly rainfall events, both in volume and spatial distribution, and demonstrated limited performance for daily events and anomalies, mainly in mountainous regions (coast and southwest). This may be related to the orographic rainfall in these regions, associating the intensity of the rain, and the topography. IMERG products can be considered as a source of precipitation data, especially on a monthly scale. Product calibrations are suggested for use on a daily scale and for time-series analysis.

Keywords: remote sensing in hydrology; precipitation; performance evaluation; GPM

1. Introduction

Precipitation plays a fundamental role in the hydrological cycle. It is considered the main water source input in the soil water balance and runoff and is used as an input in hydrological and climatological modeling. In the management of water resources, knowledge of the volume and intensity of precipitation is essential for the prediction of floods and droughts, the distribution of water for urban and industrial uses, and the planning of irrigation in agriculture and hydraulic infrastructure.

Precipitation can be measured by gauges, sensors onboard satellites, and radars [1–3]. Precipitation gauges are fundamental instruments, and their observations are considered as a reference in many studies [4]. However, to represent spatiotemporal variability of intensity and type of occurrence of precipitation, a dense measuring network is necessary with long-period information, which unfortunately is not the reality in many regions of the world [5]. In South American countries, monitoring by gauges is limited in terms of infrastructure, maintenance, density, and frequency of observations [6,7].

Regarding indirect methods, weather radars provide precipitation estimates with high spatial and temporal resolution but have limited accuracy in mountainous regions and cold climates [5,8]. On the other hand, satellite estimates of precipitation provide vast spatial and temporal coverage and are freely available. Over the last two decades, several satellite precipitation products have been developed, such as Tropical Rainfall Measuring Mission (TRMM) [9]; Climate Prediction Center Morphing Method (CMORPH) [10]; Global Satellite Mapping of Precipitation (GSMaP) [11]; Climate Hazards Group Infrared Precipitation with Stations (CHIRPS) [12]; and Multi-Source Weighted-Ensemble Precipitation (MSWEP) [13].

The TRMM Multi-satellite Precipitation Analysis (TMPA) algorithm combines precipitation estimates from satellite systems with data measured on the Earth's surface, to provide a calibrated final product and results in the "best" satellite estimate [14]. A successor to TRMM, the Global Precipitation Measurement (GPM) was launched in 2014, on a joint mission between NASA (National Aeronautics and Space Administration) and JAXA (Japan Aerospace Exploration Agency) and continues to offer products to this day. The GPM constellation consists of the first Dual-frequency phased array Precipitation Radar and a GPM Microwave Imager, which represent the most advanced versions compared to the Precipitation Radar (PR) and the TRMM Microwave Imager (TMI), onboard the TRMM satellite [15]. Relevant improvements in the GPM products include an increase in latitudinal coverage (global coverage of 60° N/S) and the detection of heavy rain, light rain, and snow [5,6]. In the era of GPM, the Integrated Multi-satellite Retrievals for GPM (IMERG) algorithm operates, intending to calibrate, unite, and interpolate satellite precipitation estimates with data from gauges [16].

Many studies have analyzed and compared the TRMM data to those obtained by gauge stations [6,17–20], which allowed for advances in the application of remote sensing in determining the volume and behavior of precipitation in several countries. Given that precipitation represents the most important parameter of the hydrologic cycle, it can directly affect physical, chemical, and biological processes. The spatial and temporal distribution of the precipitation is key for the understanding of hydrologic responses in watersheds, for example, runoff, streamflow, and flooding. Thus, the accuracy of precipitation data is essential for the good performance of hydrological models and inconsistencies in its estimates can negatively affect the hydrologic investigation.

Previous studies have reported some limitations of the IMERG products [21] related to its low performance in estimating the precipitation over North China, where snowfall events can affect the precipitation estimates from satellite products. The authors also reported the lower accuracy of the satellite precipitation estimates in areas of high altitudes or in arid and semiarid climates. Evaluating the IMERG precipitation compared to ground-based data, [22] observed a tendency for restimating the precipitation using the sub-daily products and overestimating the maximum rainfall in monthly variations over the Indian subcontinent. In a seasonal analysis, [21,23] found that IMERG performed better in warm and wet seasons. In a subtropical climate, [24] concluded that IMERG performed well for detecting precipitation events with a limitation in representing the amount of rainfall at sub-daily scales, while in tropical climates [25,26] demonstrated that it is essential to calibrate the IMERG products to reduce random and systematic errors.

On the other hand, studies comparing the IMERG to different satellite products, highlighted the better accuracy of IMERG in estimating precipitation. [27] found better performance of IMERG in estimating precipitation on a global scale compared to TMPA products. The better performance of IMERG's products was also observed by [28] concerning the TMPA in estimating precipitation in all regions of Brazil. Recently, [29] observed better performance of IMERG-Late version 6 products compared to IMERG-Early, GSMaP-NRT, GSMaP-MVK, TMPA-RT, and PERSIANN-CCS products on a global scale The better performance of IMERG in relation to other remote sensing products, reported in previous studies, supports the analysis of the performance of its products in different climates and geographic scales.The quality of IMERG products was improved over time by increasing the number of passive microwave samples [30]. The latest version of the IMERG algorithm (version 6), made available to the public in October 2019, combines the reanalysis of precipitation estimated by satellites between 2000–2014 by TMPA and in the subsequent period by GPM, resulting in 19 years of information so far. Its products have a spatial resolution of 0.1° and a 30-min temporal resolution [31]. In this way, trend analysis and analysis of extreme events could be carried out with greater accuracy. In addition, the performance of climatological and hydrological models could be improved with greater detail of recent precipitation information. Thus, studies evaluating the performance of IMERG version 6 products are important and promising at regional scales. Brazil is a country of continental dimensions, with different micro-climates and rainfall patterns throughout its territory. Among its states, Paraná, located in the south of the country in the Paraná River Basin region, is the most important socio-economically hydrologic region in Brazil [32]. This basin has the largest hydroelectric infrastructure in the country, which is responsible for approximately 44.6% of the electric energy production and transmission system in Brazil [33]. Thus, the determination of precipitation in this region is essential for forecasting with hydrological and climatological monitoring models. Also, it is important for this region to detect anomalies related to excess or deficit of precipitation, which can compromise the supply of energy and hydraulic structures.

Thus, the purpose of this study is to provide a comprehensive assessment of IMERG (version 6) precipitation estimates over a subtropical region, specifically the objectives of the study are (1) to evaluate the performance of IMERG's daily and monthly products, and (2) to evaluate IMERG's performance in detecting monthly anomalies in Paraná using observations from a dense network of precipitation gauges. This study is expected to provide a reference for the use of IMERG products in monthly and daily temporal resolutions and further contribute to improvements of the satellite precipitation algorithm. The remainder of this article is organized into the following topics: Section 2 describes the study area, the precipitation data sets used, the metrics used, and the detailed methodology; Section 3 presents the main results and the discussion, and Section 4 reports the main conclusions of the study.

2. Material and Methods

2.1. Study Region

The study area is the state of Paraná in the south region of Brazil (Figure 1). Paraná occupies an area of 199,315 km² and covers 399 municipalities [34]. According to the Köppen classification, carried out by [35], the state is in the transition from tropical and subtropical climates where the humid subtropical Cfa (hot summer) and Cfb (warm summer) predominate across 61.7% and 37.0% of the state, respectively. Because of its extensive area, there is a great diversity in terms of climate, soil types, vegetation, and agricultural use. The main biomes that constitute the state are the Atlantic Forest and the Cerrado. The predominant crops are maize, soy, and sugar cane. Most of Paraná's relief is found at altitudes above 600 m (Figure 1), subdivided into four Morpho-sculptural Units [36].

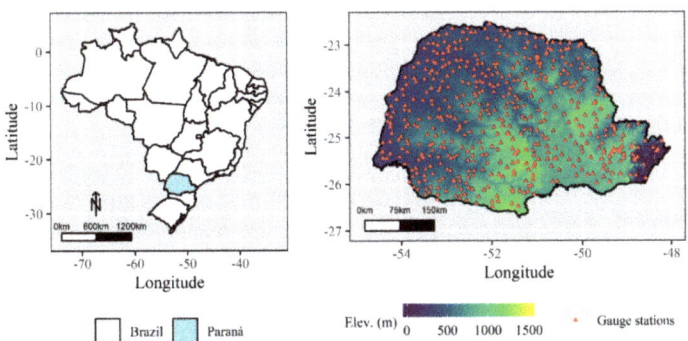

Figure 1. Study area and gauges used to validate the Integrated Multi-satellite Retrievals for GPM (IMERG) estimates of precipitation at daily and monthly time-steps.

Precipitation in Paraná varies spatially, with an annual average between 1300–2200 mm. Summer is the season with the highest rainfall in South America, including the subtropical region [37,38]. The geographic mesoregions of Paraná, demarcated by the Brazilian Institute of Geography and Statistics (IBGE) and shown in Figure 2, were considered for the regional analysis of the performance of the IMERG products over the study area.

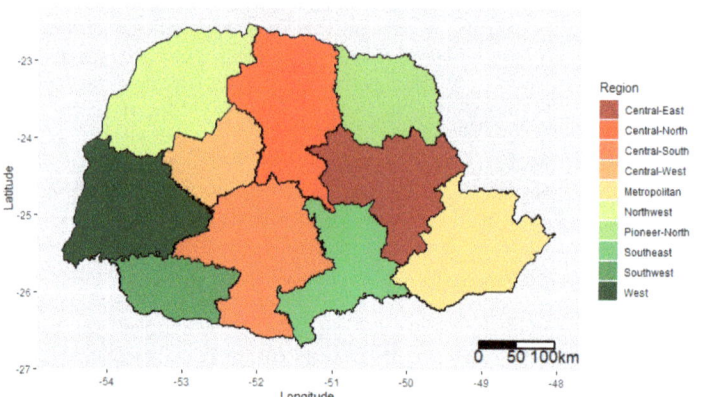

Figure 2. Mesoregions of Paraná state demarcated by the Brazilian Institute of Geography and Statistics (IBGE).

2.2. Data

2.2.1. Observed Data: Ground Gauge

The precipitation data used were acquired from 511 gauges (Figure 1) of the National Water Agency (ANA), through the Hidroweb Portal (http://www.snirh.gov.br/hidroweb/serieshistoricas, accessed on 1 March 2020), which is available to the public after consistency checks. The analysis of consistency of precipitation data is aimed at the identification and correction of errors related to data collection, as well as filling gaps in the monthly precipitation series. All available daily data were used for analysis on a daily scale, while, for analysis on a monthly scale, monthly totals were excluded when there were more than 5% of daily failures in the corresponding month. The daily data were accumulated (totaled) to produce the monthly information. The analyzed time series was from 1 June 2000 to 31 December 2018.

2.2.2. Satellite Data: IMERG

Precipitation data from remote sensing were acquired as daily and monthly temporal resolution and 0.1° spatial resolution, from the satellite constellation of the GPM, IMERG version 6 product, distributed by Goddard Earth Sciences Data and Information Services Center Distributed Active Archive Center (GES DISC DAAC), available online on http://mirador.gsfc.nasa.gov/com, accessed on 15 March 2020. The daily and monthly products "Final Run" were used, and the time series analyzed was coincident with that of the gauges (1 June 2000 to 31 December 2018). The IMERG algorithm operates to intercalibrate, merge, and interpolate all satellite microwave precipitation estimates, microwave-calibrated infrared estimates, gauge observations, and other data from potential sensors from the TRMM and GPM eras [31]. The "Final Run" product includes microwave-infrared estimates without gauge adjustment and the calibrated product based on the Global Precipitation Climatology Centre monthly gauge analysis [30]. In general, the "Final Run" products present bias correction and more accurate results than the other products supplied in near real-time (Early and Late Run) [39].

2.2.3. Performance Analyses

The IMERG products were compared to ground-based gauge using statistical indices. For this, the IMERG data were sampled at the exact location of the gauge stations. The performance of the IMERG products assessed using the coefficient of determination (R^2), the Kling-Gupta efficiency index (KGE), the error skewness (SK), the mean error (MBE), the mean absolute error (MAE), and the root of the mean square error (RMSE). The data qualitative assessment was performed using the categorical skills metrics: probability of detection (POD), false alarm ratio (FAR), and critical success index (CSI). Such metrics are used in several studies to assess the performance of satellite products [19,39–41]. The equations for the metrics used are shown in Table 1. The rainfall threshold was considered as amounts higher than 1 mm day^{-1}, as used by [42].

Table 1. Summary of statistical indices used to evaluate the satellite precipitation products.

Index	Unit	Equation *	Best Value
Determination coefficient (R^2)	-	$R^2 = \frac{\left\{\sum_{i=1}^{n}(P_i - \overline{P})(O_i - \overline{O})\right\}^2}{\sum_{i=1}^{n}(P_i - \overline{P})^2 \sum_{i=1}^{n}(O_i - \overline{O})^2}$	1
Mean error (MBE)	mm	$MBE = \frac{1}{n}\sum_{i=1}^{n}(P_i - O_i)$	0
Mean absolute error (MAE)	mm	$MAE = \frac{1}{n}\sum_{i=1}^{n}|P_i - O_i|$	0
Root of the mean square error (RMSE)	mm	$RMSE = \sqrt{\frac{1}{n}\sum_{i=1}^{n}(P_i - O_i)^2}$	0
Kling–Gupta Eficiency (KGE)	-	$KGE = 1 - \sqrt{(r-1)^2 + (\alpha-1)^2 + (\beta-1)^2}$	1
Coefficient of skewness (SK)	-	$SK = \frac{3(\overline{P} - M_o)}{\sigma}$	0
Probability of detection (POD)	-	$POD = \frac{Hits}{Hits + Misses}$	1
Critical success index (CSI)	-	$CSI = \frac{Hits}{Hits + FalseAlarm + Misses}$	1
False alarm ratio (FAR)	-	$FAR = \frac{FalseAlarm}{Hits + FalseAlarm}$	0

* where, Oi is the observed data of gauges of order i; Pi is the estimated order data (IMERG) of order i; r = rPearson, α is the ratio between simulated variance and observed variance, and β is the ratio between simulated mean and observed mean; $\overline{P}$ is the average of the estimated data (IMERG); $\overline{O}$ is the average of the observed data (gauges); M_o is the median error of satellite precipitation estimates; Hits are the days when IMERG and the station recorded rain; FalseAlarm are the days when IMERG recorded rain, but the gauges did not; Misses are the days when IMERG did not register rain, but the gauges did.

R^2 measures how much of the variability of one variable can be explained by the other variable. KGE is an objective index which assesses error in terms of correlation, variability, and bias. MBE is a simple average of the error and informs if the estimate on average over- or underestimate the observed data. MAE and RMSE are error metrics in the same units as the observed variable and represent the average magnitude of the error. In this case, RMSE penalizes large errors by squaring the differences between those observed and those estimated. SK represents the distributions of errors. A positive [negative] SK value indicates a median error is smaller [larger] than the mean error, i.e., there is a higher frequency of errors smaller [larger] than the mean error. If the mean error (MBE) is centered close to 0, a positive [negative] SK value can indicate a higher frequency of underestimations [overestimations]. Errors were calculated by subtracting observed from predicted values. POD indicates the fraction of rain events detected correctly with the total number of events detected by satellite; FAR measures the fraction of occurrences of unreal events among the total number of events detected by satellite; CSI denotes the proportion of rain events correctly detected by satellite to the total number of observed events.

Additionally, the IMERG's performance for estimating the precipitation was evaluated using a confusion matrix, for the daily and monthly products. The columns in this matrix shows the frequency distribution of IMERG precipitation amount within each ground gauge precipitation amount class. In the perfect scenario, the matrix should be antidiagonal and present a value equal to 1 along the antidiagonal line and 0 for the all the other elements [43].

2.2.4. Analysis of Anomalies

For the analysis of anomalies, the monthly values of the gauges and the monthly products of IMERG were used. The investigation of the volume of precipitation for each month concerning its average (2000–2018), was carried out by calculating the normalized anomalies of precipitation with standard deviation (Equation (1)) [44].

$$X_{Anomaly} = \frac{(X_i - \overline{X}_{2000-2018})}{\sigma_{2000-2018}} \quad (1)$$

where, Xi is the month of the year analyzed, X is the monthly average of the 2000–2018 series, and σ is the monthly standard deviation of the time series. The mean ($\overline{X}$) and standard deviation (σ) were calculated using Equations (2) and (3), respectively.

$$\sigma = \sqrt{\frac{\sum_{i=1}^{n}(X - \overline{X})^2}{n}} \quad (2)$$

$$\overline{X} = \frac{\sum_{i=1}^{n}(X_i)}{n} \quad (3)$$

Significant anomalies (95% confidence interval) were $X_{Anomaly}$ values ranging between ±1.96 as adopted by [45–47]. Thus, a $X_{Anomaly} > |1.96|$ gives us a 95% confidence to assume the observations as an anomaly, i.e., different than the historical mean.

A schematic diagram is presented in Figure 3, with the data and performance for the overall evaluation process of this study.

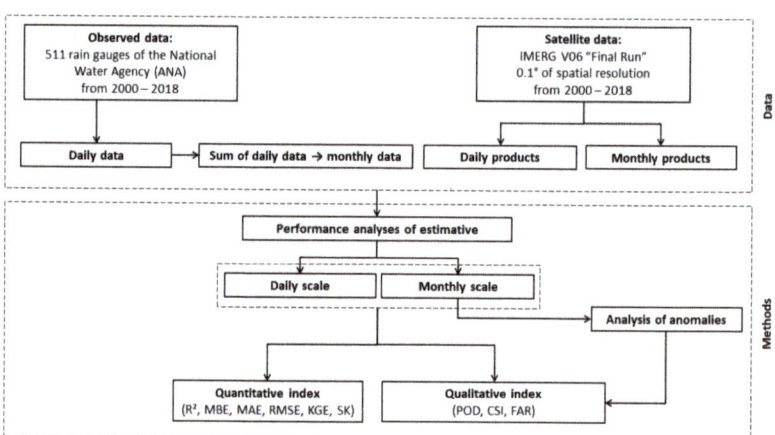

Figure 3. Schematic diagram of the source, resolution, and period of the data and metrics used for evaluation of the precipitation estimative by IMERG products.

3. Results

3.1. Temporal and Spatial Distribution of Precipitation

Figure 4 shows the average monthly precipitation (mm month^{-1}) observed by gauges and estimated by IMERG, between June 2000 and December 2018. The volume and spatial distribution of observed and estimated precipitation were consistent over all months of the year. The rainfall distribution density curve, which relates the precipitated volume to the observation frequency, had a similar distribution of the observed data from the gauges and those estimated by IMERG.

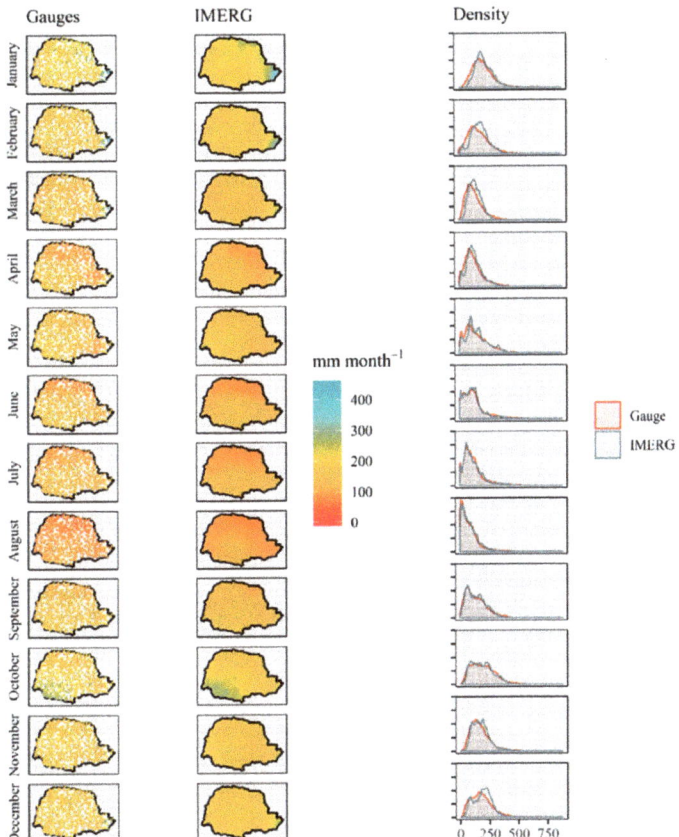

Figure 4. Average monthly rainfall observed during the study period for gauges and by IMERG, and density curve of monthly observations.

The highest frequency of precipitation observations occurs between 125–200 mm month^{-1} from October to March which is the wet season and is spatially well distributed in Paraná (Figure 4). In January and February, precipitation above 300 mm month^{-1} occurs in the coastal area. In October, the same behavior is observed in the southwest of the state. The IMERG data overestimated the high values of monthly precipitation recorded by the gauges, presenting a higher frequency of the monthly precipitation peaks in the wet season (Figure 4).

The dry season occurs between April and September, during which the precipitation decreases from the south to the north across the state, towards the Tropic of Capricorn, with a greater frequency observed between 0–125 mm month^{-1}. In this period, the months of July and August stand out as the driest months of the year, reaching a frequency of observation close to 0 mm month^{-1}.

3.2. Daily and Monthly Evaluation of IMERG Products

The relationship between IMERG daily and monthly precipitation and the data observed by gauges are presented in Figure 5. For the daily values, it is observed that IMERG overestimates the values, with a low coefficient of determination ($R^2 = 0.19$). At the monthly scale, precipitation values are closer to those observed, with R^2 of 0.73.

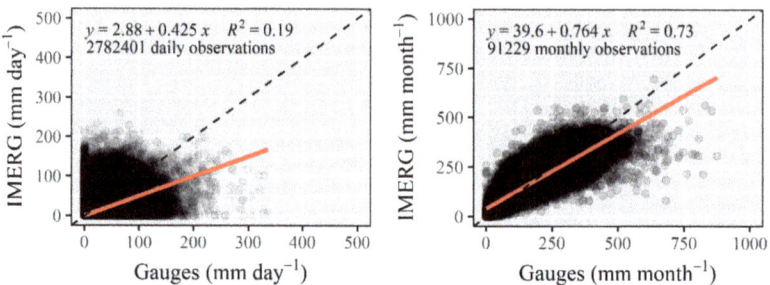

Figure 5. Regressions between daily and monthly data observed by gauges and IMERG.

Based on the general summaries of the metrics used in this study, presented in Figure 6, IMERG showed better performance for estimating monthly precipitation, compared to the daily product. The high KGE value (0.81) indicates a strong correlation between the monthly products of IMERG and the precipitation gauge data, which shows its ability to quantify monthly precipitation in the humid subtropical region. A lower accuracy is observed for daily products (KGE = 0.43), indicating a lower correlation between satellite precipitation data and pluviometers. This is due mainly to the high variability of precipitation over small areas on a daily scale. Our findings indicate that both IMERG products (daily and monthly) overestimate the observed precipitation, with a bias (MBE) of 0.19 and 5.98 mm, in the daily and monthly products, respectively. Regarding the errors, IMERG presented MAE of 5.64 and 35.90 mm and accuracy (RMSE) of 13.10 and 50.12 mm for daily and monthly products, respectively.

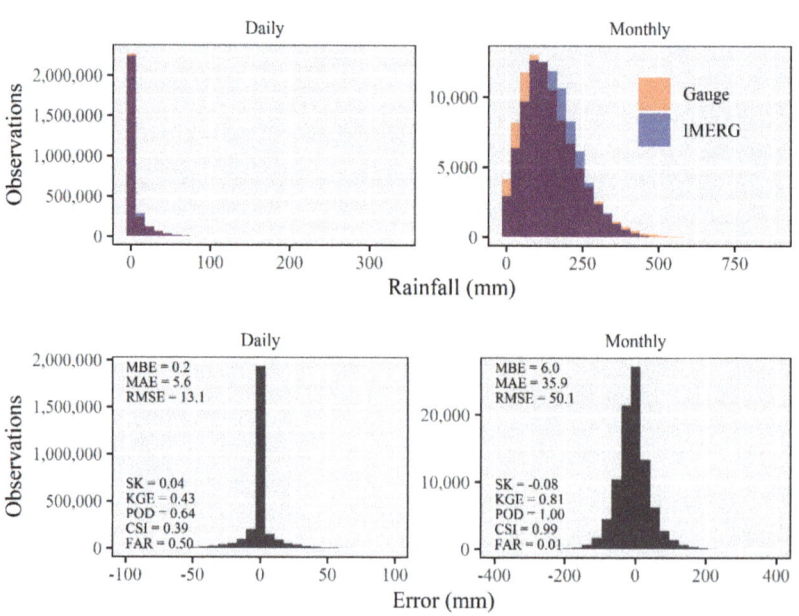

Figure 6. Error distribution and performance criteria of the IMERG products at daily and at monthly time scales.

Regarding the ability to detect rain at the monthly resolution, IMERG had a nearly perfect performance, with POD equal to 1, CSI of 0.99, and FAR of 0.01 (very close to ideal, 0). In a daily resolution, IMERG demonstrated a limited performance to detect rain events, with CSI of 0.39, a detection probability of 64%, and the risk of false alarm of 50%.

To better understand the accuracy of the IMERG products in different ranges of precipitation observations, a confusion matrix is shown in Figure 7. At the daily scale, the IMERG generally underestimate rainfall events, as seen by a large number of rainfall events greater than 1 mm) predicted in the range 0–1. This agrees with the positive SK value at daily scale (Figure 6). Considering the positive values of the error metrics showed in Figure 6 (MBE, MAE and RMSE), the positive errors of IMERG daily products could offset its negative errors.

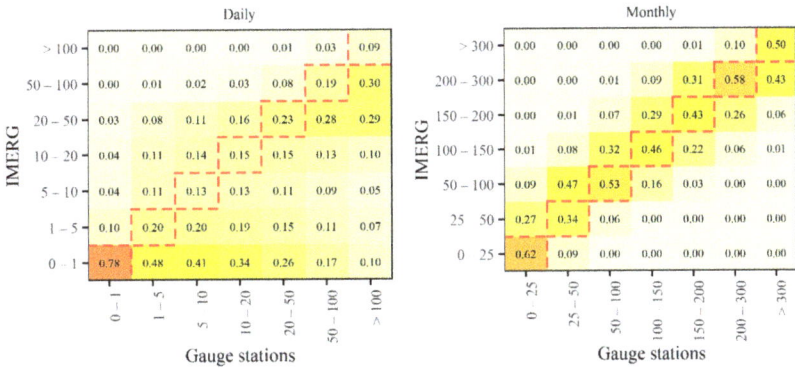

Figure 7. Confusion matrix for different precipitation ranges at daily and monthly scales.

In contrast, monthly precipitation estimates of the IMERG generally overestimate events as observed in gauge stations (negative SK in Figure 6). Only for months with higher rainfall rates (>200 mm) the IMERG product underestimates observed data. The underprediction at both daily and monthly scale can be attributed to the IMERG spatial and temporal resolution. Heavy rainfall events usually occur in shorter time intervals and small areas. Thus, sparse heavy rainfall events can be underestimated in the daily scale and not be accounted for in the monthly scale, even though the "Final run" relies on the calibration using monthly observations.

The spatial distribution of the metrics for the Paraná regions (Figure 2) at daily and monthly scale are summarized in Table 2. Less accuracy was observed in the Southwest, West, Central-West, and Metropolitan regions in daily and monthly products.

Table 2. Summary of error metrics on a daily and monthly scale of the satellite precipitation products in Paraná regions. The metrics were calculated based on mean areal precipitation.

Region	RMSE	MAE	MBE	RMSE	MAE	MBE
	(mm day^{-1})			(mm month^{-1})		
Central-South	13.70	6.19	0.02	48.50	36.30	0.83
Central-West	13.40	5.79	0.14	51.20	37.80	4.44
Central-East	11.90	5.24	0.18	43.20	31.90	5.52
Metropolitan	13.10	5.99	0.44	56.90	41.80	13.20
Northwest	12.40	5.14	0.31	50.50	36.90	9.37
Central-North	12.20	5.17	0.14	44.40	32.10	4.25
Pioneer-North	11.30	4.66	0.26	46.90	33.40	7.75
West	14.30	6.12	0.08	55.40	39.50	2.38
Southeast	12.60	5.58	0.31	41.60	31.10	9.41
Southwest	14.80	6.50	0.17	48.20	35.80	5.54

The spatial distribution of the error metrics for estimating daily and monthly rainfall by IMERG, for each gauge in the state of Paraná, are shown in Figure 8. Corroborating the results presented in Table 2, the MBE and MAE values are spatially well distributed in the study area with close values to the averages shown in Figure 5, for daily and monthly data. However, less accurate metrics were observed in the coastal areas (eastern part of the state), where IMERG presents, greater disagreements in some stations. Since the IMERG pixel covers an extensive area, large variability in precipitation is masked in areas where orographic effects are prevalent. This is evident near the coastal region for the daily monthly products and over the southwest region for daily products where there are abrupt changes in elevation (Figure 1). The SK presented homogeneous spatial distribution in a daily scale, and a greater number of negative values over the west region (ranging between 0 and −0.5), which suggest an overestimation of precipitation in this area. The KGE was also homogeneous across Paraná, with values close to 0.43 for the daily and 0.81 for the monthly products, as shown in Figure 8.

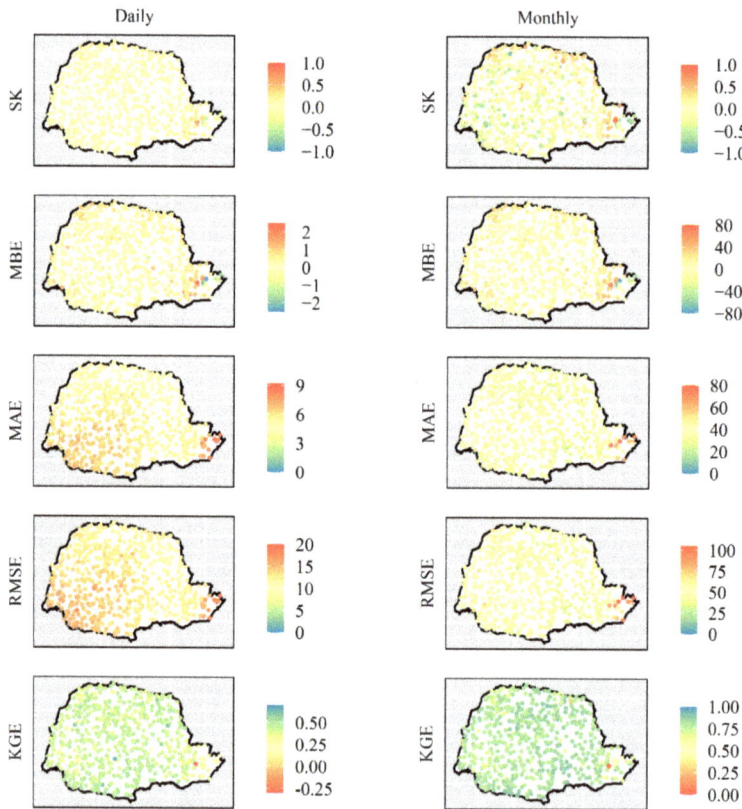

Figure 8. Spatial distribution of performance metrics on a daily and monthly scale.

The distribution of POD, CSI, and FAR (Figure 9) showed good performance of IMERG's products in detecting monthly rain events throughout Paraná, with values very close to ideals in the entire area. The performance of IMERG's products in estimating daily rainfall also was homogeneous across the state area, with the worst performance on the coast, in agreement with the statistical metrics.

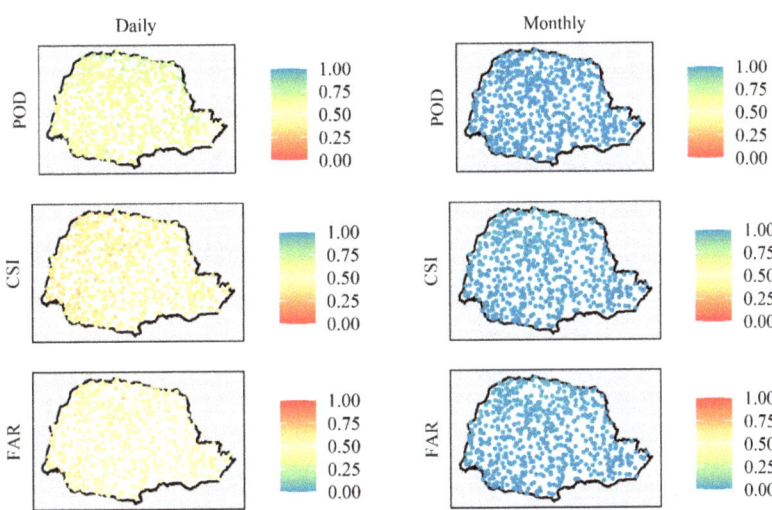

Figure 9. Spatial distribution of the qualitative metrics of IMERG performance on a daily and monthly scale.

Concerning the daily and monthly RMSE, higher values are found in the southwest and coastal areas of Parana, which correspond to areas with high volumes of rain in the autumn and summer, respectively, and the highest volumes of annual rainfall in Paraná (Figure 4). Thus, the precipitation estimates by IMERG performed better in the drier areas of the state. This behavior is better confirmed by a seasonal assessment of the errors. In Figure 10 the performance metrics for the monthly IMERG product grouped by seasons are presented. For example, IMERG showed better metrics in the winter (JJA = June, July, and August), which is the period with lowest rainfall rates. Likewise, the monthly predictions during autumn (MAM = March, April, and May) and spring (SON = September, October, and November) showed better scores than during summer (DJF = December, January, and February).

The orographic effect of the coastal region is not captured in any of the seasons. Only in winter (JJA) the errors were smaller because of the lower rainfall rates. Higher RMSEs were already expected during the wet season, because of the higher magnitude in rainfall rates. However, summer (DJF), also showed the highest positive bias, i.e., IMERG overestimates rainfall observed in gauge stations. The SK presented an overall tendency of median errors higher than the mean error (negative SK). This can be attributed to the IMERG product generally overestimating monthly rainfall rates (Figure 7) and resulting in negative SK, especially in seasons or regions with MBE closer to 0.

3.3. Rainfall Anomalies between 2000 and 2018

The spatial distribution of the performance of IMERG's monthly resolution products in detecting anomalies observed by the gauges is shown in Figure 11. In general, IMERG showed a limited performance for detecting anomalies across the state, considering +/−1.96 monthly standard deviation. The best performance was observed in the south-central region of Paraná, with POD above 0.75, CSI above 0.50, and FAR below 0.50. The worst performance occurred in the northeast region, the region with the lowest annual rainfall, and in the coastal and southwest regions, which correspond to the regions with the highest annual rainfall in the state (Figure 1), agreeing with the worst performance of the daily and monthly product metrics (Figure 8).

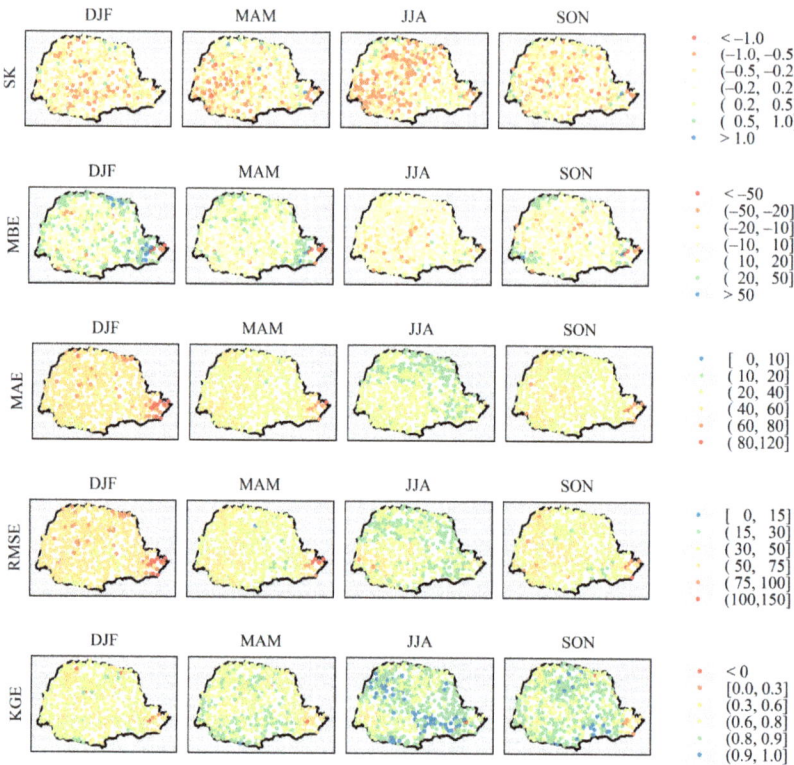

Figure 10. Seasonal distribution of the performance metrics for IMERG at a monthly scale.

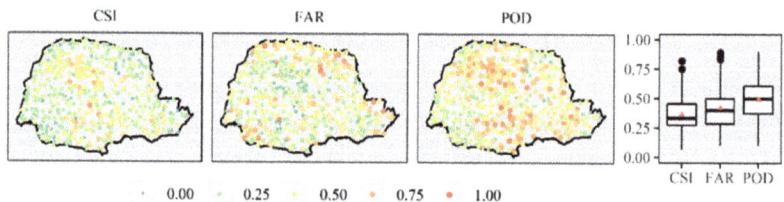

Figure 11. Spatial distribution of performance metrics for anomaly detection.

In the analysis of the boxplot of the metrics (POD, CSI, and FAR) used to assess the performance of IMERG in detecting anomalies, a relative deviation from the mean and extreme values of CSI and FAR were observed (Figure 11). Therefore, eight stations were selected randomly, in different regions of the state, to detect anomalies by the gauges and by IMERG, between the years 2000 and 2018, shown in Figure 12.

Positive anomalies were detected by IMERG monthly products and gauges stations in all regions. In general, the gauges stations detected the anomalies with greater magnitude compared to those estimated by remote sensing. On the other hand, only IMERG detected negative anomalies, for example in the gauge stations 2349036, 2449040, and 2452062.

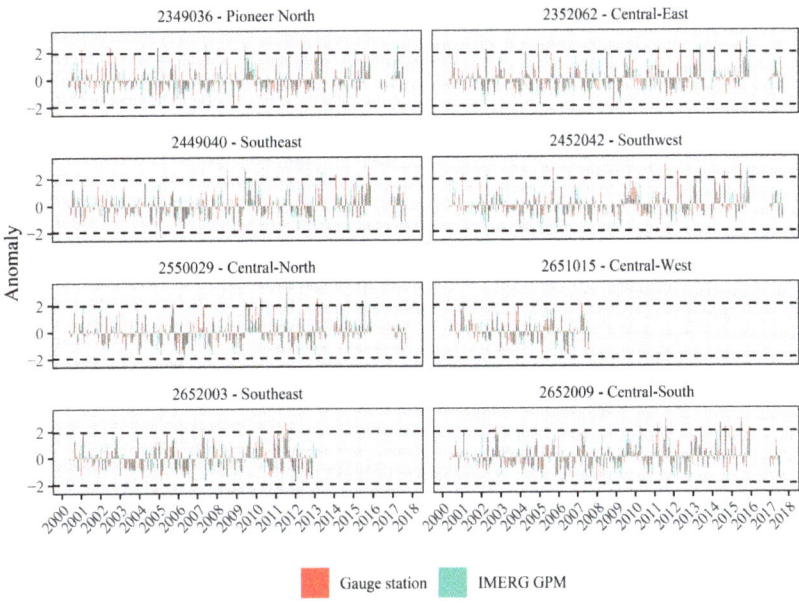

Figure 12. Anomalies for 9 stations selected at random. The dashed lines represent the values of −1.96 and 1.96 σ (monthly standard deviation).

4. Discussion

4.1. Temporal and Spatial Distribution of Precipitation

During the summer months, the Paraná State region typically receives a humidity air mass that moves from the Amazon to the southwestern Atlantic region, defined as the South Atlantic Convergence Zone (SACZ) [48], which is directly connected to the South American monsoon system along a northwest–southeast axis. The precipitation observed in the southwest region in October, on the other hand, is related to convective and frontal complexes [32,49]. The definition of the summer as the wet season and the spatial and temporal distribution in the study area presented here is corroborated by previous studies [32,49–51], confirming the precipitation estimates by IMERG through remote sensing approaches.

4.2. Daily and Monthly Evaluation of IMERG Products

Intensity and volume on a daily scale provide important information in hydrological applications, such as frequency analysis, daily precipitation event detection, and irrigation planning. The overestimation by IMERG daily products compared to ground data was observed by [29] when evaluating IMERG, GSMaP, and PERSIANN products over the whole globe (~26, 44, and 23% of bias, respectively). However, the authors observed higher correlation for IMERG products (~0.6) compared to GSMaP and PERSIANN (~0.5 and 0.4, respectively). Also, [28] related the overestimation of TMPA, IMERG-F, and GSMaP over South Brazil, but a slightly lower values of RMSE and mean error (0.86 and 0.09 mm day^{-1}, respectively) for GSMaP compared to TMPA (1.31 and 0.99 mm day^{-1}, respectively) and IMERG (1.31 and 1.01 mm day^{-1}, respectively). Regarding the underestimation of light precipitation and overestimation of heavy precipitation by IMERG presented in this study, [22] observed a tendency of IMERG daily products underestimate the frequency of rainfall events (<1 mm day^{-1}) and overestimate the frequency of intense rainfall events (>10 mm day^{-1}) across the Tibetan Plateau region. In this way, due to the limited performance of the daily products, the use of this data requires attention by the user and previous calibrations of the products at this temporal scale.

Regarding the comparison of the precipitation estimate on a monthly and daily scale, [18] also observed better performance of the monthly estimates of the precipitation when analyzing the TRMM products in Brazil. According to the authors, the monthly estimates are less affected by systematic errors than daily estimates. The IMERG products are calibrated using monthly data of in-situ gauging stations of the Global Precipitation Climatology Centre (GPCC) network [52], which can also explain the better performance of products on a monthly basis as compared to the daily scale. The better performance of the monthly precipitation product throughout Brazil also was confirmed by [20]. The authors observed an average of CC of 0.93 and RMSE of 23.20 mm when comparing IMERG version 5 in the state of Paraná. The better performance observed by [20], compared to that found in this study, may be due to the methods used for interpolation of observed capture data when assimilating them to the IMERG forecast fields. Typically, interpolation methods do not capture a large spatial variability of rain and the estimation is complex due to the spatial discontinuity [53]. The interpolation leads to the smoothing of high and low peaks of precipitation in a region, improving the relationship between observed and modeled values. With respect to the performance of IMERG for estimating seasonal precipitation, [22] showed different performance of the estimative in dry and wet seasons over sub-regions of Mainland China. The authors showed greater correlation of IMERG's products and ground gauges over the dry season for two sub-regions, as observed in this study, and the opposite for the other six. Thus, the investigation of the performance of satellites products in a regional scale is very relevant.

However, the IMERG overestimated the rainfall over regions of heavy rainfall as well as moderate rainfall. In coastal areas, even the difference of a short distance between the ocean and coastal mountains can induce failure of the satellite sensor to discriminate between the adjacent pixels of land and water, generating signal contamination and resulting in poor performance in estimating precipitation [17]. The limitation of the precipitation estimates by satellites (Global Precipitation Climatology Project–GPCP) in orographic regions was also observed in the Andes by [54]. According to the authors, the spatial resolution (2.5°) and dependence on passive microwaves and infrared precipitation recoveries used by satellites may imply the worst performance of representing precipitation in locations with orographic-type precipitation. Despite presenting higher spatial resolution (0.1°), IMERG products also showed less accuracy for estimating rainfall in orographic regions.

The estimative of the precipitation in the mountainous region tends to be underestimated towards the ocean (east area), where orographic rains occur. On the opposite side, after the sudden elevation change, it tends to overestimate the precipitation. Corroborating these results, [55] also observed trends of underestimation of precipitation by the TRMM (2A25 version 7) product in regions of orographic rainfall and overestimation in regions of valleys or flat areas in the southeastern Appalachians. According to the authors, this behavior occurs due to the spatial resolution and the correction of soil disorder made by the satellite. Another possibility for underestimating precipitation in the coastal region is that precipitation occurs while the top of the cloud is still relatively warm. Satellites are unable to fully identify rain, as heat exceeds infrared thresholds and the lower amount of ice in the air makes detection by passive microwave sensors difficult, and thus satellite products detect only part of the precipitation [2,56,57].

In addition, in mountainous regions, precipitation is extremely variable and there are changes in rainfall distribution over short distances [58], which can result in a representation of precipitation with less accuracy in these areas, since satellite products have the limitation of estimating precipitation considering the pixel size, when compared to gauges that measure the precipitation in-situ.

In the coastal area, the highest peak of precipitation occurs in the summer and is related to the predominant role of the Atlantic Tropical Mass [59], which finds the mountain as a physical barrier, culminating in orographic rains, with great volume over a short duration interval. Such an event may not be fully captured by satellites, as observed by [60] on the west coast of the United States, in the analysis of the precipitation of six satellite products (AFWA, TMPA–3B42, TMPA–3B42RT, CMORPH, PERSIANN, and NRL) and [33] in China, with IMERG version 5 products. Despite notable improvements in version 6 of IMERG's algorithm over previous versions as detailed by [30,58] observed lower accuracy of IMERG version 6 for estimating rainfall on the coastline of the Adriatic Sea in Europe, due to complex orography terrain. Additionally, [61] related the limitation of estimating precipitation in the Ebro River basin in Spain, in an area where the weather was dominated by the advection of wet maritime air masses. Thus, the measurement of the precipitation over coastal locations continues to be a challenge and deserves further research.

4.3. Rainfall Anomalies between 2000 and 2018

The northeastern area of the state is concentrated in the central part of the Paraná River basin, close to the climate transition line (subtropical and tropical) that separates several active climate systems in an area of greater atmospheric instability [32]. Thus, the precipitation estimate by the satellite may present a greater limitation (overestimation or underestimation) in this area due to atmospheric conditions.

As previously mentioned, high-intensity orographic rainfall occurs along the coast, which is estimated with less accuracy by satellite-based algorithms, resulting in poor performance in detecting anomalies. The southwestern region of Paraná has favorable conditions for the formation of severe storms and hail, which occur very quickly [62–64]. In September and October, the highest amounts of hail formation are observed in days per month [64], which exactly coincides with the period of greatest rainfall in the region (Figure 3). Thus, as on the coast of Paraná, the sensitivity of IMERG in detecting anomalies in this region may have lower performance compared to other regions.

According to [38], positive anomalies in rainfall can occur during the summer under conditions of El Niño Southern Oscillation (ENSO) in the south of Brazil. Over the years considered in this study, two El Niño events classified as "moderate," between 2002–2003 and 2009–2010, and one classified as "strong," between 2015–2016 (Golden Gate Weather Services, 2020) occurred. During these periods, positive anomalies were detected by IMERG, and gauges in all stations shown in Figure 12.

In the other years, IMERG and gauges detected some anomalies in all nine stations analyzed. However, only IMERG detected negative anomalies. [39] reported that IMERG version 5 products tend to underestimate precipitation amounts for rainfall rates 40–75 mm day^{-1}, but overestimate precipitation amounts for high rainfall rates (>80 mm day^{-1}). Thus, the anomalies detected in this study by the satellite could occur due to the under or overestimation of rain events. Thus, the use of IMERG products in anomaly studies should consider their variable performance for this purpose, requiring calibrations and prior data assessments.

5. Conclusions

In this study, the performance of IMERG version 6 products in estimating daily and monthly precipitation were evaluated in comparison with the data observed by 511 gauges distributed in the state of Paraná in Brazil. The results showed better metrics for monthly precipitation. In summary, the main findings of this study were as follows:

i. The volume and spatial distribution of observed and estimated rainfall are consistent across all months of the year in the monthly products of IMERG version 6, with similar rainfall distribution density curves.

ii. IMERG version 6 has a good relationship between precipitation estimates and those observed by gauges on the monthly time scale, with high correlation and accuracy, and low errors in statistical metrics. However, a lower performance was observed in estimating rainfall in regions with abrupt changes in topography along the coast, related to the lower accuracy when estimating orographic affected rainfall.

iii. The monthly products of IMERG version 6 performed very close to perfect considering qualitative assessments for the detection of rainfall events in this time scale throughout the study area.

iv. The daily estimates of IMERG version 6 were limited in representing the rainfall observed by the gauges, with little correlation between the data and low values of rain event detection rates. Although the gauges are direct observations and considered references, it is known there is great spatial variability in daily data, which is the probable cause of the low performance.

v. The detection of anomalies by the monthly products of IMERG version 6 showed limited performance over the years analyzed and the study area, probably due to the topography and rainfall regime in the northeast, coast, and southeast.

Based on the results presented here, IMERG version 6 can be used as a source of monthly precipitation data over the territory of Paraná. However, on a daily scale, prior calibration of the product is recommended to ensure the positive performance of the estimate on this time scale, especially for mountainous areas. Future improvements to IMERG version 6 products may increase its accuracy and favor its application for the detection of rain in coastal areas and anomalies. Also, studies that consider seasonal analyses and other time scales (hourly and half-hourly), areas with complex topographies, and the other products of IMERG version 6 (Early and Late Run) are recommended.

Author Contributions: Conceptualization, all the authors; methodology, J.G.N., D.A. and H.C.B.; software and validation D.A.; writing—original draft preparation, J.G.N. D.A. and H.C.B.; writing—review and editing, all the authors. All authors have read and agreed to the published version of the manuscript.

Funding: This research received no external funding.

Informed Consent Statement: Not applicable.

Acknowledgments: This study was financed in part by the Coordenação de Aperfeiçoamento de Pessoal de Nível Superior (CAPES–In English: Coordination of Improvement of Higher Education Personnel)—Finance code 001. Additional funding from the Daugherty Water for Food Global Institute at the University of Nebraska is also appreciated.

Conflicts of Interest: The authors declare no conflict of interest.

References

1. Shen, Y.; Xiong, A. Validation and comparison of a new gauge-based precipitation analysis over mainland China. *Int. J. Climatol.* **2016**, *36*, 252–265. [CrossRef]
2. Guo, H.; Chen, S.; Bao, A.; Behrangi, A.; Hong, Y.; Ndayisaba, F.; Hu, J.; Stepanian, P.M. Early assessment of Integrated Multi-satellite Retrievals for Global Precipitation Measurement over China. *Atmos. Res.* **2016**, *176*, 121–133. [CrossRef]
3. Kucera, P.A.; Ebert, E.E.; Turk, F.J.; Levizzani, V.; Kirschbaum, D.B.; Tapiador, F.J.; Loew, A.; Borsche, M. Precipitation from Space: Advancing Earth System Science. *Bull. Am. Meteorol. Soc.* **2013**, *94*, 365–375. [CrossRef]
4. Tapiador, F.J.; Turk, F.; Petersen, W.; Hou, A.Y.; García-Ortega, E.; Machado, L.A.; Angelis, C.F.; Salio, P.; Kidd, C.; Huffman, G.J.; et al. Global precipitation measurement: Methods, datasets and applications. *Atmos. Res.* **2012**, *104*, 70–97. [CrossRef]
5. Hou, A.Y.; Kakar, R.K.; Neeck, S.; Azarbarzin, A.A.; Kummerow, C.D.; Kojima, M.; Oki, R.; Nakamura, K.; Iguchi, T. The Global Precipitation Measurement Mission. *Bull. Am. Meteorol. Soc.* **2014**, *95*, 701–722. [CrossRef]
6. Hobouchian, M.P.; Salio, P.; Skabar, Y.G.; Vila, D.; Garreaud, R. Assessment of satellite precipitation estimates over the slopes of the subtropical Andes. *Atmos. Res.* **2017**, *190*, 43–54. [CrossRef]
7. Salio, P.; Hobouchian, M.P.; Skabar, Y.G.; Vila, D. Evaluation of high-resolution satellite precipitation estimates over southern South America using a dense rain gauge network. *Atmos. Res.* **2015**, *163*, 146–161. [CrossRef]
8. Falck, A.; Maggioni, V.; Tomasella, J.; Diniz, F.; Mei, Y.; Beneti, C.; Herdies, D.; Neundorf, R.; Caram, R.; Rodriguez, D. Improving the use of ground-based radar rainfall data for monitoring and predicting floods in the Iguaçu river basin. *J. Hydrol.* **2018**, *567*, 626–636. [CrossRef]
9. Kummerow, C.; Barnes, W.; Kozu, T.; Shiue, J.; Simpson, J. The tropical rainfall measuring mission (TRMM) sensor pack-age. *J. Atmos. Ocean. Technol.* **1998**, *15*, 809–817. [CrossRef]
10. Joyce, R.J.; Janowiak, J.E.; Arkin, P.A.; Xie, P. CMORPH: A method that produces global precipitation estimates from passive microwave and infrared data at high spatial and temporal resolution. *J. Hydrometeorol.* **2004**, *5*, 487–503. [CrossRef]
11. Mega, T.; Ushio, T.; Takahiro, M.; Kubota, T.; Kachi, M.; Oki, R. Gauge-Adjusted Global Satellite Mapping of Precipitation. *IEEE Trans. Geosci. Remote Sens.* **2018**, *57*, 1928–1935. [CrossRef]

12. Funk, C.; Peterson, P.; Landsfeld, M.; Pedreros, D.; Verdin, J.; Shukla, S.; Husak, G.; Rowland, J.; Harrison, L.; Hoell, A.; et al. The climate hazards infrared precipitation with stations—A new environmental record for monitoring extremes. *Sci. Data* **2015**, *2*, 1–21. [CrossRef] [PubMed]
13. Beck, H.E.; Van Dijk, A.I.J.M.; Levizzani, V.; Schellekens, J.; Miralles, D.G.; Martens, B.; De Roo, A. MSWEP: 3-hourly 0.25° global gridded precipitation (1979–2015) by merging gauge, satellite, and reanalysis data. *Hydrol. Earth Syst. Sci.* **2017**, *21*, 589–615. [CrossRef]
14. Huffman, G.J.; Bolvin, D.T.; Nelkin, E.J.; Wolff, D.B.; Adler, R.F.; Gu, G.; Hong, Y.; Bowman, K.P.; Stocker, E.F. The TRMM Multisatellite Precipitation Analysis (TMPA): Quasi-Global, Multiyear, Combined-Sensor Precipitation Estimates at Fine Scales. *J. Hydrometeorol.* **2007**, *8*, 38–55. [CrossRef]
15. Wang, C.; Tang, G.; Han, Z.; Guo, X.; Hong, Y. Global intercomparison and regional evaluation of GPM IMERG Version-03, Version-04 and its latest Version-05 precipitation products: Similarity, difference and improvements. *J. Hydrol.* **2018**, *564*, 342–356. [CrossRef]
16. Huffman, G.J.; Bolvin, D.T.; Nelkin, E.J. Integrated Multi-SatellitE Retrievals for GPM (IMERG) Technical Documentation. NASA/GSFC Code. Available online: http://pmm.nasa.gov/sites/default/files/document_files/IMERG_doc.pdf (accessed on 2 February 2020).
17. El Kenawy, A.M.; Lopez-Moreno, J.I.; McCabe, M.F.; Vicente-Serrano, S.M. Evaluation of the TMPA-3B42 precipitation product using a high-density rain gauge network over complex terrain in northeastern Iberia. *Glob. Planet. Chang.* **2015**, *133*, 188–200. [CrossRef]
18. Melo, D.D.C.D.; Xavier, A.C.; Bianchi, T.; Oliveira, P.T.S.; Scanlon, B.R.; Lucas, M.C.; Wendland, E. Performance evaluation of rainfall estimates by TRMM Multi-satellite Precipitation Analysis 3B42V6 and V7 over Brazil. *J. Geophys. Res. Atmos.* **2015**, *120*, 9426–9436. [CrossRef]
19. Fang, J.; Yang, W.; Luan, Y.; Du, J.; Lin, A.; Zhao, L. Evaluation of the TRMM 3B42 and GPM IMERG products for extreme precipitation analysis over China. *Atmos. Res.* **2019**, *223*, 24–38. [CrossRef]
20. Gadelha, A.N.; Coelho, V.H.R.; Xavier, A.C.; Barbosa, L.R.; Melo, D.C.; Xuan, Y.; Huffman, G.J.; Petersen, W.A.; Almeida, C.D.N. Grid box-level evaluation of IMERG over Brazil at various space and time scales. *Atmos. Res.* **2019**, *218*, 231–244. [CrossRef]
21. Chen, F.; Li, X. Evaluation of IMERG and TRMM 3B43 Monthly Precipitation Products Over Mainland China. *Remote Sens.* **2016**, *8*, 472. [CrossRef]
22. Krishna, U.V.M.; Das, S.K.; Deshpande, S.M.; Doiphode, S.L.; Pandithurai, G. The assessment of Global Precipitation Measurement estimates over the Indian subcontinent. *Earth Space Sci.* **2017**, *4*, 540–553. [CrossRef]
23. Moazami, S.; Najafi, M. A comprehensive evaluation of GPM-IMERG V06 and MRMS with hourly ground-based precipitation observations across Canada. *J. Hydrol.* **2021**, *594*, 125929. [CrossRef]
24. Islam, A. Statistical comparison of satellite-retrieved precipitation products with rain gauge observations over Bangladesh. *Int. J. Remote Sens.* **2018**, *39*, 2906–2936. [CrossRef]
25. Bhuiyan, A.E.; Yang, F.; Biswas, N.K.; Rahat, S.H.; Neelam, T.J. Machine Learning-Based Error Modeling to Improve GPM IMERG Precipitation Product over the Brahmaputra River Basin. *Forecasting* **2020**, *2*, 248–266. [CrossRef]
26. Oliveira, R.; Maggioni, V.; Vila, D.; Porcacchia, L. Using Satellite Error Modeling to Improve GPM-Level 3 Rainfall Estimates over the Central Amazon Region. *Remote Sens.* **2018**, *10*, 336. [CrossRef]
27. Liu, Z. Comparison of Integrated Multisatellite Retrievals for GPM (IMERG) and TRMM Multisatellite Precipitation Analysis (TMPA) Monthly Precipitation Products: Initial Results. *J. Hydrometeorol.* **2016**, *17*, 777–790. [CrossRef]
28. Rozante, J.R.; Vila, D.A.; Barboza Chiquetto, J.; Fernandes, A.D.A.; Souza Alvim, D. Evaluation of TRMM/GPM blended daily products over Brazil. *Remote Sens.* **2018**, *10*, 882. [CrossRef]
29. Chen, H.; Yong, B.; Shen, Y.; Liu, J.; Hong, Y.; Zhang, J. Comparison analysis of six purely satellite-derived global precipitation estimates. *J. Hydrol.* **2020**, *581*, 124376. [CrossRef]
30. Tang, G.; Clark, M.P.; Papalexiou, S.M.; Ma, Z.; Hong, Y. Have satellite precipitation products improved over last two decades? A comprehensive comparison of GPM IMERG with nine satellite and reanalysis datasets. *Remote Sens. Environ.* **2020**, *240*, 111697. [CrossRef]
31. Huffman, G.J.; Bolvin, D.T.; Nelkin, E.J.; Tan, J. Integrated Multi-Satellite Retrievals for GPM (IMERG) Technical Documentation. Available online: https://docserver.gesdisc.eosdis.nasa.gov/public/project/GPM/IMERG_doc.06.pdf (accessed on 2 February 2020).
32. Zandonadi, L.; Acquaotta, F.; Fratianni, S.; Zavattini, J.A. Changes in precipitation extremes in Brazil (Paraná River Basin). *Theor. Appl. Clim.* **2016**, *123*, 741–756. [CrossRef]
33. ANA. Conjuntura dos Recursos Hídricos no Brasil 2017. Relatório Pleno/Agência Nacional de Águas. Brasília. Available online: http://www.ana.gov.br (accessed on 18 March 2020).
34. IBGE-Instituto Brasileiro de Geografia e Estatística. Sinopse do Censo Demográfico Rio de Janeiro. Available online: https://www.in.gov.br/en/web/dou/-/resolucao-n-3-de-26-de-agosto-de-2019-212912380 (accessed on 8 May 2020).
35. Alvares, C.A.; Stape, J.L.; Sentelhas, P.C.; Gonçalves, J.L.D.M.; Sparovek, G. Köppen's climate classification map for Brazil. *Meteorol. Z.* **2013**, *22*, 711–728. [CrossRef]
36. Santos, L.J.C.; Oka-Fiori, C.; Canali, N.E.; Fiori, A.P.; Da Silveira, C.T.; Da Silva, J.M.F.; Ross, J.L.S. Mapeamento Geomorfológico do Estado do Paraná. *Rev. Bras. Geomorfol.* **2006**, *7*, 3–12. [CrossRef]

37. Grimm, A.M.; Pal, J.S.; Giorgi, F. Connection between Spring Conditions and Peak Summer Monsoon Rainfall in South America: Role of Soil Moisture, Surface Temperature, and Topography in Eastern Brazil. *J. Clim.* **2007**, *20*, 5929–5945. [CrossRef]
38. Grimm, A.M. Interannual climate variability in South America: Impacts on seasonal precipitation, extreme events, and possible effects of climate change. *Stoch. Environ. Res. Risk Assess.* **2010**, *25*, 537–554. [CrossRef]
39. Su, J.; Lü, H.; Zhu, Y.; Cui, Y.; Wang, X. Evaluating the hydrological utility of latest IMERG products over the Upper Huaihe River Basin, China. *Atmos. Res.* **2019**, *225*, 17–29. [CrossRef]
40. Tang, G.; Ma, Y.; Long, D.; Zhong, L.; Hong, Y. Evaluation of GPM Day-1 IMERG and TMPA Version-7 legacy products over Mainland China at multiple spatiotemporal scales. *J. Hydrol.* **2016**, *533*, 152–167. [CrossRef]
41. Chen, C.; Chen, Q.; Duan, Z.; Zhang, J.; Mo, K.; Li, Z.; Tang, G. Multiscale Comparative Evaluation of the GPM IMERG v5 and TRMM 3B42 v7 Precipitation Products from 2015 to 2017 over a Climate Transition Area of China. *Remote Sens.* **2018**, *10*, 944. [CrossRef]
42. Tan, M.L.; Duan, Z. Assessment of GPM and TRMM Precipitation Products over Singapore. *Remote Sens.* **2017**, *9*, 720. [CrossRef]
43. Shi, J.; Yuan, F.; Shi, C.; Zhao, C.; Zhang, L.; Ren, L.; Zhu, Y.; Jiang, S.; Liu, Y. Statistical Evaluation of the Latest GPM-Era IMERG and GSMaP Satellite Precipitation Products in the Yellow River Source Region. *Water* **2020**, *12*, 1006. [CrossRef]
44. Aragão, L.E.O.C.; Malhi, Y.; Roman-Cuesta, R.M.; Saatchi, S.; Anderson, L.O.; Shimabukuro, Y.E. Spatial patterns and fire response of recent Amazonian droughts. *Geophys. Res. Lett.* **2007**, *34*, 7. [CrossRef]
45. Junior, C.H.L.S.; Almeida, C.T.; Santos, J.R.N.; Anderson, L.O.; Aragão, L.E.O.C.; Silva, F.B. Spatiotemporal rainfall trends in the Brazilian legal amazon between the years 1998 and 2015. *Water* **2018**, *10*, 1220. [CrossRef]
46. Anderson, L.O.; Malhi, Y.; Aragão, L.E.O.C.; Ladle, R.; Arai, R.; Barbier, N.; Phillips, O. Remote sensing detection of droughts in Amazonian forest canopies. *New Phytol.* **2010**, *187*, 733–750. [CrossRef] [PubMed]
47. Lee, J.; Wong, D.W.S. *Statistical Analysis with ArcView GIS*; John Wiley and Sons, Inc.: Hoboken, NJ, USA, 2001.
48. Hirata, F.E.; Grimm, A.M. The role of synoptic and intraseasonal anomalies in the life cycle of summer rainfall ex-tremes over South America. *Clim. Dyn.* **2016**, *46*, 3041–3055. [CrossRef]
49. Boulanger, J.P.; Leloup, J.; Penalba, O.; Rusticucci, M.; Lafon, F.; Vargas, W. Observed precipitation in the Paraná-Plata hydrological basin: Long-term trends, extreme conditions, and ENSO teleconnections. *Clim. Dyn.* **2005**, *24*, 393–413. [CrossRef]
50. Grimm, A.M.; Ferraz, S.E.; Gomes, J. Precipitation anomalies in southern Brazil associated with El Niño and La Niña events. *J. Clim.* **1998**, *11*, 2863–2880. [CrossRef]
51. Terassi, P.M.D.B.; Galvani, E. Identification of Homogeneous Rainfall Regions in the Eastern Watersheds of the State of Paraná, Brazil. *Climate* **2017**, *5*, 53. [CrossRef]
52. Anjum, M.N.; Ding, Y.; Shangguan, D.; Ahmad, I.; Ijaz, M.W.; Farid, H.U.; Yagoub, Y.E.; Zaman, M.; Adnan, M. Performance evaluation of latest integrated multi-satellite retrievals for Global Precipitation Measurement (IMERG) over the northern highlands of Pakistan. *Atmos. Res.* **2018**, *205*, 134–146. [CrossRef]
53. Hewitson, B.C.; Crane, R.G. Gridded Area-Averaged Daily Precipitation via Conditional Interpolation. *J. Clim.* **2005**, *18*, 41–57. [CrossRef]
54. Schumacher, V.; Justino, F.; Fernández, A.; Meseguer-Ruiz, O.; Sarricolea, P.; Comin, A.; Venancio, L.P.; Althoff, D. Comparison between observations and gridded data sets over complex terrain in the Chilean Andes: Precipitation and temperature. *Int. J. Clim.* **2020**, *40*, 5266–5288. [CrossRef]
55. Duan, Y.; Wilson, A.M.; Barros, A.P. Scoping a field experiment: Error diagnostics of TRMM precipitation radar estimates in complex terrain as a basis for IPHEx. *Hydrol. Earth Syst. Sci.* **2015**, *19*, 1501–1520. [CrossRef]
56. Dinku, T.; Chidzambwa, S.; Ceccato, P.; Connor, S.J.; Ropelewski, C.F. Validation of high-resolution satellite rainfall products over complex terrain. *Int. J. Remote Sens.* **2008**, *29*, 4097–4110. [CrossRef]
57. Karaseva, M.O.; Prakash, S.; Gairola, R.M. Validation of high-resolution TRMM-3B43 precipitation product using rain gauge measurements over Kyrgyzstan. *Theor. Appl. Clim.* **2011**, *108*, 147–157. [CrossRef]
58. Navarro, A.; García-Ortega, E.; Merino, A.; Sánchez, J.L.; Kummerow, C.; Tapiador, F.J. Assessment of IMERG Precipitation Estimates over Europe. *Remote Sens.* **2019**, *11*, 2470. [CrossRef]
59. Vanhoni, F.; Mendonça, F. O Clima Do Litoral Do Estado Do Paraná. *Rev. Bras. de Clim.* **2008**, *3*, 3. [CrossRef]
60. Tian, Y.; Peters-Lidard, C.D.; Eylander, J.B.; Joyce, R.J.; Huffman, G.J.; Adler, R.F.; Hsu, K.-L.; Turk, F.J.; Garcia, M.; Zeng, J. Component analysis of errors in satellite-based precipitation estimates. *J. Geophys. Res. Space Phys.* **2009**, *114*, 24. [CrossRef]
61. Navarro, A.; García-Ortega, E.; Merino, A.; Sánchez, J.L.; Tapiador, F.J. Orographic biases in IMERG precipitation estimates in the Ebro River basin (Spain): The effects of rain gauge density and altitude. *Atmos. Res.* **2020**, *244*, 105068. [CrossRef]
62. Brooks, H.E.; Lee, J.W.; Craven, J.P. The spatial distribution of severe thunderstorm and tornado environments from global reanalysis data. *Atmos. Res.* **2003**, *67*, 73–94. [CrossRef]
63. Martins, J.A.; Brand, V.S.; Capucim, M.N.; Felix, R.R.; Martins, L.D.; Freitas, E.D.; Gonçalves, F.L.; Hallak, R.; Dias, M.A.F.S.; Cecil, D.J. Climatology of destructive hailstorms in Brazil. *Atmos. Res.* **2017**, *184*, 126–138. [CrossRef]
64. Beal, A.; Hallak, R.; Martins, L.D.; Martins, J.A.; Biz, G.; Rudke, A.P.; Tarley, C.R. Climatology of hail in the triple border Paraná, Santa Catarina (Brazil) and Argentina. *Atmos. Res.* **2020**, *234*, 104747. [CrossRef]

Article

Historic and Simulated Desert-Oasis Ecotone Changes in the Arid Tarim River Basin, China

Fan Sun [1,2], Yi Wang [1,3,*], Yaning Chen [1], Yupeng Li [1], Qifei Zhang [1,2], Jingxiu Qin [1,2] and Patient Mindje Kayumba [1,2]

1. State Key Laboratory of Desert and Oasis Ecology, Xinjiang Institute of Ecology and Geography, Chinese Academy of Sciences (CAS), Urumqi, Xinjiang 830011, China; sunfan18@mails.ucas.ac.cn (F.S.); chenyn@ms.xjb.ac.cn (Y.C.); liyupeng@ms.xjb.ac.cn (Y.L.); zhangqifei15@mails.ucas.ac.cn (Q.Z.); qinjingxiu17@mails.ucas.ac.cn (J.Q.); patientestime001@mails.ucas.ac.cn (P.M.K.)
2. College of Resources and Environment, University of Chinese Academy of Sciences, Beijing 100049, China
3. School of Water Resources and Hydropower Engineering, North China Electric Power University, Beijing 102206, China
* Correspondence: 51102473@ncepu.edu.cn; Tel.: +86-991-7823175

Abstract: The desert-oasis ecotone, as a crucial natural barrier, maintains the stability of oasis agricultural production and protects oasis habitat security. This paper investigates the dynamic evolution of the desert-oasis ecotone in the Tarim River Basin and predicts the near-future land-use change in the desert-oasis ecotone using the cellular automata–Markov (CA-Markov) model. Results indicate that the overall area of the desert-oasis ecotone shows a shrinking trend (from 67,642 km^2 in 1990 to 46,613 km^2 in 2015) and the land-use change within the desert-oasis ecotone is mainly manifested by the conversion of a large amount of forest and grass area into arable land. The increasing demand for arable land for groundwater has led to a decline in the groundwater level, which is an important reason for the habitat deterioration in the desert-oasis ecotone. The rising temperature and drought have further exacerbated this trend. Assuming the current trend in development without intervention, the CA-Markov model predicts that by 2030, there will be an additional 1566 km^2 of arable land and a reduction of 1151 km^2 in forested area and grassland within the desert-oasis ecotone, which will inevitably further weaken the ecological barrier role of the desert-oasis ecotone and trigger a growing ecological crisis.

Keywords: Tarim River Basin; desert-oasis ecotone; land-use change; CA-Markov model

1. Introduction

The spatial interface between two or more ecological regions and their material, energy, and structural and functional systems is called an ecotone or an ecological ecotone [1]. The desert-oasis ecotone is distributed along the periphery of an oasis and is characterized by a zone of desert vegetation that separates the extensive desert from the oasis [2]. The ecotone records the interaction and mutual transformation between the desert and oasis ecosystems [3] and serves as an ecological link connecting the two. A desert-oasis ecotone is a unique ecosystem between a desert and an oasis, usually characterized by low diversity, sparse cover, and dominance by perennial herbaceous grasses and semi-shrubs, such as *Phragmites australis*, *Tamarix ramosissima*, *Karelinia caspia*, and *Alhagi sparsifolia*. The ecotone can be used for ranching (of both livestock and wild animals); its vegetation can also increase the roughness of the underlying ground surface, thereby hindering the development of desertification and protecting the oasis from wind erosion and sand deposition [4–7]. At the same time, the desert-oasis ecotone is the interface between the oasis ecosystem and the desert ecosystem where energy, material, and information exchange occurs [8], which is highly sensitive to external environmental and human disturbances, affected easily by human activities, including the expansion of cultivated land

and urbanization and precipitation [9]. A desert-oasis ecotone being a natural ecological barrier that prevents the desert from expanding into the oasis, its analysis provides an important indicator and an early warning of ecological changes. The desert-oasis ecotone also plays a large role in the development of the oasis economy. Therefore, the ecotone has been a topic of significant research in recent years. Previous research on this ecotone primarily focuses on characterizing and classifying the vegetation diversity, microclimate, and soil moisture, among other parameters in the ecotone and/or oasis [8,10–12]. The comparison of these studies has partially revealed the causes of formation and the ecological environment of ecotones in deserts and oases of different arid areas, thus providing the empirical basis for the ecological protection of desert-oasis ecotones.

The observed changes and development of ecotones are closely related to the dynamic evolution of the overall environment and climate of the region, as well as human influencing factors. Maintaining the stability and development of oases requires a comprehensive analysis of long-term trends and of the causes of the changes in the ecotone as well as within the entire basin. However, such studies are generally lacking. Previous studies have also shown that changes in the ecotone are closely related to changes in land-use type, which provides research ideas for predicting future changes in the ecotone. Thus, the future dynamic evolution of the ecotone can be inferred by predicting land-use changes.

There are many methods of simulating and predicting the evolution of land-use patterns, such as system dynamics, CLUE-S, artificial neural network (ANN), and cellular automata–Markov chain (CA-Markov) methods [13–16]. The system dynamics model is based on cybernetics, information theory, and system theory to analyze the drivers of land-use change. At present, the system dynamics simulation software STELLA has not been fully combined with the spatial analysis function of GIS to implement land-use change simulation and has not played its role as a powerful dynamics system. [17]. The CLUE-S model has limitations, and the setting of some parameters in the model mainly relies on expert knowledge, which will bring a high degree of subjectivity [13]. The ANN simulation of land-use change requires long simulation times, and the method does not provide the user with a specific evolution formula and contains large errors [18]. The CA-Markov model not only retains the advantages of the Markov model for long-term prediction but also integrates the ability of the cellular automata (CA) model to simulate complex spatiotemporal system changes. Thus, the CA-Markov model can better simulate land-use changes in time and space and has been widely used [19–21].

In the Tarim River Basin, artificial oasis [22] and desertification processes are increasing [23]. As a result, the desert-oasis ecotone has rapidly decreased in size and ecological concerns have become increasingly prominent. Meanwhile, the rapid advancement of urbanization in the Tarim River Basin, as well as the continuous development of the social economy, has led to significant changes in the pattern of land use, which have produced a series of impacts on the ecological environment of the ecotone.

The main purposes of this study are (1) to analyze the spatial and temporal variability and driving forces of the desert-oasis ecotone in the Tarim River Basin from 1990 to 2015, (2) to evaluate the applicability of the CA-Markov model, and (3) to further predict the near-future land-use changes in the Tarim River Basin. This study will deepen our knowledge of the evolution of the desert-oasis ecotone, which has important implications for the protection of the ecological environment in the arid zone and the construction of an ecological civilization in the Silk Road Economic Belt.

2. Study Area

The Tarim River Basin is an inland basin located far from the ocean in northwest China (Figure 1). The area is characterized by a temperate arid continental climate with scarce precipitation and strong evaporation. In the study area, the average annual precipitation is about 53.14 mm, while the annual potential evaporation is much higher, about 2196 mm. The average annual temperature is about 3.9 °C, which is typical of an inland arid climate. The basin covers an area of 1.02×10^6 km^2 and is the largest inland river basin in China

(Figure 1). In response to the global climate change and increasing human activities, the natural ecosystems in the basin are facing a series of crises and challenges. Its fragile ecological environment possesses abundant natural resources [24]. The drainage network in the basin consists of the main stream of the Tarim River and 144 drainage systems associated with nine major tributary basins: the Yarkand River, the Aksu River, the Kaidu-Kongque River, the Hotan River, the Kaxgar River, the Weigan River, the Dina River, the Keriya River, and the Qarqan River. The tributaries to the main stream of the Tarim River form a centripetal shape around the Tarim Basin [25]. The Tarim River is a dissipative inland river whose runoff is mainly supplied by meltwater from glaciers and snow. The sources of the Tarim River runoff include glacial melt water, accounting for 48.2%; precipitation in the form of rain and snow, accounting for 27.4%; and river base flow, accounting for 24.4% of the total [26,27]. The Tarim River Basin is a typical oasis agricultural production area in China; the oasis area in the basin has been increasing, and arable land has increased during the past 30 years. This has led to a shortage of water in the basin, which is mainly used for agriculture. The demand for agricultural irrigation water is large and accounts for about 96% of the total water use in the Tarim River Basin [28].

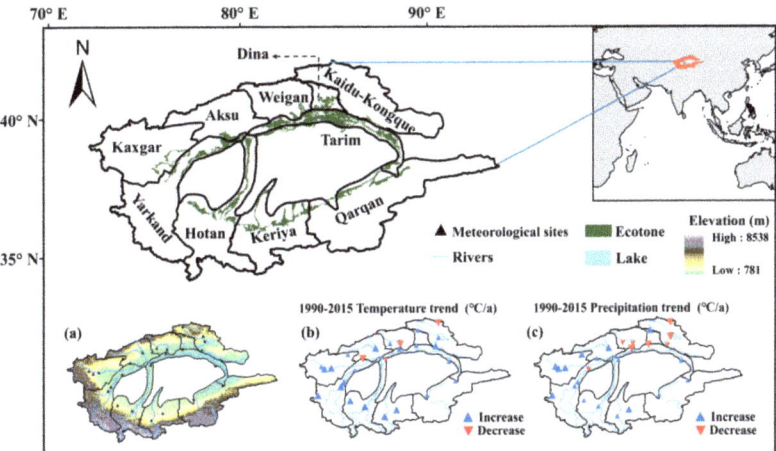

Figure 1. Map of the study area showing the Tarim River Basin and its nine main tributary river basins: the Kaidu-Kongque, Aksu, Weigan, Yarkand, Qarqan, Dina, Hotan, Kaxgar, and Keriya river basins; (**a**) the elevation of this area ranges from 781 to 8538 m above sea level; the spatial and temporal distribution of (**b**) temperature and (**c**) precipitation changes from 1990 to 2015. The black triangles represent the meteorological stations. Blue and red represent the increase and decrease, respectively, and the size of the triangle represents the magnitude of the change.

3. Materials and Methods

3.1. Materials

To analyze the dynamic evolution of the Tarim River Basin and its ecotone, and the factors controlling the observed changes, this article mainly uses remote sensing, land use, and meteorological data collected between 1990 and 2015. The analyzed historical trends in such parameters as the area of land-use transfer were subsequently used to extrapolate and predict the potential future changes in the study area.

3.1.1. Remote Sensing Data

Landsat 5 Thematic Mapper (TM)/Landsat 7 Enhanced Thematic Mapper (ETM) +/Landsat 8 Operational Land Imager (OIL) satellite imagery from 1990, 2000, and 2015 (a total of 78 pieces, Table S1) was used in this study. The acquisition time was from June to September of each year. The cloud cover was less than 10%, and the pixel resolution of the data set was 30 m × 30 m. The ENVI 5.3 software was used to perform radiometric calibration, atmospheric correction, and Normalized Difference Vegetation Index (NDVI) extraction calculations on the remote sensing images.

3.1.2. LUCC Data

The land-use data set for the Tarim River Basin was obtained from the existing remote sensing monitoring data set of land use in China. It was provided by the Data Center for Resources and Environmental Sciences, Chinese Academy of Sciences (http://www.resdc.cn). In this paper, we used three periods of Chinese land-use data (30 m × 30 m), collected in 1990, 2000, and 2015. The database offers the most comprehensive coverage of China's land use/land cover and has been used in a number of published studies [29,30]. The land-use types were classified into six categories: arable land, forest, grassland, water, built-up land, and unused land (Table S2).

3.1.3. Meteorological Data

Monthly scale meteorological information on temperature, precipitation, wind speed, humidity, and pressure from 1990 to 2015 for 26 meteorological stations in the Tarim River Basin were used to describe the recent changes in climatic conditions. The meteorological data were obtained from the China Meteorological Science Data Sharing Service Network, which have good continuity and has been tested for consistency. Station selection required the following: (1) the station was a national ground meteorological station, and (2) missing data accounted for less than 1% of the total data.

3.1.4. Groundwater Data

We selected groundwater-level data from the Yarkand, Kaxgar, and Weigan river basins. Three groundwater monitoring wells were selected for each basin. Groundwater data for the Yarkand River Basin (2004–2010) were obtained from the Kaxgar Hydrological Bureau, and the groundwater data for the Weigan River Basin (2000–2012) and the Kaxgar River Basin (2004–2010) were obtained from groundwater monitoring wells deployed by the Xinjiang Institute of Ecology and Geography of the Chinese Academy of Sciences for multi-year actual measurements.

3.2. Methods

3.2.1. NDVI Calculation

Vegetation indices are often used for vegetation analysis [31–33] and are typically formed by combining certain bands of image spectral data that possess vegetation-sensitive properties. Widely used vegetation indices include simple vegetation indices, ratio vegetation indices, NDVI, and transformed normalized vegetation indices [33,34]. The NDVI was used as the analysis index. It was calculated as follows:

$$NDVI = (NIR - R)/(NIR + R) \quad (1)$$

where NIR is the TM near-infrared band value and R is the TM visible-red band value. The range of the image element values was $-1 \leq NDVI \leq 1$. Negative values indicate that the ground cover consists of clouds, water, snow, etc.; 0 indicates the occurrence of rock or bare soil; positive values indicate that there is vegetation cover, and the values increase with increasing coverage.

ENVI 5.3 was used to perform radiometric calibration, atmospheric correction, and mosaic and other processing of the Landsat images. The NDVI calculation tools were used to calculate and output images from the three selected years separately. The calculation results were then loaded into ArcGIS10.6, which was used to eliminate outliers. The GIS raster calculator was used to extract the desert-oasis ecotone.

Sun et al. (2020) indicated that when the NDVI is between 0.05 and 0.35, the scope of the delineated ecotone can match the actual scope of the ecotone to a greater extent [35]. On this basis, we also combined visual interpretation and field verification to exclude the artificial protection forest and arable land in this range, and with this criterion, the desert-oasis ecotones of the Tarim River Basin in 1990, 2000, and 2015 were obtained.

3.2.2. Land-Use Transfer Matrix

This research quantitatively studied the conversion between land-use types in different years. ArcGIS and MATLAB were used to process the land-use TIFF data from the study area and to calculate the land-use transfer matrix from 1990 to 2015.

The land-use transition matrix was derived from system analysis and was used to quantitatively describe the mutual feedback relationship between the system state and the state transition. Put differently, the matrix describes the change in conditions of the system from time T to time T + 1 [36]. At present, it is widely used to describe the internal characteristics of the transfer structure and transfer direction of land-use/land-cover changes in basins. This method can not only describe the structural characteristics of the land-use area within a period of time but also effectively describes the transfer area and transfer direction of various land-use types at the beginning and the end of the period, as follows:

$$S_{ij} = \begin{bmatrix} S_{11} & \cdots & S_{1n} \\ \vdots & \ddots & \vdots \\ S_{n1} & \cdots & S_{nn} \end{bmatrix} \quad (2)$$

where S is the area; i and j (i, j = 1,2..., n) are the land-use type before and after transfer, respectively; S_{ij} is the area where land use changes from type i to type j; and n is the number of land-use types before and after the transfer.

3.2.3. The Standardized Precipitation Evapotranspiration Index

The Standardized Precipitation Evapotranspiration Index (SPEI) is widely used in global and regional drought detection and characterization [37–40]. This study used monthly site data and the Penman–Monteith formula to calculate differences in potential evapotranspiration, which were needed to calculate the SPEI. Specifics of the calculation can be found in Allen et al. [2].

A positive SPEI value indicates a relatively wet condition, whereas a negative SPEI value indicates a dry state. An SPEI value between -0.5 and 0.5 indicates normal condition. In this study, SPEIs at 3-month, 6-month, 9-month, and 12-month timescales were used for analysis. The ranges of SPEI values were divided into five classes according to the national meteorological drought scale (Table S3).

3.2.4. The CA-Markov Model

The CA-Markov model was used herein to simulate and predict future land use in the Tarim River Basin. The Markov model is a method for predicting the probability of occurrence at a specified time based on the Markov chain process theory. It is often used for the prediction of geographic events with no after-effect characteristics [41,42]. The evolution of land use has the nature of the Markov process. In fact, land-use type corresponds to the "possible state" of the Markov process, and the area or ratio of the conversion between land-use types can be represented as a state transition probability matrix [43]. The model is expressed as follows:

$$S_{(t+1)} = P_{ij} \times S_t \quad (3)$$

$$P_{ij} = \begin{bmatrix} P_{11} & \cdots & P_{1n} \\ \vdots & \ddots & \vdots \\ P_{n1} & \cdots & P_{nn} \end{bmatrix} \quad (4)$$

$$[0 \leq P_{ij} < 1 \text{ and } \sum_{j=1}^{n} P_{ij} = 1 (i.j = 1, 2 \cdots n)] \quad (5)$$

In Equation (3), $S_{(t+1)}$ is the system state at $t+1$ and P_{ij} is the state transition probability matrix.

The cellular automata (CA) model is a lattice dynamics model with discrete time and space states. It focuses on the interaction of different temporal and spatial characteristic cells and has powerful spatial calculation and simulation capabilities [44,45]. In terms of land-use prediction, the Markov model focuses on the prediction of the amount of land-use change but it cannot spatially express the spatial distribution of the various types of land-use changes. The cellular automata model can express the spatiotemporal dynamic evolution of complex spatial systems, thereby making up for the deficiency in the Markov model [46].

This study applied the CA-Markov model in the Idrisi 17.0 software to land-use data collected in 1990, 2000, and 2015. The first step in the approach used the 2000 data as the starting year and predicted the land use in 2015. The land-use predictions were then compared with the actual land use in 2015 to verify the reliability of the CA-Markov model simulation. Once verified, the spatial patterns in land use in 2030 were predicted using 2015 as the starting year (Figure 2). The images involved in the processing of the Idrisi software are portrayed as raster data, and the land raster size used in this analysis was 30×30 m. The spatial data processing was completed using ArcGIS software.

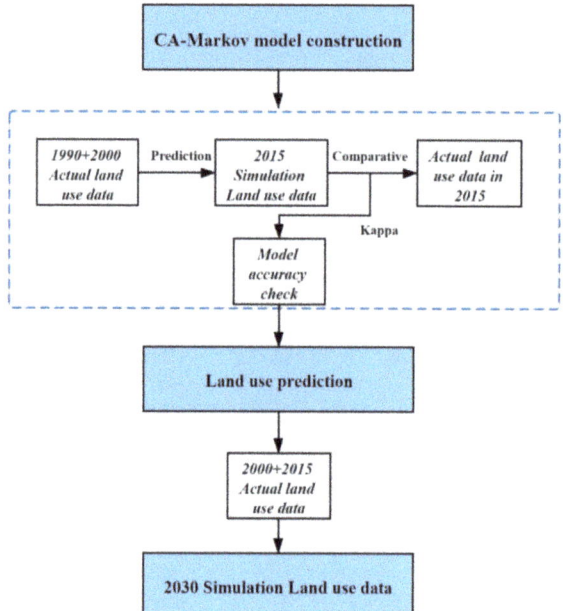

Figure 2. Flow chart showing primary steps in the simulation process.

3.2.5. The Kappa Index

The Kappa index is often used to interpret remote sensing accuracy and to evaluate the similarity of two spatial maps. The Kappa index was used in this study to verify the accuracy of the CA-Markov model for simulating the evolution of land use in each tributary basin [47]. The index is calculated as follows:

$$kappa = \frac{P_o - P_c}{P_p - P_c} \tag{6}$$

$$P_o = \frac{n_1}{n}, P_c = \frac{1}{N} \tag{7}$$

where P_o is the proportion of the raster that is correctly simulated, P_c is the desired proportion of correctly simulated raster grid cells, P_p is the proportion of correctly simulated grids for an ideal classification, n is the total number of grids, n_1 is the number of grids that are correctly simulated, and N is the number of land-use types (N = 6 in this study). The degree of consistency is weak when the Kappa index is less than 0.2 and significant when the Kappa index is greater than 0.8. A higher Kappa index means a better model simulation. (The detailed relationship between the Kappa index and consistency is shown in Table S4.)

4. Results
4.1. Desert-Oasis Ecotone and Land-Use Changes in the Tarim River Basin
4.1.1. Desert-Oasis Ecotone Changes in the Tarim River Basin

Figure 3 shows the changes in the extent of the desert-oasis ecotone in the Tarim River Basin and its various sub-basins in 1990, 2000, and 2015. The desert-oasis ecotone of the Tarim River Basin declined during this period, decreasing from 67,642 km² in 1990 to 46,613 km² in 2015. At the same time, the area of the desert in the study area expanded, with the proportion of desert area increasing from 59.88% in 1990 to 63.36% in 2015.

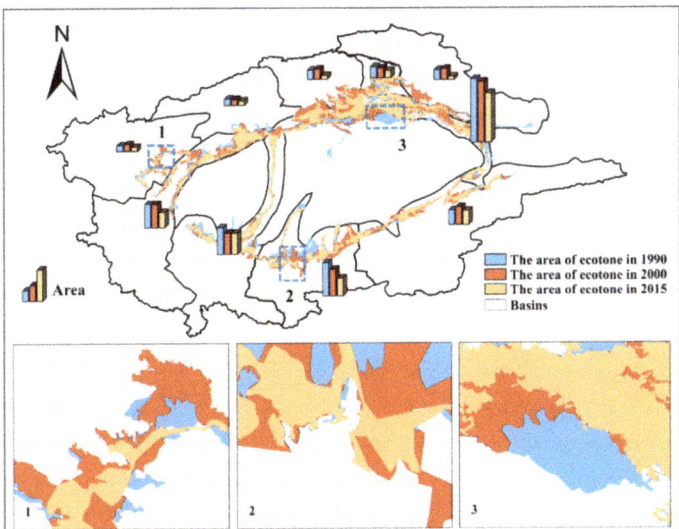

Figure 3. The areal extent of the ecotone in 1990, 2000, and 2015 in each basin. The lower maps show changes in the local ecotone in (**1**) the Weigan River Basin, (**2**) the Keriya River Basin, and (**3**) the main stream of the Tarim River.

At the sub-basin scale, the area of the ecotone of each sub-basin also decreased by different degrees, among which the three basins with the most significant area reduction were the Kaidu-Kongque, Weigan and Keriya river basins; the areas of reduction were 2223 km², 1704 km², and 5090 km², respectively (Table S5). While the ecotone decreased in size, the vegetation coverage also significantly decreased, as shown by a drop in the NDVI from 0.142 to 0.127. This indicates a deterioration in the desert-oasis ecotone in terms of area and quality between 1990 and 2015.

4.1.2. Land-Use Changes in the Tarim River Basin

The shrinkage of the desert-oasis ecotone in the Tarim River Basin is closely related to the strong land-use changes in the basin in recent decades. Table 1 further describes the area and proportion of each land-use type in the Tarim River Basin in 1990, 2000, and 2015. Overall, from 1990 to 2015, the area of arable land, water bodies, industrial land, and unused land increased. In contrast, the area of grassland decreased, whereas the area of forest land did not significantly change. Unused land was the dominant land-use type in this area (accounting for about 53%), increasing by 1419 km² in the preceding two decades. Grassland also represented a dominant land-use type in the area. Grasslands decreased from 35.97% in 1990 to 34.60% in 2015; the areal decrease was 8911 km². These land-use patterns reflect the intensity and nature of human activities; in fact, the expansion of arable land far exceeds the increase in the area of industrial land. Overall, the arable land area expanded drastically from 1990 to 2015, increasing by 7125 km², while the area of industrial land increased by only 66 km². The area of water bodies increased from 1990 to 2000 and then slightly decreased from 2000 to 2015. Over the entire period, the area of water bodies increased by 283 km². Forest land, which occupies a relatively small proportion of the total, exhibited insignificant changes in the area; its relative proportion was stable, at about 1.90%.

Table 1. Areas and proportions of land-use types in the study area in 1990, 2000, and 2015.

Type	1990		2000		2015	
	Area (km²)	Ratio (%)	Area (km²)	Ratio (%)	Area (km²)	Ratio (%)
Arable land	24,522.41	3.79	26,725.11	4.13	31,647.51	4.89
Forest land	12,055.43	1.86	12,688.41	1.96	12,062.48	1.87
Grassland	232,629.10	35.97	226,322.97	35.00	223,717.63	34.60
Water	34,774.43	5.38	35,508.45	5.49	35,057.50	5.42
Industrial land	1563.65	0.24	1497.20	0.23	1630.12	0.25
Unused land	341,124.92	52.75	343,917.04	53.18	342,543.93	52.97

Figure 4 shows the spatial distribution of the interconversion between different land uses in the Tarim River Basin during the period 1990–2015. The increase in arable land in 1990–2015 was mainly in the periphery of the original arable land and oasis and extended to the unused land. The increase in the area of arable land was mainly at the expense of grass land, unused land, and forest land. The increase in unused land was mainly distributed near the location of arable land and original grassland, and its areal expansion mainly came from the degradation of some arable land, forest land, and grassland. The increase in water bodies was mainly distributed in the foothills of the southern edge of the Tarim Basin, such as in the upstream areas of the Hotan and Yarkand river basins, and it was mainly derived from the transformation of grassland (Table S6).

4.2. Driving Force Analysis

Changes in the desert-oasis ecotone of the Tarim River Basin are inextricably linked to natural and anthropogenic factors. Therefore, this study investigated the intrinsic causes and drivers of the changes in the desert-oasis ecotone by changes in climatic parameters and anthropogenic activities.

4.2.1. Meteorological Factors

A total of 26 meteorological stations in the Tarim River Basin were selected, and temporal trends in mean annual temperature and annual precipitation were analyzed. Of the total stations, 22 stations exhibited an increase in temperature and 18 stations exhibited an increase in precipitation (Figure 1). Changes in dry and wet conditions were analyzed by calculating the SPEI for different time scales of drought in the Tarim River Basin from 1990 to 2015. At the 3-month, 6-month, 9-month, and 12-month time scales, average SPEI values exhibited a decrease, indicating enhanced aridification in the study area (Figure 5). Moderate droughts occurred in 2006 and 2008. Since 2000, the frequency and severity of droughts have become stronger, suggesting that droughts are an important reason for the accelerated decline in the ecotone after 2000. The results of the analysis also demonstrate that multi-scale SPEI can effectively show the degree of drought and drought duration in the Tarim Basin. SPEI of different scales show different degrees of interannual oscillations and interannual variability, but the overall direction of change was the same; the study area became more arid after 1990.

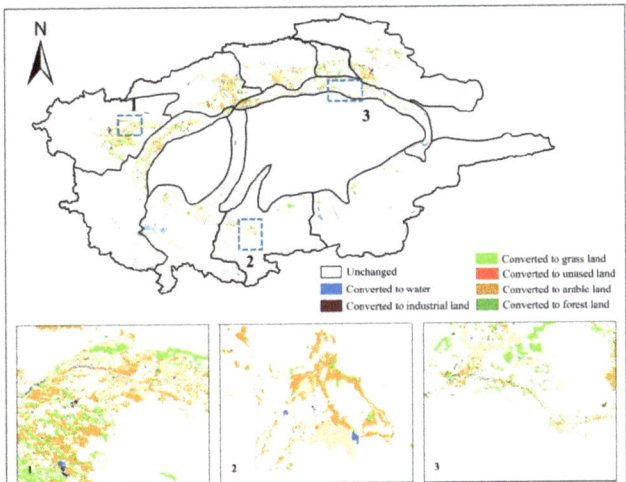

Figure 4. Land-use transfer map of the Tarim River Basin from 1990 to 2015. The lower maps show the local changes in (**1**) the Weigan River Basin, (**2**) the Keriya River Basin, and (**3**) the main stream of the Tarim River.

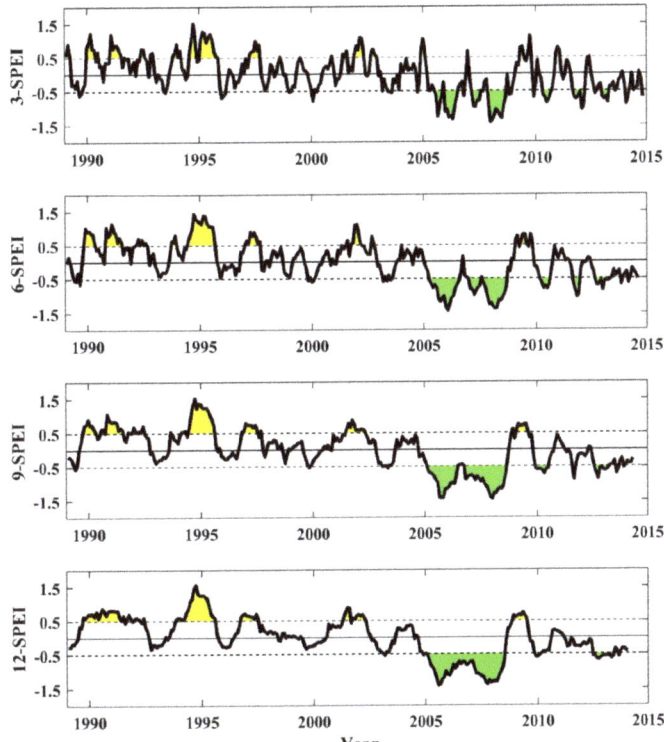

Figure 5. Time series of the 3-, 6-, 9-, and 12-month SPEI values in Tarim River Basin from 1990 to 2015.

4.2.2. Human Factors: Groundwater Changes

Within the Tarim River Basin, where precipitation is extremely low, groundwater is an important source of irrigation water. With the increase in arable land in the basin, the exploitation of groundwater has also increased. For example, groundwater monitoring data collected in the Yarkand, Kaxgar, and Weigan river basins show that groundwater levels have decreased during the 21st century (Figure 6). The most significant decline has been in the Weigan River Basin, where declines in groundwater levels of up to 2.48 m, 4.93 m, and 3.97 m have been observed. The expansion of oasis cultivation and irrigation in the basin has presumably caused a significant decrease in groundwater levels, which is a key hydrological element for the survival of natural vegetation and directly affects the growth and maintenance of natural vegetation in the desert-oasis ecotone.

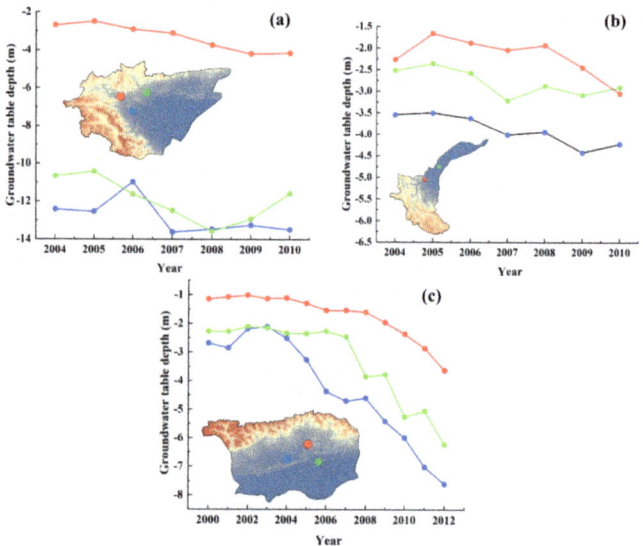

Figure 6. Changes in the groundwater table in (**a**) the Kaxgar River Basin, (**b**) the Yarkand River Basin, and (**c**) the Weigan River Basin. The dots represent the observation points of the groundwater level, and the fold lines represent the change in the trend of the groundwater level. The color of the fold line is consistent with the color of the observation point of the groundwater level.

4.3. Simulation and Prediction of Land Use in the Ecotone and Its Basin in 2030

4.3.1. Accuracy Verification

Since the shrinkage of the ecotone is closely related to land-use type changes, we also predicted future land-use changes based on the CA-Markov model in order to understand the future changes of the transition zone. To simulate future land-use changes in the basin, we first validated the accuracy of the 2015 land-use data. Since the desert-oasis ecotone of the Tarim River Basin is included in the whole Tarim River Basin area, we directly validated the accuracy of the simulation of the entire basin. The land-use structure of the study area was simulated and predicted based on the CA-Markov model. Land-use maps in 1990 and 2000 were defined as input data to simulate land use in 2015. Effectiveness of the simulation was assessed using spatial raster contrast in which the land-use type in specific spatial locations was compared to the actual 2015 land-use map.

Figure 7 shows that the simulation error, as determined by inconsistent land-use locations, was 3.62% of the total number of raster cells. Most inconsistencies appeared adjacent to water bodies and forest land. A total of 96.38% of the regions were consistent with the actual map in 2015. The Kappa index of the simulation result was 0.9551, also indicating the high reliability of the result and the CA-Markov model in predicting land-use types. The Kappa index of each sub-basin in the Tarim River Basin exceeded 0.80, which meets the accuracy requirement of the Kappa index.

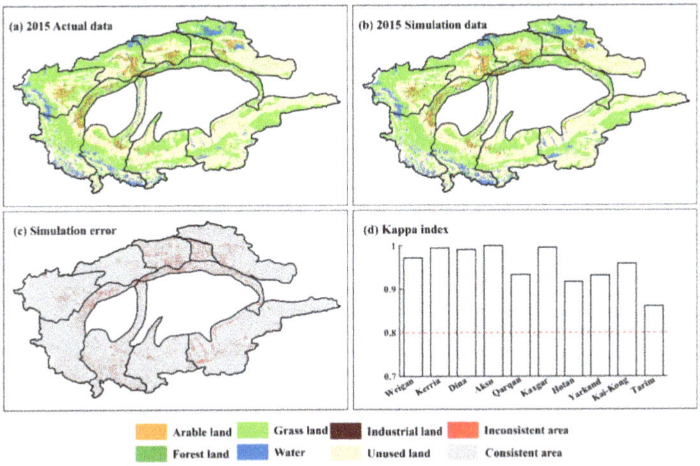

Figure 7. Comparisons of measured (**a**) and simulated (**b**) land-use maps of the Tarim River Basin in 2015, along with verification of forecast accuracy; (**c**) The simulation error between 1990 and 2015; (**d**) the Kappa index for each river basin.

The quantitative accuracy of the area of each land type in the simulation was also evaluated by comparing it with the actual area in 2015 (Table 2). The prediction error in 2015 was expressed as the absolute value of the error between the predicted and actual values of each land-use type area. Except for forest land and water bodies, where the error was 6% and 8.56%, respectively, the error in predicting the land-use types was within 5%. The error associated with industrial land and unused land was less than 1%, indicating that the simulation method had a high precision and credibility. Therefore, the CA-Markov model was able to effectively simulate the land-use changes in the study area and can be used to simulate future land use.

Table 2. Comparison between simulated and actual land use within the study area in 2015.

Type	Predicted Area (km²)	Ratio (%)	Actual Area (km²)	Ratio (%)	Quantitative Accuracy Error (%)
Arable land	33,073.94	5.04	31,647.51	4.89	4.51
Forest land	12,786.20	1.27	12,062.48	1.87	6.00
Grassland	216,813.78	33.36	223,717.63	34.60	3.09
Water	38,060.14	6.14	35,057.50	5.42	8.56
Industrial land	1638.14	0.26	1630.12	0.25	0.49
Unused land	344,293.60	53.93	342,543.93	52.97	0.51

4.3.2. Forecast of Changes in the Desert-Oasis Ecotone in the Tarim River Basin

Figure 8 predicts the spatial distribution of land-use change in the Tarim River Basin in 2030 using the CA-Markov model. The prediction suggests that past land-use trends in the Tarim River Basin will continue in 2030 (Table S7). The land type with the greatest change will be arable land, whose area will increase from 31,647 km² in 2015 to 34,909 km² in 2030, an increase of 10.31%. The land-use type exhibiting the largest decrease will be grassland (it will decrease by 12,497 km² compared to 2015), while forest land area, industrial land, water bodies, and unused land will exhibit a small increase. Future projections and simulations of land types within the Tarim River desert-oasis ecotone can infer its ecological status and development trends. This study simulated and predicted land-use changes within the ecotone by 2030, using the 2015 ecotone extent as the boundary. Land-use changes within the ecotone show a similar trend to the entire area. The simulation found that the arable land area in the ecotone will increase significantly, from 1033 km² in 2015 to 2599 km² in 2030, while the areas of forest and grassland will decrease by 318 km² and 833 km², respectively. As the area of natural vegetation decreases and the arable land increases within the ecotone, the quality of the future ecotone habitats will further deteriorate.

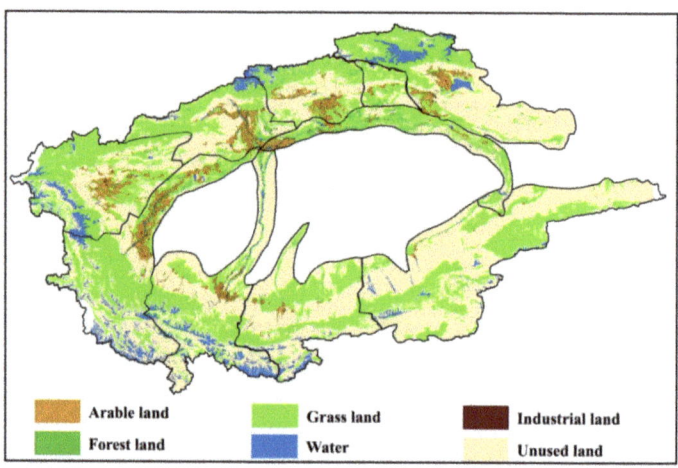

Figure 8. Land-use in Tarim River Basin in 2030.

5. Discussion

5.1. Criteria for the Classification of the Desert-Oasis Ecotone

There are different approaches to the current delineation of the desert-oasis ecotone. Most scholars define the ecotone as a zone of limited width along the edge of the oasis [48], while others extend the ecotone to the entire foothills or define it as terrain without a clear spatial location [49]. In this study, we found that the usual criteria for delineating the desert-oasis ecotone do not work well in the Tarim River Basin, which consists of a large area composed of many watersheds. As a result, the ecotone is not uniformly distributed. Therefore, the NDVI contour data calculated from the TM images (30 m resolution) were used here, in conjunction with the remote sensing images of land-use/land-cover change in 1990, 2000, and 2015. The land-use type corresponding to the NDVI values in the range between 0.05 and 1 were interpreted to be forest and grassland, whereas the land-use type corresponding to NDVI values between 0.35 and 0.95 were mostly artificial oases organized in neat patterns of blocks and strips. Thus, NDVI values in the range between 0.05 and 0.35 were selected as the desert-oasis ecotone in the Tarim Basin. The results were verified by using ENVI and ACRGIS to interpret, classify, and digitize the images from the three analyzed years, and the results were verified by fieldwork.

5.2. Combined Effect of Climate Change and Human Activities on the Desert-Oasis Ecotone

Analysis of temperature and precipitation in the study area shows that both parameters have increased since 1990 (Figure 1) but the increase in temperature has been much greater than the increase in precipitation. By calculating the SPEI values of the basin for different time scales, we found that there has been an increase in drought conditions in the study area since 2005, especially during the years of 2005, 2006, and 2008, when droughts reaching the moderate level occurred year-round. The increase in droughts, and external climatic changes in general, may be an important reason for the accelerated decrease in the area of the ecotone after 2000. Runoff (flow) in the Tarim River Basin is primarily generated in high mountain areas where glaciers and snowmelt recharge dominate. The input from these water sources is very sensitive to global climate change. The increased hydrological volatility and water resource uncertainty caused by climate change may lead to more prominent conflicts between water supply and demand in the oasis economy and desert ecosystem in the basin [50].

Because of the expansion of arable land, agricultural water consumption remains high. Agricultural water has long been the main form of water in this area. The proportion of agricultural water is too large, and the water structure is seriously imbalanced. The proportion of agricultural water in the basin has long been as high as about 95%, which is much higher than the Chinese average (65%) and the world average (70%). At present, the development of water resources in the Tarim River Basin has greatly exceeded the carrying capacity of regional water resources [28]. For example, the groundwater level in the Kaxgar River Basin dropped by nearly 1 m between 2004 and 2010 and the water crisis has become more prominent [25]. Groundwater overexploitation has led to the degradation of desert vegetation and damage to ecosystems [51]. For instance, the 321 km river cutoff in the downstream reach of the Tarim River has caused shrinkage and even disappearance of oases [52,53]. The reduction in ecological water has also led to a decrease in surface vegetation cover, NDVI, and the area of the desert-oasis ecotone.

5.3. Applicability of the Land-Use Change Model and Future Work

Human influence on land use reflects not only natural factors but also economic and social factors. Therefore, predicting land-use changes is extremely important for promoting natural, economic, and sustainable development and protecting ecological balance. Relevant data show that the irrigated area of the Tarim River Basin has increased by 67% in the last 30 years [53] and the future expansion of cultivated land area will decrease ecological space and ecological water, leading to a shrinkage in the desert-oasis ecotone and a decline in its function as an ecological barrier. Therefore, it is necessary to perform quantitative prediction and analysis of future land-use and pattern changes in the Tarim River Basin. The CA-Markov model is widely used in urban land-use pattern simulation and in the assessment of watershed land-use change; however, the model has rarely been applied in arid areas, especially in areas with complex geographical features combining desert-oasis characteristics. This study experimentally applied and optimized the CA-Markov model using relevant, scientifically selected factors, and improved the simulation accuracy to obtain reliable prediction results. As a case study, it provides a good example and basis for the prediction of land-use change in arid areas.

The model also has a few shortcomings related to the quantification of some factors. For example, the influence of road distance, water body distance, and various administrative policies were not considered and the analysis of the social, geographical, economic, and resource environment that affects land-use changes was not fully characterized. Therefore, to improve model accuracy, future research should consider the influence of natural and human factors on the change in the geospatial system. In addition, future research can attempt to establish a model of land-use/land-cover change under the joint action of several different influencing factors and different decision makers by adding weighting elements.

6. Conclusions

This paper investigated the dynamic evolution of the desert-oasis ecotone in the Tarim River Basin and predicted the near-future land-use change in the desert-oasis ecotone using the CA-Markov model. The main findings of this paper are as follows:

With the decrease in the NDVI (from 0.142 in 1990 to 0.127 in 2015) of the desert-oasis ecotone, the area of the ecotone also shrank from 67,642 km^2 in 1990 to 46,613 km^2 in 2015. In the context of global climate change, the temperature showed a significant increase compared with precipitation, which led to an obvious increase in aridity in the study area.

Meanwhile, the increase in arable land area led to a decrease in the groundwater table. The above factors have led to the shrinkage of the desert-oasis ecotone in the Tarim River Basin.

The CA-Markov model was verified to have good applicability in this study area, which was used to predict and simulate the future land dynamics of the study basin. Assuming the present development trend continues without intervention, the arable land area in the ecotone will increase from 1033 km^2 in 2015 to 2599 km^2 in 2030 and the woodland area and grassland area will decrease from 318 km^2 to 833 km^2, respectively. The main land-use types in the Tarim River Basin in 2030 will be arable land, unused land, and grassland.

In light of the above conclusions, it is necessary to establish reasonable management countermeasures for land-use planning in Tarim River Basin development to achieve sustainable development and protect the ecology of the basin.

Supplementary Materials: The following are available online at https://www.mdpi.com/2072-4292/13/4/647/s1: Table S1: The detailed information on Landsat images used in this study; Table S2: Land cover classification system; Table S3: SPEI categories; Table S4: Kappa index and consistency relationship; Table S5: Changes in the ecotone area of each basin in 1990, 2000, and 2015; Table S6: Transfer matrix of land-use area between 1990 and 2015 in the Tarim River Basin; and Table S7: Area of land-use types in the study area in 1990, 2000, 2015, and 2030.

Author Contributions: F.S. and Y.W. conceived the original design of this paper. Y.C. and Y.L. improved the structure of the paper. Q.Z., J.Q., and P.M.K. provided comments on this paper. All authors have read and agreed to the published version of the manuscript.

Funding: The research was supported by the National Youth Thousand Talents Project (Y771071001).

Institutional Review Board Statement: Not applicable.

Informed Consent Statement: Not applicable.

Data Availability Statement: The data presented in this study are available on request from the corresponding author.

Acknowledgments: We thank LetPub (www.letpub.com) for the linguistic assistance and scientific consultation provided during the preparation of this manuscript.

Conflicts of Interest: The authors declare no conflict of interest.

References

1. Chen, Y.; Chen, Y.; Zhu, C.; Li, W. The concept and mode of ecosystem sustainable management in arid desert areas in northwest China. *Acta Ecol. Sin.* **2019**, *39*, 7410–7417.
2. Allen, R.G. *Crop Evapotranspiration-Guidelines for Computing Crop Water Requirements*; FAO Irrigation and Drainage Paper No.56; FAO: Rome, Italy, 1998; p. 300.
3. Zhou, X.; Tao, Y.; Wu, L.; Li, Y.; Zhang, Y. Divergent Responses of Plant Communities under Increased Land-Use Intensity in Oasis-Desert Ecotones of Tarim Basin. *Rangel. Ecol. Manag.* **2020**, *73*, 811–819. [CrossRef]
4. Rosenfeld, D.; Rudich, Y.; Lahav, R. Desert dust suppressing precipitation: A possible desertification feedback loop. *Proc. Natl. Acad. Sci. USA* **2001**, *98*, 5975–5980. [CrossRef] [PubMed]
5. Chen, Y.; Zhang, X.; Fang, G.; Li, Z.; Wang, F.; Qin, J.; Sun, F. Potential risks and challenges of climate change in the arid region of northwestern China. *Reg. Sustain.* **2020**, *1*, 20–30. [CrossRef]
6. Gosz, J.R. Ecological functions in a biome transition zone: Translating local responses to broad-scale dynamics. In *Landscape Boundaries*; Springer: New York, NY, USA, 1992; pp. 55–75.
7. Traut, B.H. The role of coastal ecotones: A case study of the salt marsh/upland transition zone in California. *J. Ecol.* **2005**, *93*, 279–290. [CrossRef]
8. Li, X.; Yang, K.; Zhou, Y. Progress in the study of oasis-desert interactions. *Agric. Meteorol.* **2016**, *230–231*, 1–7. [CrossRef]
9. Hou, J.; Du, L.; Liu, K.; Hu, Y.; Zhu, Y. Characteristics of vegetation activity and its responses to climate change in desert/grassland biome transition zones in the last 30 years based on GIMMS3g. *Theor. Appl. Climatol.* **2019**, *136*, 915–928. [CrossRef]
10. Fan, Z.; Li, J.; Yue, T. Land-cover changes of biome transition zones in Loess Plateau of China. *Ecol. Model.* **2013**, *252*, 129–140. [CrossRef]
11. Ji, S.; Bai, X.; Qiao, R.; Wang, L.; Chang, X. Width identification of transition zone between desert and oasis based on NDVI and TCI. *Sci. Rep.* **2020**, *10*, 8672. [CrossRef]

12. Wang, J.; Gao, Y.; Sheng, W. Land use/cover change impacts on water table change over 25 years in a desert-oasis transition zone of the Heihe River basin, China. *Water* **2015**, *8*, 11. [CrossRef]
13. Verburg, P.H.; Soepboer, W.; Veldkamp, A.; Limpiada, R.; Espaldon, V.; Mastura, S.S.A. Modeling the spatial dynamics of regional land use: The CLUE-S model. *Environ. Manag.* **2002**, *30*, 391–405. [CrossRef] [PubMed]
14. Buckland, C.E.; Bailey, R.M.; Thomas, D.S.G. Using artificial neural networks to predict future dryland responses to human and climate disturbances. *Sci. Rep.* **2019**, *9*, 3855. [CrossRef]
15. Memarian, H.; Balasundram, S.; Talib, J.; Teh, C.; Sood, A.; Abbaspour, K. Validation of CA-Markov for simulation of land use and cover change in the Langat Basin, Malaysia. *J. Geogr. Inf. Syst.* **2012**, *44*, 542–554. [CrossRef]
16. Huang, Q.; Shi, P.; He, C.; Li, X. Modelling land use change dynamics under different aridification scenarios in Northern China. *Acta Geogr. Sin.* **2006**, *61*, 1299.
17. Shen, Q.; Chen, Q.; Tang, B.; Yeung, S.; Hu, Y.; Cheung, G. A system dynamics model for the sustainable land use planning and development. *Habitat Int.* **2009**, *33*, 15–25. [CrossRef]
18. Civco, D.L. Artificial neural networks for land-cover classification and mapping. *Int. J. Geogr. Inf. Syst.* **1993**, *7*, 173–186. [CrossRef]
19. Nouri, J.; Gharagozlou, A.; Arjmandi, R.; Faryadi, S.; Adl, M. Predicting urban land use changes using a CA–Markov model. *Arab. J. Sci. Eng.* **2014**, *39*, 5565–5573. [CrossRef]
20. Sang, L.; Zhang, C.; Yang, J.; Zhu, D.; Yun, W. Simulation of land use spatial pattern of towns and villages based on CA–Markov model. *Math. Comput. Model.* **2011**, *54*, 938–943. [CrossRef]
21. Kamusoko, C.; Aniya, M.; Adi, B.; Manjoro, M. Rural sustainability under threat in Zimbabwe–Simulation of future land use/cover changes in the Bindura district based on the Markov-cellular automata model. *Appl. Geogr.* **2009**, *29*, 435–447. [CrossRef]
22. Chen, H.; Chen, Y. Changes of desert riparian vegetation along the main stream of Tarim River, Xinjiang. *Chin. J. Ecol.* **2015**, *34*, 3166.
23. Zhao, R.; Chen, Y.; Shi, P.; Zhang, L.; Pan, J.; Zhao, H. Land use and land cover change and driving mechanism in the arid inland river basin: A case study of Tarim River, Xinjiang, China. *Environ. Earth Sci.* **2013**, *68*, 591–604. [CrossRef]
24. Chen, Y.; Hao, X.; Chen, Y.; Zhu, C. Study on water system connectivity and ecological protection countermeasures of Tarim River Basin in Xinjian. *Bull. Chin. Acad. Sci.* **2019**, *34*, 1156–1164.
25. Fang, G.; Yang, J.; Chen, Y.; Li, Z.; Ji, H.; De Maeyer, P. How hydrologic processes differ spatially in a large basin: Multi-site and multi-objective modeling in the Tarim River Basin. *J. Geophys. Res. Atmos.* **2018**, *123*, 7098–7113.
26. Chen, Y.; Xu, C.; Hao, X.; Li, W.; Chen, Y.; Zhu, C.; Ye, Z. Fifty-year climate change and its effect on annual runoff in the Tarim River Basin, China. *Quat. Int.* **2009**, *208*, 53–61.
27. Xu, Z.; Chen, Y.; Li, J. Impact of climate change on water resources in the Tarim River basin. *Water Resour. Manag.* **2004**, *18*, 439–458. [CrossRef]
28. Wang, F.; Chen, Y.; Li, Z.; Fang, G.; Li, Y.; Xia, Z. Assessment of the irrigation water requirement and water supply risk in the Tarim River Basin, Northwest China. *Sustainability* **2019**, *11*, 4941. [CrossRef]
29. Lai, L.; Huang, X.; Yang, H.; Chuai, X.; Zhang, M.; Zhong, T.; Chen, Z.; Chen, Y.; Wang, X.; Thompson, J.R. Carbon emissions from land-use change and management in China between 1990 and 2010. *Sci. Adv.* **2016**, *2*, e1601063. [CrossRef] [PubMed]
30. Zhang, W.; Lu, Y.; van der Werf, W.; Huang, J.; Wu, F.; Zhou, K.; Deng, X.; Jiang, Y.; Wu, K.; Rosegrant, M.W. Multidecadal, county-level analysis of the effects of land use, Bt cotton, and weather on cotton pests in China. *Proc. Natl. Acad. Sci. USA* **2018**, *115*, E7700. [CrossRef]
31. Ma, M.; Veroustraete, F. Reconstructing pathfinder AVHRR land NDVI time-series data for the Northwest of China. *Adv. Space Res.* **2006**, *37*, 835–840. [CrossRef]
32. Piao, S.; Wang, X.; Park, T.; Chen, C.; Lian, X.; He, Y.; Bjerke, J.W.; Chen, A.; Ciais, P.; Tømmervik, H. Characteristics, drivers and feedbacks of global greening. *Nat. Rev. Earth Environ.* **2020**, *1*, 14–27. [CrossRef]
33. Zhang, Y.; Song, C.; Band, L.E.; Sun, G.; Li, J. Reanalysis of global terrestrial vegetation trends from MODIS products: Browning or greening? *Remote Sens. Environ.* **2017**, *191*, 145–155. [CrossRef]
34. Deng, Y.; Wang, S.; Bai, X.; Luo, G.; Wu, L.; Chen, F.; Wang, J.; Li, C.; Yang, Y.; Hu, Z. Vegetation greening intensified soil drying in some semi-arid and arid areas of the world. *Agric. For. Meteorol.* **2020**, *292*, 108103. [CrossRef]
35. Sun, F.; Wang, Y.; Chen, Y. Dynamic changes of the desert-oasis ecotone and its influencing factors in Tarim Basin. *Chin. J. Ecol.* **2020**, *39*, 3397–3407.
36. Long, H.; Liu, Y.; Hou, X.; Li, T.; Li, Y. Effects of land use transitions due to rapid urbanization on ecosystem services: Implications for urban planning in the new developing area of China. *Habitat Int.* **2014**, *44*, 536–544. [CrossRef]
37. Li, Y.; Chen, Y.; Li, Z. Dry/wet pattern changes in global dryland areas over the past six decades. *Glob. Planet. Chang.* **2019**, *178*, 184–192. [CrossRef]
38. Stagge, J.H.; Tallaksen, L.M.; Gudmundsson, L.; Van Loon, A.F.; Stahl, K. Candidate distributions for climatological drought indices (SPI and SPEI). *Int. J. Climatol.* **2015**, *35*, 4027–4040. [CrossRef]
39. Beguería, S.; Vicente-Serrano, S.M.; Reig, F.; Latorre, B. Standardized precipitation evapotranspiration index (SPEI) revisited: Parameter fitting, evapotranspiration models, tools, datasets and drought monitoring. *Int. J. Climatol.* **2014**, *34*, 3001–3023. [CrossRef]

40. Manzano, A.; Clemente, M.A.; Morata, A.; Luna, M.Y.; Beguería, S.; Vicente-Serrano, S.M.; Martín, M.L. Analysis of the atmospheric circulation pattern effects over SPEI drought index in Spain. *Atmos. Res.* **2019**, *230*, 104630. [CrossRef]
41. Veldkamp, A.; Lambin, E.F. Predicting land-use change. *Agric. Ecosyst. Environ.* **2001**, *85*, 1–6. [CrossRef]
42. Fischer, G.; Sun, L. Model based analysis of future land use development in China. *Agric. Ecosyst. Environ.* **2001**, *85*, 163–176. [CrossRef]
43. Pijanowski, B.; Brown, D.; Shellito, B.; Manik, G. Using neural networks and GIS to forecast land use changes: A land transformation model. *Comput. Environ. Urban Syst.* **2002**, *26*, 553–575. [CrossRef]
44. Van Vliet, J.; Hurkens, J.; White, R.; van Delden, H. An activity-based cellular automaton model to simulate land-use dynamics. *Environ. Plan. B Plan. Des.* **2011**, *39*, 198–212. [CrossRef]
45. Li, X.; Liu, X. Defining agents' behaviors to simulate complex residential development using multicriteria evaluation. *J. Environ. Manag.* **2007**, *85*, 1063–1075. [CrossRef] [PubMed]
46. Wood, E.C.; Lewis, J.E.; Tappan, G.G.; Lietzow, R.W. The development of a land cover change model for southern Senegal. In *Land Use Modeling Workshop*; EROS Data Center: Sioux Falls, SD, USA, 1997; pp. 5–6.
47. Pontius, R.; Huffaker, D.; Denman, K. Useful techniques of validation for spatially explicit land-change models. *Ecol. Model.* **2004**, *179*, 445–461. [CrossRef]
48. Jiang, P.; Cheng, L.; Li, M.; Zhao, R.; Duan, Y. Impacts of LUCC on soil properties in the riparian zones of desert oasis with remote sensing data: A case study of the middle Heihe River basin, China. *Sci. Total Environ.* **2015**, *506*, 259–271. [CrossRef]
49. Li, J.; Zhao, C.; Zhu, H.; Li, Y.; Wang, F. Effect of plant species on shrub fertile island at an oasis–desert ecotone in the South Junggar Basin, China. *J. Arid Environ.* **2007**, *71*, 350–361. [CrossRef]
50. Chen, Z.; Chen, Y.; Li, B. Quantifying the effects of climate variability and human activities on runoff for Kaidu River Basin in arid region of northwest China. *Theor. Appl. Climatol.* **2013**, *111*, 537–545. [CrossRef]
51. Shen, Y.; Chen, Y. Global perspective on hydrology, water balance, and water resources management in arid basins. *Hydrol. Process.* **2009**, *24*, 129–135. [CrossRef]
52. Chen, Y.; Ye, Z.; Shen, Y. Desiccation of the Tarim River, Xinjiang, China, and Mitigation Strategy. *Quat. Int.* **2011**, *244*, 264–271. [CrossRef]
53. Fang, G.; Chen, Y. Variation in agricultural water demand and its attributions in the arid Tarim River Basin. *J. Agric. Sci.* **2018**, *156*, 1–11. [CrossRef]

Article

Implementation of an Improved Water Change Tracking (IWCT) Algorithm: Monitoring the Water Changes in Tianjin over 1984–2019 Using Landsat Time-Series Data

Xingxing Han [1,2], Wei Chen [1,2], Bo Ping [1,2] and Yong Hu [3,*]

1 Institute of Surface-Earth System Science, School of Earth System Science, Tianjin University, Tianjin 300072, China; hanxingxing@tju.edu.cn (X.H.); chenwei19@tju.edu.cn (W.C.); pingbo@tju.edu.cn (B.P.)
2 Tianjin Key Laboratory of Earth Critical Zone Science and Sustainable Development in Bohai Rim, Tianjin University, Tianjin 300072, China
3 Chongqing Institute of Surveying and Monitoring for Planning and Natural Resources, Chongqing 400120, China
* Correspondence: rihor@sina.com; Tel.: +86-023-6771-2435

Citation: Han, X.; Chen, W.; Ping, B.; Hu, Y. Implementation of an Improved Water Change Tracking (IWCT) Algorithm: Monitoring the Water Changes in Tianjin over 1984–2019 Using Landsat Time-Series Data. *Remote Sens.* 2021, 13, 493. https://doi.org/10.3390/rs13030493

Academic Editors: Weili Duan, Shreedhar Maskey, Pedro Luiz Borges Chaffe, Pingping Luo, Bin He, Yiping Wu and Jingming Hou
Received: 31 December 2020
Accepted: 25 January 2021
Published: 30 January 2021

Publisher's Note: MDPI stays neutral with regard to jurisdictional claims in published maps and institutional affiliations.

Copyright: © 2021 by the authors. Licensee MDPI, Basel, Switzerland. This article is an open access article distributed under the terms and conditions of the Creative Commons Attribution (CC BY) license (https://creativecommons.org/licenses/by/4.0/).

Abstract: Tianjin is the largest open city along the coastline in Northern China, which has several important wetland ecosystems. However, no systematic study has assessed the water body changes over the past few decades for Tianjin, not to mention their response to human activities and climate change. Here, based on the water change tracking (WCT) algorithm, we proposed an improved water change tracking (IWCT) algorithm, which could remove built-up shade noise (account for 0.4%~6.0% of the final water area) and correct omitted water pixels (account for 1.1%~5.1% of the final water area) by taking the time-series data into consideration. The seasonal water product of the Global Surface Water Data (GSWD) was used to provide a comparison with the IWCT results. Significant changes in water bodies of the selected area in Tianjin were revealed from the time-series water maps. The permanent water area of Tianjin decreased 282.5 km^2 from 1984 to 2019. Each time after the dried-up period, due to government policies, the land reclamation happened in Tuanbo Birds Nature Reserve (TBNR), and, finally, 12.6 km^2 of the lake has been reclaimed. Meanwhile, 488.6 km^2 of land has been reclaimed from the sea along the coastal zone in the past 16 years at a speed of 28.74 km^2 yr^{-1} in the Binhai New Area (BHNA). The method developed in this study could be extended to other sensors which have similar band settings with Landsat; the products acquired in this study could provide fundamental reference for the wetland management in Tianjin.

Keywords: inland water; IWCT; Tianjin; Landsat data

1. Introduction

Inland water systems often provide critical ecosystem functions, i.e., water and food provision, local climate regulation, conservation of biological diversity [1–3]. However, due to intensive human activities (e.g., land reclamation and water conservancy projects), the area of inland water decreased sharply in China [4,5]. Thus, accurate monitoring the long-term changes of inland water bodies is important to both scientific community for water research and local governments for water-resource planning and management [6–11].

Remote sensing has become one of the most efficient methods in water area monitoring with synoptic and repeated observations. Various methods have been developed to discriminate and map water using multi-spectral, hyperspectral and radar images [12–14]. The most simple and popular method was basis on threshold segmentation by using a single band, or a ratio of two bands of data (e.g., Normalized Difference Vegetation Index (NDVI), Normalized Difference Water Index (NDWI) [15] and the Modified Normalized Difference Water Index (MNDWI) [16,17]). However, as the spectral characteristics of lakes vary temporally and spatially, the single threshold may not be suitable for large areas and long time-interval due to varied band configurations and illumination conditions [14]. The supervised classification, decision tree and machine learning algorithms were also used to classify the water body [18,19], while they need training data from different regions and different seasons to serve as prior knowledge or auxiliary data.

The daily global water map data set from 2001 to 2016 was generated based on 500-m resolution time-series daily MODIS (Moderate Resolution Imaging Spectroradiometer) reflectance data [20,21]. The MODIS 250-m resolution data collected between 2000 and 2010 were used by Feng et al. [13] to document the temporal inundation changes of Poyang Lake with Floating Algae Index (FAI) and a gradient method. Han and Niu [21] constructed a new Global Surface Water Extent Dataset (GSWED) covering 2000–2018 with a temporal resolution of 8 days and a spatial resolution of 250 m based on MODIS data and Google Earth Engine (GEE) cloud computing platform. Although, the 250-m and 500-m resolution MODIS data can be used to effectively analyze the long-term water body changes in large areas, the results obtained from the MODIS images may misinterpret the small-scale inland water bodies due to a mixing pixels problem [22]. The publicly accessible Landsat archive makes the historical water body mapping feasible at 30-m level [23]. Based on classification tree models and top of atmosphere (TOA) reflectance of 3.4 million Landsat scenes, Pickens et al. [19] generated the global water maps and assessed the accuracy using sub-pixel analysis based on samples from 5-m resolution RapidEye imagery. Hou, et al. [24] delineated the lake boundaries using a thresholding method based on Landsat-derived NDWI to track reclamation-induced changes in the Yangtze Plain lakes. The Function of mask (FMASK) was developed to detect cloud, cloud shadow, water and snow for time-series Landsat images [25,26]. With Landsat time-series, the global inland water dynamic was documented by different methods (multiple indices and thresholds, the expert system based on a procedural sequential decision tree, etc.) [19,27,28]. The multi-source remote sensing data were also used for small-coverage inland water detection [29–32]. These methods need a large number of auxiliary data and thus cannot be fully automated employed to extract long-term water areas.

Recently, a water change tracking (WCT) algorithm based on Minimum Normalized Water Score (MNWS) for accurate inland water mapping was proposed and has been proven to be a promising method for Landsat images to extract water bodies automatically. Compared with FMASK and G1WBM methods, the WCT algorithm seems to be more robust for inland water extraction. However, while the WCT algorithm has been validated on large lakes, non-urban areas and wetland water extraction in China, its ability for small lakes and urban-dominated area water extraction has not been discussed. When this method was applied to extract water bodies in urban-dominant area in Tianjin, lots of very bright water pixels were omitted and the algorithm performed poorly in urban areas with high buildings. Hence, given the availability of the time-series Landsat images and lack of systematic assessment of the Tianjin water bodies, this study has two objectives:

(1) Improve the existing WCT algorithm to extract the water bodies in Tianjin using Landsat images;
(2) Document water changes in Tianjin over 1984–2019 and understand their linkage with human activities and climate changes.

The IWCT can remove the built-up shades and correct the omitted pixels using the time-series remote sensing data. Furthermore, the IWCT will be fully validated in the urban dominated areas.

2. Materials and Methods

2.1. Study Area

Tianjin, located at the downstream of Haihe river Basin (116°42′7.5″–118°3′37.6″E and 38°33′23″–40°15′7.5″N, Figure 1), is the largest open city along the coastline in northern China. Influenced by temperate monsoon climate, Tianjin has two seasons: dry season (September to June) and wet season (July to August), and more than 60% of annual precipitation occurs in the wet season [33]. The wetland resources in Tianjin serve as the main resting places for migratory birds from East Asia to Australia. To conserve the rare and endangered migratory birds, the government has constructed several nature reserves: Beidagang Wetland Nature Reserve (BWNR), Dahuangpu Wetland Nature Reserve (DWNR), Ancient Coast and Wetland National Nature Reserve (ACWNNR) and Tuanbo Birds Nature Reserve (TBNR). The other major water bodies of Tianjin distributed in the Yuqiao Reservoir (YR) which is the important source of drinking water for Tianjin, and Binhai New Area (BHNA).

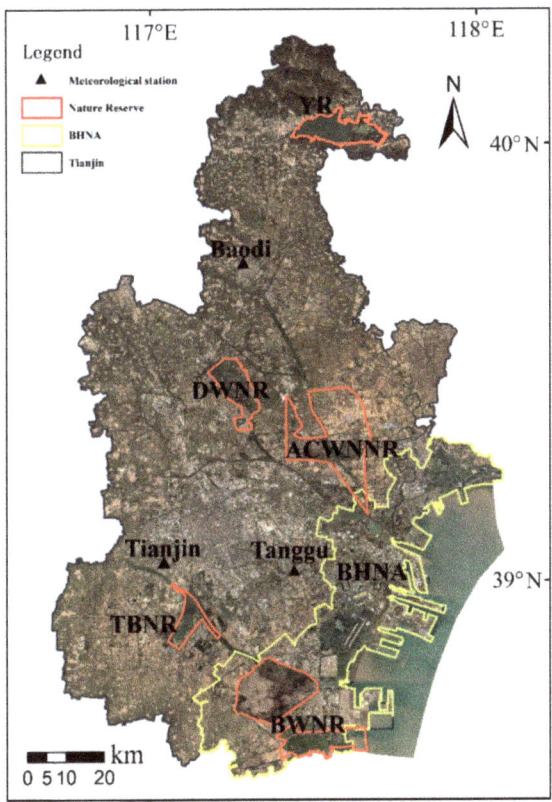

Figure 1. The location of study area. The true-color image is the Landsat-8 OLI image collected on 12 September 2017 and 26 October 2018. The boundaries of the four nature reserves (Beidagang Wetland Nature Reserve (BWNR), Dahuangpu Wetland Nature Reserve (DWNR), Ancient Coast and Wetland National Nature Reserve (ACWNNR), Tuanbo Birds Nature Reserve (TBNR)), Binhai New Area (BHNA) and Yuqiao Reservoir (YR) of Tianjin are delineated with red lines and yellow lines. Note that the two red boundaries near BWNR all belong to BWNR.

2.2. Dataset

Landsat surface reflectance data, including Landsat 8 Operational Land Imager (OLI), Landsat 7 Enhanced Thematic Mapper Plus (ETM+), Landsat 5 Thematic Mapper (TM) and Landsat 4 Thematic Mapper (TM) with 30-m resolution were downloaded from the U.S. Geological Survey (USGS) (http://glovis.usgs.gov/) and used to map Tianjin water bodies and their spaito-temporal changes. The entire study region is covered by four Landsat footprints (path 122 row 032, path 122 row 033, path 123 row 032 and path 123 row 033). After excluding images with significant cloud covers based on visual inspection, 1087 Landsat images between 1984 and 2019 were kept. The temporal distribution of the data is shown in the bubble chart (Figure 2). The size of the bubbles represents the valid observation area after removing the "bad pixels" (snow, ice or cloud covered areas in the satellite images), and arranges from small (the smallest valid observation area, 500 km²) gradually to large (the largest valid observation area, 13,000 km²).

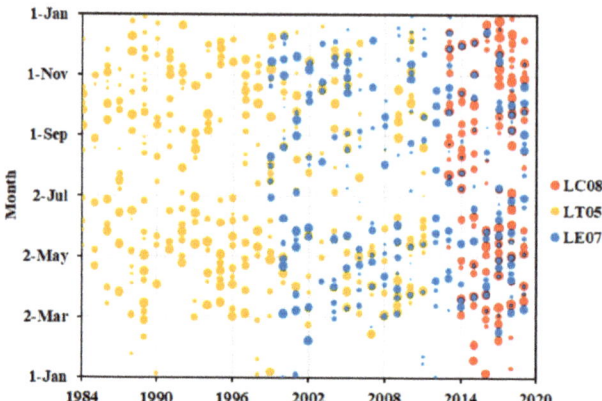

Figure 2. Scene acquisition dates for the Landsat images from 1985 to 2019. The size of the bubbles indicates useful observation areas which masked the "bad pixels" for the whole study area. The size of bubbles represents the valid data areas changed from 500 (the smallest size) to 13,000 km² (the largest size).

The high-resolution images on Google Earth could be used as ground truth for the water classification. Two high-resolution images (acquisition on 21 November 2005 and 27 May 2018) that 1 day ahead and on the same day with the objective Landsat images (20051122123LE07 and 20180527122LE07) were used to discriminate the water and non-water pixels. The seasonality dataset of Global Surface Water-Data (GSWD) for Tianjin was downloaded from the website (https://global-surface-water.appspot.com/) [3]. The Seasonality map provides the intra-annual changes of water surface in the period of October 2014 to October 2015 and shows the number of months water was present. The images were used to select ground truth water samples to validate the Landsat extract water maps. The Shuttle Radar Topography Mission (SRTM) DEM data of Tianjin were downloaded from USGS. The urban boundary was downloaded from the website (http://data.ess.tsinghua.edu.cn/) [34]. The GaoFen (GF)-2 PMS data were downloaded from China Centre for Resources Satellite Data and Application (http://www.cresda.com/CN/). Precipitation and temperature data were obtained from the three meteorological stations (Baodi, Tianjin and Tanggu, shown in Figure 1) through China Meteorological Data Service Center (CMDC) (http://data.cma.cn/en).

Furthermore, the boundary of YR used in this study was calculated based on the largest inundation of YR during the whole study period. The boundary of four nature reserves and BHNA was obtained from Tianjin Institute of Surveying and Mapping (http://tjch.com.cn/).

2.3. Method

The WCT algorithm is designed to automatically extract multi-temporal water maps for large area. The processing steps could be summarized as follows:

1. Collection of Reliable Water Samples (RWS): The RWS was extracted by the criteria that MNDWI was larger than 0.3 and the minimum reflectance of the red, green, and near infrared bands (MGRN) lower than 0.15 [27,35].
2. Water sample clustering for the Reliable Water Samples: The k-means clustering method was applied to the visible bands of the RWS to cluster the water bodies into

eight classes. The spectral statistical parameters (the mean and standard deviation values) for each band of each cluster were calculated.

3. Calculation of the Minimum Normalized Water Score (MNWS): The Normalized Water Score (NWS) were calculated for all the images and the MNWS was calculated by the Equations (1) and (2).

$$NWS_i = \sqrt{\frac{1}{m}\sum_{j=1}^{m}\left(\frac{(x_j - \overline{x_{i,j}})}{\sigma_{i,j}}\right)^2}, \quad (1)$$

$$MNWS = \min_{i=1}^{n}\{NWS_i\}, \quad (2)$$

$\overline{x_{i,j}}$ and $\sigma_{i,j}$ is the mean and standard deviation of cluster i and band j, m is the number of the band, n is the number of the cluster.

4. Extraction of Water Body: The water body extracted by the criteria that the MNWS < 2.5.

When applied the WCT algorithm to the study region for the Landsat time-series images of 1984–2019, we found that the WCT algorithm has some defects. The defects of the WCT could be summarized as follows:

(1) Water pixels with very high reflectance may be misclassified as background. The k-means clustering method was applied to the RWS to divide water bodies into different clusters according to the local environment. The WCT is highly dependent on the RWS selected from the images. The water pixels in the sun glint and in shallow areas affected by sand bottoms may be observed as high reflectance data. The water samples of these regions always did not meet the criteria of RWS and would be misclassified to non-water pixels.

(2) Urban building shadows may be misclassified as water. The built-up shade noise account for 0.4%~6.0% of the final water areas from June to December/next January. The spectral characteristics of the materials under the shadows were not represented and the shadows share similar spectral characters with water. If the shadows in the images were not masked, they may be misinterpreted as water.

(3) The time-series data were used to detect the changes of the water in the WCT algorithm, rather than extracting water. The omitted water pixels account for 1.1%~5.1% of the final water areas.

Based on the WCT algorithm, we proposed an improved water change tracking (IWCT) algorithm to overcome the defects of the WCT. The flow chart (Figure 3) shows the details of the data processing steps. The practical processing steps show as follows:

1. The WCT algorithm: Implied the WCT algorithm described in the Section 2.3 to the Landsat (TM/ETM+ and OLI) images to extract preliminary water body map [12].

2. "Bad pixel" mask: When the water pixel is covered by snow, cloud or ice, the spectral characteristic of the pixel changes greatly. The automatic cloud detection methods may be not suitable in this study because of their limits, such as ① The ground objects with high reflectance may be misclassified as clouds; ② It is hard to determine the boundary of optically thin clouds and their shadows; ③ Water may be misclassified to cloud shadows. In this study, visual interpretation method was used to remove bad pixels. Red-Green-Blue "true-color" composite images were generated using three bands (Red Band: 655, Green Band: 536, Blue Band: 480 nm) from the Landsat data. The regions where the RGB images showed "bad pixels" were manually delineated using the region of interest (ROI) Tool. The pixels within the ROI will be set as non-valid observations and discarded.

3. Shadow mask: The spectral characteristics of the terrain and building shadows are similar to water and are always misinterpreted as water in the original algorithms. The water bodies and terrain shadows could be separated using the DEM data with the criteria that if the slopes of the pixels are greater than 5° [27]. In the urban-dominated areas, the shadows of high buildings will affect the water extraction accuracy. Building

shadows is seasonally dependent, and will be related to the changes in the solar angle. When the sun is directly over the Tropic of Cancer on 21 June of the year, the sun has highest elevation angle in the northern hemisphere. Considering the geographic coordinates of Tianjin area, the images collected in June–July were less affected by the urban shade, and thus can be used to distinguish urban shades in the urban areas delineated by the urban boundary and water bodies for the rest data of the year.
4. Omitted water pixels correction: Using only one index combination often leads to omission errors for those water pixels with high reflectance. With the repeat monitoring, the remote sensing not only provides spatial information, it also provides the temporal coverage of water bodies. Thus, the time-series water classification results and Landsat reflectance data were used to correct the water pixels which were omitted in the WCT algorithm. Images in the same month within a four-year window (i.e., two years before and later) were selected, if the data were classified as water within four years, and the data meet the criteria (MNDWI > 0 or NDWI > 0 or NDVI < 0 and NDVI < 0.3) it will set as omitted water.
5. Water frequency and water area mapping: For each pixel, the water frequency map is calculated as the ratio of the number of pixels detected as water and the number of valid observations within one year. To quantify the water area changes of Tianjin during the last three decades, the water areas of the six typical regions (ACWNNR, BWNR, DWNR, TBNR, BHNA and YR) were calculated for all the cloud-free Landsat images of these specific regions. Pixels with water frequency equals 1 were classified as permanent water.

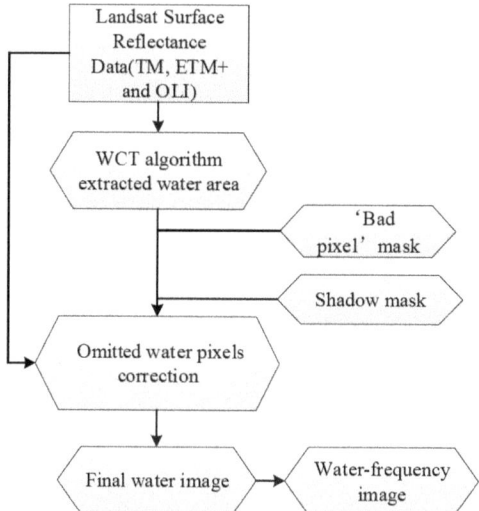

Figure 3. The flow chart of the Landsat-based water extraction method improved water change tracking (IWCT) algorithm developed in this study.

2.4. Accuracy Assessment

The omission and commission errors at pixel scale were used to evaluate the performance of the IWCT. For the selected two Landsat images, 2000 random points for water and non-water pixels in the IWCT images were generated within the overlapped region between Google earth high-resolution images and Landsat data (20051122123LE07 and 20180527122LE07). The points were visually classified as water/non-water on the Google Earth high-resolution image, and were used for validation of the result.

As both of the seasonal GSWD products and IWCT-based water map were calculated with the entire archive of the Landsat images and have the same spatial resolution, the

GSWD product can be used to assess the accuracy of IWCT-based results. The seasonality map of the water surfaces for a single year (2014–2015) was available on the website, and was chosen to make a comparison with the data calculated in this study. If all the pixels in the valid observations are water pixels, the data will be marked as permanent water in the GSWD. As the water frequency was calculated based on the ratio between the number of pixels detected as water and the number of valid observations, the pixels with water frequency equal to 1 would be treated as permanent water for IWCT. The permanent water in the GSWD and the permanent water in IWCT were calculated based on the same principle. The permanent water data of the two images were resampled to the same resolution as 3 km × 3 km, and then the correlation analysis was conducted for all the points in the data. To test the applicability of proposed algorithm, it was applied to a FLAASH atmospheric corrected GF-2 PMS image. The extraction result based on 3.9-m resolution GF-2 PMS was then compared with Landsat result.

3. Results

The water bodies were extracted using the IWCT method excluding the bad pixels and shadows. The omitted water pixels were corrected using the time-series Landsat images and primary water results. To assess the accuracy of IWCT product, the commission and omission errors of the WCT and IWCT were estimated using the same validation samples. The results show that the IWCT method was much better than the WCT by providing lower commission errors and omission errors (Table 1). The IWCT presented lower commission and omission errors for 20051122123LE07 and 20180527122LE07 images than WCT algorithm (commission error: 0.8–5.1% for IWCT versus 0.8–9.5% for WCT, omission error: 0.9–3.2% for IWCT versus 2.4–4.3% for WCT). Scatterplots of permanent water area from IWCT, corresponding to 3 km × 3 km resolution, were made against GSWD. The result indicated that the two data have extremely high consistency with a determination coefficient (R^2) of 0.97 and RMSE of 0.46 ($p < 0.05$) (Figure 4).

Table 1. Classification accuracy of the WCT and IWCT.

	WCT		IWCT	
Image	Commission Error	Omission Error	Commission Error	Omission Error
20051122123LE07	9.5%	4.3%	5.1%	3.2%
20180527122LE07	0.8%	2.4%	0.8%	0.9%

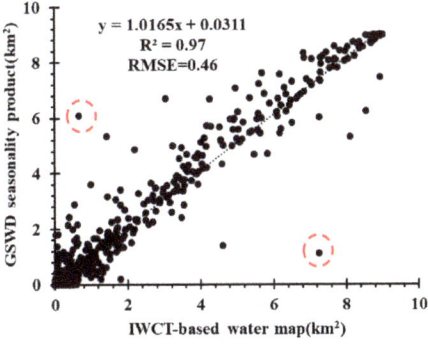

Figure 4. Validation of the IWCT-based water map using existed water bodies image from GSWD seasonality product. The two red circled points showed great differences, and the corresponding water area will be discussed in the Discussion section below.

Figure 5 shows the dynamic water body maps of Tianjin between 1984 and 2019. Each map was generated using the Landsat time-series images in the corresponding year. The water bodies with higher frequencies are presented in dark blue and seasonal water with lower frequencies are shown in lighter blue. The spatial distributions of the water bodies were clearly demonstrated within each panel, and the long-term water frequency changes were also revealed across each panel. In general, the water bodies with high frequencies located mainly in the BHNA, YR and TBNR. The water frequencies are more variable in the ACWNNR, BWNR and DWNR in different years. The water landscape shows the trend of fragmentation from 1984 to 2019. The water area of TBNR shows significant variability in different years, with a larger area of water has been reclaimed in after 2002. In addition, land has been reclaimed from sea along the coastal zone after 2003 in the BHNA.

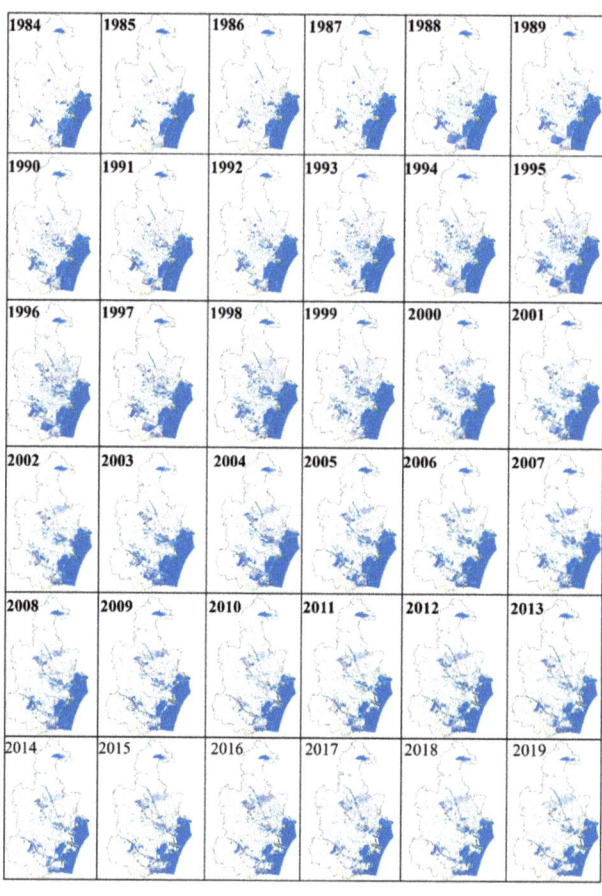

Figure 5. Annual water frequency map of Tianjin during 1984–2019.

To present the details of the water changes in Tianjin, six typical regions were selected from Tianjin to show the long-term changes of water area. Figure 6 plots the estimated areas of water and permanent water bodies between 1984 and 2019, Figure 6a–g represent the results of ACWNNR, BWNR, DWNR, TBNR, BHNA, YR and Tianjin, respectively. All the cloud-free images in the corresponding area were considered when calculating the long-term changes. The coverages of the water area showed a seasonal and annual fluctuation trend during 1984–2019 for the ACWNNR, BWNR and DWNR (Figure 6a–c). The water area within TBNR showed a remarkable seasonal and annual fluctuation from 1984 to 2019, which almost dried up in the periods (1992–1994, 1999–2004, and 2008–2011) and remained at a fluctuated level (46.2~54.6 km^2) after 2012 (Figure 6d). Specifically, the water area showed a fluctuation trend in the period (1984–2003) and a significant shrinking trend (-28.7 km^2 yr^{-1}, $p < 0.05$) after 2003 for the BHNA (Figure 6e). The water area of YR had a significant seasonal variation (Figure 6f). The permanent water area (including permanent shallow marine water and inland water) of Tianjin decreased 282.5 km^2 from 1984 to 2019.

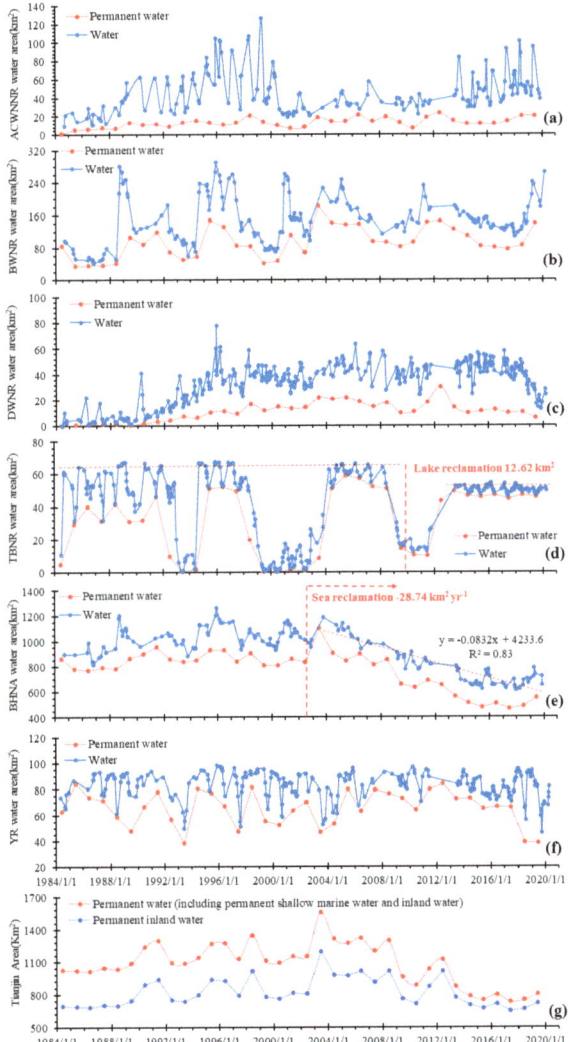

Figure 6. Long-term water area (blue dots) and permanent water area (red dots) changes for ACWNNR (**a**), BWNR (**b**), DWNR (**c**), TBNR (**d**), BHNA (**e**), YR (**f**) and Tianjin (**g**) during 1984–2019. The permanent water refers to pixels which were water for the duration of a calendar year. Note that the vertical dashed red lines represent the year when large area of land reclamation activities begins. The horizontal dashed red line in (**d**) represents mean largest water area before and after 2009. The oblique dashed red line in (**e**) represents the regression line for the water area after 2003.

4. Discussion

4.1. Driving Forces

To test whether the area of water bodies was driven by precipitation, the relationship between water area in each selected study area and local precipitation from the nearest meteorological station was examined. Correlation analysis showed a statistically significant correlation between the total precipitation of three months before the image acquisition dates and water area during July to September for four out of six sites, with the determination coefficients (R^2) of 0.25 (N = 35, $p < 0.01$) for the BWNR, 0.07 (N = 71, $p < 0.05$) for DWNR, 0.16 (N = 72, $p < 0.01$) for TBNR and 0.29 (N = 61, $p < 0.01$) for YR, which are statistically significant. Specifically, the water area of the BHNA and precipitation showed a statistically significant correlation ($R^2 = 0.71$, $p < 0.01$) if the result in 1984 (red circled points) was not considered due to very low sea level in 1984. Note that the red data points in the TBNR and BHNA were collected after the land reclamation activities happened in the local study area, were not taken into consideration when making the correlation analysis (Figure 7). The precipitation appears to negatively correlate with the water area in ACWNNR ($R^2 = 0.04$, $p > 0.05$) and DWNR ($R^2 = 0.07$, $p < 0.05$). After checking the land use map, we found that the main vegetation types are dry land crop and paddy field crop for ACWNNR and DWNR, respectively (http://www.dsac.cn/DataProduct/Detail/20081103). The water bodies could be affected by the surface vegetation types. Thus, the changes in local precipitation might be partially associated to water area changes in July to September for parts of study regions.

The long-term change of water area or water quality could be taken as the indicators characterizing the ecosystem change. With historical surface water changing trends and ancillary data, we can document the impact of the human activities and climate change on the ecosystem. The water bodies of TBNR was considerably fluctuated in different years and remained at a low coverage level during the periods (1992–1994, 1999–2004, 2008–2011) (Figures 6d and 8d,h,l) cloud be linked to important government policies. The reservoir was built in 1978, the storage capacity of the reservoir was 0.98 hundred million m^3. The running out of water happened in 1992 is due to a project which increased the storage capacity of the reservoir (Figure 8c). The dike of the reservoir faced some safety problem after operation for a long time, the reservoir was repaired and carefully maintained after 2008 (http://www.docin.com/p-525234617.html); thus, the water ran out after 2008 (Figure 8k). Each time after the running out of water period, parts of the water area were reclaimed (Figure 8e,i,m) and finally 12.6 km^2 lake has been reclaimed. The decreasing of the water area after 2003 in the BHNA was attributed to the sea reclamation activities (Figure 9) [36]. Figure 9 showed the coastal line of the Tianjin in 2004, 2009, 2014 and 2019, indicating the land expanded towards sea, with an average trend of 28.7 km^2 yr^{-1} from 2004 to 2019. The decrease of permanent water area in Tianjin was obtained in accordance with the decrease of permanent water area in BHNA. The permanent shallow marine waters gradually reduced due to sea reclamation activities in the BHNA after 2004. The permanent inland water areas fluctuated in different years. The decrease of permanent water area (including permanent shallow marine water and inland water) in Tianjin was due to sea reclamation activities.

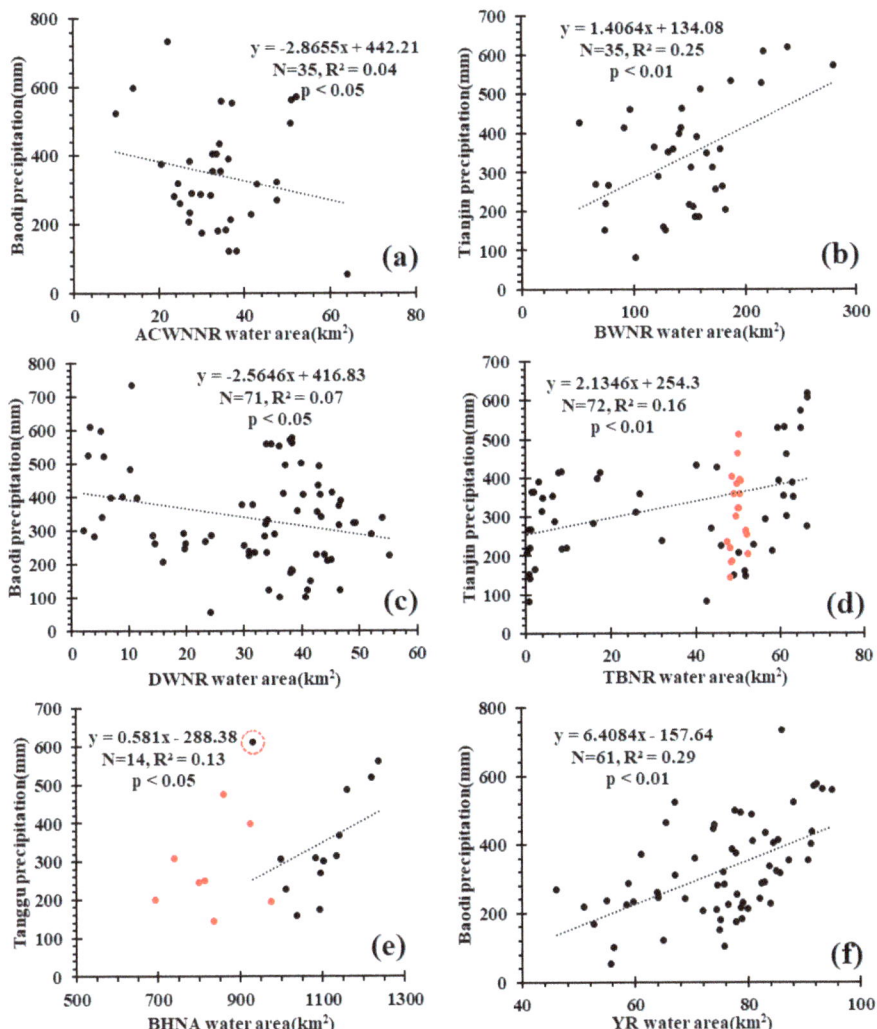

Figure 7. Relationship between the water area of the six study sites and accumulated precipitation of three months before the image acquisition dates in the nearest meteorological station. The red dots data were collected after the land reclamation and were not considered when calculated the regression functions. (**a**) Relationship between ACWNNR water area and precipitation of Baodi meteorological station, (**b**) Relationship between BWNR water area and precipitation of Tianjin meteorological station, (**c**) Relationship between DWNR water area and precipitation of Baodi meteorological station, (**d**) Relationship between TBNR water area and precipitation of Tianjin meteorological station, (**e**) Relationship between BHNA water area and precipitation of Tanggu meteorological station, (**f**) Relationship between YR water area and precipitation of Baodi meteorological station.

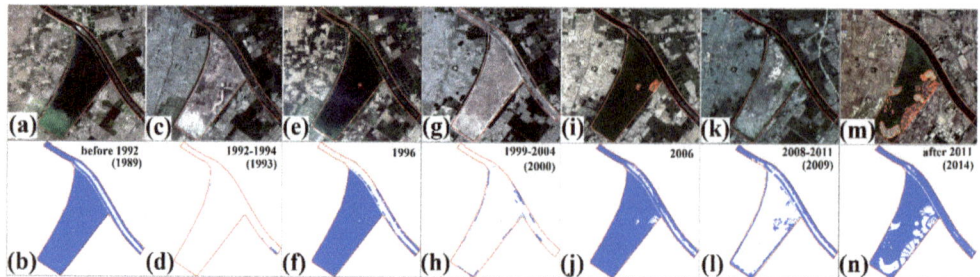

Figure 8. Illustration of land reclamation and images of running out water period in the TBNR from Landsat images of different periods. (**a,c,e,g,i,k,m**) represent true-color Landsat image of TBNR, the corresponding water bodies are shown in (**b,d,f,h,j,l,n**). Note, the years in the brackets represent the image selected year.

Figure 9. Sea reclamation in Tianjin coastal area for 2004, 2009, 2014 and 2019. The coastal line spatially extended towards the sea.

Additionally, changes in other meteorological factors (i.e., temperature) might influence the evapotranspiration of water and thus the water coverage, no significant relationship was found between the temperature and water area over the six study regions.

4.2. Validity of the Results and Future Applications

Long-term changes of water in Tianjin were clearly revealed and quantified via Landsat observation. However, are these Landsat derived water areas valid?

IWCT method achieved much better performance than the WCT by providing lower commission errors and omission errors. The correlation analysis between the maximum water frequency maps from IWCT result (a) and GSWD result shows the two results are highly consistent except for several points (Figure 4). The red circled points located in the YR (Figure 10) show great difference were analyzed. There is a big difference for the water map between IWCT result (Figure 10a) and GSWD result (Figure 10b) for the two points (Figure 4). After inspecting all the water classification results and the corresponding RGB images from October 2014 to October 2015, we found that the IWCT-based water frequency map is more reliable. Although lots of cloud mask algorithms had been developed before, the algorithms still have issues that cannot be ignored for the time-series images and will induce uncertainties to the final water map [26]. After a more robust cloud mask algorithm was developed in the further, the algorithm will be used to automatically extract water map.

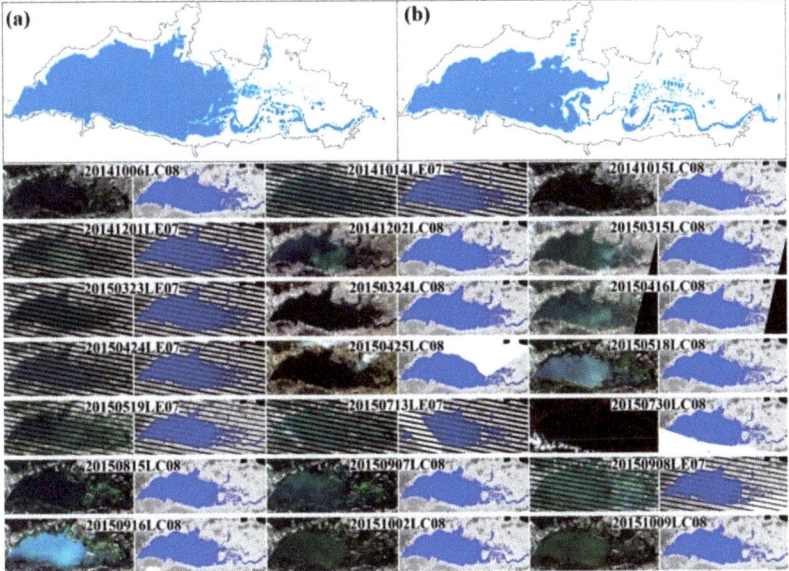

Figure 10. The maximum water frequency maps of YR for IWCT result (**a**) and GSWD result (**b**). The rest images are the Table 2014. to October 2015.

Several points classified as omitted water were used to carry out the sensitivity analysis in this study. We kept the criteria (MNDWI > 0 or NDWI > 0 or NDVI < 0) stable, and then changed the NDVI values from 0 to 0.5. The sensitivity analysis results can reflect the impact of different NDVI criteria on the omitted water detection. Results show that the classification error rate decreases with the increasing NDVI values, but increases as the NDVI > 0.3. The reason is that when NDVI > 0.3, the vegetation will be more probably misclassified as omitted water (Figure 11). Overall, the criteria (NDVI < 0.3) give satisfied results in the sensitivity analysis section.

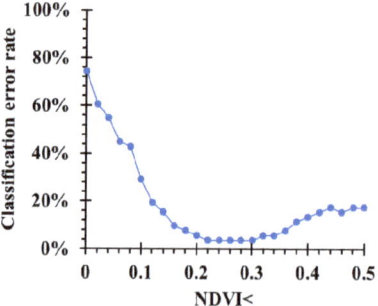

Figure 11. The classification error rate with different values for the omitted water.

The Chinese government has launched a series of high-resolution (GF) mapping satellites in the last seven years. The sensors onboard have similar spectral bands with the Landsat images and higher resolutions. We applied the proposed algorithm to 3.9-m resolution GF-2 PMS data which were collected on 29 May 2019 and on the same day as the OLI image. The water bodies for the two images were upscaled to 0.015° latitude and 0.007° longitude, respectively (Figure 12). A comparison showed that the water bodies extracted from GF-2 is consistent with the OLI-based result. The result shows that the algorithm can be used for GF-2 image water extraction and the high-resolution GF-2 image could be used to extract more small area of water (narrow rivers, small lakes).

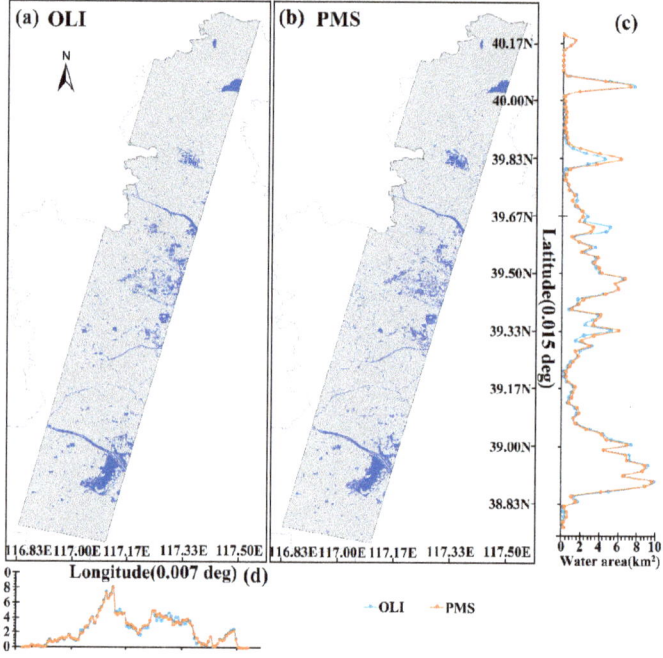

Figure 12. Comparison of water classification result and water area for 30-m 29 May 2019 OLI and 3.9-m resolution GF-2 PMS images using IWCT algorithm. IWCT-based water maps for Landsat OLI (**a**), IWCT-based water maps for GF-2 PMS (**b**), water body area for each 0.015° latitude (**c**) and water body area for each 0.007° longitude (**d**).

5. Conclusions

In this study, we developed the IWCT method which could remove built-up shade noise and correct omitted water pixels by taking the time-series data into consideration. The algorithm shows better performance in the urban dominated areas. With the IWCT algorithm, we documented the time-series water body changes of Tianjin using observations during 1984–2019. The annual water frequency map shows significantly decreasing trends after the land reclaimed activities. In addition, 488.6 km^2 of land has been reclaimed from the sea along the coastal zone in the last 16 years at the speed of 28.74 km^2 yr^{-1} in the BHNA. Overall, the change of water bodies in July to September could be partly attributed to precipitation except for the dry land crop and paddy field crop dominated areas, and the decrease of the water coverage appears to be a result of human activities.

IWCT method achieved much better performance than the WCT by providing lower commission and omission errors. However, this study still has some limitations. Although the urban boundary has been used to correct the IWCT algorithm in urban-dominated areas, parts of the urban shades (account for 2.9% of the randomly selected water pixels in the validation) were still misclassified as water. In the summer, the shadow of the skyscraper was misclassified as water. Very dark surface (the ground covered by coal, account for 0.1% of the randomly selected water pixels in the validation) will be misclassified as water. A more robust urban shade detection model is needed in the further research to remove the urban shades. The inclusion of manual clouds removal in the algorithm seriously limits its automatic application in large-scale dynamic inland water mapping. After developing a more robust cloud mask algorithm, the algorithm will be used to automatically extract water map. As several high-resolution sensors share similar bands with Landsat data, the method developed in this study can be extended to other sensors to study the long-term changes of even smaller water bodies in response to human activities and climate variability.

Author Contributions: Conceptualization, X.H. and Y.H.; methodology, X.H. and Y.H.; software, X.H.; validation, X.H. and Y.H.; formal analysis, W.C.; investigation, X.H.; resources, X.H.; data curation, X.H.; writing—original draft preparation, X.H.; writing—review and editing, X.H., W.C. and B.P.; visualization, X.H.; supervision, Y.H. and W.C.; project administration, Y.H.; funding acquisition, Y.H. All authors have read and agreed to the published version of the manuscript.

Funding: This research was funded by the National Natural Science Foundation of China (Nos. 41901386) and the Natural Science Foundation of ChongQing (Nos. cstc2019jcyj-msxmX0548).

Institutional Review Board Statement: Not applicable.

Informed Consent Statement: Not applicable.

Data Availability Statement: The data that support the findings of this study are available on request from the first author (X.H.) and the corresponding author (Y.H.).

Acknowledgments: We thank the USGS and China Centre for Resources Satellite Data and Application for providing Landsat and GF-2 data. We are indebted to CMDC for providing meteorological data. We also thank the anonymous reviewers for their valuable comments for the improvement of the manuscript.

Conflicts of Interest: The authors declare no conflict of interest.

References

1. Sheng, Y.; Song, C.; Wang, J.; Lyons, E.A.; Knox, B.R.; Cox, J.S.; Gao, F. Representative lake water extent mapping at continental scales using multi-temporal Landsat-8 imagery. *Remote Sens. Environ.* **2016**, *185*, 129–141. [CrossRef]
2. Palmer, S.C.J.; Kutser, T.; Hunter, P.D. Remote sensing of inland waters: Challenges, progress and future directions. *Remote Sens. Environ.* **2015**, *157*, 1–8. [CrossRef]
3. Pekel, J.; Cottam, A.; Gorelick, N.; Belward, A.S. High-resolution mapping of global surface water and its long-term changes. *Nature* **2016**, *540*, 418–422. [CrossRef] [PubMed]
4. Ma, R.; Duan, H.; Hu, C.; Feng, X.; Li, A.; Ju, W.; Jiang, J.; Yang, G. A half-century of changes in China's lakes: Global warming or human influence? *Geophys. Res. Lett.* **2010**, *37*. [CrossRef]

5. Gong, P.; Niu, Z.; Cheng, X.; Zhao, K.; Zhou, D.; Guo, J.; Liang, L.; Wang, X.; Li, D.; Huang, H.; et al. China's wetland change (1990–2000) determined by remote sensing. *Sci. China-Earth Sci.* **2010**, *53*, 1036–1042.
6. Piao, S.; Ciais, P.; Huang, Y.; Shen, Z.; Peng, S.; Li, J.; Zhou, L.; Liu, H.; Ma, Y.; Ding, Y. The impacts of climate change on water resources and agriculture in China. *Nature* **2010**, *467*, 43–51. [CrossRef]
7. Oki, T.; Kanae, S. Global Hydrological Cycles and World Water Resources. *Science* **2006**, *313*, 1068–1072. [CrossRef]
8. Duan, W.; Zou, S.; Chen, Y.; Nover, D.; Fang, G.; Wang, Y. Sustainable water management for cross-border resources: The Balkhash Lake Basin of Central Asia, 1931–2015. *J. Clean. Prod.* **2020**, *263*, 121614. [CrossRef]
9. Duan, W.; Chen, Y.; Zou, S.; Nover, D. Managing the water-climate-food nexus for sustainable development in Turkmenistan. *J. Clean. Prod.* **2019**, *220*, 212–224. [CrossRef]
10. Liu, D.; Chen, W.; Menz, G.; Dubovyk, O. Development of integrated wetland change detection approach: In case of Erdos Larus Relictus National Nature Reserve, China. *Sci. Total Environ.* **2020**, *731*, 139166. [CrossRef]
11. Chen, W.; Cao, C.; Liu, D.; Tian, R.; Wu, C.; Wang, Y.; Qian, Y.; Ma, G.; Bao, D. An evaluating system for wetland ecological health: Case study on nineteen major wetlands in Beijing-Tianjin-Hebei region, China. *Sci. Total Environ.* **2019**, *666*, 1080–1088. [CrossRef] [PubMed]
12. Chen, X.; Liu, L.; Zhang, X.; Xie, S.; Lei, L. A Novel Water Change Tracking Algorithm for Dynamic Mapping of Inland Water Using Time-Series Remote Sensing Imagery. *IEEE J. Sel. Top. Appl. Earth Obs. Remote Sens.* **2020**, *13*, 1661–1674. [CrossRef]
13. Feng, L.; Hu, C.; Chen, X.; Cai, X.; Tian, L.; Gan, W. Assessment of inundation changes of Poyang Lake using MODIS observations between 2000 and 2010. *Remote Sens. Environ.* **2012**, *121*, 80–92. [CrossRef]
14. Khandelwal, A.; Karpatne, A.; Marlier, M.E.; Kim, J.; Lettenmaier, D.P.; Kumar, V. An approach for global monitoring of surface water extent variations in reservoirs using MODIS data. *Remote Sens. Environ.* **2017**, *202*, 113–128. [CrossRef]
15. McFeeters, S.K. The use of the Normalized Difference Water Index (NDWI) in the delineation of open water features. *Int. J. Remote Sens.* **1996**, *17*, 1425–1432. [CrossRef]
16. Xu, H. Modification of normalised difference water index (NDWI) to enhance open water features in remotely sensed imagery. *Int. J. Remote Sens.* **2006**, *27*, 3025–3033. [CrossRef]
17. Huang, W.; Duan, W.; Nover, D.; Sahu, N.; Chen, Y. An integrated assessment of surface water dynamics in the Irtysh River Basin during 1990–2019 and exploratory factor analyses. *J. Hydrol.* **2021**, *593*, 125905. [CrossRef]
18. Wang, S.; Zhang, L.; Zhang, H.; Han, X.; Zhang, L. Spatial–Temporal Wetland Landcover Changes of Poyang Lake Derived from Landsat and HJ-1A/B Data in the Dry Season from 1973–2019. *Remote Sens.* **2020**, *12*, 1595. [CrossRef]
19. Pickens, A.H.; Hansen, M.C.; Hancher, M.; Stehman, S.V.; Tyukavina, A.; Potapov, P.; Marroquin, B.; Sherani, Z. Mapping and sampling to characterize global inland water dynamics from 1999 to 2018 with full Landsat time-series. *Remote Sens. Environ.* **2020**, *243*, 111792. [CrossRef]
20. Ji, L.; Gong, P.; Wang, J.; Shi, J.; Zhu, Z. Construction of the 500-m resolution daily global surface water change database (2001–2016). *Water Resour. Res.* **2018**, *54*, 10–270. [CrossRef]
21. Han, Q.; Niu, Z. Construction of the Long-Term Global Surface Water Extent Dataset Based on Water-NDVI Spatio-Temporal Parameter Set. *Remote Sens.* **2020**, *12*, 2675. [CrossRef]
22. Carroll, M.L.; Townshend, J.R.G.; Dimiceli, C.; Noojipady, P.; Sohlberg, R.A. A new global raster water mask at 250 m resolution. *Int. J. Digit. Earth* **2009**, *2*, 291–308. [CrossRef]
23. Zhu, Z.; Wulder, M.A.; Roy, D.P.; Woodcock, C.E.; Hansen, M.C.; Radeloff, V.C.; Healey, S.P.; Schaaf, C.B.; Hostert, P.; Strobl, P. Benefits of the free and open Landsat data policy. *Remote Sens. Environ.* **2019**, *224*, 382–385. [CrossRef]
24. Hou, X.; Feng, L.; Duan, H.; Chen, X.; Sun, D.; Shi, K. Fifteen-year monitoring of the turbidity dynamics in large lakes and reservoirs in the middle and lower basin of the Yangtze River, China. *Remote Sens. Environ.* **2017**, *190*, 107–121. [CrossRef]
25. Zhu, Z.; Woodcock, C.E. Object-based cloud and cloud shadow detection in Landsat imagery. *Remote Sens. Environ.* **2012**, *118*, 83–94. [CrossRef]
26. Qiu, S.; Zhu, Z.; He, B. Fmask 4.0: Improved cloud and cloud shadow detection in Landsats 4–8 and Sentinel-2 imagery. *Remote Sens. Environ.* **2019**, *231*, 111205. [CrossRef]
27. Yamazaki, D.; Trigg, M.A.; Ikeshima, D. Development of a global ~90 m water body map using multi-temporal Landsat images. *Remote Sens. Environ.* **2015**, *171*, 337–351. [CrossRef]
28. Aires, F.; Prigent, C.; Fluet-Chouinard, E.; Yamazaki, D.; Papa, F.; Lehner, B. Comparison of visible and multi-satellite global inundation datasets at high-spatial resolution. *Remote Sens. Environ.* **2018**, *216*, 427–441. [CrossRef]
29. Sun, F.; Ma, R.; He, B.; Zhao, X.; Zeng, Y.; Zhang, S.; Tang, S. Changing Patterns of Lakes on The Southern Tibetan Plateau Based on Multi-Source Satellite Data. *Remote Sens.* **2020**, *12*, 3450. [CrossRef]
30. Schwatke, C.; Dettmering, D.; Seitz, F. Volume Variations of Small Inland Water Bodies from a Combination of Satellite Altimetry and Optical Imagery. *Remote Sens.* **2020**, *12*, 1606. [CrossRef]
31. Chen, T.; Song, C.; Ke, L.; Wang, J.; Yao, F.; Liu, K.; Wu, Q. Estimating seasonal water budgets in global lakes by using multi-source remote sensing measurements. *J. Hydrol.* **2020**, *593*, 125781. [CrossRef]
32. Zhang, W.; Pan, H.; Song, C.; Ke, L.; Wang, J.; Ma, R.; Deng, X.; Liu, K.; Zhu, J.; Wu, Q. Identifying emerging reservoirs along regulated rivers using multi-source remote sensing observations. *Remote Sens.* **2019**, *11*, 25. [CrossRef]
33. Zhao, N.; Yue, T.; Li, H.; Zhang, L.; Yin, X.; Liu, Y. Spatio-temporal changes in precipitation over Beijing-Tianjin-Hebei region, China. *Atmos. Res.* **2018**, *202*, 156–168. [CrossRef]

34. Liu, H.; Gong, P.; Wang, J.; Clinton, N.; Bai, Y.; Liang, S. Annual Dynamics of Global Land Cover and its Long-term Changes from 1982 to 2015. *Earth Syst. Sci. Data Discuss.* **2019**, *12*, 1217–1243. [CrossRef]
35. Irish, R.R. Landsat 7 automatic cloud cover assessment. In Proceedings of the Algorithms for Multispectral, Hyperspectral, and Ultraspectral Imagery VI, Orlando, FL, USA, 24–26 April 2000; pp. 348–355.
36. Duan, H.; Zhang, H.; Huang, Q.; Zhang, Y.; Hu, M.; Niu, Y.; Zhu, J. Characterization and environmental impact analysis of sea land reclamation activities in China. *Ocean Coast. Manag.* **2016**, *130*, 128–137. [CrossRef]

Article

Spatial Allocation Method from Coarse Evapotranspiration Data to Agricultural Fields by Quantifying Variations in Crop Cover and Soil Moisture

Zonghan Ma [1,2], Bingfang Wu [1,2,*], Nana Yan [1], Weiwei Zhu [1], Hongwei Zeng [1] and Jiaming Xu [1,2]

[1] State Key Laboratory of Remote Sensing Science, Aerospace Information Research Institute, Chinese Academy of Sciences, Beijing 100101, China; mazh@aircas.ac.cn (Z.M.); yannn@aircas.ac.cn (N.Y.); zhuww@aircas.ac.cn (W.Z.); zenghw@aircas.ac.cn (H.Z.); xujm@aircas.ac.cn (J.X.)
[2] College of Resources and Environment, University of Chinese Academy of Sciences, Beijing 100101, China
* Correspondence: wubf@aircas.ac.cn; Tel.: +86-10-6484-2375

Abstract: Cropland evapotranspiration (ET) is the major source of water consumption in agricultural systems. The precise management of agricultural ET helps optimize water resource usage in arid and semiarid regions and requires field-scale ET data support. Due to the combined limitations of satellite sensors and ET mechanisms, the current high-resolution ET models need further refinement to meet the demands of field-scale ET management. In this research, we proposed a new field-scale ET estimation method by developing an allocation factor to quantify field-level ET variations and allocate coarse ET to the field scale. By regarding the agricultural field as the object of the ET parcel, the allocation factor is calculated with combined high-resolution remote sensing indexes indicating the field-level ET variations under different crop growth and land-surface water conditions. The allocation ET results are validated at two ground observation stations and show improved accuracy compared with that of the original coarse data. This allocated ET model provides reasonable spatial results of field-level ET and is adequate for precise agricultural ET management. This allocation method provides new insight into calculating field-level ET from coarse ET datasets and meets the demands of wide application for controlling regional water consumption, supporting the ET management theory in addressing the impacts of water scarcity on social and economic developments.

Keywords: agricultural water management; crop water consumption; remote sensing model; evapotranspiration allocation

1. Introduction

Land evapotranspiration (ET) is a major component of terrestrial water cycling and groundwater consumption [1]. For basin-scale water balance, the major component of basin water output is ET, followed by surface runoff and infiltration. It is estimated that nearly 70% of the total land precipitation and inflow is returned to the air by terrestrial ET. From an agricultural water management perspective, ET is an important indicator of cropland water consumption and water resource investment [2]. In arid and semiarid regions, social and economic development is limited by the amount of available fresh water [3], and water quotas for agricultural irrigation systems need judicious management, where the balance between irrigation water supply and necessary crop water demand is optimally controlled to maintain normal production [4]. The accurate and timely acquisition of crop ET is critical for reflecting the status of cropland water and determining irrigation strategies [5]. Moreover, views on water management have shifted from the conventional focus on increasing water income and cutting expenditures to the current focus on controlling water consumption. The applications of ET-based water management and a water consumption-oriented water rights allocation system have been demonstrated to be effective in the Turpan Basin, China, which has raised the demand for high-accuracy and low-cost farm-level ET monitoring methods [6,7].

With its broad spatial coverage and high temporal resolution, remote sensing (RS) technology provides a feasible approach to frequently monitor regional ET at low costs and with high efficiency [8]. Accurate surface vegetation cover, albedo, and temperature data provide a solid foundation for the simulation of ET-related physical and physiological processes [9]. Many RS-based ET models have been developed and applied at global and regional scales depending on multiple theories, including the ETWatch model based on the parametrized surface energy balance theory [10], the MOD16 series product based on the global MODIS satellite data and meteorological data [11], the GLEAM product based on the water balance theory [12], and the PML ET product focused on the plant water carbon stomata mechanism [13]. These RS-based ET products provide regional and global ET results at daily or eight-day intervals with moderate spatial resolutions, usually at 500-m to one-km scales. Such ET data have been adopted for studying regional ecological impacts on water resources [14–16], modeling land-surface processes [17], analyzing water cycling, and evaluating regional available water resources [16]. However, the spatial scales of these moderate-resolution ET products are too coarse for agricultural water management, especially within small irrigation districts that usually occupy tens to hundreds of square meters, and detailed spatial information is omitted from km-level ET data. Ground-level field ET measurement methods, such as lysimeters, sap flows, and eddy covariances (ECs), are high in cost and require careful management [18]. The combination of precise agricultural ET management and RS technology has developed a strong demand for ET models that show detailed cropland water consumption at the field scale.

Due to the limitations of satellite sensors, high-resolution spatial data usually have long temporal intervals, such as the Landsat series at 30-m resolution and 16-day intervals and the Sentinel-2 data at 10–60-m resolutions and five-day intervals. For agricultural ET monitoring and water consumption applications, the daily scale is ideal for timely evaluations and adjustments of agricultural activities [8]. Considering the impacts of clouds on optical RS data, the temporal gap could be 10 days or even longer. Such long temporal intervals are not satisfactory in many RS-based ET models, and daily ET variations could accumulate and lead to large errors in the monthly and crop ET estimates during growing seasons. The current km-level RS-based ET models are widely applied and highly accurate, and these models help address the limitations of field ET estimation and provide insights into agricultural water applications. Restrained by the inability of high-resolution satellite sensors to meet temporal requirements, the calculation of fine-resolution ET inevitably relies on data downscaling methods that combine high- and low-resolution RS datasets. The downscaling methods mainly focus on the pixel level of satellite images [19].

Most ET downscaling is conducted by downscaling the major parameters in the ET calculation procedure and is combined with other available fine-resolution input data to directly calculate high-resolution ET results. The DisALEXI model is an example of using high-resolution surface temperature to produce a downscaled surface ET [20]. DisALEXI has developed a framework of ET downscaling algorithm, which is to build up a linkage of key parameters between coarse and high-resolution RS data, such as land surface temperature (LST) and normalized difference vegetation index (NDVI) or other vegetation indexes [20–23]. The downscaling of LST has benefited from multiple research developments on the downscaling of Landsat thermal bands combined with MODIS LST data [24–26], such as the thermal sharpening (TSP), land surface temperature disaggregation (DLST), and temperature unmixing (TUM) methods [27,28]. LST downscaling methods are abundantly developed using multiple spatial, temporal, and spectral resolution RS data. However, the scale effects on the LST data and the unstable relationship of intermediate parameters with LST have limited the improvement of the input data accuracy and, thus, limited the LST-based downscaling model [29]. Since crop transpiration processes are accompanied by variability in crop photosynthetic processes [13] and can be monitored with high-resolution optical satellite data, crop transpiration can be estimated with stomatal behavior and biophysical conditions combined with meteorological data. The surface resistance (r_s) is an important factor in calculating daily ET and can be downscaled using one-km r_s combined with 30-m satellite data and then input into the ET calculations. The downscaling of r_s provides fine-resolution information for the ET calculations and is regarded as the basic approach to the spatial monitoring of ET for field-scale management. Studies have shown that the r_s-based ET downscaling approach requires the underlying surfaces to be covered with some amount of plants, since its feasibility depends on the biophysical activities of plants' transpiration through the stomata. For sparsely vegetated regions, the accuracy of the r_s-based ET downscaling approach is limited [21]. Those algorithms focus on the relationship of each coarse resolution (m^3) pixel and fine resolution (m^1) pixels within it; the relationships of adjacent fine resolutions from different coarse pixels are hardly considered, which could bring in some "edges", as shown in Tan, Wu and Yan's [21] downscaling results.

Distinguished from pixel-level ET downscaling, the agricultural fields can be regarded as objects that have homogeneous ET properties within each field and heterogeneous properties between the fields. Studies on the ET mechanism and impact analysis of factors on ET have revealed that ET is highly correlated with vegetation cover, soil moisture, air temperature, and vapor pressure deficiency [14,15]. ET can vary among fields due to differences in the soil moisture from irrigation, the transpiration abilities of different crop breeds, fertilization, and growth conditions; thus, it is possible to use the differences in soil water conditions and crop conditions as indicators of the ET differences between fields. At different scales, ET impact factors can vary, and some of these factors can be acquired from high-resolution RS data; this suggests possible approaches in which the field ET within coarse ET pixels could be allocated by means of building relationships between impact factors and ET at the field scale. Combining high-resolution RS data and moderate-resolution actual ET products, field-scale actual ET can be calculated and applied for crop water monitoring. We chose the ETWatch model as the moderate-resolution ET data, as ETWatch is reliable for daily and monthly ET estimations across different climate types and surface characteristics, especially in cropland regions. Its accuracy has been proven by more than 50 research groups for water consumption structure and agricultural irrigation management, including our research region, Haihe Basin and Heihe Basin [7,8,14,16,30]. For ET calculations at high resolution, ETWatch has also been used as input data in several researches, including some downscaling methods [19,21].

To explore the appropriate methods of field ET monitoring and irrigation evaluation, we discuss whether the field ET allocation method combining moderate-resolution ET data and high-resolution satellite data is applicable. The model results are compared with ground ET observations from two typical cropland stations in the North China Plain and Northeast China, both of which are located in arid to semiarid regions. In this research, a total of two years of Sentinel-2 optical data and coarse ET products from the ETWatch model are combined with observations from two ground stations: Guantao from the Haihe Basin, North China Plain and Daman from the Heihe Basin, Northwest China. In Section 2, the data, proposed model, and other research methods are described in detail. Section 3 presents the model results and evaluation. Section 4 discusses the performance of the model and its potential application in field-scale agricultural water management.

2. Materials and Methods

2.1. Method

Figure 1 shows the flowchart of the allocation method from 1 km ET data to agricultural field scale. The crop growth conditions and field moistures are associated with the crop transpiration capacity based on the stomatal behavior and soil water status, and both influence the field ET. In our research, we neglected the possible transport of the horizontal water vapor of adjacent fields; currently, horizonal vapor is analyzed on a large scale, such as the global and continent levels [31]. Most of the current ET RS models have not taken into account that the vapor amount moves horizontally, possibly due to the difficulty in monitoring and estimating horizonal vapors at the regional level. Thus, we assumed that the 1-km ET input data are accurate enough for allocation. One of our research regions,

Daman, is located in the Zhangye Oasis, Heihe Basin. The oasis effect of the meteorological parameter differences between an oasis and desert may affect the accuracy of allocation results in heterogeneous regions [32]. To avoid the potential impact of the oasis effect, we limited our research region to a cropland area close to the center of the oasis.

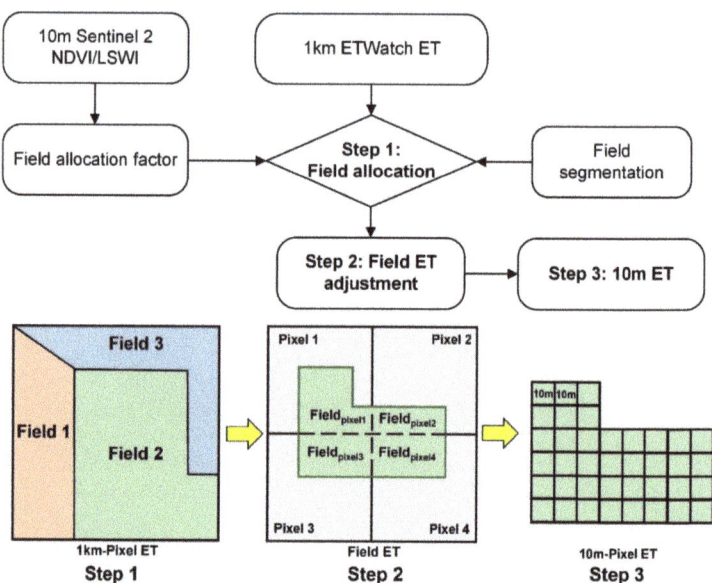

Figure 1. Flowchart of the evapotranspiration (ET) allocation method. The upper figure is the calculation algorithm, and the below figure is the three steps of field ET allocation from 1-km pixel ET to 10-m pixel ET. NDVI: normalized difference vegetation index and LSWI: land surface water index.

Based on the two classes of virtually unchanged and field-variant parameters that influence the ET, the ET capacity of each field can be evaluated. Thus, we can assume that, between adjacent fields within a coarse ET pixel, some allocation factors derived from the field-variant parameters can be regarded as equivalent to the ET capacity of each field. The relation above can be summarized using the following equation:

$$ET_{field_1} : ET_{field_2} : \cdots ET_{field_i} = AF_{field_1} : AF_{field_2} : \cdots AF_{field_i} \quad (1)$$

where ET_{field_i} represents the ET of the ith field ($i = 1, 2, \cdots n$) within each coarse ET pixel, and AF_{field_i} represents the allocation factor of the field i ($i = 1, 2, \cdots n$) within each coarse ET pixel. Additionally, the relationship between the ET of field i and the coarse ET pixel can be expressed as follows:

$$\frac{ET_{field_i}}{ET_{coarse}} = \frac{AF_{field_i}}{AF_{coarse}} \quad (2)$$

where AF_{coarse} represents the mean allocation factor of the coarse pixel. With Equation (2), each field ET within the coarse ET pixel can be calculated based on the allocation factor, which can also be regarded as the allocation of the coarse pixel-level ET (approximately 1-km level) to the farm-level (10 m level) ET. In the actual practice of field ET allocation, the allocation factor should be calculated from high-resolution RS data that can show the inner coarse-pixel farm-level spatial heterogeneity. The following section will introduce how the allocation factor is derived.

Conventional ET direct calculation methods focus on the estimation of the latent heat flux, including the Penman-Monteith (PM) equation [33] and the Priestley-Taylor equation

(PT equation) [34–36]. Recently, several studies have revealed that the PT equation is suitable for conditions with sufficient soil moisture, such as irrigated cropland. The PT equation is independent from surface and aerodynamic resistances and is calculated with meteorological data and the coefficient α. The PT equation has been developed primarily for potential ET estimation; for actual ET applications, the α coefficient should be adjusted with several constraint functions, such as surface wetness (w), temperature (t), and crop fraction of absorbed photosynthesis active radiation (fapar) [36]. Here, we used the vegetation cover fraction (FVC) to represent the fapar limit, as many studies have done [11,13]. We used the RS-based land surface water index (LSWI) as the soil water constraint. The LSWI is the normalization of the near-infrared (NIR) and shortwave-infrared (SWIR) bands [37–39].

The PT equation can be expressed as follows:

$$LE = f(lswi, fvc, t)\alpha \frac{\Delta}{\Delta + \gamma}(R_n - G) \quad (3)$$

where α is the PT coefficient and is initially set to 1.26, Δ and γ are the same as in the PM equation, $f(lswi)$ represents the surface wetness, and low surface moisture conditions limit the ET volume, $f(t)$ represents the temperature constraint, and the optimal temperature of the crops is the optimal temperature for transpiration, $f(fvc)$ represents the radiation constraint, and a low FVC limits the net radiation for the ET process.

As we focused on the cropland, crop transpiration is the major component of ET. We proposed an assumption that, at the field scale within coarse pixels (1-km level), the daily environmental conditions (air temperature, relative humidity, wind speed, and net radiation) are virtually unchanged, and the spatial differences in the meteorological factors are compared with two automatic weather source (AWS) matrices. Based on this assumption, some parts of the PT equation can be neglected at the field scale: Δ and γ are calculated from the relative humidity and air temperature, R_n is mostly concerned with the solar shortwave radiation calculated from the sunshine duration and surface longwave radiation derived from the surface emissivity and air temperature, and these climatic parts can be neglected, since cropland fields are homogeneous in vegetation type and the emissivity differences are small between fields. We neglected the R_n part of the field allocation factor, and G can be regarded as a constant part of R_n [40], so G is also neglected. By these terms, we deduced the calculation of the field allocation factors using ET constraint functions.

We proposed our field allocation factor calculation equation is as follows:

$$AF_{field_i} = 1 - \max\left(\min\left(\frac{FVC_{max} - FVC}{FVC_{max} - FVC_{min}}, 1\right), 0\right) \cdot \max\left(\min\left(\frac{LSWI_{max} - LSWI}{LSWI_{max} - LSWI_{min}}, 1\right), 0\right) \quad (4)$$

$$FVC = FVC_{max} \times \max\left(\min\left(\frac{NDVI - 0.1}{0.9 - 0.1}, 1\right), 0\right) \quad (5)$$

$$LSWI = \frac{band_{nir} - band_{swir}}{band_{nir} + band_{swir}} \quad (6)$$

where $LSWI_{max}$ indicates the high moisture condition, $LSWI_{min}$ indicates the dry condition, and the FVC_{max} is set as 0.95. This method assumes that the pixel has no vegetation cover with a NDVI < 0.1 and is fully covered with a NDVI > 0.9. $band_{nir}$ is the near-infrared band reflectance, and $band_{swir}$ is the shortwave-infrared band (SWIR) reflectance. We used the Sentienl-2 satellite SWIR band 11 (1613.7 nm) to calculate the LSWI, which is sensitive to the surface water status; the Sentinel 2 band 11-based LSWI has recently been widely applied with good performances for reflecting the surface moisture condition in agricultural applications [41], such as crop intensity mapping [42], plantation dynamics [43], leaf area index, and aboveground biomass estimation [44].

The field ET allocation from coarse ET pixels involves the allocation of a single pixel between fields and the adjustment of fields that span pixels. Figure 2 shows the flowchart of the field ET allocation process. Based on Equation (2), the allocation is conducted with the following procedures:

Step 1 is the allocation of fields at each coarse ET pixel. Field ET within the same coarse ET pixel is allocated using the allocation factor:

$$\mathrm{ET}_{\mathrm{field}_i} = \mathrm{ET}_{1\mathrm{km}} \frac{AF_{field_i}}{AF_{1km}} \quad (7)$$

where $\mathrm{ET}_{1\mathrm{km}}$ is the coarse ET pixel value at the 1-km level, and AF_{field_i} is the allocation factor of field i. AF_{1km} is the mean allocation factor of the area that corresponds to the 1-km ET pixel calculated from high-resolution RS data.

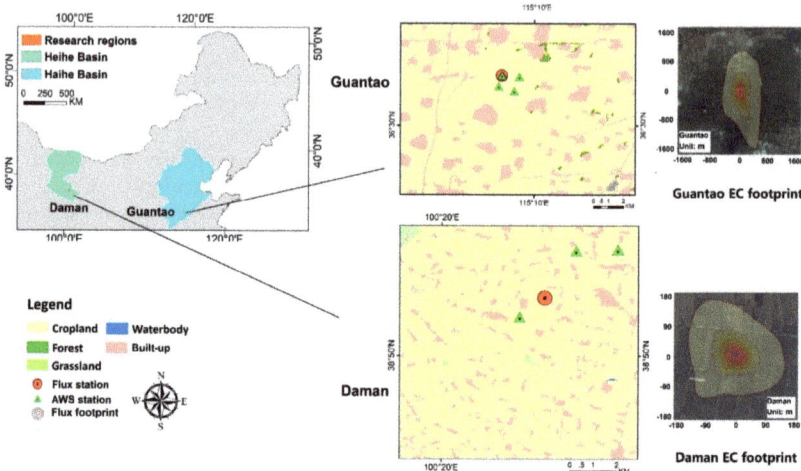

Figure 2. Locations and land cover types of the research regions. The flux station locations are marked on the maps, with photos of the flux footprints shown below. The flux observation footprints are calculated using the flux footprint prediction model [45]. EC: eddy covariance and AWS: automatic weather source.

Step 2 is the adjustment of fields that span pixels. Some large fields may cover different coarse pixels, and each part of the field is calculated with a different allocated ET value with different 1-km pixels. The final field ET result is adjusted and unified based on the area and allocated ET value of every part of the field from different 1-km pixels using the following equation:

$$\mathrm{ET}_{\mathrm{field}_i} = \frac{\sum \mathrm{ET}_{\mathrm{field}_i, \mathrm{pixel}_j} A_{field_i pixel_j}}{A_{field_i}} \quad (8)$$

where i indicates the field, and j indicates the covered 1-km pixel of *field i*. The ET amount is accumulated from the allocated ET multiplied by the field area in *pixel j*. Then, the summation is divided by the total field area to give the final field ET result.

For Step 3, the final ET data of the field can be derived using a transformation of Equation (6), with the mean allocation factor of the field and the high-resolution allocation factor, and the final ET data can be acquired:

$$\mathrm{ET}_{10m} = \mathrm{ET}_{\mathrm{field}} \frac{AF_{10m}}{AF_{field}} \quad (9)$$

2.2. Data

2.2.1. Research Region

The model was developed and tested at two cropland research stations: Guantao Station (115.1274 E, 36.5150 N) located in the Haihe Basin, North China Plain and Daman Station (100.3722 E, 38.8555 N) located in the Heihe Basin, Northwest China. The locations and land cover types of the two stations are shown in Figure 2. Both stations are cultivated with maize, and the climate types between the two sites are different. Guantao Station is located in a temperate monsoon climate in the south with a winter wheat and summer maize crop rotation, and the research region was chosen as the 10 km × 12 km cropland region. The annual average precipitation of Guantao is 560 mm, and the average temperature is 13 °C. Daman Station is located in a temperate continental climate that is typical of semiarid and semi-humid regions in the Heihe River Basin, Northwest China with a 10 km × 10 km cropland region. The annual average precipitation of Daman is 110 mm, and the annual mean temperature is 7 °C; only April–October, with high temperatures, is suitable for crop cultivation. The precipitation of both regions is not sufficient for crop growth and requires additional irrigation.

2.2.2. Remote-Sensing Data

In this research, the coarse ET data were calculated by ETWatch. The input data were mainly MOD09GA reflectance data, MOD11A1 surface temperature data [46], and MCD43B1 BRDF data. All MODIS datasets were acquired from the NASA Land Processes Distributed Active Archive Center (LP DAAC) (https://lpdaac.usgs.gov/) and then processed for the research region at a 1-km spatial resolution after geometric correction, radiance calibration, and atmospheric correction. Considering the impact of clouds on the surface reflectance, we used the Savitzky-Golay filter method (S-G filter) to extend the cloud-free days albedo to a daily scale at each pixel. The SG filter method has been widely used in the temporal extension of the NDVI [47] and albedo [48] in many researches.

For the field-scale ET model, we used Sentinel 2 satellite data from the European Commission's Copernicus program [49]. The data were downloaded using the Google Earth engine and were already processed with atmospheric correction and transformed into bottom of air (BOA) reflectance data. The bands used were mainly the 10-m resolution red band (red, 664.6 nm), near-infrared band (NIR, 832.8 nm), and shortwave-infrared band (SWIR, 1613.7 nm), which are sensitive to vegetation and ground water dynamics. The three bands were then applied in the NDVI and LSWI calculations.

The cropland distribution map in this research was extracted from the 30-m resolution ChinaCover dataset developed by the Aerospace Information Research Institute, Chinese Academy of Sciences [50]. The ChinaCover land use map divided the land cover into six major classes: forest, grassland, cropland, built-up, waterbody, and bare land.

2.2.3. ETWatch Model Data

In this research, we used the ETWatch model as the coarse ET input. The ETWatch model was developed by Wu [51]. The ETWatch model is based on the surface energy balance theory that the energy for evapotranspiration, latent heat flux, is regarded as the residual of the surface net radiation (R_n) which is the input energy from solar radiation, sensible heat flux (H), which is a major output energy from the energy balance for heating the air, soil heat flux (G_0), which takes up a small proportion of the net radiation for heating the soil. Thus, the calculation of ET is based on the precise estimation of the R_n, H, and G_0. Considering the cloud impact on satellite images to retrieve land surface characteristics, temporal extension methods are needed to extend the ET results from cloud-free days to a daily scale. The calculation processes of the ETWatch model involves several steps. First, on cloud-free days, the instantaneous latent heat and sensible heat from a satellite pass by moment are calculated with the surface energy balance theory and parametric models; then, the instantaneous ratio of latent heat (LE) to sensible heat (H) is acquired. Second, the instantaneous LE result is extended to a daily scale using the evaporation fraction, which assumes the ratio of LE to H remains little changed during the day, and the daily net radiation is calculated using meteorological data and the sunshine duration data. Based on the daily LE results, the r_s on cloud-free days can be retrieved with an inverse form of the Penman-Monteith (PM) equation. Third, the daily r_s is extended from the cloud-free day r_s results by a time scale extension model developed from the Jarvis model [52]. Finally, the daily ET is calculated based on the PM equation with the daily r_s and daily net radiation. After tens of years of development, ETWatch is robust when applied, and researchers have developed many parametric models [53], such as the net radiation model, sunshine duration model from stationary satellite data, sensible heat flux model [54], aerodynamic roughness length model [55], and the r_s time scale extension model. ETWatch has been validated for various climates and land cover types, including the semi-humid semiarid Haihe Basin [10], arid Loess Plateau area [14], Heihe Basin [54], and extremely arid Turpan Basin, Xinjiang [7]. These applications have proven the quality of the ETWatch model and shown that it is reliable for this research [8].

2.2.4. Crop Field Segmentation Map

In this research, we used the simple linear iterative clustering (SLIC) method for cropland field detection [56]. The SLIC method is a common method applied in medical image processing, RS segmentation, and computer object identification [57]. SLIC is developed based on an iterative algorithm with the k-means clustering theory. The cluster that best fits each pixel is chosen from the neighboring cluster cores instead of applying a cluster calculation with all cores in the image. By this method, SLIC can conduct k-means clustering at high speeds with low computing resources and can be applied in image segmentation. In agricultural cropland fields, the image values within the field are quite similar to each other and are divided by field boundaries, roads, and built-up human facilities. The different growing conditions from different agricultural activities can be seen with satellite NDVI. In this research, we used the temporal series NDVI files of maize growing, peaking, and mature periods to perform cropland segmentation with the SLIC method. Figure 3 shows the cropland segmentation results. The cropland segmentation is further used in field-scale ET allocation as the basic cropland field.

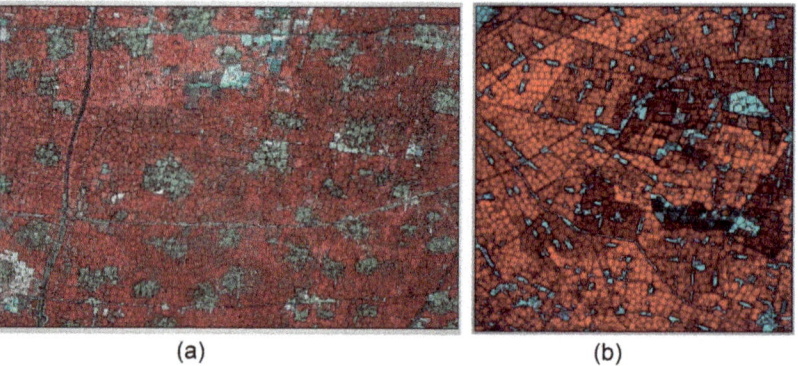

Figure 3. Field extraction from the simple linear iterative clustering (SLIC) method in the research regions: (**a**) Guantao field and (**b**) Daman field.

2.2.5. Site Observation Data

For model validation, ground observational EC data were gathered from two stations. Daman observations were acquired from the National Tibetan Plateau Data Center (https://data.tpdc.ac.cn/en/) under the HiWATER eco-hydrological experiment with multiple ground observation instruments [58–60]. Guantao observational data were collected from our experimental site [5,10]. The equipment of each of the two EC stations included a 3D

Sonic Anemometer and Infrared Gas Analyzer, and the data were processed under standard approaches [8,61,62]. The EC stations were accompanied by automatic weather observation instruments (AWSs), and the meteorological observations included air temperature, relative humidity, wind velocity and direction, air pressure, precipitation, and four-component radiation. Quality control of meteorological observational data was applied. The footprints of the EC observations were calculated using the flux footprint prediction model [45]. For analysis of the spatial heterogeneity of meteorological parameters and solar radiation, four separate AWSs from the Guantao observation matrix and three separate AWSs from Daman were chosen for observations of the air temperature, relative humidity, wind speed, and net radiation. The locations of the AWSs are marked in Figure 2. The Guantao AWSs were located at distances of 1 to 2 km, and the Daman AWSs were located at distances of 2–5 km, which matched the coarse km-scale ET pixel and subpixel levels.

To evaluate the spatial heterogeneity at the field level of the major meteorological parameters that impact the ET, we compared the growing season's air temperature (T_a), relative humidity (RH), wind speed (WS), and net radiation (R_n) using two site AWS observation matrices. At the Daman site, the AWS was installed at 4 m high and the radiometer at 4.5 m. At the Guantao site, the installment height of the AWS was 10 m, and the radiometer was 15 m. The results showed that, during the main crop-growing period (June–September), at a daily scale on average, the Guantao AWS observation matrix differences in the T_a, RH, WS, and R_n were 0.07–0.26 °C, 0.5–2%, 0.4–1.0 m/s, and 6.6 W/m^2, respectively, and the Daman AWS observation matrix differences in the T_a, RH, WS, and R_n were <0.3 °C, <3%, 0.4–1.0 m/s, and 10 W/m^2, respectively. The relatively small differences in the AWS observation matrix results indicated that the variations in the T_a, RH, WS, and R_n were small at the subpixel 1–5 km level, which covered the scope of the fields in this research and demonstrated that our neglect of the meteorological and radiation differences in the PT equation was acceptable for deriving the ET allocation factors.

2.3. Model Evaluation

The accuracy assessment was based on three statistical indicators: correlation factor (R), root mean square error (RMSE), and William's index of differences (d). Detailed calculation methods of the statistical indicators are shown below:

$$R = \frac{\sum_{i=1}^{n}[(Y_i - \overline{Y})(X_i - \overline{X})]}{\sqrt{\sum_{i=1}^{n}(Y_i - \overline{Y})^2}\sqrt{\sum_{i=1}^{n}(X_i - \overline{X})^2}} \quad (10)$$

$$\text{RMSE} = \sqrt{\frac{1}{n}\sum_{i=1}^{n}(Y_i - X_i)^2} \quad (11)$$

$$d = 1 - \frac{\sum_{i=1}^{n}(Y_i - X_i)^2}{\sum_{i=1}^{n}(|Y_i - \overline{X}| + |X_i - \overline{X}|)^2} \quad (12)$$

where X_i is the model result, Y_i is the validation data, and $\overline{X}$ and $\overline{Y}$ represent the average values of the model result and validation data, respectively.

3. Results

3.1. Field-Scale ET Allocation Factor

In this research, we developed an allocation factor that indicates the ET capacity variations of different fields. Combining the coarse ET dataset (1 km), the field-scale ET can be allocated at a high resolution (10 m). The effectiveness of the allocation factor in representing the ET differences is essential, and we performed a correlation analysis between the site ET observations and RS-based allocation factors in Daman and Guantao. The correlation results are shown in Figure 4. We compared the eight-day average ET observations and allocation factors. Based on the previous definition, the allocation factor can be regarded as an indicator of long-term ET trends, and an eight-day average can decrease the daily variations for comparison. We found that the allocation factor has a satisfactory correlation with the EC ET data. At the Daman and Guantao Stations, the correlation coefficients of the allocation factors and site EC ET observations are greater than 0.77, indicating significant correlations ($p < 0.01$). As shown in Figure 4, the ET amount can be reduced during rainy days, during which the RH is saturated and the vapor pressure deficit (VPD) between the air and leaf surfaces is small, which slows the transpiration speed. The allocation factor cannot reflect the differences in rainfall events. Considering that the daily meteorological variations such as VPD and temperature strongly influence the ET variations and are not considered when composing the allocation factors, the correlation result is sufficient for field-level allocation. The daily site ET variations are presented with the allocation factors in Figure 4a,c. Since daily ET is sensitive to daily meteorological changes, the site ET is more variable than the allocation factor, which is smoothed between days with the S-G filter method. Both at the daily scale and at eight-day intervals, the allocation factor corresponds well with the ET variation and can be used for further field ET modeling.

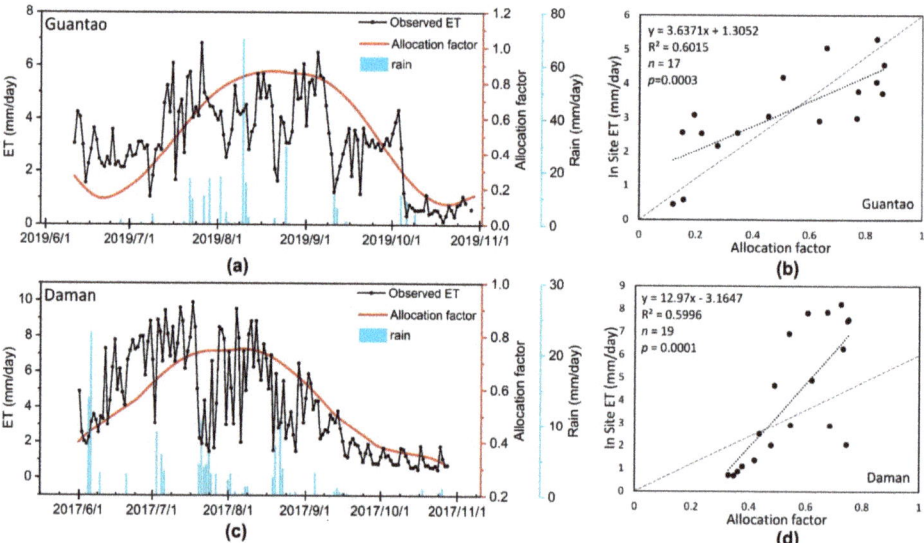

Figure 4. The correlation between the allocation factors and site ET observations. (**a,c**) The daily comparisons of the proposed allocation factor and observed ground ET in Guantao and Daman; the ET observations are derived from the site EC towers, and rainfall data are derived from the site rain gauge observations. (**b,d**) The correlations between the allocation factor and site ET at 8-day intervals. The correlation factor R was greater than 0.77 at both sites and statistically significant ($p < 0.01$).

3.2. Field ET Allocation Performances Based on the ETWatch Model

The accuracy of the field model ET was evaluated using the site ET observations from the ground EC instruments. Detailed evaluation indicator values are exhibited in Table 1. Figure 5 shows the model validation of the allocated field-scale ET results. Compared with the site ET observations, the allocated ET is very accurate. At Daman and Guantao Stations, the correlation coefficients R^2 of the field model ET are higher (0.954 for Guantao and 0.941 for Daman) and the root mean square errors are lower (0.98 mm/day for Guantao and 1.59 mm/day for Daman) than those of the ETWatch performances (R^2 0.949 for Guantao and 0.890 for Daman and RMSE 0.946 mm/day for Guantao and 1.67 mm/day for Daman). Figure 6 shows the temporal variations in the field model ET and site observations. The field model ET is well-correlated with the site observations during the peak growing periods of maize (June–August) and can precisely exhibit the variations in daily ET in the field. These model validation results show that the field-scale ET model allocated from the ETWatch 1-km-level dataset has achieved high accuracy according to the site observations and with the appropriate coarse ET data. Field-scale allocation can provide improved accuracy and present ET distributions with more resemblances to the actual field conditions than those of lower-resolution models, and the spatial distribution of the field model ET is presented in Figures 7 and 8.

Table 1. Site evaluation of the field model evapotranspiration (ET) allocated from the ETWatch data. R^2: correlation coefficient, RMSE: root mean square error, and d: William's index of differences.

Site	ETWatch			Field Model ET		
	Adj. R^2	RMSE	d	Adj. R^2	RMSE	d
Guantao	0.949	0.946	0.915	0.954	0.981	0.916
Daman	0.890	1.67	0.874	0.941	1.50	0.931

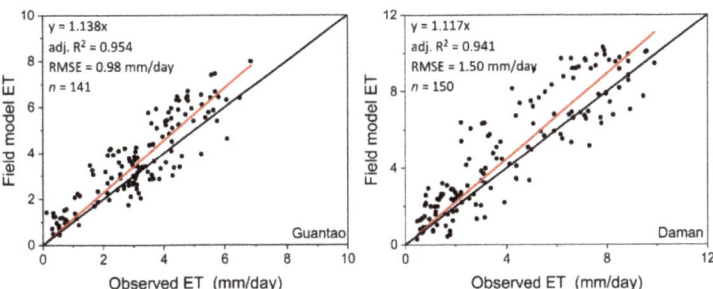

Figure 5. Model validation of the field model ET with the site observations. R^2: correlation coefficient and RMSE: root mean square error.

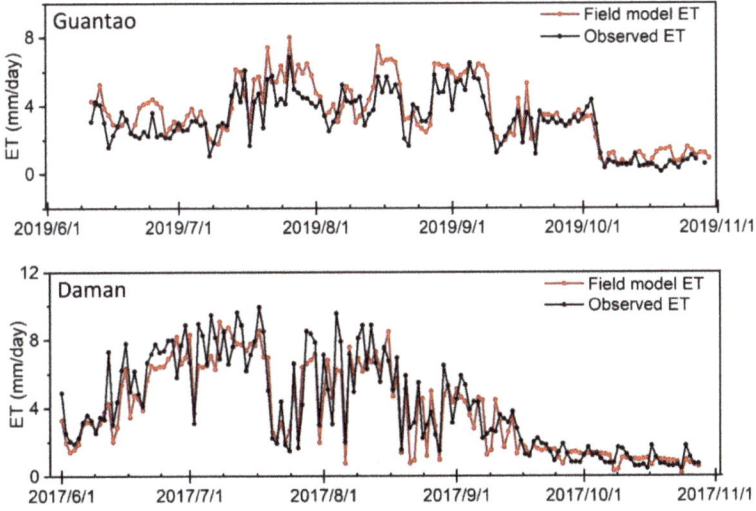

Figure 6. Temporal development of the field model ET at the two research stations. In each panel, the red line indicates the field model ET, and the black line indicates the site observed ET from the eddy covariance instruments.

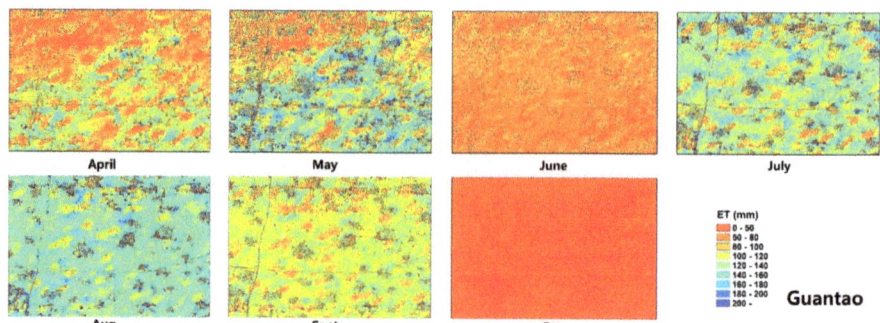

Figure 7. The spatial distribution of the ET allocation results in the Guantao research region.

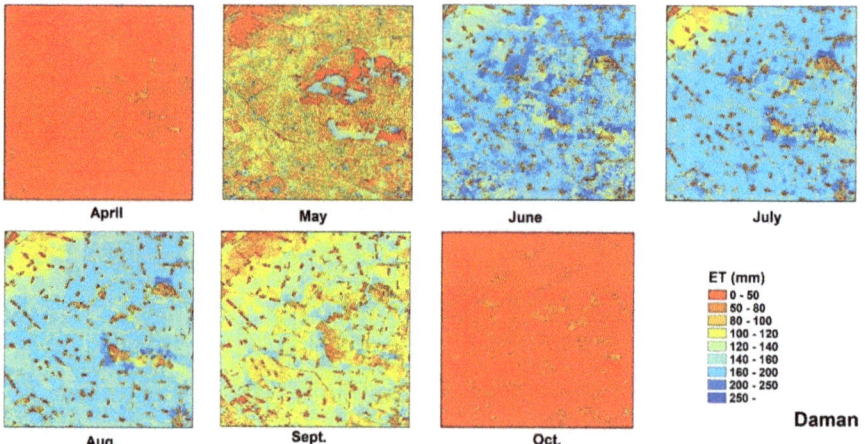

Figure 8. The spatial distribution of the ET allocation results in the Daman research region.

The spatial distribution of the ET allocation is presented in Figure 7 for Guantao and Figure 8 for Daman. Compared with the coarse ET data, the field-scale allocated ET has more spatial details that can show the temporal development of the water consumption results in each field. Since the coarse ET pixel is computed based on the comprehensive characteristics of the 1-km^2 region, the high- and low-ET areas are averaged in the coarse ET results; however, the allocated ET data revealed the high- and low-ET regions and remain consistent in the total ET amount between the coarse ET and allocated field ET. The concept of water balance was initially adopted with the ET allocation approaches. At the monthly level, the field model ET can show the spatiotemporal development of maize in each field at the farm level and be used for agricultural water consumption management. ET is the major water consumption pathway of irrigation systems, and precise ET monitoring is essential in achieving water usage cuts, especially in arid, water-scarce regions. The field ET map provides foundational data for water management.

3.3. Comparison with the Pixel-Level Downscaling Method

Figure 9 shows the spatial distribution map comparison between the original coarse ET datasets and the allocated ET results. The spatial distribution of a pixel-level ET downscaling approach based on surface resistance is also shown. The results show that the spatial texture is improved, and more spatial ET information can be extracted from the allocated ET. Since the coarse ET pixel can be regarded as the average value of the inner pixel regions, the differences in the field-level ET are erased, and the spatial information of the coarse-resolution ET is limited. The allocated ET can finely capture the spatial differences in ET capacities between fields under different irrigation and fertilization strategies and other agricultural activities. Comparing different coarse ET data sources, the allocated ET values are dependent on the original ET data, and the total regional ET amounts of the original coarse ET and the allocated ET are equivalent based on the water balance; thus, with sufficiently accurate coarse ET models, the accuracy of the allocated ET results is guaranteed.

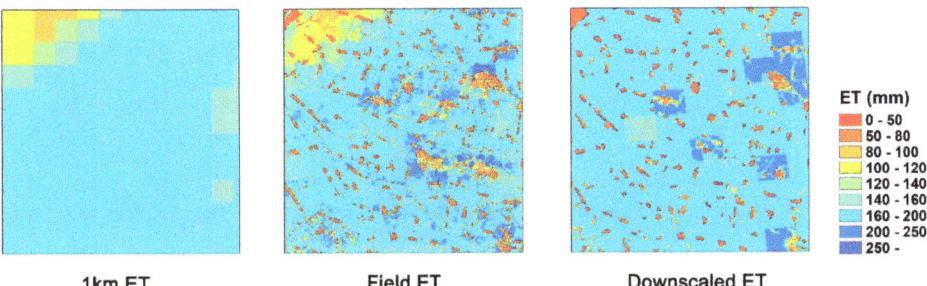

Figure 9. The spatial distribution comparison of monthly fine-resolution ET data with the surface resistance (rs)-based downscaling ET method [21] in July. The 1-km ET data are calculated with the ETWatch model, and the field ET is derived with the allocation method from this research. The downscaled ET is based on the surface resistance downscaling approach.

The downscaling methods are currently aimed at the pixel levels, for which the process flow can be described as follows. The values of a chosen factor at each high-resolution pixel within a coarse-pixel area are compared against the average value of the coarse pixel. Each factor can be directly correlated with the ET. Despite the differences in downscaling methods, with pixel-level calculations, the small differences between adjacent coarse pixels can be magnified in the downscaling process and show differences similar to "boundaries" in downscaled fine-resolution ET results, as shown in Figure 9. The ET based on downscaled surface resistance shows apparent differences at the boundaries of the coarse ET dataset. Since our allocation method is based on agricultural fields, the field is regarded as the independent object of ET allocation. We consider the allocation between the cropland fields and the ET within the fields; thus, our results are more reasonable in spatial distributions, with the ET differences mainly located at the agricultural field boundaries and in line with the actual situations.

4. Discussion

4.1. The ET Allocation Method Performance

In this research, we proposed a new method to calculate high-resolution ET from coarse ET datasets that can better fit the agricultural field scale for applications in water management. We viewed agricultural fields as the basic ET objects for ET allocation from coarse resolutions to high resolutions and developed ET allocation factors by quantifying the ET capacity based on the surface moisture and crop physiological parameters. Our model was practical to apply, because it utilized the abundant ET datasets at the km-resolution level. The model performance was evaluated using two cropland stations, Guantao in the Haihe Basin and Daman of the Huailai Station, both of which are maize cultivation research stations. The validation results were satisfactory, and the evaluation indicators (R^2, RMSE, and d) had better performances in the allocated field model ET than in the original km-level ET (Table 1 and Figure 9).

Based on the derivation of the PT equation, our field ET allocation model captured the variations between fields in the main ET parameters, which reflected the crop's capacity to absorb solar radiation, represented by the FVC from the NDVI and ground water status from the LSWI. These variation parameters were combined to form what we called the field allocation factor in this research and acted as the driving parameter of the ET field allocation model. The mechanism of the allocation model was that between different crop fields; the ET capacity of the fields relied on the proposed variation parameters, which were highly correlated with the ET and present large spatial diversity. The ET allocation model acceptably neglected the meteorological conditions, which were uniformly distributed in the homogeneous cropland fields at the coarse-km level. It can be seen in Figure 4 that the proposed allocation factors had similar trends but smaller fluctuations than those of the observed ET variations. One reason is that the NDVI and LSWI input data were processed with the S-G filter method to acquire the daily-scale data, which applied a smoothing effect to the five-day interval data. Another major reason was that NDVI and LSWI were not sensitive to temporal variations, since they reflected the surface variables, such as chlorophyll, crop growth conditions, and water contents; this is only the reason that the S-G filter method is suitable for temporal extension. As the S-G filter method has been commonly adopted in the temporal extension of remote sensing-based indexes, the errors and residuals in the daily smoothed data were acceptable. Since the allocation is conducted at the spatial level, the temporal variations in climate data do not participate in the ET allocation processes and have small spatial effects at the field level. Thus, the temporal differences in the variability between the allocation factor and actual ET have minor effects on the allocation model. For the allocation factor, we used the NDVI and LSWI; during the dense canopy period, such as the peak growth period of maize, the LSWI reflected more of the water content in vegetation instead of the soil water content. At this period, the crop transpiration takes the majority of the ET with high FVC and low radiation for soil evaporation, FVC plays the major role of allocation, and the influence of the LSWI on the ET allocation is limited.

4.2. Improvement to Pixel-Level ET Downscaling Methods

When applying ET methods to precise agricultural water management, the lack of stable, accurate, high-resolution ET data has heavily restrained the attention on ET in water management. Our method is different from previous ET downscaling methods, such as the LST or vegetation index disaggregation, which usually concentrate on the spatial resolution refinement of several ET model parameters and input them into the original ET models for high-resolution ET calculations. Up to a point, this kind of ET downscaling approach compensates for the absence of critical high-resolution input data in ET models, especially in surface energy-related models. Restrained by the band characteristics and current sensor technology, the daily temporal resolution of thermal infrared sensors is incompatible with a 10-m spatial resolution. Most of the LST downscaling methods combine MODIS daily 1-km LST data with Landsat monthly 30-m LST data, and the large temporal gap between MODIS and Landsat inevitably leads to errors in the spatial information and larger uncertainties in the ET calculations, which limit their usefulness for product applications.

Another ET downscaling method is by directing the correlation of the ET with remote sensing-based parameters such as the NDVI and LST and distributing the ET within coarse pixels. This downscaling approach can be regarded as a pixel-oriented method, and the downscaled results can be influenced by the coarse-pixel scale effect that the adjacent coarse pixel brings into distinct edges, as shown in Figure 9. These models are validated with high accuracy at the point level; however, the spatial distribution of the boundaries is not similar to real cropland ET situations. Our model is distinguished from the pixel-oriented downscaling method in that we regarded the agricultural field as the basic object of the ET distribution units. Based on the water balance theory, the statistical results of coarse ET and fine-resolution ET should remain equivalent, and ET can be allocated to each field. The cropland field can avoid the impacts of coarse-pixel differences on the ET results, and the spatial texture has more resemblance to the actual field ET distribution than that of the lower resolution models. After ET allocation among fields, the field ET is then downscaled based on pixels within the field and is capable of reflecting more precise spatial information of the inner field ET conditions. With the attributions of more mechanisms for ET variations instead of linear correlations from downscaling methods, more emphasis on the essential role of the field in km-level inner pixel ET allocation, and reasonable allocation approaches between and within the fields, our field ET allocation model performs better than the previous methods at estimating the ET with a refined resolution.

4.3. Future Application in Agricultural Water Management

At present, our model achieved satisfactory accuracy in the cropland. The calculation of the ETWatch one-km ET required spatially interpolated ground meteorological parameters, which sustain an adequate accuracy at the km level while, at 10-m level, the values may differ far from the real conditions under the influence of horizontal flow eddies. We neglected the spatial differences of the meteorological parameters in relative homogeneous and plane underlaying surfaces such as croplands, as the ground observation matrix showed that the meteorological parameters were little changed at the one-km level. However, when applying the allocation method in rugged terrain, the differences of wind speed, relative humidity, and air temperature should be concerned; the spatial diversity of the meteorological parameters should be preliminarily analyzed before applying the allocation model, as the meteorological condition homogeneity is the precondition of applying the allocation method. The movement of horizontal flow of the eddies carrying vapors between fields is another potential impact factor on the accuracy of field-scale ET. Wang and Dickinson's [63] review on the Monin-Obukhov Similarity Theory (MOST) showed that the MOST theory has potential in reflecting the turbulent fluxes at horizontally homogeneous and stationary surface layers, which may shed light on solving horizontal vapor estimations in future researches.

Modern water resource management requires advanced management theories to harmonize the requirements of economic and social developments with limited water resources. In agricultural water management, recent studies have added more emphasis to the control of ET from cropland and irrigation systems. Fine-resolution ET that matches the scale of cropland fields is needed to support policy decision-makers and set ET targets for water consumption cuts. One potential application of the field-level ET model is in the determination of water rights. The conventional determination of water rights involves setting limits on farmers' water withdrawal amounts, with no emphasis on how farmers use the water or the return flows of irrigation water. However, these water rights management methods neglect the water cycling mechanism where, with soil infiltration and ground runoff, some of the withdrawn water can return to the regional water system, and ET is the real water consumption that did not return to the local river or groundwater systems. Thus, an ET-oriented water rights allocation system is more reasonable, and since different crops have various ET amounts, the crop-planting structure should also be considered. In the Turpan Basin, China, the local government has experimented with using ET to set farmers' withdrawal amounts and has achieved considerable success in relieving the local water use crisis by changing the crop-planting structure to low-ET crops and reducing the use of high-ET irrigation methods [7,64]. These applications require accurate ET datasets that match each farmer's field, and our ET allocation model can provide the data foundation for farm-level water management.

5. Conclusions

In this research, we proposed a method to calculate ET at the field scale by allocating coarse ETWatch ET data to fields based on the allocation factor derived from high-resolution satellite data. The model achieved satisfactory accuracy compared with ground observations from two maize-growing cropland stations and improvement in the spatial representation and accuracy compared with coarse ET. The field model ET data are capable of field-scale water management in agricultural systems to precisely monitor the crop ET status, providing insights into water management approaches based on ET and water consumption. This allocation method can calculate field-scale ET with accuracy, stability, and speed, the exact characteristics that meet the demands of a wide application based on using ET data to control the regional water consumption, supporting ET management theory in addressing the impacts of water scarcity on societal and economic developments.

Author Contributions: Conceptualization, B.W. and Z.M.; methodology, B.W. and Z.M.; writing—original draft preparation, Z.M.; and writing—review and editing, B.W., N.Y., W.Z., H.Z. and J.X. All authors have read and agreed to the published version of the manuscript.

Funding: This research was funded by National Natural Scientific Foundations of China, grant number 41991232, Advanced Science Foundation Research Project of the Chinese Academy of Sciences, grant number QYZDY-SSW-DQC014, and the National Key Research & Development Program of China, grant number 2016YFC0501601.

Acknowledgments: The authors would like to thank the National Tibetan Plateau/Third Pole Environmental Data Center (https://data.tpdc.ac.cn/en/) for providing the ground observation data of Daman Station.

Conflicts of Interest: The authors declare no conflict of interest.

References

1. Oki, T.; Kanae, S. Global hydrological cycles and world water resources. *Science* **2006**, *313*, 1068–1072. [CrossRef]
2. Sellers, P.; Dickinson, R.; Randall, D.; Betts, A.; Hall, F.; Berry, J.; Collatz, G.; Denning, A.; Mooney, H.; Nobre, C. Modeling the exchanges of energy, water, and carbon between continents and the atmosphere. *Science* **1997**, *275*, 502–509. [CrossRef]
3. Mekonnen, M.M.; Hoekstra, A.Y. Four billion people facing severe water scarcity. *Sci. Adv.* **2016**, *2*, e1500323. [CrossRef]
4. Gebbers, R.; Adamchuk, V.I. Precision agriculture and food security. *Science* **2010**, *327*, 828–831. [CrossRef]
5. Wu, B.; Jiang, L.; Yan, N.; Perry, C.; Zeng, H. Basin-wide evapotranspiration management: Concept and practical application in Hai Basin, China. *Agric. Water Manag.* **2014**, *145*, 145–153. [CrossRef]
6. The World Bank. *Design of Water Consumption Based Water Rights Administration System for Turpan Prefecture of Xinjiang China*; Water Partnership Program; World Bank Group: Washington, DC, USA, 2012. Available online: http://documents.worldbank.org/curated/en/588081468216268772/Design-of-water-consumption-based-water-rights-administration-system-for-Turpan-prefecture-of-Xinjiang-China (accessed on 5 December 2020).
7. Tan, S.; Wu, B.; Yan, N.; Zeng, H. Satellite-Based Water Consumption Dynamics Monitoring in an Extremely Arid Area. *Remote Sens.* **2018**, *10*, 1399. [CrossRef]
8. Wu, B.; Zhu, W.; Yan, N.; Xing, Q.; Xu, J.; Ma, Z.; Wang, L. Regional Actual Evapotranspiration Estimation with Land and Meteorological Variables Derived from Multi-Source Satellite Data. *Remote Sens.* **2020**, *12*, 332. [CrossRef]
9. Chen, J.M.; Liu, J. Evolution of evapotranspiration models using thermal and shortwave remote sensing data. *Remote Sens. Environ.* **2020**, *237*, 111594. [CrossRef]
10. Wu, B.; Yan, N.; Xiong, J.; Bastiaanssen, W.; Zhu, W.; Stein, A. Validation of ETWatch using field measurements at diverse landscapes: A case study in Hai Basin of China. *J. Hydrol.* **2012**, *436*, 67–80. [CrossRef]
11. Mu, Q.; Zhao, M.; Running, S.W. Brief introduction to MODIS evapotranspiration data set (MOD16). *Water Resour. Res.* **2005**, *45*, 1–4.
12. Martens, B.; Gonzalez Miralles, D.; Lievens, H.; Van Der Schalie, R.; De Jeu, R.A.; Fernández-Prieto, D.; Beck, H.E.; Dorigo, W.; Verhoest, N. GLEAM v3: Satellite-based land evaporation and root-zone soil moisture. *Geosci. Model. Dev.* **2017**, *10*, 1903–1925. [CrossRef]
13. Zhang, Y.; Kong, D.; Gan, R.; Chiew, F.H.; McVicar, T.R.; Zhang, Q.; Yang, Y. Coupled estimation of 500 m and 8-day resolution global evapotranspiration and gross primary production in 2002–2017. *Remote Sens. Environ.* **2019**, *222*, 165–182. [CrossRef]
14. Ma, Z.; Yan, N.; Wu, B.; Stein, A.; Zhu, W.; Zeng, H. Variation in actual evapotranspiration following changes in climate and vegetation cover during an ecological restoration period (2000–2015) in the Loess Plateau, China. *Sci. Total Environ.* **2019**, *689*, 534–545. [CrossRef]
15. Yan, N.; Tian, F.; Wu, B.; Zhu, W.; Yu, M. Spatiotemporal Analysis of Actual Evapotranspiration and Its Causes in the Hai Basin. *Remote Sens.* **2018**, *10*, 332. [CrossRef]
16. Wu, B.; Zeng, H.; Yan, N.; Zhang, M. Approach for Estimating Available Consumable Water for Human Activities in a River Basin. *Water Resour. Manag.* **2018**, *32*, 2353–2368. [CrossRef]
17. Liang, S.; Wang, K.; Zhang, X.; Wild, M. Review on estimation of land surface radiation and energy budgets from ground measurement, remote sensing and model simulations. *IEEE J. Sel. Top. Appl. Earth Obs. Remote Sens.* **2010**, *3*, 225–240. [CrossRef]
18. Zhang, Z.; Tian, F.; Hu, H.; Yang, P. A comparison of methods for determining field evapotranspiration: Photosynthesis system, sap flow, and eddy covariance. *Hydrol. Earth Syst. Sci.* **2014**, *18*, 1053–1072. [CrossRef]

19. Tan, S.; Wu, B.; Yan, N.; Zhu, W. An NDVI-Based Statistical ET Downscaling Method. *Water* **2017**, *9*, 995. [CrossRef]
20. Norman, J.; Anderson, M.; Kustas, W.; French, A.; Mecikalski, J.; Torn, R.; Diak, G.; Schmugge, T.; Tanner, B. Remote sensing of surface energy fluxes at 10^1-m pixel resolutions. *Water Resour. Res.* **2003**, *39*, 1221–1261. [CrossRef]
21. Tan, S.; Wu, B.; Yan, N. A method for downscaling daily evapotranspiration based on 30-m surface resistance. *J. Hydrol.* **2019**, *577*, 123882. [CrossRef]
22. Hong, S.-h.; Hendrickx, J.M.; Borchers, B. Down-scaling of SEBAL derived evapotranspiration maps from MODIS (250 m) to Landsat (30 m) scales. *Int. J. Remote Sens.* **2011**, *32*, 6457–6477. [CrossRef]
23. Wang, Y.Q.; Xiong, Y.J.; Qiu, G.Y.; Zhang, Q.T. Is scale really a challenge in evapotranspiration estimation? A multi-scale study in the Heihe oasis using thermal remote sensing and the three-temperature model. *Agric. For. Meteorol.* **2016**, *230*, 128–141. [CrossRef]
24. Weng, Q.; Fu, P.; Gao, F. Generating daily land surface temperature at Landsat resolution by fusing Landsat and MODIS data. *Remote Sens. Environ.* **2014**, *145*, 55–67. [CrossRef]
25. Tian, F.; Qiu, G.; Lü, Y.; Yang, Y.; Xiong, Y. Use of high-resolution thermal infrared remote sensing and "three-temperature model" for transpiration monitoring in arid inland river catchment. *J. Hydrol.* **2014**, *515*, 307–315. [CrossRef]
26. Mukherjee, S.; Joshi, P.; Garg, R. A comparison of different regression models for downscaling Landsat and MODIS land surface temperature images over heterogeneous landscape. *Adv. Space Res.* **2014**, *54*, 655–669. [CrossRef]
27. Zhan, W.; Chen, Y.; Zhou, J.; Wang, J.; Liu, W.; Voogt, J.; Zhu, X.; Quan, J.; Li, J. Disaggregation of remotely sensed land surface temperature: Literature survey, taxonomy, issues, and caveats. *Remote Sens. Environ.* **2013**, *131*, 119–139. [CrossRef]
28. Chen, Y.; Zhan, W.; Quan, J.; Zhou, J.; Zhu, X.; Sun, H. Disaggregation of remotely sensed land surface temperature: A generalized paradigm. *IEEE Trans. Geosci. Remote Sens.* **2014**, *52*, 5952–5965. [CrossRef]
29. Fu, P.; Weng, Q. Consistent land surface temperature data generation from irregularly spaced Landsat imagery. *Remote Sens. Environ.* **2016**, *184*, 175–187. [CrossRef]
30. Feng, X.; Fu, B.; Piao, S.; Wang, S.; Ciais, P.; Zeng, Z.; Lü, Y.; Zeng, Y.; Li, Y.; Jiang, X. Revegetation in China's Loess Plateau is approaching sustainable water resource limits. *Nat. Clim. Chang.* **2016**, *6*, 1019–1022. [CrossRef]
31. Ploeger, F.; Günther, G.; Konopka, P.; Fueglistaler, S.; Müller, R.; Hoppe, C.; Kunz, A.; Spang, R.; Grooß, J.U.; Riese, M. Horizontal water vapor transport in the lower stratosphere from subtropics to high latitudes during boreal summer. *J. Geophys. Res. Atmos.* **2013**, *118*, 8111–8127. [CrossRef]
32. Ruehr, S.; Lee, X.; Smith, R.; Li, X.; Xu, Z.; Liu, S.; Yang, X.; Zhou, Y. A mechanistic investigation of the oasis effect in the Zhangye cropland in semiarid western China. *J. Arid Environ.* **2020**, *176*, 104120. [CrossRef]
33. Monteith, J.L. Evaporation and environment. In *Symposia of the Society for Experimental Biology*; Cambridge University Press (CUP): Cambridge, UK, 1965; Volume 19, pp. 205–234.
34. Hao, Y.; Baik, J.; Choi, M. Developing a soil water index-based Priestley–Taylor algorithm for estimating evapotranspiration over East Asia and Australia. *Agric. For. Meteorol.* **2019**, *279*, 107760. [CrossRef]
35. Priestley, C.H.B.; Taylor, R. On the assessment of surface heat flux and evaporation using large-scale parameters. *Mon. Weather Rev.* **1972**, *100*, 81–92. [CrossRef]
36. Fisher, J.B.; Tu, K.P.; Baldocchi, D.D. Global estimates of the land–atmosphere water flux based on monthly AVHRR and ISLSCP-II data, validated at 16 FLUXNET sites. *Remote Sens. Environ.* **2008**, *112*, 901–919. [CrossRef]
37. Chandrasekar, K.; Sesha Sai, M.; Roy, P.; Dwevedi, R. Land Surface Water Index (LSWI) response to rainfall and NDVI using the MODIS Vegetation Index product. *Int. J. Remote Sens.* **2010**, *31*, 3987–4005. [CrossRef]
38. Xiao, X.; Zhang, Q.; Saleska, S.; Hutyra, L.; De Camargo, P.; Wofsy, S.; Frolking, S.; Boles, S.; Keller, M.; Moore III, B. Satellite-based modeling of gross primary production in a seasonally moist tropical evergreen forest. *Remote Sens. Environ.* **2005**, *94*, 105–122. [CrossRef]
39. Zheng, Y.; Zhang, M.; Zhang, X.; Zeng, H.; Wu, B. Mapping winter wheat biomass and yield using time series data blended from PROBA-V 100-and 300-m S1 products. *Remote Sens.* **2016**, *8*, 824. [CrossRef]
40. Pereira, L.S.; Allen, R.G.; Smith, M.; Raes, D. Crop evapotranspiration estimation with FAO56: Past and future. *Agric. Water Manag.* **2015**, *147*, 4–20. [CrossRef]
41. Zhang, T.; Su, J.; Liu, C.; Chen, W.-H.; Liu, H.; Liu, G. Band selection in Sentinel-2 satellite for agriculture applications. In Proceedings of the 23rd International Conference on Automation and Computing (ICAC), Huddersfield, UK, 7–8 September 2017; pp. 1–6.
42. Liu, L.; Xiao, X.; Qin, Y.; Wang, J.; Xu, X.; Hu, Y.; Qiao, Z. Mapping cropping intensity in China using time series Landsat and Sentinel-2 images and Google Earth Engine. *Remote Sens. Environ.* **2020**, *239*, 111624. [CrossRef]
43. You, N.; Dong, J. Examining earliest identifiable timing of crops using all available Sentinel 1/2 imagery and Google Earth Engine. *ISPRS J. Photogramm. Remote Sens.* **2020**, *161*, 109–123. [CrossRef]
44. Wang, J.; Xiao, X.; Bajgain, R.; Starks, P.; Steiner, J.; Doughty, R.B.; Chang, Q. Estimating leaf area index and aboveground biomass of grazing pastures using Sentinel-1, Sentinel-2 and Landsat images. *ISPRS J. Photogramm. Remote Sens.* **2019**, *154*, 189–201. [CrossRef]
45. Kljun, N.; Calanca, P.; Rotach, M.W.; Schmid, H.P. A simple two-dimensional parameterisation for Flux Footprint Prediction (FFP). *Geosci. Model. Dev.* **2015**, *8*, 3695–3713. [CrossRef]

46. Wan, Z. *MODIS Land Surface Temperature Products Users' Guide*; Institute for Computational Earth System Science, University of California: Santa Barbara, CA, USA, 2006.
47. Chen, J.; Jönsson, P.; Tamura, M.; Gu, Z.; Matsushita, B.; Eklundh, L. A simple method for reconstructing a high-quality NDVI time-series data set based on the Savitzky–Golay filter. *Remote Sens. Environ.* **2004**, *91*, 332–344. [CrossRef]
48. Bian, J.-h.; Li, A.; Song, M.; Ma, L.; Jiang, J. Reconstruction of NDVI time-series datasets of MODIS based on Savitzky-Golay filter. *J. Remote Sens.* **2010**, *14*, 725–741.
49. Drusch, M.; Del Bello, U.; Carlier, S.; Colin, O.; Fernandez, V.; Gascon, F.; Hoersch, B.; Isola, C.; Laberinti, P.; Martimort, P. Sentinel-2: ESA's optical high-resolution mission for GMES operational services. *Remote Sens. Environ.* **2012**, *120*, 25–36. [CrossRef]
50. Wu, B.; Qian, J.; Zeng, Y.; Zhang, L.; Yan, C.; Wang, Z.; Li, A.; Ma, R.; Yu, X.; Huang, J. *Land Cover Atlas of the People's Republic of China (1: 1 000 000)*; China Map Publishing House: Beijing, China, 2017.
51. Wu, B.-F.; Xiong, J.; Yan, N.-N.; Yang, L.-D.; Du, X. ETWatch for monitoring regional evapotranspiration with remote sensing. *Adv. Water Sci.* **2008**, *19*, 671–678.
52. Xu, J.; Wu, B.; Yan, N.; Tan, S. Regional daily ET estimates based on the gap-filling method of surface conductance. *Remote Sens.* **2018**, *10*, 554. [CrossRef]
53. Wu, B.; Xiong, J.; Yan, N. ETWatch: Models and methods. *J. Remote Sens* **2010**, *15*, 224–230.
54. Zhuang, Q.; Wu, B.; Yan, N.; Zhu, W.; Xing, Q. A method for sensible heat flux model parameterization based on radiometric surface temperature and environmental factors without involving the parameter KB−1. *Int. J. Appl. Earth Obs. Geoinf.* **2016**, *47*, 50–59. [CrossRef]
55. Yu, M.; Wu, B.; Yan, N.; Xing, Q.; Zhu, W. A Method for Estimating the Aerodynamic Roughness Length with NDVI and BRDF Signatures Using Multi-Temporal Proba-V Data. *Remote Sens.* **2016**, *9*, 6. [CrossRef]
56. Kim, K.S.; Zhang, D.; Kang, M.C.; Ko, S.J. Improved simple linear iterative clustering superpixels. In Proceedings of the IEEE International Symposium on Consumer Electronics (ISCE), Hsinchu, Taiwan, 3–6 June 2013; pp. 259–260.
57. Choi, K.S.; Oh, K.W. Subsampling-based acceleration of simple linear iterative clustering for superpixel segmentation. *Comp. Vis. Image Underst.* **2016**, *146*, 1–8. [CrossRef]
58. Liu, S.; Xu, Z.; Song, L.; Zhao, Q.; Ge, Y.; Xu, T.; Ma, Y.; Zhu, Z.; Jia, Z.; Zhang, F. Upscaling evapotranspiration measurements from multi-site to the satellite pixel scale over heterogeneous land surfaces. *Agric. For. Meteorol.* **2016**, *230*, 97–113. [CrossRef]
59. Li, X.; Cheng, G.; Liu, S.; Xiao, Q.; Ma, M.; Jin, R.; Che, T.; Liu, Q.; Wang, W.; Qi, Y. Heihe watershed allied telemetry experimental research (HiWATER): Scientific objectives and experimental design. *Bull. Am. Meteorol. Soc.* **2013**, *94*, 1145–1160. [CrossRef]
60. Xu, Z.; Liu, S.; Li, X.; Shi, S.; Wang, J.; Zhu, Z.; Xu, T.; Wang, W.; Ma, M. Intercomparison of surface energy flux measurement systems used during the HiWATER-MUSOEXE. *J. Geophys. Res. Atmos.* **2013**, *118*, 13–140. [CrossRef]
61. Liu, S.; Xu, Z.; Zhu, Z.; Jia, Z.; Zhu, M. Measurements of evapotranspiration from eddy-covariance systems and large aperture scintillometers in the Hai River Basin, China. *J. Hydrol.* **2013**, *487*, 24–38. [CrossRef]
62. Liu, S.M.; Xu, Z.W.; Wang, W.; Jia, Z.; Zhu, M.; Bai, J.; Wang, J. A comparison of eddy-covariance and large aperture scintillometer measurements with respect to the energy balance closure problem. *Hydrol. Earth Syst. Sci.* **2011**, *15*, 1291–1306. [CrossRef]
63. Wang, K.; Dickinson, R.E. A review of global terrestrial evapotranspiration: Observation, modeling, climatology, and climatic variability. *Rev. Geophys.* **2012**, *50*. [CrossRef]
64. The World Bank. *Summary of Water Consumption Management Technology in Turpan City Based on Remote Sensing Technology: Innovation and Highlights*; World Bank: Washington, DC, USA, 2017. (In Chinese)

Article

Error Correction of Multi-Source Weighted-Ensemble Precipitation (MSWEP) over the Lancang-Mekong River Basin

Xiongpeng Tang [1,2,3], Jianyun Zhang [1,2,3], Guoqing Wang [1,2,3,*], Gebdang Biangbalbe Ruben [1], Zhenxin Bao [1,2,3], Yanli Liu [1,2,3], Cuishan Liu [1,2,3] and Junliang Jin [1,2,3]

1. State Key Laboratory of Hydrology-Water Resources and Hydraulic Engineering, Nanjing 210098, China; xptang@hhu.edu.cn (X.T.); jyzhang@nhri.cn (J.Z.); rubengebdang@hhu.edu.cn (G.B.R.); zxbao@nhri.cn (Z.B.); ylliu@nhri.cn (Y.L.); csliu@nhri.cn (C.L.); jljin@nhri.cn (J.J.)
2. Nanjing Hydraulic Research Institute, Nanjing 210029, China
3. Research Center for Climate Change, Ministry of Water Resources, Nanjing 210029, China
* Correspondence: gqwang@nhri.cn

Citation: Tang, X.; Zhang, J.; Wang, G.; Ruben, G.B.; Bao, Z.; Liu, Y.; Liu, C.; Jin, J. Error Correction of Multi-Source Weighted-Ensemble Precipitation (MSWEP) over the Lancang-Mekong River Basin. *Remote Sens.* **2021**, *13*, 312. https://doi.org/10.3390/rs13020312

Received: 18 December 2020
Accepted: 15 January 2021
Published: 18 January 2021

Publisher's Note: MDPI stays neutral with regard to jurisdictional claims in published maps and institutional affiliations.

Copyright: © 2021 by the authors. Licensee MDPI, Basel, Switzerland. This article is an open access article distributed under the terms and conditions of the Creative Commons Attribution (CC BY) license (https://creativecommons.org/licenses/by/4.0/).

Abstract: The demand for accurate long-term precipitation data is increasing, especially in the Lancang-Mekong River Basin (LMRB), where ground-based data are mostly unavailable and inaccessible in a timely manner. Remote sensing and reanalysis quantitative precipitation products provide unprecedented observations to support water-related research, but these products are inevitably subject to errors. In this study, we propose a novel error correction framework that combines products from various institutions. The NASA Modern-Era Retrospective Analysis for Research and Applications (AgMERRA), the Asian Precipitation Highly-Resolved Observational Data Integration Towards Evaluation of Water Resources (APHRODITE), the Climate Hazards group InfraRed Precipitation with Stations (CHIRPS), the Multi-Source Weighted-Ensemble Precipitation Version 1.0 (MSWEP), and the Precipitation Estimation from Remotely Sensed Information using Artificial Neural Networks-Climate Data Records (PERSIANN) were used. Ground-based precipitation data from 1998 to 2007 were used to select precipitation products for correction, and the remaining 1979–1997 and 2008–2014 observe data were used for validation. The resulting precipitation products MSWEP-QM derived from quantile mapping (QM) and MSWEP-LS derived from linear scaling (LS) are evaluated by statistical indicators and hydrological simulation across the LMRB. Results show that the MSWEP-QM and MSWEP-LS can better capture major annual precipitation centers, have excellent simulation results, and reduce the mean BIAS and mean absolute BIAS at most gauges across the LMRB. The two corrected products presented in this study constitute improved climatological precipitation data sources, both time and space, outperforming the five raw gridded precipitation products. Among the two corrected products, in terms of mean BIAS, MSWEP-LS was slightly better than MSWEP-QM at grid-scale, point scale, and regional scale, and it also had better simulation results at all stations except Strung Treng. During the validation period, the average absolute value BIAS of MSWEP-LS and MSWEP-QM decreased by 3.51% and 3.4%, respectively. Therefore, we recommend that MSWEP-LS be used for water-related scientific research in the LMRB.

Keywords: Lancang-Mekong river basin; MSWEP; AgMERRA; APHRODITE; CHIRPS; PERSIANN; error correction

1. Introduction

Precipitation is a key element associated with terrestrial–atmospheric circulation. Thus, it governs terrestrial renewable water resources that affect urban development, ecological water storage, and agricultural irrigation [1,2]. On the other hand, precipitation is a complex natural phenomenon affected by various natural and anthropogenic factors, and its characteristics have significant variability both on a spatial and temporal scale. Thus, it is essential to obtain more accurate precipitation with a higher temporal and spatial resolution for various purposes, such as climate change research [3], analysis of temporal and spatial evolution of precipitation [4], and streamflow simulation [5]. For an extended period until the launch of the Tropical Rainfall Measuring Mission satellite in 1997, gauge observation was the only means to obtain actual precipitation values with a point scale. However, traditional gauge observation is often limited by sparse gauge distribution and poor spatial and temporal representation [6,7].

Benefitting from the development of remote sensing and computer technology in today's era, an increasing number of satellite-based precipitation products and reanalysis precipitation products have been developed, which have provided unprecedented data support for global and regional hydrometeorological research. To some extent, the development of remote sensing has also made up for the shortcomings of insufficient spatial and temporal gauge observation data, especially for areas without long series observations [8]. An increasing number of satellite-based and reanalysis precipitation products are now available with an extended period, daily or sub-daily scale, such as the Tropical Rainfall Measuring Mission 3B42 (TMPA) [9], Multi-Source Weighted-Ensemble Precipitation (MSWEP) [10], Precipitation Estimation from Remotely Sensed Information using Artificial Neural Networks-Climate Data Record (PERSIANN-CDR) [11], and the Asian Precipitation Highly-Resolved Observational Data Integration Towards Evaluation of Water Resources (APHRODITE) [12]. Unfortunately, due to the fact of satellite sampling errors, indirect measurement, and data fusion technology deviations, these products inevitably have specific errors compared with gauge observations [13,14]. Furthermore, as pointed out by Worqlul et al. [15], error correction processes or data fusion technology based on gauge observations should be conducted before using remote sensing precipitation for hydrological simulations. Therefore, error correction of long-term satellite-based or reanalysis precipitation products is vital before conducting hydrometeorological-related research.

Bias correction methods are often used to correct precipitation of global climate models (GCMs) and satellite-based or reanalysis estimates. An increasing number of bias correction methodologies have proven to improve the accuracy of raw remotely sensed effectively or reanalysis precipitation products [16–19]. In general, there are two commonly used correction mechanisms, one is global average correction, and the other is the more widely used method, that is, a local correction method that considers the accuracy of the temporal and spatial performance of precipitation estimation [20]. In the set of local correction methods, the more commonly used methods include objective analysis (OA) [21], optimal interpolation method (OI) [22], distribution fitting (DF) [20], geographically weighted regression (GWR) [23], linear scaling (LS) [24], and quantile mapping (QM) [25]. Among the methods mentioned above, the LS and the QM methods are widely used because they do not require the available time series of collected gauged observations to be completely consistent with the remote sensed or reanalysis precipitation estimates. Liu et al. [26] indicated that the linear regression model improves satellite precipitation accuracy at both monthly and annual scales over China. Ghimire et al. [18] also concluded that quantile mapping could significantly improve the hydrological simulation performance. However, these bias correction methods often require precipitation observation data of the same sequence length as remote sensing precipitation as a baseline. For regions with insufficient data observed, such as the Lancang-Mekong River, these methods often have certain limitations. Therefore, a new bias correction framework that requires a short series of observational precipitation as a reference needs to be proposed and verified.

The Lancang-Mekong River Basin (LMRB) is located on the Indo-China Peninsula. Its biodiversity and floodplains feed more than 60 million people. Its climate ranges from the upper temperate plateau to the lower tropical monsoon climate, resulting in significant spatial and temporal variability of precipitation within the basin [3,27]. Meanwhile, the LMRB flows through six countries, making the collection and unified management of gauge observation data challenging within the basin [7,28,29]. In the past decade, increasing studies have focused on the evaluation and error correction of multiple satellite-based and reanalysis precipitation products over the LMRB [3,6,7,28,30–33]. Among these studies, Tang et al. [3] used statistical analysis and hydrological simulations to assess the accuracy of the AgMRRA, MSWEP, PERSIANN, and TMPA (Tropical Rainfall Measuring Mission 3B42 Version7) compared with gauge observations from 1998 to 2007 over the LMRB. They concluded that the MSWEP and TMPA have good performance with a higher correlation coefficient (CC = 0.86) and lower mean error (ME is –0.32 mm/day and –0.01 mm/day, respectively) in terms of watershed average precipitation; for each gauge in different spaces, the four products show different pros and cons. For example, PERSIANN has the relative smallmean error (ME = 0.25 mm/day) but does not have a high correlation coefficient (CC = 0.81); the MSWEP has the opposite performance in some stations, it has a high correlation but with a higher mean error. Chen et al. [31] compared and evaluated TMPA, PERSIANN, MERRA2 (Modern-Era Retrospective Analysis for Research and Applications), ERA-Interim (European Centre for Medium-Range Weather Forecasts Interim Reanalysis), and CFSR (Climate Forecast System Reanalysis) with APHRODITE, which was developed based on gauge observations over the LMRB, and they found that both PERSIANN and TMPA have high reliability. In terms of error correction of satellite-based precipitation in LMRB, Chen et al. [33] used APHRODITE with 0.25 degree resolution as a reference and then reconstructed the CMORPH (CPC MORPHing technique) with 0.05 degree resolution; finally, they obtained daily-scale precipitation with a 0.05 degree resolution over the whole LMRB. However, the results of Lauri et al. [28], Tang et al. [13], and Chen et al. [33] show that although the APHRODITE has a high correlation coefficient compared with gauge observations, there is a severe underestimation of precipitation in the lower LMRB. On the other hand, since APHRODITE only provides daily precipitation from 1951 to 2007, this also significantly limits its application. According to our literature search, the current research on precipitation bias correction in the LMRB is mainly focused on downscaling remotely sensed precipitation at a larger spatial scale to higher-precision precipitation data based on certain auxiliary data (such as terrain data, Normalized Difference Vegetation Index (NDVI)). Li et al. [23] used the GWR method, taking the Lancang River Basin as the study area, combined with the relationship between precipitation and NDVI, land surface temperature (LST), and digital elevation model (DEM) to spatially downscale the Tropical Rainfall Measuring Mission (TMPA) 3B43 Version 7 precipitation product (2001–2015) with a resolution of 0.25° to 1 km. It was found that the downscaled TMPA precipitation had a better performance than the original TMPA data. Zhang et al. [34] selected the LMRB as the study area, used the random forest regression method, combined with the correlation between precipitation and longitude, latitude, elevation, NDVI, etc., to downscale the TMPA and PERSIANN products with a spatial resolution of 0.25° in 2001 (wet year), 2005 (normal year), and 2009 (dry year) to 1 km. They found that in terms of the mean square error (RMSE) and the mean absolute error (MAE), the downscaled precipitation performed better than the original products. In general, the current precipitation bias correction in the LMRB is concentrated in the shorter available time series (limited by the length of the available observation data series); however, for research related to climate change, it is necessary to provide meteorological data for at least 30 years according to the recommendation of the World Meteorological Organization. At present, there is currently relatively little literature focusing on improving the accuracy of these precipitation products with long available time series in LMRB. To fill this research gap, in this study, we proposed a novel correction frame to improve the accuracy of long-term daily satellite-based precipitation, and we expect this frame can be used in other ungauged or poorly gauged areas. This framework first evaluates the prediction accuracy of five sets of precipitation products on whether rainfall event occurs or not, selects the precipitation product with higher accuracy as the benchmark, then the mean error (ME) between each grid point of the five sets of precipitation products and the observed precipitation after interpolation is evaluated at monthly scale, the product with the smallest ME in each grid is used as the "true" precipitation value, and then the error correction is performed on the benchmark precipitation product. Therefore, this study focuses on improving long-term daily satellite-based precipitation in LMRB, where there is a lack of sufficient gauge observations. This study's primary objective is to use limited gauge precipitation observations to combine the advantages of five precipitation products (i.e., AgMERRA, APHRODITE, CHIRPS, MSWEP, and PERSIANN) in both time and space. Additionally, the precipitation data will be used to generate long-term daily precipitation data with high accuracy that is suitable for the LMRB.

2. Study Area and Data Sets
2.1. Study Area

The Lancang-Mekong River (LMRB) originates from the northeast slope of the Tanggula Mountains on the Tibetan Plateau in China, with an approximate total catchment area of 795,000 km^2, with elevation ranges from −5 to 5580 m. It flows 4909 km in length through six countries from north to south (i.e., China, Myanmar, Thailand, Lao PDR, Cambodia, and Vietnam), and then empties into the South China Sea in Ho Chi Minh City, Vietnam. The river is generally called Lancang River (LRB) in China, which is ~2140 km length (~43.5% of the total length), has a catchment area of ~195,000 km^2 (~24.5% of the total catchment area), and flows through China's Qinghai Province, Tibet Autonomous Region, and Yunnan Province. After flowing out of China, it is called the Mekong River (MRB), is ~2709 km in length (~56.5% of the total length), and has a catchment area of ~600,000 km^2 (~75.5% of the total catchment area) (Figure 1).

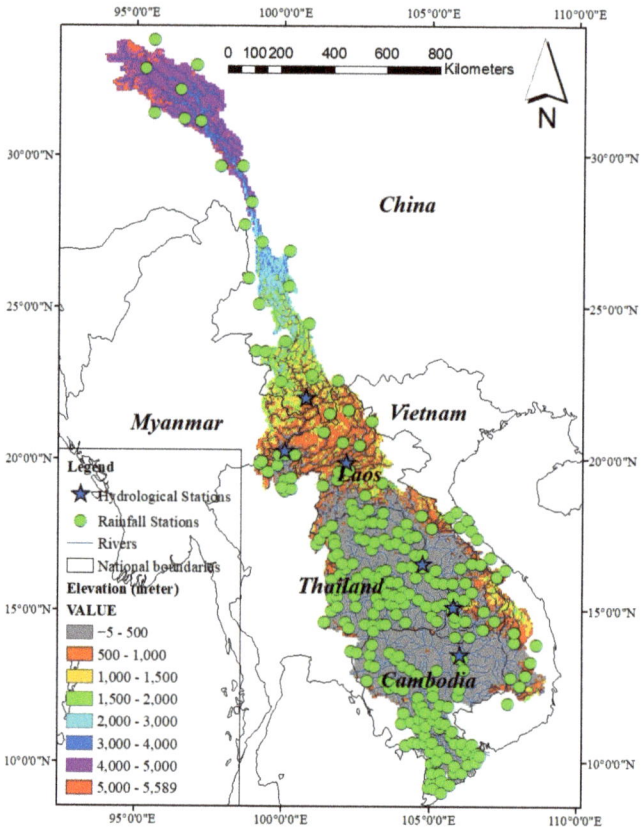

Figure 1. Location of the Lancang-Mekong− River in Asia and its topographical map. Distribution of rainfall gauge stations and six hydrological stations: from north to south, the hydrological stations are Yunjinghong, Chiang Saen, Luang Prabang, Mukdahan, Pakse, and Stung Treng).

The LMRB consists of seven different natural geographical areas with diverse topography, drainage patterns, and landforms, of which the Tibet Plateau (TP), Three River Area (TRA), and Lancang Basin (LB) form the Lancang River Basin in China. The other four areas include the Northern Highlands (NHs), Khorat Plateau (KP), Tonle Sap Lake Basin (TSLB), and Mekong Delta Basin (MDB) that make up the Mekong River Basin. The temperature climate of the LRB and the tropical monsoon climate of the MRB together lead to extremely uneven spatial and temporal distributions of precipitation in the LMRB, which also divide the precipitation and streamflow processes into the wet season (from May to October) and dry season (from November to April) [35]. The annual precipitation in the upper LRB can be as little as 600 mm, while in some mountainous areas of the lower MRB, the annual precipitation even exceeds 3000 mm [36,37].

2.2. Data

2.2.1. Precipitation Products

In this study, two reanalysis precipitation products include AgMERRA [38] and APHRODITE [12], three satellite-based precipitation products, namely, CHIRPS [39], MSWEP [10], and PERSIANN [40] were used. The main reasons we chose these five precipitation products were as follows: (1) they all have the same spatial resolution (0.25 degree), which can avoid the additional errors caused by resampling; (2) they all provide more than 30 years of daily records to make them more representative for precipitation prediction; (3) according to the results of published research [3,13,28,33], these five products have different optimal performance across the LMRB; some have high correlation coefficients and a large mean error against gauge observations, while some have a small mean error but with low correlation coefficients. This section will briefly introduce these five precipitation products, and the necessary information for these five precipitation products is presented in Table 1.

Table 1. Characteristics of the precipitation products used in this study. CMA = China Meteorological Administration; MRC = Mekong River Commission.

Precipitation	Temporal Resolution	Spatial Resolution	Date	Date Sources
Gauge	Daily	Point	1979–2014	CMA and MRC
AgMERRA	Daily	0.25°	1980–2010	https://data.giss.nasa.gov/impacts/agmipcf/agmerra/
APHRODITE	Daily	0.25°	1951–2007	http://www.chikyu.ac.jp/precip/english/products.html
CHIRPS	Daily	0.25°	1981–present	https://chc.ucsb.edu/data/chirps
MSWEP	Daily	0.25°	1979–2016	https://platform.princetonclimate.com/PCA_Platform/index.html
PERSIANN	Daily	0.25°	1983–present	https://climatedataguide.ucar.edu/climate-data/persiann-cdr-precipitation-estimation-remotely-sensed-information-using-artificial

(AgMERRA: NASA Modern-Era Retrospective Analysis for Research and Applications, APHRODITE: Asian Precipitation Highly-Resolved Observational Data Integration Towards Evaluation of Water Resources, CHIRPS: Climate Hazards group InfraRed Precipitation with Stations, MSWEP: Multi-Source Weighted-Ensemble Precipitation, PERSIANN: Precipitation Estimation from Remotely Sensed Information using Artificial Neural Networks-Climate Data Records).

The AgMERRA precipitation estimates were developed at the National Aeronautics and Space Administration (NASA) as one meteorological element of the Agricultural Model Inter-comparison and Improvement Project (AgMIP) to provide daily-scale, consistent time series [41]. This product incorporates the MERRA-Land product, which has significantly improve the spatial resolution of daily precipitation and the accuracy of extreme precipitation compared with other climate forcing data sets. It provides daily precipitation with a 0.25° (~25 km) horizontal resolution from 1980 to 2010 [42,43].

The APHRODITE data sets are provided at a 0.25 resolution with the Asia coverage (extends Himalayas, South, and Southeast Asia, and mountainous areas in Middle Asia) and daily precipitation values from 1951 to 2007 [12,44]. The APHRODITE was developed by the Japan Meteorological Agency (JMA) [13]. This product was developed based on the daily precipitation data provided by dense surface rainfall stations (between 5000 and 12,000) in Asia. The results of published studies in the Lancang-Mekong River Basin show that the APHRODITE has a high correlation coefficient, but the amount of precipitation is underestimated in the lower Mekong Basin [28,33].

The CHIRPS product provides land-only daily precipitation with a high resolution ($0.05° \times 0.05°$ and $0.25° \times 0.25°$) from 1980 to the present. This data set incorporates monthly precipitation climatology from Climate Hazards Group Precipitation Climatology (CHPClim) from Tropical Rainfall Measuring Mission's 3B42 product (TRMM 3b42). In addition, ground gauge precipitation observations from various sources from global and regional meteorological systems amd atmospheric model precipitation fields from NOAA (National Oceanic and Atmospheric Administration) [6,39] were added to the data set.

The MSWEP version 1.1 precipitation estimates provide daily time series from 1979 to 2015 with a 0.25 degree resolution developed by Beck et al. [10]. The monthly data set of CHPClim and gauge observations were used to calculate the long-term mean of MSWEP. Basin-scale average precipitation inferred from streamflow observations at 13,762 stations across the globe were used to remove the orographic effects and gauge under-catch. A weighted average of seven satellite/reanalysis precipitation data sets was used to correct the temporal variability of MSWEP, which included CMORPH, GsMAP-MVK, CPC Unified, GPCC, TMPA 3B42RT, ERA-Interim, and JRA-55 [3,10].

The PERSIANN is a multi-satellite-based precipitation data set that provides a near-global (60°S to 60°N) daily precipitation estimate, with a 0.25-degree spatial resolution from 1983 to near present [40,45]. The PERSIANN algorithm was used to develop the daily precipitation estimate. The hourly precipitation data from the National Centers for Environmental Prediction (NCEP) stage IV was used to train the artificial neural network. Finally, the Global Precipitation Climatology Project's (GPCP) monthly products were used to adjust the daily estimate.

As baseline data, we collected daily precipitation data from 267 stations within or around the LMRB. These stations are mainly run and maintained by the China Meteorological Administration and the Mekong River Commission. To access the precipitation data provided by the China Meteorological Administration (CMA), we needed to register an account first, and then apply to download the data set the data managed by the downstream Mekong River Commission (MRC) can be downloaded on its official website or one can refer to the supplementary data provided by Wang et al. [7]. It should be noted that the most extended sequence of these observations is from 1979 to 2014; that is, these gauges have different series lengths. Since the spatial resolution of the remote sensing and reanalysis precipitation products collected in this study are all 0.25 degrees, for the case where there are multiple gauges in a $0.25° \times 0.25°$ grid, we use the arithmetic mean of several gauges as the observed values within this grid. The available period of these gauges is shown in Figure 2. As shown in Figure 2, the precipitation data provided by most stations were from 1991 to 2007. Few stations provided recent observational precipitation data, which is one of the reasons we implemented this study.

2.2.2. Auxiliary Data

The construction of the SWAT (Soil and Water Assessment Tool) model required daily-scale meteorological forcing data (daily maximum and minimum temperatures, daily relative humidity, daily wind speeds, and daily solar radiation). In addition, topographical data (digital elevation model), soil data, land-use data, and daily-scale discharge data were required for calibration.

The NCEP-CFSR reanalysis data provided by NCEP (National Centers for Environmental Prediction) were also used in this study for hydrological modeling. The NCEP-CFSR reanalysis data assimilated a 6 h grid point statistical interpolation system (GSI) using the GEOS-5 (Goddard Earth Observing System) model and data assimilation system [46,47]. This reanalysis data had a 38 km spatial resolution. It has been applied to the Mekong River Basin [3] and the Lancang River basin [13] with excellent simulation results (with a Nash–Sutcliffe Efficiency coefficient (NSE) greater than 0.75).

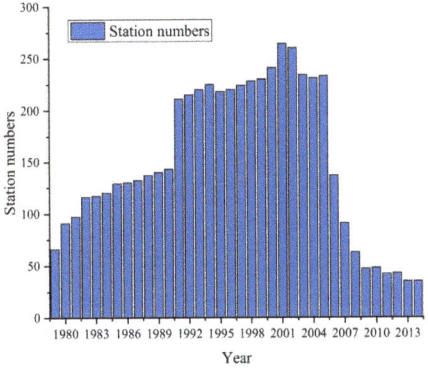

Figure 2. Station numbers of available data from 1979 to 2014.

The Digital Elevation Model (DEM) database with a 90 m resolution used in this study was downloaded from the NASA Shuttle Radar Topographical Mission (SRTM) (http://srtm.csi.cgiar.org/). The soil database with a ~1 km resolution was collected from the Food and Agriculture Organization (FAO) of the United Nations (http://www.fao.org/land-water/databases-and-software/hwsd/en/), and it contained two soil layers. The land use database was obtained from the Global Land Cover 2000 Project (GLC 2000), and it also had a 1 km spatial resolution (https://ec.europa.eu/jrc/en/scientific-tool/global-land-cover).

The daily streamflow data of six hydrological stations on the Mekong mainstream were collected from the Information Center of the Ministry of Water Resources of China (ICMWR) and the Mekong River Commission (MRC) to calibrate the model with the available period from 1998 to 2007. These hydrological data are subject to strict quality control by both departments. The selected six stations were Yunjinghong (China), Chiang Saen (Thailand), Luang Prabang (Laos), Mukdahan (Thailand), Pakse (Laos), and Stung Treng (Cambodia). The basic information on the six hydrological stations is shown in Table 2.

Table 2. Basic information on the six hydrological stations used in this study.

Station	Country	Latitude (Degree)	Longitude (Degree)	Elevation (Meter)	Period
Yunjinghong	China	100.78	22.03	592	1998–2007
Chiang Saen	Myanmar	100.08	20.27	372	1998–2007
Luang Prabang	Laos	102.14	19.89	316	1998–2007
Mukdahan	Thailand	104.74	16.54	133	1998–2007
Pake	Laos	105.8	15.12	102	1998–2007
Stung Treng	Cambodia	106.02	13.55	51	1998–2007

3. Methodology

3.1. Statistical Criteria of Performance Comparison

Before the error correction, we first compared the precipitation products (i.e., AgMERRA, APHRODITE, CHIRPS, MSWEP, and PERSIANN) with the site-observed precipitation pixel scale (point to pixel). An evaluation was conducted at a daily scale covering the period from 1979 to 2014, following Zhu et al. [46], Gumindoga et al. [20], and Tang et al. [3] who conducted a point-pixel evaluation in multiple basins worldwide. To qualitatively evaluate the performance of precipitation products with gauged precipitation observations, the following statistical indices were used: the correlation coefficient (CC), mean error (ME), relative bias (BIAS), and the probability of precipitation detection (POD01). These equations were calculated as shown in Equations (1)–(4).

$$CC = \frac{\sum_i^n (P_{o,i} - \overline{P}_{o,i})(P_{s,i} - \overline{P}_{s,i})}{\sqrt{\sum_i^n (P_{o,i} - \overline{P}_{o,i})^2} \sqrt{\sum_i^n (P_{s,i} - \overline{P}_{s,i})^2}} \quad (1)$$

$$ME = \frac{\sum_i^n (P_{s,i} - P_{o,i})}{n} \quad (2)$$

$$BIAS = \frac{\sum_i^n P_{s,i} - \sum_i^n P_{o,i}}{\sum_i^n P_{o,i}} \times 100\% \quad (3)$$

$$POD01 = \frac{H_{00} + H_{11}}{n} \quad (4)$$

where $P_{o,i}$ and $\overline{P}_{o,i}$ are the individual and averages observed precipitation provided by ground gauges, respectively. $P_{s,i}$ and $\overline{P}_{s,i}$ are, respectively, the daily and averages of precipitation products, and n is the total number of data series. The H_{00} represents the number of days when no precipitation occurred both in the observation data and the remote sensed or reanalysis precipitation product, while H_{11} represents the total number of days when precipitation occurred both in the observation data and the remote sensed or reanalysis precipitation product.

3.2. Framework of Precipitation Error Correction

In the proposed frame, the process of error correction for long-term satellite-based precipitation over poorly gauged areas can be divided into three steps, and the flow chart of this framework is shown in Figure 3. First, collect multiple sets of precipitation products with a long available period. Then, select a set of precipitation products with a high correlation coefficient and POD01 compared with gauged precipitation observation of all ground stations. The second step compares the IDW-interpolated (Inverse Distance Weighted) gauged precipitation observations with all precipitation products at each grid point, select the smallest ME product as the benchmark, and corrects the precipitation product selected the first step. The last step is to validate the corrected precipitation products using the remaining gauge observations. The detailed steps for error correction of daily precipitation are shown below:

(1) Select multiple sets of long-term daily-scale precipitation products with high resolution.
(2) Compare precipitation products with observed precipitation from all gauge stations, and select a set of precipitation products with a higher correlation coefficient and POD01 for correction.
(3) Select gauged precipitation data with a certain period (1998 to 2007 in this study) containing more stations (Figure 4), and these gauges should have better spatial representation. Then monthly grid-scale precipitation data with the same spatial resolution as the precipitation products are obtained through IDW interpolation.
(4) Compare the IDW monthly scale precipitation data with monthly scale gauged precipitation. The precipitation product with the smallest ME at each grid point in each month is obtained as the actual rainfall value for correction.
(5) The precipitation data obtained in the fourth step are used to correct the product selected in the second step at each grid point every month. Then the daily-scale rainfall products with higher accuracy are obtained.
(6) Statistical indicators and hydrological simulation are used to assess the accuracy of the corrected precipitation product. In this study, the SWAT model was used for streamflow simulation.

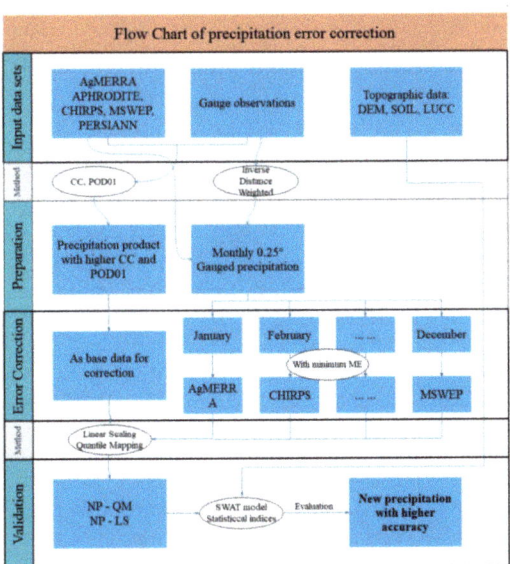

Figure 3. Flow chart of precipitation error correction for this study (CC: correlation coefficient, POD01: the probability of detection for precipitation and without precipitation, SWAT: Soil and Water Assessment Tool, NP: new precipitation products, QM: quantile mapping, LS: linear scaling).

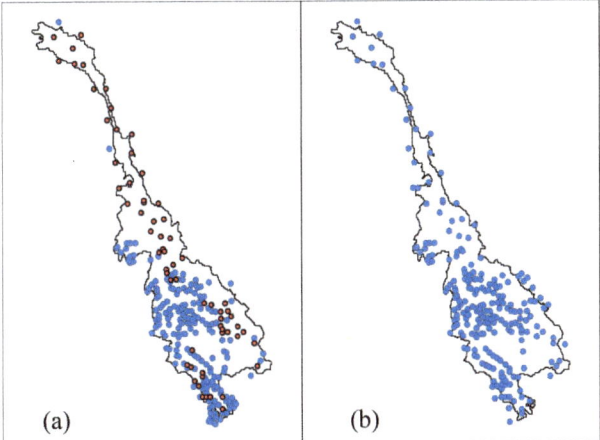

Figure 4. (a) The spatial distribution of precipitation stations with observation data from 1998 to 2007, and the red point means that there were no missing observation data at this station. (b) The spatial distribution of precipitation stations with observation data from 1979 to 1997 and 2008 to 2014.

In this study, two error correction techniques were used to remove the satellite-based precipitation product's bias. The first technique was nonparametric empirical quantile mapping (QM) (Equation (5)) and the another was linear scaling (LS) (Equation (6)). These two techniques were used because they are easy to implement and have been proven to effectively correct daily-scale precipitation data [18,20,48]. It should be noted that the corrections in this study were performed on a monthly scale, which means that we needed to perform 12 corrections for each grid point. Following the study of Reiter et al. [49], which evaluated the rainfall corrections at multiple time scales, they found that corrections at a monthly scale were most effective in removing daily-scale precipitation bias. The calculation formulas for the two correction methods are shown below:

$$P_{raw} = F_{raw}^{-1}(F_{corr}(P_{corr})) \tag{5}$$

$$P_{corr} = P_{raw} \times Scale \qquad Scale = \frac{mean_{per}}{mean_{raw}} \tag{6}$$

where P_{corr}, P_{raw}, $mean_{per}$ and $mean_{raw}$ mean precipitation of corrected, precipitation products selected in step 2, the mean value of precipitation product selected in step 4, and the mean value of precipitation product selected in step 2, respectively. F_{corr} is the cumulative distribution function (CDF) of P_{corr}, and F_{raw}^{-1} is the inverse CDF (or quantile function) corresponding to P_{raw}.

3.3. Brief Description of the SWAT Model

The SWAT is a semi-distributed hydrological model developed by the the Agricultural Research Service of the United States Department of Agriculture (USDA-ARS). It has been widely applied in various watersheds associated with climate change assessment, soil erosion, and non-point pollution [3,46,50,51]. The SWAT version 2012, coupled with the ArcGIS interface, was used in this study to evaluate the precipitation products' performance. This model first divides the study area into several sub-basins based on the topography data sets (i.e., DEM data, mask data). Each sub-basin was discretized into multiple hydrological response units (HRUs), which are the most fundamental computational unit according to the soil type, land use data, and slope data [50,52]. The water cycle calculated by the model was simulated on each HRU. This production flow then converged to the corresponding sub-basin. Finally, the total streamflow of the study area was calculated from the output of each sub-basin. The calibration of the SWAT model was done using a separate software named SWAT-CUP (SWAT Calibration and Uncertainty Program), which can be used for calibration, validation, and uncertainty analysis of the model [52–54]. The SUFI-2 (Sequential Uncertainty Fitting Version 2) within the SWAT-CUP was used in this study to calibrate the model [55].

In order to evaluate the performance of the model, the Nash–Sutcliffe Efficiency coefficient (NSE) and relative bias (BIAS) (Equation (3)) were used [56]. The NSE calculation formula is shown below:

$$NSE = 1 - \frac{\sum_i^n (Q_o^i - Q_s^i)^2}{\sum_i^n (Q_o^i - \overline{Q}_o)^2} \tag{7}$$

where Q_o^i and Q_s^i represent the observed and simulated streamflow, respectively. $\overline{Q}_o$ means the average of observed streamflow, and n is the total number of streamflow data.

4. Results

4.1. Evaluation of Five Precipitation Products with Gauged Observations

Spatial distributions and boxplots of the correlation coefficient for AgMERRA, APHRODITE, CHIRPS, MSWEP, and PERSIANN compared with gauged observations at a daily scale over the LMRB are shown in Figure 5. Overall, for the whole basin, APHRODITE had the best linear correlation with the gauged observations compared with the other four precipitation products (0.61 in the whole basin), followed by the MSWEP (0.49) and AgMERRA (0.39). In contrast, the CHIRPS and PERSIANN had the smallest CC (0.35). There were 204 gauges for APHRODITE with CCs more significant than 0.5, while MSWEP, AgMERRA, PERSIANN, and CHIRPS had 107, 48, 10, and 9 gauges with CC exceeding 0.5, the highest CCs for CHIRPS and PERSIANN products among all gauges were 0.54 and 0.59, respectively. In terms of the spatial distribution, gauges in China (i.e., within the Lancang River Basin) generally have higher CCs than those located in the lower LMRB. This also indicates that these five sets of precipitation products included more observation information from the gauges in the upper LRMB when they were developed.

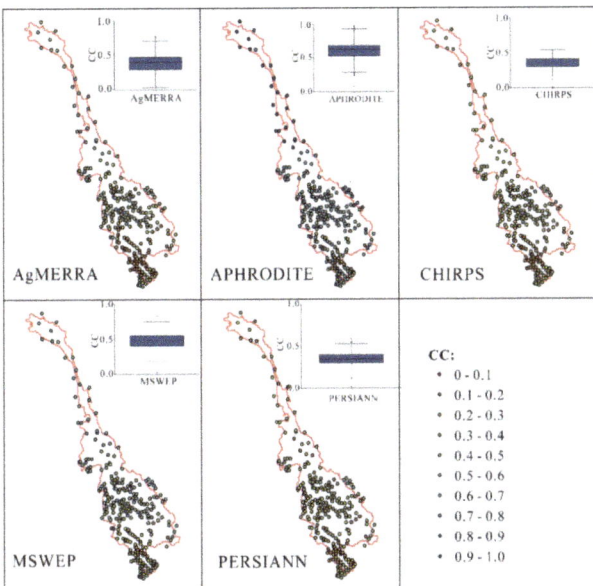

Figure 5. Correlation coefficient (CC) of five precipitation products against gauged observations at the daily scale. The boxplot of the correlation coefficients for all stations are shown in the upper right corner of the picture.

Figure 6 shows the POD01 of five precipitation products compared with gauged observations on a daily scale. Consistent with the CC results, CHIRPS also had the smallest POD01 coefficient than the other four products. From the boxplots' results, we can see that MSWEP was the best performing product with the highest POD01, which means that this product can most accurately predict precipitation over the LMRB. From the perspective of spatial distribution, MSWEP had the largest POD01 (higher than 0.9) in the entire LMRB, and the performance of the APHRODITE product was slightly lower than MSWEP. In contrast to the spatial performance of correlation coefficients, the POD01 of stations in the upper LMRB were lower than those in the lower LMRB, which may be affected by the complex terrain of the Qinghai–Tibet Plateau.

Figures 5 and 6 show that APHRODITE and MSWEP had higher CCs and POD01 over the entire LMRB. At the same time, the prediction of precipitation occurrence was worse than MSWEP. However, considering that APHRODITE only provides daily-scale data until 2007, and Lauri et al. [28] pointed out that although APHRDOTE had a high correlation coefficient in the Mekong River Basin, there is still an underestimation of precipitation in the downstream regions. Therefore, this study chose MSWEP as the corrected precipitation product because it had a relatively high correlation coefficient compared with gauge observations and can best predict precipitation occurrence.

Figure 7 shows the spatial distribution of five precipitation products compared with IDW interpolation monthly with the smallest ME from 1998 to 2007. It can be seen from Figure 7 that no particular product can perform well in all 12 months. In general, MSWEP in the upper Qinghai–Tibet Plateau region had a better performance from January to March, May, and June. The PERSIANN performed well in April and from June to November in the middle region of the river basin. While in the downstream region, AGMERRA had a better performance from March to September. For other months and regions, no one product performed significantly better than other products. Table 3 presents the number of grids for the five precipitation products with the smallest ME over the whole basin compared with IDW-derived gauge observations at a monthly scale. We can see that from March to September, which included the rainy season of the study area, APHRODITE had the fewest grids with the smallest ME. This also means that although APHRODITE had the most massive CC compared to the other for products, its estimation of precipitation in the whole basin was not accurate. Compared with the other four products, MSWEP had the most grids with the smallest ME in February, from May to July, November, and December. In other months, AgMERRA, APHRODITE, CHIRPS, and PERSIANN performed better in September and October, January, April, and August, respectively.

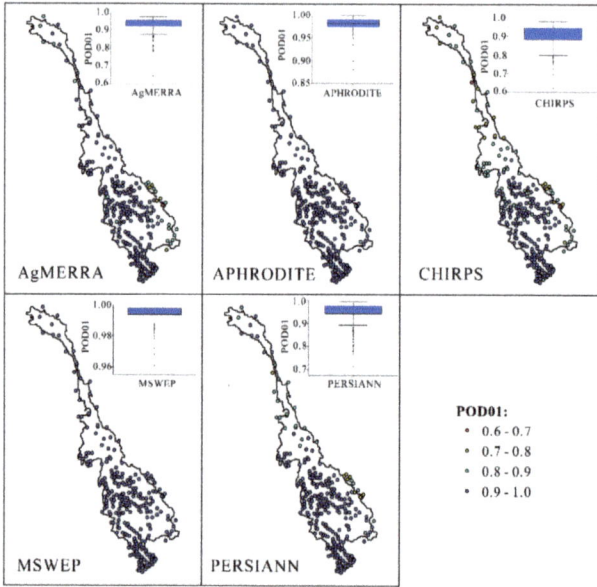

Figure 6. Same as Figure 5 but for POD01.

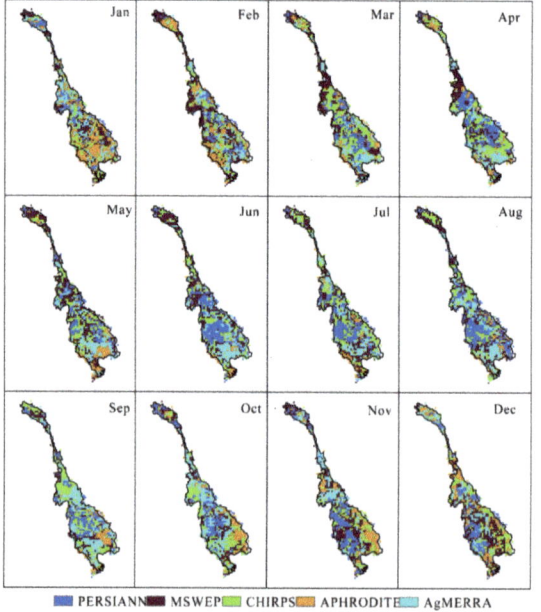

Figure 7. Monthly minimum ME for five precipitation products against inverse distance weighted (IDW)-derived gauge observation during 1998 to 2007.

Table 3. The number of grid points over the whole basin that had the smallest ME compared with the IDW-interpolated precipitation at a monthly scale.

Product	Jan	Feb	Mar	Apr	May	Jun	Jul	Aug	Sep	Oct	Nov	Dec
AgMERRA	241	205	202	200	229	268	220	252	313 *	292 *	194	177 #
APHRODITE	382 *	300	156 #	125 #	143 #	87 #	110 #	79 #	121 #	156 #	217	286
CHIRPS	150	172	288 *	281 *	222	196	282	209	307	258	172 #	203
MSWEP	239	322 *	288	274	364 *	307 *	283 *	287	192	235	313 *	295 *
PERSIANN	127 #	140 #	205	259	181	281	244	312 *	206	198	243	178

(Jan: January, Feb: February, Mar: March, Apr: April, Jun: June, Jul: July, Aug: August, Sep: September, Oct: October, Nov: November, Dec: December. The maximum number are marked with "*" and the minimum number are marked with "#").

4.2. Grid-Scale Evaluation of Corrected Precipitation with Gauge Observations from 1998 to 2007

Based on the results in Section 4.1 and the correction mechanism introduced in Section 3.2, we selected daily-scale MSWEP from 1979 to 2014 as the product to be corrected and used the QM and LS methods to correct the MSWEP grid by grid and month to month. The corrected precipitation is called MSWEP-QM and MSWEP-LS, respectively.

Figure 8 shows the spatial distribution of annual average precipitation derived from gauge observation, AgMERRA, APHRODITE, CHIRPS, MSWEP, PERSIANN, and two corrected products (i.e., MSWEP-QM, MSWEP-LS) from 1998 to 2007. In general, the corrected precipitation (MSWEP-QM and MSWEP-LS) can correctly represent the spatial distribution of annual-scale precipitation in the entire LMRB compared with gauge observations. The corrected products can better predict the precipitation centers of the LMRB (including the Khorat Plateau in northeastern Thailand and The Northern Highlands downstream), which is influenced by the Indian Ocean monsoon and southeast monsoon. From the spatial distribution perspective, in the upper Qinghai–Tibet Plateau, all products except PERSIANN performed well, while AgMERRA and APHRODITE underestimated annual precipitation in the middle area. In the downstream area of the LMRB, CHIRPS, MSWEP, and PERSIANN overestimated the precipitation compared to the gauge observations. For corrected precipitation, we can see that both MSWEP-QM and MSWEP-LS can better remove the precipitation error on the multi-year average scale.

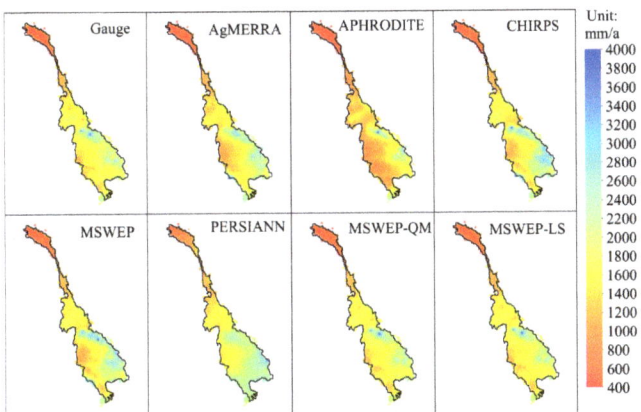

Figure 8. Spatial patterns of annual average precipitation derived from gauge observation, AgMERRA, APHRODITE, CHIRPS, MSWEP, PERSIANN, MSWEP-QM, and MSWEP-LS from 1998 to 2007.

Figure 9 shows the scatterplots of annual precipitation comparisons for the whole LMRB between AgMERRA, APHRODITE, CHIRPS, MSWEP, PERSIANN, MSWEP-QM, MSWEP-LS, and gauges. As shown in Figure 9, annual MSWEP-QM and MSWEP-LS agree well with gauge observations over the entire LMRB (CC = 0.97 and 0.98, respectively).

Generally, for the five raw precipitation products before correction, APHRODITE underestimated the precipitation (with BIAS equals −19.01%) but had the highest CC (0.91). The CC of CHIRPS and MSWEP were also very satisfying (0.85 and 0.87, respectively) but slightly overestimated the precipitation (with BIAS equals 2.44% and 1.98%, respectively). The PERSIANN had the smallest CC, the phenomenon of overestimating the light precipitation and underestimating the heavy one existed. It should be pointed out that the original five sets of precipitation products all had a certain number of grid points that had a tendency to underestimate (overestimate) the precipitation compared with gauge observations, while the corrected precipitation products (MSWEP-QM and MSWEP-LS) effectively reduced these anomalies point (Figure 9). In conclusion, the correction mechanism proposed in this study can effectively remove the precipitation error at an annual scale, and MSWEP-QM performed slightly better than MSWEP-LS.

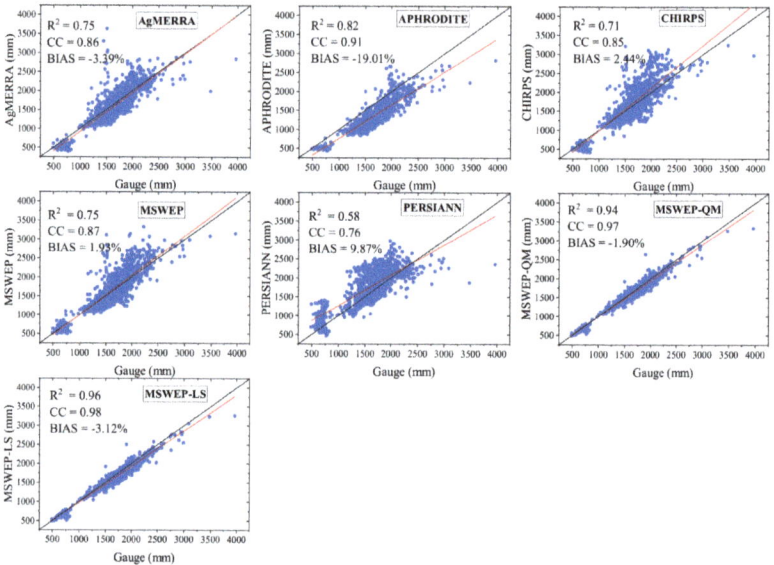

Figure 9. Scatterplot of annual average precipitation of gauge observations against AgMERRA, APHRODITE, CHIRPS, MSWEP, PERSIANN, MSWEP-QM, and MSWEP-LS from 1998 to 2007. The diagonal line is black, and the best linear fit line is red which using the least squares method.

4.3. Point-Scale Evaluation of Corrected Precipitation with Gauge Observations from 1998 to 2007

In this section, we first compare the performance of two corrected precipitation products (i.e., MSWEP-QM and MSWEP-LS) with MSWEP on a daily scale from 1998 to 2007. Then the accuracy of all precipitation products with gauge observations at basin scale is conducted. Figure 10 illustrates the BIAS of (a) MSWEP, (b) MSWEP-QM, and (c) MSWEP-LS. From these three sub-figures, in general, the two corrected precipitation can effectively remove the BIAS at a daily scale in the entire LMRB. However, there is still relatively large BIAS in high-altitude areas in the southeast of the basin and the Mekong Delta region; this may be due to the short data sequence of the gauges in these two regions to some errors in the IDW interpolation precipitation. By comparing the two corrected rainfall products, we can see that MSWEP-LS performed slightly better than MSWEP-QM in the upper Qinghai–Tibet Plateau and performed exceptionally well in other parts the river basin. Among 246 evaluated gauges, MSWEP-LS and MSWEP-QM had 214 and 211 gauges with BIAS between −20% and 20%, while MSWEP had 183 gauges. Figure 10d,e show the changes in the BIAS of MSWEP-QM and MSWEP-LS compared to MSWEP products. The MSWEP-QM and MSWEP-LS products had 165 and 171 gauges of BIAS, showing a decreasing trend, respectively. For the remaining stations, the amount of BIAS change was also less than 10%. From the average BIAS of all gauges, the average BIAS of MSWEP-QM and MSWEP-LS reached 0.17% and –0.5%, respectively, which significantly reduced the average BIAS of MSWEP (2.48%). The average absolute BIAS of all gauges of MSWEP-QM and MSWEP-LS also decreased from 16.57% to 11.21% and 10.69%. Generally, the corrected precipitation products can perform better than MSWEP at most stations, and MSWEP-LS performs slightly better than MSWEP-QM.

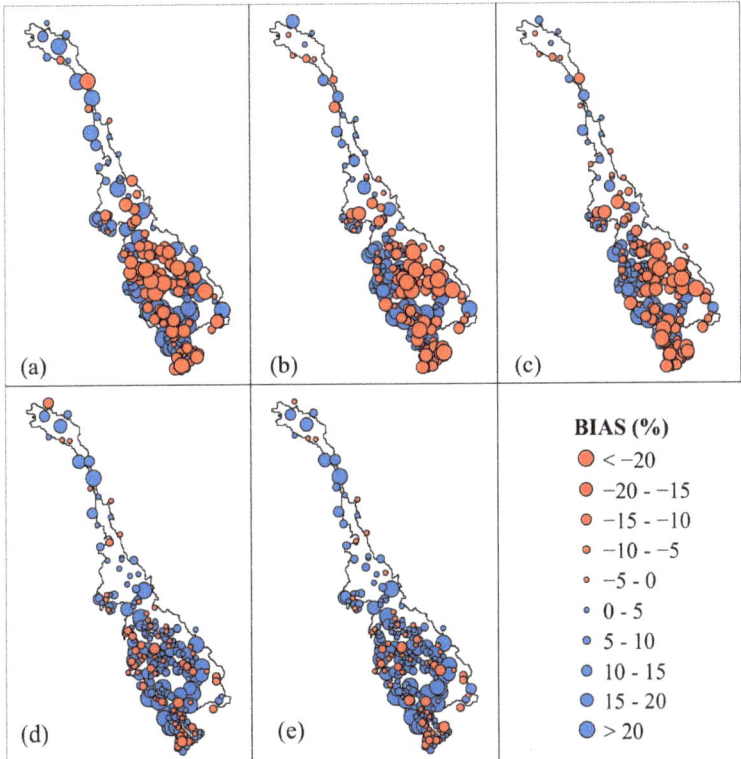

Figure 10. BIAS of MSWEP (**a**), MSWEP-QM (**b**), MSWEP-LS (**c**) compared with gauge observations from 1998 to 2007 at daily scale; the second line shows the change in BIAS between MSWEP-QM (**d**), MSWEP-LS (**e**), and MSWEP, in which blue points mean BIAS decreased, and red means the opposite.

Figure 11 shows scatterplots of precipitation estimate comparisons for the entire LMRB at a daily scale between AgMERRA, APHRODITE, CHIRPS, MSWEP, PERSIANN, MSWEP-QM, MSWEP-LS, and gauge observations from 1998 to 2007. Precipitation values were calculated from the arithmetic mean of the precipitation at the grid points that contain those stations, and gauge observations over the LMRB at a daily scale, respectively. As shown in Figure 11, APHRODITE has the largest correlation coefficient than gauge observations. However, it also has the most extensive BIAS (–15.5%), consistent with the results shown in Figures 8 and 9. The PERSIANN had the smallest CC (0.79) and R-Square (0.62), and relatively large BIAS (10.44%), which is also consistent with the results in Figure 5 and Table 3, which means that there are a relatively large number of grids points for accurate precipitation estimation. For corrected precipitation products, MSWEP-QM had a smaller BIAS and almost equal CC and R-Square than MSWEP, and MSWEP-LS had slightly larger CC and R-Square than MSWEP. However, the absolute value of its BIAS increased slightly. The higher BIAS was probably caused by the cancelation of positive and negative values. Overall, the corrected precipitation products can slightly increase the CC and reduce the BIAS at the whole river basin scale.

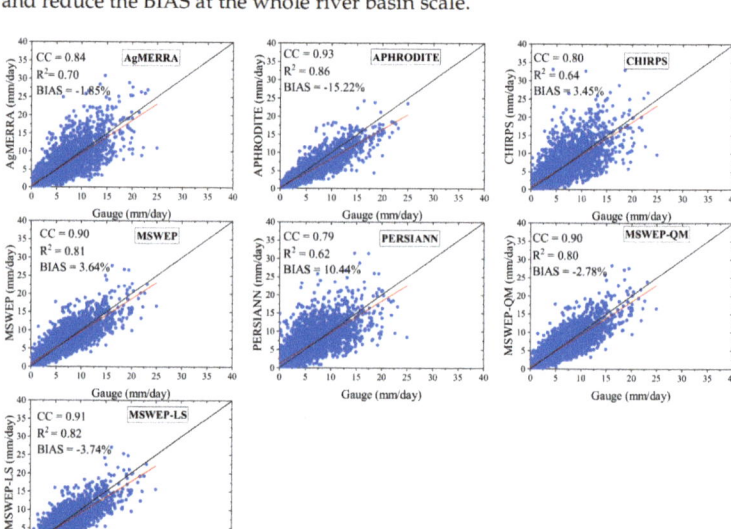

Figure 11. Scatterplots of precipitation comparisons for the entire Lancang-Mekong River Basin at daily scale between AgMERRA, APHRODITE, CHIRPS, MSWEP, PERSIANN, MSWEP-QM, MSWEP-LS, and gauge observations from 1998 to 2007. (The diagonal line is black, and the best linear fit line is red which uses the least squares method).

4.4. Hydrological and Regional Evaluation of Corrected Precipitation from 1998 to 2007

Hydrological and regional evaluations are conducted in this section to evaluate the performance of corrected precipitation products. We first used the gauge observations from 1998 to 2007 to calibrate the SWAT model, and then the calibrated model was used to assess the accuracy of all precipitation products. The regional evaluation was conducted in seven zones which were divided into six hydrological stations (Figure 1), namely, Y (above Yunjinghong station), YC (from Yunjinghong to Chiang Saen), CL (from Chiang Saen to Luang Prabang), LM (from Luang Prabang to Mukdahan), MP (from Mukdahan to Pakse), PS (from Pakse to Stung Treng), and SD (from Stung Treng to Mekong Delta).

Table 4 shows the SWAT model's simulation results at six selected hydrological stations with all eight precipitation inputs (including gauge observations). The results indicate that this model has good adaptability in the entire LMRB. For the five precipitation products, in general, the simulation results at the Yunjinghong station were the worst compared to other stations. Even though the NSE coefficient of MSWEP product reached 0.8, it had a relatively large BIAS (15.07%). For the other five stations, APHRODITE mostly underestimated the streamflow (except the Luang Prabang station), which was also consistent with the results presented in Figures 8 and 11. The AgMERRA, CHIRPS, MSWEP, and PERSIANN all had large negative BIAS at the Mukdahan station. Through Figure 8, we can see that these four products all underestimated the precipitation. As for the corrected precipitation products, we can see that the MSWEP-LS showed better performance than the MSWEP product in five of the six stations, while MSWEP-QM had better simulation results than MSWEP in just two stations. In summary, the corrected MSWEP-LS product can better simulate the daily streamflow processes in the entire LMRB (with all NSE greater than 0.8 and all BIAS lower than 11.4). To further explain the simulation results presented in Table 4 and evaluate the corrected rainfall product's accuracy, the regional evaluation was conducted next.

Table 4. Streamflow simulation results of six hydrological stations: NSE and BIAS (%).

Station	Gauge		AgMERRA		APHRODITE		CHIRPS	
	NSE	BIAS	NSE	BIAS	NSE	BIAS	NSE	BIAS
Yunjinghong	0.83	1.94	0.72	−10.66	0.52	−23.58	0.51	9.92
Chiang Saen	0.88	3.34	0.9	−2.8	0.87	−11.57	0.76	8.31
Luang Prabang	0.89	16.96	0.88	10.66	0.9	4.89	0.82	18.25
Mukdahan	0.93	6.61	0.87	−17.49	0.76	−27.73	0.80	−19.58
Pakse	0.97	7.79	0.95	−7.24	0.92	−14.47	0.95	−4.88
Stungtreng	0.98	−0.94	0.96	−2.67	0.96	−8.83	0.96	2.27
Station	MSWEP		PERSIANN		MSWEP-QM		MSWEP-LS	
	NSE	BIAS	NSE	BIAS	NSE	BIAS	NSE	BIAS
Yunjinghong	0.8	15.07	0.54	49.51	0.79	16.22	**0.83**	**7.32**
Chiang Saen	0.89	1.75	0.82	−3.33	0.88	3.35	**0.89**	**−0.4**
Luangprabang	0.89	12.5	0.86	8.9	0.88	16.99	**0.89**	**11.4**
Mukdahan	0.87	−17.6	0.86	−17.8	**0.91**	**−11.81**	0.9	−10.33
Pakse	0.95	−7.93	0.96	9.32	**0.97**	**4.83**	0.97	3.41
StungTreng	0.97	−2.8	0.96	−2.01	**0.97**	**−6.2**	0.96	−7.75

(Performance of corrected precipitation products which are better than MSWEP are shown in bold).

Table 5 shows the BIAS of five precipitation products (i.e., AgMERRA, APHRODITE, CHIRPS, MSWEP, and PERSIANN) and two corrected precipitation products (MSWEP-QM and MSWEP-LS) against gauge observations over seven sub-regions at a daily scale from 1998 to 2007. Overall, the BIAS of the two corrected precipitation products is within ±9%, which means that the framework proposed in this study can effectively remove the precipitation errors on a sub-regional scale. From a spatial perspective, the two corrected products all have smaller BIAS in Y, YC, PS, and SD regions than the original five precipitation products. In contrast, in the other two regions, the BIAS was slightly larger than the MSWEP product. From the performance of different rainfall products and their impact on hydrological simulation results, we can see that APHRODITE underestimates the precipitation in all seven sub-regions. The underestimation of precipitation led to the underestimated flow process of this product's hydrological simulation results (Table 4). The PERSAINN had a 20.99% precipitation error in the Y region, resulting in a larger BIAS (49.51%) for its streamflow simulation. In general, the two corrected products showed a smaller BIAS in the four sub-regions (Y, YC, PS, and SD) than the original five rainfall products, while the BIAS in the other three sub-regions shows a small increase. This may be related to the offset of the positive and negative BIAS values at different gauges. The results in Figure 10 indicate that the corrected precipitation data decreased BIAS at most gauges.

Table 5. BIAS (%) of five precipitation products and two corrected precipitation products against gauge observations over seven sub-regions at daily scale.

Region	AgMERRA	APHRODITE	CHIRPS	MSWEP	PERSIANN	MSWEP-QM	MSWEP-LS
Y	−2.10	−7.78	3.13	9.41	20.99	2.35	**1.22**
YC	−5.20	−11.59	4.20	1.28	−8.32	**0.88**	−0.81
CL	−1.57	−12.73	−2.61	**−0.72**	0.88	−0.96	−2.39
LM	−11.49	−24.63	−9.79	**−7.12**	−3.39	−8.46	−8.91
MP	−9.70	−24.50	−5.36	**−4.24**	5.73	−7.77	−8.18
PS	−1.67	−14.45	12.20	7.49	5.29	**−4.86**	−5.12
SD	8.89	−20.16	13.31	10.14	42.16	**−0.16**	−1.88

(where Y, YC, CL, LM, MP, PS, and SD means region over Yunjinghong station, the region between Yunjinghong station and Chiang Saen station, the region between Chiang Saen station and Luang Prabang station, the region between Luang Prabang station and Mukdahan station, the region between Mukdahan station and Pakse station, the region between Pakse station and Stung Treng station, the region between Stung Treng station and Mekong Delta Region, respectively; product with the smallest BIAS in each region are shown in bold)

4.5. Validation of Corrected Precipitation with Gauge Observations from 1979 to 1997 and 2008 to 2014

In the previous sections, we evaluated the corrected precipitation products at the point scale and grid-scale from 1998 to 2007. This section evaluates the corrected precipitation products on the remaining observations (i.e., from 1979 to 1997 and 2008 to 2014) to assess the performance of the bias correction framework proposed in this study.

Figure 12a–c show the BIAS of MSWEP, MSWEP-QM, and MSWEP-LS compared with gauge observation on a daily scale from 1979 to 1997 and 2008 to 2014. From these three subfigures, we can see that the two sets of corrected precipitation products have smaller BIAS compared to MSWEP in the upper Mekong River Basin. However, most stations downstream still showed a tendency to underestimate precipitation. The cause of this phenomenon may be that these sets of precipitation products only contained information on the limited ground observation gauges in the downstream area during the generation process. We further compared the BIAS changes at a daily scale between the MSWEP-QM (d), MSWEP-LS (e), and MSWEP. The results showed that the corrected precipitation products can effectively reduce BIAS at most stations, especially in the upstream and middle reaches. However, BIAS in some high-altitude areas in the southeast of the basin and some Mekong Delta stations showed a larger trend. The average absolute BIAS of MSWEP was 21.4%, and the corrected MSWEP-QM and MSWEP-LS were 17.98% and 17.87%, which were reduced by 3.4% and 3.51%, respectively.

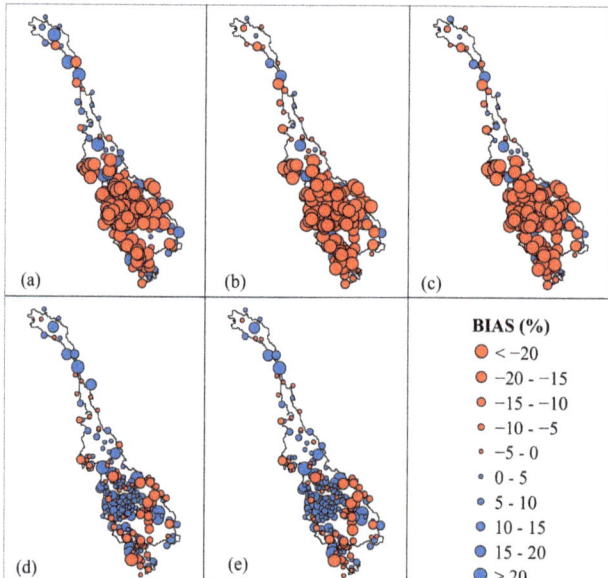

Figure 12. BIAS of MSWEP (**a**), MSWEP-QM (**b**), MSWEP-LS (**c**) compared with gauge observations from 1979 to 1997 and 2008 to 2014 at daily scale; the second line shows the change in BIAS between MSWEP-QM (**d**), MSWEP-LS (**e**), and MSWEP, in which the blue points mean BIAS decreased, and red means the opposite.

5. Discussion

5.1. Performance of Different Precipitation Products

In this study, before error correction of the precipitation products, we first evaluated five precipitation products (i.e., AgMERRA, APHRODITE, CHIRPS, MSWEP, and PERSIANN). Then, we compared them with gauge observations in the Lancang-Mekong River Basin mainly from three aspects (correlation coefficient, probability of detection, and mean error of each month). We found that APHRODITE had the largest correlation coefficient among the five products (followed by MSWEP, AgMERRA, PERSIANN, and CHIRPS) in the LMRB (Figure 5), and this conclusion is consistent with the previous research results of the LMRB [28,33]. However, at the same time, we also found that APHRODITE had severely underestimated precipitation (Figures 8 and 11), especially in the lower LMRB. The underestimated precipitation also led to underestimating the hydrological simulation of the runoff processes (Table 5), and this conclusion was rarely mentioned in published studies. Although the development of the APHRODITE precipitation product included 5000–12,000 verification stations in Asia, the monthly precipitation data in this study area only included ground gauges in Thailand and the Lancang River area in China [44]. This may be one of the reasons for its poor performance in the downstream area of LMRB. On the other hand, Figure 6 in the original article published on the development of the APHRODITE [44] compared the annual average precipitation of APHRODITE and Global Precipitation Climatology Centre (GPCC) from 1961 to 2004, and the result also indicates that the precipitation in the lower LMRB was underestimated. However, due to the fact tjat surface observation data in the LMRB are scarce and difficult to collect, the APHRODITE has always been selected as the actual precipitation values in many published studies without considering its errors [28,31]. As these References [46,57] have shown, any small precipitation error may be amplified by hydrological simulations and affect the entire water resource allocation, sustainable development strategy formulation, etc. For the accuracy of precipitation event detection (including precipitation equal to 0), we can see that MSWEP had the best performance over the other four products, followed by APHRODITE. This is because the MSWEP product uses a weighted average of seven rainfall products (i.e., CPC Unified, GPCC, CMORPH, GSMaP-MVK, TMPA 3B42V7, ERA-Interim, and JRA-55) during the development process [10]. It also uses the average precipitation and streamflow of 13,762 watersheds worldwide to remove the impact of terrain on precipitation. The AgMERRA mainly uses data sets of MERRA-Land and wet days of CRU to correct the precipitation days [42,58]. Whereas CHIRPS uses the CHPclim model to integrate the FAO and Global Historical Climate Network (GHCN) stations, remote sensing data TMPA and CMORPH [59]. The PERSIANN mainly uses GPCP monthly average precipitation to correct its product [40]. That means that the MSWEP incorporates more information about precipitation products and ground stations, which may be an important reason for its more accurate detection of rainfall events. We understand why MSWEP has the most grids in more months (February, May, June, July, November, and December) to estimate precipitation more accurately than the other four products (Table 3). However, as shown in Figures 7 and 8, MSWEP still has large errors in some areas of the LMRB. Therefore, in this study, we propose a bias correction mechanism based on various remote sensing and reanalysis precipitation products, hoping to provide a set of long-range precipitation products with high accuracy for the LMRB.

5.2. Applicability of the Error Correction Framework

The error correction framework proposed in this study using Quantile Mapping (QM) and Linear Scaling (LS) both fit quite well with smaller ME and BIAS in most stations, and it can also better spatially predict the precipitation in the entire LMRB (Figures 6 and 7). Because we first select the precipitation product based on the CC and rainfall events estimates (POD01), and then further corrects the precipitation errors. Therefore, after correction, we use BIAS, which focuses on its estimation of quantity for the precipitation assessment. From the corrected rainfall products' evaluation results, MSWEP-LS is slightly better than MSWEP-QM, mainly due to the two correction methods' different internal mechanisms. Because the QM method uses the cumulative precipitation distribution function (CDF) to correct the selected product (MSWEP in this study) by one percentile. However, in this study, we selected the precipitation product with the smallest ME at each grid point in each month as the benchmark. There may be errors between the cumulative distribution curve and those of station observations. The LS method only uses a scaling factor to correct the precipitation. MSWEP has a higher correlation coefficient than the site-observed precipitation in the LMRB, which may be one reason why the MSWEP-LS method performs slightly better than MSWEP-QM. Ghimire et al. [18] compared and analyzed the effects of LS and QM methods on streamflow simulation after global climate models (GCMs) precipitation product bias correction, and they found that the LS method showed better performance than QM. As shown in Figure 9, the BIAS of MSWEP-QM was slightly smaller than that of MSWEP-LS. However, the results shown in Figure 8 indicate that the BIAS of MSWEP-LS had more stations showing a decreasing trend, which is probably caused by the cancelation of positive and negative values [60]. We have not achieved perfect results in LM (from Luang Prabang to Mukdahan) and MP (from Mukdahan to Pakse) regions from the regional evaluation results. This setback is partly due to the shorter available data length of the stations we collected in these two regions, which led to our IDW interpolation results in these areas may be insufficient. In general, although this study was conducted in the Lancang-Mekong River Basin, the proposed framework could also be applicable in other areas, especially for those with limited gauge observations.

5.3. Limitations and Future Directions of This Study

Although the framework proposed in this study can effectively remove the BIAS of precipitation on a daily scale over the entire LMRB (Figures 8, 10 and 12), there are also limitations. In this study, the IDW interpolation method was used to obtain the grid-scale (0.25°) monthly precipitation with observed precipitation from 1998 to 2007. However, there is no doubt that interpolation will bring some errors [33], especially in the lower LMRB where the gauges have shorter data sequences available. In future research, we should collect more ground radar and gauge observation data to reduce further the errors caused by interpolation. Second, due to the difficulty of collecting the observation data of LMRB and the available data series are always short, in this study, the product with the smallest mean error was selected as the actual value at each grid point each month, in other words, the accuracy of the remote sensed or reanalysis precipitation product at a specific grid point has a greater contribution to the accuracy of the corrected precipitation than the gauge observations. However, it is clear that each product inevitably has specific errors compared to gauge observations [8,61], making it impossible for us to remove the precipitation error using the available data we collected. Third, the correction result is also related to the spatial distribution of the collected rainfall stations. From the Figure 8, we can see that in the eastern mountainous area of the lower reaches of the LMRB, the corrected precipitation products can better present the annual rainfall distribution of the basin compared with the gauge observations, but in the lower reaches of Cambodia, the observation data we collected by the ground rainfall stations in this area is very limited, therefore, there may still be some uncertainty in the corrected precipitation in this area (Figure 8). In the last aspect, our study used mean error as the unique indicator to select the precipitation products as the actual value; as concluded by Wang et al. [62], extreme precipitation is more meaningful for the prediction of water-related disasters and the sustainable use of water resources in the watershed. Therefore, in the next research, we should consider extreme precipitation indicators when selecting precipitation products. However, as we have shown in our introduction and results, our proposed framework can reduce the BIAS of MSWEP at most gauges. According to Figures 10–12, both the MSWEP-QM and MSWEP-LS performed better than the raw precipitation products (i.e., AgMERRA, APHRODITE, CHIRPS, MSWEP, and PERSIANN), of which MSWEP-LS performs better. Therefore, we recommend that MSWEP-LS be used for related studies such as hydrological simulation in the Lancang-Mekong River Basin.

6. Conclusions

In this study, we proposed and implemented a novel daily-scale precipitation bias correction framework based on multiple long-term remote sensing and reanalysis precipitation products in Lancang-Mekong River Basin, which also can be used in other poorly gauged areas. We first compared the five rainfall products (i.e., AgMERRA, APHRODITE, CHIRPS,

MSWEP, and PERSIANN) with the observed precipitation. The resulting precipitation products MSWEP-QM derived from quantile mapping and MSWEP-LS derived from linear scaling were evaluated in calibration (from 1998 to 2007) and validation (1979 to 1997 and 2008 to 2014) periods. The main conclusions are summarized in the following points:

1. The APHRODITE showed the highest CC (0.61) with gauge observations at a daily scale but greatly underestimated the precipitation (with BIAS equals −15.5%), especially in the downstream areas. This means that we should carefully choose APHRODITE as the actual value of the LMRB for related research. The average probability of precipitation detection (POD01) estimated by MSWEP was 0.99, which was the highest among the five raw precipitation products.
2. The monthly grid-scale evaluation results showed that most grids of MSWEP had the smallest ME in February, from May to July, November, and December. The AgMERRA, APHRODITE, CHIRPS, and PERSIANN had the most grids with the smallest ME in September and October, January, April, and August, respectively. The variation of five precipitation products' performance over the entire LMRB was associated with the data sources included in their respective development processes and the different algorithms they adopt.
3. Grid-scale evaluation shows that two resulting precipitation products both can capture the spatial variability of multi-year average precipitation across the entire LMRB in the calibration period. The MSWEP-QM (0.97) and MSWEP-LS (0.98) have higher CC than AgMRRA (0.86), APHRODITE (0.91), CHIRPS (0.86), MSWEP (0.87), PERSIANN (0.76). The point-scale evaluation results indicate that the BIAS of MSWEP-QM (165 in 246), and MSWEP-LS (171 in 246) have more gauges showing a downward trend.
4. Hydrological and regional revaluation shows that MSWEP-LS and MSWEP-QM achieved better simulation results in five regions (i.e., Y, YC, CL, LM, and MP regions) compared to the two regions derived from MSWEP (LM and MP). The BIAS of MSWEP-QM and MSWEP-LS in seven sub-regions all reach within ±9% on a daily scale. They also had smaller BIAS in Y, YC, PS, and SA regions than the five raw precipitation products.
5. Validation results indicated that the average absolute BIAS of MSWEP-QM and MSWEP-LS reduced by 3.4% and 3.51%, respectively, compared to MSWEP. The BIAS of MSWEP-QM and MSWEP-LS had 141 and 142 gauges showing a decreasing trend than MSWEP.

In general, the novel precipitation bias-correction framework proposed in this study is considered to provide a viable study for blending five selected precipitation products in regions with limited gauge observations. We also recommend that the MSWEP-LS can be used for further water-related research in LMRB.

Author Contributions: Conceptualization, X.T. and G.W.; methodology, X.T.; writing—original draft preparation, X.T.; writing—review and editing, G.B.R., J.Z., Z.B., Y.L., C.L., and J.J. All authors have read and agreed to the published version of the manuscript.

Funding: This study was funded by the National Key Research and Development Plan of China (grant number: 2016YFA0601501), the National Natural Science Foundation of China (grant number: 41830863) and the National Key Research and Development Plan of China (grant number: 2017YFC0405604).

Institutional Review Board Statement: Not applicable.

Informed Consent Statement: Not applicable.

Data Availability Statement: The data presented in this study are available on request from the corresponding author.

Acknowledgments: The authors are very grateful to all anonymous reviewers, chief editors and associate editors for their valuable comments and suggestions, which not only greatly improved the quality of this paper but also have immense value for our future research.

Conflicts of Interest: The authors declare no competing interests.

References

1. Duan, Z.; Bastiaanssen, W. First results from Version 7 TRMM 3B43 precipitation product in combination with a new downscaling-calibration procedure. *Remote Sens. Environ.* **2013**, *131*, 1–13. [CrossRef]
2. Gao, C.; Booij, M.J.; Xu, Y.P. Impacts of climate change on characteristics of daily-scale rainfall events based on nine selected GCMs under four CMIP5 RCP scenarios in Qu River basin, east China. *Int. J. Climatol.* **2020**, *40*, 887–907. [CrossRef]
3. Zhu, Q.; Hsu, K.l.; Xu, Y.P.; Yang, T. Evaluation of a new satellite-based precipitation data set for climate studies in the Xiang River basin, southern China. *Int. J. Climatol.* **2017**, *37*, 4561–4575. [CrossRef]
4. Tang, G.; Long, D.; Hong, Y.; Gao, J.; Wan, W. Documentation of multifactorial relationships between precipitation and topography of the Tibetan Plateau using spaceborne precipitation radars. *Remote Sens. Environ.* **2018**, *208*, 82–96. [CrossRef]
5. Tang, X.; Zhang, J.; Gao, C.; Ruben, G.B.; Wang, G. Assessing the Uncertainties of Four Precipitation Products for Swat Modeling in Mekong River Basin. *Remote Sens.* **2019**, *11*, 304. [CrossRef]
6. Luo, X.; Wu, W.; He, D.; Li, Y.; Ji, X. Hydrological Simulation Using TRMM and CHIRPS Precipitation Estimates in the Lower Lancang-Mekong River Basin. *Chin. Geogr. Sci.* **2019**, *29*, 13–25. [CrossRef]
7. Wang, W.; Lu, H.; Yang, D.; Khem, S.; Yang, J.; Gao, B.; Peng, X.; Pang, Z. Modelling Hydrologic Processes in the Mekong River Basin Using a Distributed Model Driven by Satellite Precipitation and Rain Gauge Observations. *PLoS ONE* **2016**, *11*, e0152229. [CrossRef]
8. Yong, B.; Chen, B.; Gourley, J.J.; Ren, L.; Hong, Y.; Chen, X.; Wang, W.; Chen, S.; Gong, L. Intercomparison of the Version-6 and Version-7 TMPA precipitation products over high and low latitudes basins with independent gauge networks: Is the newer version better in both real-time and post-real-time analysis for water resources and hydrologic extremes? *J. Hydrol.* **2014**, *508*, 77–87. [CrossRef]
9. Huffman, G.J.; Bolvin, D.T.; Nelkin, E.J.; Wolff, D.B.; Adler, R.F.; Gu, G.; Hong, Y.; Bowman, K.P.; Stocker, E.F. The TRMM multisatellite precipitation analysis (TMPA): Quasi-global, multiyear, combined-sensor precipitation estimates at fine scales. *J. Hydrometeorol.* **2007**, *8*, 38–55. [CrossRef]
10. Beck, H.E.; Van Dijk, A.I.J.M.; Levizzani, V.; Schellekens, J.; Miralles, D.G.; Martens, B.; De Roo, A. MSWEP: 3-hourly 0.25° global gridded precipitation (1979–2015) by merging gauge, satellite, and reanalysis data. *Hydrol. Earth Syst. Sci.* **2016**, *21*, 1–38. [CrossRef]
11. Mazzoleni, M.; Brandimarte, L.; Amaranto, A. Evaluating precipitation datasets for large-scale distributed hydrological modelling. *J. Hydrol.* **2019**, *578*, 124076. [CrossRef]
12. Yatagai, A.; Kamiguchi, K.; Arakawa, O.; Hamada, A.; Yasutomi, N.; Kitoh, A. APHRODITE: Constructing a Long-Term Daily Gridded Precipitation Dataset for Asia Based on a Dense Network of Rain Gauges. *Bull. Am. Meteorol. Soc.* **2012**, *93*, 1401–1415. [CrossRef]
13. Tang, X.; Zhang, J.; Wang, G.; Yang, Q.; Yang, Y.; Guan, T.; Liu, C.; Jin, J.; Liu, Y.; Bao, Z. Evaluating Suitability of Multiple Precipitation Products for the Lancang River Basin. *Chin. Geogr. Sci.* **2019**, *29*, 37–57. [CrossRef]
14. Ma, Y.; Hong, Y.; Chen, Y.; Yang, Y.; Tang, G.; Yao, Y.; Long, D.; Li, C.; Han, Z.; Liu, R. Performance of optimally merged multisatellite precipitation products using the dynamic Bayesian model averaging scheme over the Tibetan Plateau. *J. Geophys. Res. Atmos.* **2018**, *123*, 814–834. [CrossRef]
15. Worqlul, A.W.; Ayana, E.K.; Maathuis, B.H.P.; Macalister, C.; Philpot, W.D.; Leyton, J.M.O.; Steenhuis, T.S. Performance of bias corrected MPEG rainfall estimate for rainfall-runoff simulation in the upper Blue Nile Basin, Ethiopia. *J. Hydrol.* **2017**, *556*, S0022169417300689. [CrossRef]
16. Gutjahr, O.; Heinemann, G. Comparing precipitation bias correction methods for high-resolution regional climate simulations using COSMO-CLM. *Theor. Appl. Clim.* **2013**, *114*, 511–529. [CrossRef]
17. Gudmundsson, L.; Bremnes, J.; Haugen, J.; Engen-Skaugen, T. Downscaling RCM precipitation to the station scale using statistical transformations—A comparison of methods. *Hydrol. Earth Syst. Sci.* **2012**, *16*, 3383–3390. [CrossRef]
18. Ghimire, U.; Srinivasan, G.; Agarwal, A. Assessment of rainfall bias correction techniques for improved hydrological simulation. *Int. J. Climatol.* **2019**, *39*, 2386–2399. [CrossRef]
19. Habib, E.; Haile, A.; Sazib, N.; Zhang, Y.; Rientjes, T. Effect of Bias Correction of Satellite-Rainfall Estimates on Runoff Simulations at the Source of the Upper Blue Nile. *Remote Sens.* **2014**, *6*, 6688–6708. [CrossRef]
20. Gumindoga, W.; Rientjes, T.H.; Haile, A.T.; Makurira, H.; Reggiani, P. Performance of bias-correction schemes for CMORPH rainfall estimates in the Zambezi River basin. *Hydrol. Earth Syst. Sci.* **2019**, *23*, 2915–2938. [CrossRef]
21. Boushaki, F.I.; Hsu, K.-L.; Sorooshian, S.; Park, G.-H.; Mahani, S.; Shi, W. Bias adjustment of satellite precipitation estimation using ground-based measurement: A case study evaluation over the southwestern United States. *J. Hydrometeorol.* **2009**, *10*, 1231–1242. [CrossRef]
22. Xie, P.; Xiong, A.Y. A conceptual model for constructing high-resolution gauge-satellite merged precipitation analyses. *J. Geophys. Res. Atmos.* **2011**, *116*. [CrossRef]
23. Li, Y.; Zhang, Y.; He, D.; Luo, X.; Ji, X. Spatial downscaling of the tropical rainfall measuring mission precipitation using geographically weighted regression Kriging over the Lancang River Basin, China. *Chin. Geogr. Sci.* **2019**, *29*, 446–462. [CrossRef]
24. Mahmood, R.; Jia, S.; Tripathi, N.K.; Shrestha, S. Precipitation extended linear scaling method for correcting GCM precipitation and its evaluation and implication in the transboundary Jhelum River basin. *Atmosphere* **2018**, *9*, 160. [CrossRef]
25. Maraun, D. Bias correction, quantile mapping, and downscaling: Revisiting the inflation issue. *J. Clim.* **2013**, *26*, 2137–2143. [CrossRef]
26. Liu, S.; Yan, D.; Qin, T.; Weng, B.; Li, M. Correction of TRMM 3B42V7 based on linear regression models over China. *Adv. Meteorol.* **2016**, *2016*. [CrossRef]

27. Hecht, J.S.; Lacombe, G.; Arias, M.E.; Dang, T.D.; Piman, T. Hydropower dams of the Mekong River basin: A review of their hydrological impacts. *J. Hydrol.* **2018**, *568*, 285–300. [CrossRef]
28. Lauri, H.; Räsänen, T.A.; Kummu, M. Using Reanalysis and Remotely Sensed Temperature and Precipitation Data for Hydrological Modeling in Monsoon Climate: Mekong River Case Study. *J. Hydrometeorol.* **2014**, *15*, 1532–1545. [CrossRef]
29. Winemiller, K.O.; Mcintyre, P.B.; Castello, L.; Fluetchouinard, E.; Giarrizzo, T.; Nam, S.; Baird, I.G.; Darwall, W.; Lujan, N.K.; Harrison, I. Development and environment. Balancing hydropower and biodiversity in the Amazon, Congo, and Mekong. *Science* **2016**, *351*, 128. [CrossRef]
30. Ohara, N.; Chen, Z.; Kavvas, M.; Fukami, K.; Inomata, H. Reconstruction of historical atmospheric data by a hydroclimate model for the Mekong River basin. *J. Hydrol. Eng.* **2010**, *16*, 1030–1039. [CrossRef]
31. Chen, A.; Chen, D.; Azorin-Molina, C. Assessing reliability of precipitation data over the Mekong River Basin: A comparison of ground-based, satellite, and reanalysis datasets. *Int. J. Climatol.* **2018**, *38*, 4314–4334. [CrossRef]
32. Chen, C.; Jayasekera, D.; Senarath, S. Assessing Uncertainty in Precipitation and Hydrological Modeling in the Mekong. In Proceedings of the World Environmental and Water Resources Congress, Austin, TX, USA, 17–21 May 2015; pp. 2510–2519.
33. Chen, C.J.; Senarath, S.U.S.; Dima-West, I.M.; Marcella, M.P. Evaluation and restructuring of gridded precipitation data over the Greater Mekong Subregion. *Int. J. Climatol.* **2016**, *37*, 180–196. [CrossRef]
34. Zhang, J.; Fan, H.; He, D.; Chen, J. Integrating precipitation zoning with random forest regression for the spatial downscaling of satellite-based precipitation: A case study of the Lancang-Mekong River basin. *Int. J. Climatol.* **2019**, *39*, 3947–3961. [CrossRef]
35. Jacobs, J.W. The Mekong River Commission: Transboundary Water Resources Planning and Regional Security. *Geogr. J.* **2002**, *168*, 354–364. [CrossRef] [PubMed]
36. Han, Z.; Long, D.; Fang, Y.; Hou, A.; Hong, Y. Impacts of climate change and human activities on the flow regime of the dammed Lancang River in Southwest China. *J. Hydrol.* **2019**, *570*, 96–105. [CrossRef]
37. Li, D.; Long, D.; Zhao, J.; Lu, H.; Hong, Y. Observed changes in flow regimes in the Mekong River basin. *J. Hydrol.* **2017**, *551*, 217–232. [CrossRef]
38. Ruane, A.C.; Goldberg, R.; Chryssanthacopoulos, J. Climate forcing datasets for agricultural modeling: Merged products for gap-filling and historical climate series estimation. *Agric. For. Meteorol.* **2015**, *200*, 233–248. [CrossRef]
39. Retalis, A.; Tymvios, F.; Katsanos, D.; Michaelides, S. Downscaling CHIRPS precipitation data: An artificial neural network modelling approach. *Int. J. Remote Sens.* **2017**, *38*, 3943–3959. [CrossRef]
40. Faridzad, M.; Yang, T.; Hsu, K.; Sorooshian, S.; Xiao, C. Rainfall Frequency Analysis for Ungauged Regions using Remotely Sensed Precipitation Information. *J. Hydrol.* **2018**, *563*, 123–142. [CrossRef]
41. Reichle, R.H.; Koster, R.D.; De Lannoy, G.J.M.; Forman, B.A.; Liu, Q.; Mahanama, S.P.P.; Touré, A. Assessment and Enhancement of MERRA Land Surface Hydrology Estimates. *J. Clim.* **2011**, *24*, 6322–6338. [CrossRef]
42. Rienecker, M.M.; Suarez, M.J.; Gelaro, R.; Todling, R.; Bacmeister, J.; Liu, E.; Bosilovich, M.G.; Schubert, S.D.; Takacs, L.; Kim, G.K. MERRA: NASA's Modern-Era Retrospective Analysis for Research and Applications. *J. Clim.* **2011**, *24*, 3624–3648. [CrossRef]
43. Rienecker, M.M.; Suarez, M.J.; Todling, R.; Bacmeister, J.; Takacs, L.; Liu, H.; Gu, W.; Sienkiewicz, M.; Koster, R.D.; Gelaro, R. The GEOS-5 Data Assimilation System—Documentation of Versions 5.0.1, 5.1.0, and 5.2.0. 2008. Available online: https://gmao.gsfc.nasa.gov/pubs/docs/GEOS-5.0.1_Documentation_r3.pdf (accessed on 18 January 2021).
44. Yatagai, A.; Arakawa, O.; Kamiguchi, K.; Kawamoto, H.; Nodzu, M.I.; Hamada, A. A 44Year Daily Gridded Precipitation Dataset for Asia Based on a Dense Network of Rain Gauges. *Sci. Online Lett. Atmos. Sola* **2009**, *5*, 137–140. [CrossRef]
45. Ashouri, H.; Hsu, K.L.; Sorooshian, S.; Braithwaite, D.K.; Knapp, K.R.; Cecil, L.D.; Nelson, B.R.; Prat, O.P. PERSIANN-CDR: Daily Precipitation Climate Data Record from Multisatellite Observations for Hydrological and Climate Studies. *Bull. Am. Meteorol. Soc.* **2014**, *96*, 197–210. [CrossRef]
46. Zhu, Q.; Xuan, W.; Liu, L.; Xu, Y.P. Evaluation and hydrological application of precipitation estimates derived from PERSIANN-CDR, TRMM 3B42V7, and NCEP-CFSR over humid regions in China. *Hydrol. Process.* **2016**, *30*, 3061–3083. [CrossRef]
47. Saha, S.; Moorthi, S.; Pan, H.L.; Wu, X.R.; Wang, J.D.; Nadiga, S.; Tripp, P.; Kistler, R.; Woollen, J.; Behringer, D. The NCEP climate forecast system reanalysis. *Bull. Am. Meteorol. Soc.* **2010**, *91*, 1015–1057. [CrossRef]
48. Potter, N.J.; Chiew, F.H.; Charles, S.P.; Fu, G.; Zheng, H.; Zhang, L. Bias in Downscaled Rainfall Characteristics. Available online: https://hess.copernicus.org/preprints/hess-2019-139/hess-2019-139.pdf (accessed on 18 January 2021).
49. Reiter, P.; Gutjahr, O.; Schefczyk, L.; Heinemann, G.; Casper, M. Does applying quantile mapping to subsamples improve the bias correction of daily precipitation? *Int. J. Climatol.* **2018**, *38*, 1623–1633. [CrossRef]
50. Arnold, J.G.; Srinivasan, R.; Muttiah, R.S.; Williams, J.R. Large area hydrologic modeling and assessment part I: Model development. *JAWRA J.* **1998**, *34*, 73–89. [CrossRef]
51. Abbaspour, K.C.; Vaghefi, S.A.; Srinivasan, R. A Guideline for Successful Calibration and Uncertainty Analysis for Soil and Water Assessment: A Review of Papers from the 2016 International SWAT Conference. *Water* **2017**, *10*, 6. [CrossRef]
52. Arnold, J.G.; Kiniry, J.R.; Srinivasan, R.; Williams, J.R.; Haney, E.B.; Neitsch, S.L. *Soil & Water Assessment Tool: Input/Output Documentation. Version 2012*; Texas Water Resources Institute: College Station, TX, USA, 2012; pp. 1–650.
53. Abbaspour, K.C.; Vejdani, M.; Haghighat, S. SWAT-CUP calibration and uncertainty programs for SWAT. *Modsim Int. Congr. Model. Simul. Land Water Environ. Manag. Integr. Syst. Sustain.* **2007**, *364*, 1603–1609. [CrossRef]
54. Abbaspour, K.C.; Yang, J.; Maximov, I.; Siber, R.; Bogner, K.; Mieleitner, J.; Zobrist, J.; Srinivasan, R. Modelling hydrology and water quality in the pre-alpine/alpine Thur watershed using SWAT. *J. Hydrol.* **2007**, *333*, 413–430. [CrossRef]

55. Abbaspour, K.C.; Johnson, C.A.; Genuchten, M.T.V. Estimating Uncertain Flow and Transport Parameters Using a Sequential Uncertainty Fitting Procedure. *Vadose Zone J.* **2004**, *3*, 1340–1352. [CrossRef]
56. Nash, J.E.; Sutcliffe, J.V. River flow forecasting through conceptual models part I—A discussion of principles. *J. Hydrol.* **1970**, *10*, 282–290. [CrossRef]
57. Zhao, F.; Wu, Y.; Qiu, L.; Sun, Y.; Sun, L.; Li, Q.; Niu, J.; Wang, G. Parameter uncertainty analysis of the SWAT model in a mountain-loess transitional watershed on the Chinese Loess Plateau. *Water* **2018**, *10*, 690. [CrossRef]
58. Molod, A.; Takacs, L.; Suarez, M.; Bacmeister, J. Development of the GEOS-5 atmospheric general circulation model: Evolution from MERRA to MERRA2. *Geosci. Model Dev.* **2015**, *7*, 1339–1356. [CrossRef]
59. Funk, C.; Peterson, P.; Landsfeld, M.; Pedreros, D.; Verdin, J.; Shukla, S.; Husak, G.; Rowland, J.; Harrison, L.; Hoell, A. The climate hazards infrared precipitation with stations—A new environmental record for monitoring extremes. *Sci. Data* **2015**, *2*, 150066. [CrossRef] [PubMed]
60. Tang, G.; Ma, Y.; Long, D.; Zhong, L.; Hong, Y. Evaluation of GPM Day-1 IMERG and TMPA Version-7 legacy products over Mainland China at multiple spatiotemporal scales. *J. Hydrol.* **2016**, *533*, 152–167. [CrossRef]
61. Sun, Q.; Miao, C.; Duan, Q.; Ashouri, H.; Sorooshian, S.; Hsu, K.L. A review of global precipitation datasets: Data sources, estimation, and intercomparisons. *Rev. Geophys.* **2017**, *56*, 79–107. [CrossRef]
62. Wang, G.; Wang, D.; Trenberth, K.E.; Erfanian, A.; Yu, M.; Bosilovich, M.G.; Parr, D.T. The peak structure and future changes of the relationships between extreme precipitation and temperature. *Nat. Clim. Chang.* **2017**, *7*, 268. [CrossRef]

Article

Evaluation of TMPA 3B42-V7 Product on Extreme Precipitation Estimates

Jiachao Chen [1], Zhaoli Wang [1,2], Xushu Wu [1,2,*], Chengguang Lai [1,2] and Xiaohong Chen [3]

1. School of Civil Engineering and Transportation, State Key Laboratory of Subtropical Building Science, South China University of Technology, Guangzhou 510641, China; ctjiachaochen@mail.scut.edu.cn (J.C.); wangzhl@scut.edu.cn (Z.W.); laichg@scut.edu.cn (C.L.)
2. Guangdong Engineering Technology Research Center of Safety and Greenization for Water Conservancy Project, Guangzhou 510641, China
3. Center for Water Resource and Environment, Sun Yat-Sen University, Guangzhou 510275, China; eescxh@mail.sysu.edu.cn
* Correspondence: xshwu@scut.edu.cn

Abstract: Availability of precipitation data at high spatial and temporal resolution is crucial for the understanding of precipitation behaviors that are determinant for environmental aspects such as hydrology, ecology, and social aspects like agriculture, food security, or health issues. This study evaluates the performance of 3B42-V7 satellite-based precipitation product on extreme precipitation estimates in China, by using the Fuzzy C-Means algorithm and L-moment-based regional frequency analysis method. The China Gauge-based Daily Precipitation Analysis (CGDPA) product is employed to measure the estimation biases of 3B42-V7. Results show that: (1) for most regions of China, the Generalized Extreme Value and Generalized Normal distributions are preferable for extreme precipitation estimates; (2) the extreme precipitation estimations of 3B42-V7 for different return periods have a high correlation with those of CGDPA, with biases within 25% for a majority of China on extreme precipitation estimates.

Keywords: extreme precipitation; estimation; TMPA 3B42-V7; regional frequency analysis; China

1. Introduction

The knowledge and estimation of extreme precipitation are essential for many applications such as water resources management, flood forecasting, transportation, early warning, and disaster mitigation [1–4]. Observing the physical quantity of Earth's atmosphere through satellites and using algorithms to combine multi-source remote sensing data is an effective way of estimating precipitation [5–7]. This kind of quantitative precipitation estimation product overcomes the shortcomings of gauge station-based observations such as limited coverage, uneven distribution, and poor consistency. Among the satellite-based precipitation estimation products, the Tropical rainfall measurement mission Multi-satellite Precipitation Analysis (TMPA) 3B42-V7 has received much attention [8,9]. Many studies have indicated that 3B42 has higher precision among similar products [10–12]. At present, several studies have been carried out based on the precipitation data provided by TMPA and have achieved reliable results. Due to the high quality and wide spatial coverage, Jung et al. [13] obtained the global soil evaporation trend from 1998 to 2008. In terms of runoff simulation, 3B42-V7 also performs well. Wang et al. [14] obtained a Nash–Sutcliffe coefficient of 0.83 for daily runoff in the southern humid regions of China. Even under the adverse condition of terrain and lacking data for calibration, the 3B42-V7 still has good hydrological applicability [15]. The accurate recording of no rain and light rain events also allows 3B42-V7 to be widely used in drought researches. Zhong et al. [16] compared three kinds of satellite-based precipitation products, and showed that 3B42-V7 has the best performance with the smallest deviation, and it can accurately capture the center

and range of drought events. However, there are very few studies focusing on extreme precipitation estimation using 3B42-V7. Motivated by this need, the purpose of this study is to evaluate the accuracy of estimating precipitation extremes in different return periods based on 3B42-V7.

The index flood method [17] is one of the procedures to estimate precipitation extremes. The main idea of this method is to assume that the flood distributions at each location in a homogeneous region have the same coefficient of variation and skewness, thereby estimating the quantile of any return period at any location within the region. On this basis, Hosking and Wallis [18] used the L-moments method to improve the index flooding method, and proposed the L-moments-based regional frequency analysis method. This method has been widely used in the frequency analysis of regional floods, precipitation and drought [19–24]. However, using L-moments-based regional frequency analysis to estimate precipitation extremes places higher demands on the representativeness, consistency, accuracy and sequence length of precipitation data. In regions with complex terrains, such as gorge, the vertical variation of precipitation is obvious, and the data representative of rain gauge stations tends to be poor. Limited by factors such as the level of economic development, many countries and regions in the world have problems such as sparse meteorological station network, poor data consistency, high error rate, and short data accumulation, which brings difficulties for the development of infrastructure and the study of extreme weather [25–28]. In this regard, the combination of the 3B42-V7 and L-moments method is a preferable way to explore extreme precipitation characteristics.

This study will take China as an example to explore the potential of 3B42-V7 in estimating precipitation extremes by using the L-moments-based regional frequency analysis method. The main objectives are to: (1) reveal extreme precipitation under different return periods using the 3B42-V7 data and the L-moments method together with the Fuzzy C-Means algorithm (FCM, a clustering algorithm), and (2) compare 3B42-V7 with the China Gauge-based Daily Precipitation Analysis (CGDPA) product to evaluate the performance of 3B42-V7 on extreme precipitation estimates. The study can provide a reference for the application of 3B42-V7 in extreme precipitation characterization and estimation, which potentially provides an alternative but effective way for estimating extreme precipitation particularly for regions with poor station networks or lack of long-record, consistent observed data.

2. Study Area and Data

2.1. Study Area

As shown in Figure 1, China is selected as the study area ($73.375°E \sim 135.125°E$, $18.125°N \sim 49.875°N$). On one hand, China has relatively complete precipitation data, which can provide reference data for the accuracy evaluation of 3B42-V7. On the other hand, China has a variety of terrains (plateaus, mountains, canyons, plains, hills, etc.) and multiple climate zones (The boreal, the temperate, the warm temperate, the subtropical, the tropical, and the highland climate) [29,30]. Elevation and rainfall intensity are the two main factors affecting the accuracy of the 3B42-V7 product [31,32].

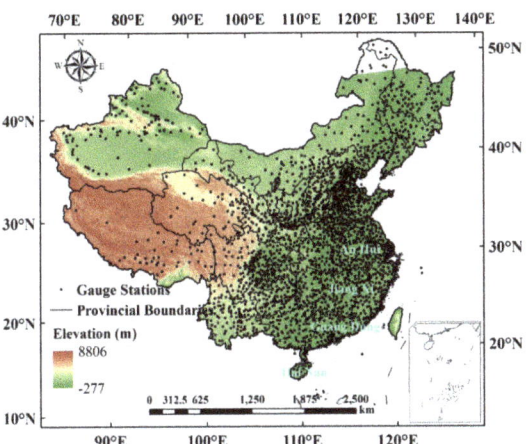

Figure 1. The geography survey of the study area.

2.2. Data

3B42-V7 multi-satellite precipitation product can provide precipitation data covering 50°N~50°S with a spatial resolution of 0.25° × 0.25°. It is the post-process product of TMPA and is calibrated by the monthly meteorological data from the Global Precipitation Climatology Centre. The calibration enhanced the accuracy of 3B42-V7 significantly comparing with the near-real-time product (3B42RT) [11,33]. Compared with the previous algorithm, the seventh version (V7) of the algorithm is considered to provide higher quality precipitation data and has better hydrological utility [34,35]. The dataset is available for download from https://pmm.nasa.gov/data-access/downloads/trmm.

In this study, CGDPA is used as a reference product. The raw precipitation data of CGDPA were collected from 2419 meteorological stations in mainland China and interpolated into raster data with a resolution of 0.25° × 0.25° by the National Meteorological Information Center using the climatology-based Optimal Interpolation method. The stations used in CGDPA are not used in 3B42-V7 and therefore CGDPA is regarded to be independent of 3B42-V7 [36]. According to the study, the results obtained based on this interpolation method can better reflect the influence of terrain on precipitation [37]. According to Shen and Xiong [38], CGDPA products have high accuracy and can capture heavy rainfall events. Currently, this dataset is widely used in the accuracy assessment of satellite precipitation products [36,39,40]. It can be downloaded from http://data.cma.cn.

The daily precipitation data of both CGDPA and 3B42-V7 were selected from 1st January 1998 to 31st December 2017. From these data, we further extracted the annual maximum consecutive 1-day, 3-day, and 5-day precipitation (RX1DAY, RX3DAY, and RX5DAY, respectively) as the extreme precipitation indices. On one hand, these three indices can reflect the characteristics of extremes. On the other hand, these are the concerns of designers when applying such as designing infrastructure, strength designing and checking. In addition, the 90 m resolution elevation data are used in this study, which comes from the Shuttle Radar Topography Mission (SRTM) [41] that is in Geotiff format and can be extracted by means of ArcGIS software.

3. Methodology

3.1. Fuzzy C-Means Algorithm

Given that the spatial distribution pattern of extreme precipitation is not solely related to terrain or climate, it is not advisable to use traditional methods such as basin, climate or administrative boundaries to divide homogeneous regions. Therefore, five factors are considered, including latitude, longitude, elevation, and average annual rainfall, to divide the homogeneous region in this study by using the FCM method. The FCM is a fuzzy clustering algorithm derived from the K-means method [42,43]. At present, several studies have effectively applied it to regional frequency analysis [44,45]. This method allows one data point to belong to multiple clusters, and each data point has a corresponding membership degree to each cluster. The sum of all memberships of a data point is 1. According to the principle of "the smallest square of the weighted error in the same cluster", each cluster center is iterated and adjusted until the center is not changed. Finally, according to the value of the membership degree, which clusters the data point belongs to is determined. The above principle can be expressed by the following formula

$$J_m = \sum_{i=1}^{N} \sum_{j=1}^{M} u_{ij}^m \|x_i - c_i^2\| \tag{1}$$

$$u_{ij} = \frac{1}{\sum_{k=1}^{M} \left(\frac{\|x_i - C_j\|}{\|x_i - C_k\|} \right)^{\frac{2}{m-1}}} \tag{2}$$

$$c_j = \frac{\sum_{i=1}^{N} u_{ij}^m \cdot x_i}{\sum_{i=1}^{N} u_{ij}^m} \tag{3}$$

where N is the group number of measured data, M is the sum of clusters, m is any real number greater than 1, u_{ij} is the membership degree of x_i in the cluster j, x_i is the ith n-dimensional measured data, c_j is the n-dimension center of cluster j, $||*||$ is any norm, and k represents the iteration steps.

3.2. L-Moments-Based Region Frequency Analysis

Describing the characteristics of precipitation can be carried out by using the frequency distribution curve, and a curve is described by several statistical parameters. The L-moments is a method for estimating the parameters of the frequency distribution curve [18,46,47]. Compared with the conventional methods, the L-moments method has small estimation bias, good unbiasedness and robustness [48].

Ordering a n independent samples of variable X, which are arranged in ascending to obtain {$X_{1:n}, X_{2:n}, \ldots, X_{n:n}$}, and the subscript i and n represent the ith minimum number in the sample of length n. The r-order L-moment (λ_r) is defined as follows:

$$\lambda_r = \frac{1}{r} \sum_{i=0}^{r-1} (-1)^i \binom{r-1}{i} E(X_{r-i:r}) \tag{4}$$

$$E(X_{r:n}) = \frac{n!}{(r-1)!(n-r)!} \int_0^1 X[F(X)]^{r-1}[1 - F(X)]^{n-r} dF(X) \tag{5}$$

To better describe the statistical characteristics of the distribution curve, Hosking proposed L-Moment ratios are used defined as follows:

$$\tau_2 = \lambda_2 / \lambda_1 \tag{6}$$

$$\tau_r = \lambda_r / \lambda_2, \ r = 3, 4 \ldots \tag{7}$$

where τ_2 is the L-coefficients of variation (L-CV) reflecting the scale characteristics, τ_3 is the L-skewness of the reflecting skewness characteristics, and τ_4 is the L-kurtosis reflecting the kurtosis characteristics.

To perform L-moments-based regional frequency analysis, several steps are required, including region division with the same precipitation characteristics, checking the discordancy of data from the same region, region homogeneity test, and selection of appropriate distributions and estimation of precipitation quantile. Among them, the division of regions can be initially obtained by the FCM algorithm. If the homogeneity test is not passed, the corresponding region needs to be adjusted or subdivided.

In order to prevent outliers in the region that are obviously wrong or that differ greatly from other sites, it is necessary to check the data discordancy. It is generally measured in D_i and is defined as follows:

$$D_i = \frac{N}{3}(u_i - \overline{u})^T A^{-1}(u_i - \overline{u}) \tag{8}$$

$$\overline{u} = \frac{\sum_{i=1}^{N} u_i}{N} \tag{9}$$

$$A = \sum_{i=1}^{N}(u_i - \overline{u})(u_i - \overline{u})^T \tag{10}$$

$$u_i = \left[\tau_2^{(i)}, \tau_3^{(i)}, \tau_4^{(i)}\right]^T \tag{11}$$

where N represents the total number of sites in the same region, T represents the transpose of a matrix, and $\tau_2^{(i)}, \tau_3^{(i)}, \tau_4^{(i)}$ i, respectively. When the number of sites in the region is greater than 15, Hosking and Wallis suggest treating $D_i > 3$ as discordant [18].

In order to ensure that sites in the same region have the same precipitation frequency distribution curve theoretically, it is necessary to use H for homogeneity testing. The formula is as follows:

$$H = \frac{V - \mu_v}{\sigma_v} \tag{12}$$

$$V = \frac{\sum_{i=1}^{N} n_i \sqrt{\left(\tau_3^{(i)} - \tau_3^R\right)^2 + \left(\tau_4^{(i)} - \tau_4^R\right)^2}}{\sum_{i=1}^{N} n_i} \tag{13}$$

$$\tau_r^R = \frac{\sum_{i=1}^{N} n_i \tau_r^{(i)}}{\sum_{i=1}^{N} n_i}, \quad r = 3, 4, \tag{14}$$

where n_i is the length of the historical precipitation data from the site i; μ_v and σ_v are the mean and standard deviation of the V values calculated from 1000 Monte Carlo simulations, respectively. A region can be regarded as "acceptably homogeneous" if $H < 1$, "possibly heterogeneous" if $1 < H < 2$, and "definitely heterogeneous" if $H > 2$.

Six alternative distributions were selected for this study: Generalized Extreme Value (GEV), Generalized Logistic (GLO), Generalized Normal (GNO), Generalized Pareto (GPA), Pearson type III (PE3), and Wakeby (WAK). Using a goodness-of-fit measurement (Z) to judge the feasibility of the hypothesized distribution:

$$Z = \frac{\tau_4^{Dist} - \tau_4^R + \beta_4}{\sigma_4} \tag{15}$$

where τ_4^{Dist} is the L-kurtosis of the candidate distribution function; β_4 and σ_4 are the deviation and standard deviation of the regional average L-kurtosis (computed from 1000 Monte Carlo simulations and measured samples), respectively. When $|Z| \leq 1.64$, it indicates that the hypothesized distribution has a 90% confidence level, and the closer $|Z|$ is to 0, the hypothesized distribution is more suitable. When $|Z| > 1.64$, it recommends selecting the more robust WAK distribution [18].

The precipitation extremes under different return periods can be calculated by the following formula:

$$Q_{Tij} = q_{Ti}\overline{x}_{ij} \tag{16}$$

where $\overline{x}_{ij}$ is the average of the samples from site j in region i; q_{Ti} is the regional growth curve, the value of which depends on the distribution function selected for region i and the return period T.

3.3. Evaluation Metrics

In order to quantitatively describe the difference between the precipitation extremes estimating by different precipitation data, the correlation coefficient (R), root mean square error (RMSE) and relative error (BIAS) are used (Li et al., 2020c):

$$R = \frac{\sum_{i=1}^{n}(X_i - \overline{X})(Y_i - \overline{Y})}{\sqrt{\sum_{i=1}^{n}(X_i - \overline{X})^2}\sqrt{\sum_{i=1}^{n}(Y_i - \overline{Y})^2}} \quad (17)$$

$$\text{RMSE} = \sqrt{\frac{1}{n}\sum_{i=1}^{n}(Y_i - X_i)^2} \quad (18)$$

$$\text{BIAS} = \frac{Y - X}{X} \quad (19)$$

where X is the reference sequence and Y is the sequence to be evaluated. The precipitation extremes estimation results of 3B42-V7 and CGDPA are organized according to the same extreme precipitation index and return period. R and RMSE are calculated using the organized sequence. Its purpose is to reflect the overall performance of 3B42-V7 (correlation and error with CGDPA results). The spatial distribution of the error is obtained by calculating the BIAS of each grid.

4. Results

4.1. Region Division

Using 3B42-V7 as the precipitation input, China was divided into 60 regions with similar precipitation conditions based on the FCM algorithm. Since there may be a slight error in the clustering result, it is possible that several grids inside the region belong to another region. Therefore, manual inspections should also be carried out to properly adjust the interior and boundaries of each region.

The division results are shown in Figure 2, from which it can be seen that each sub-region is spatially continuous, without fleck or stripe. This somewhat implies that the division is reasonable. Moreover, according to the climate zones over China, it is found that many of the sub-region boundaries are along the boundaries between different climate zones (Figure S1 in the Supplementary Material). A distinctive example can be seen for the Middle Temperate zone in which the boundary coincides with the boundaries of some sub-regions. Therefore, from the climatic viewpoint, the region division conducted by FCM is meaningful and rational. When looking into RX1Day, RX3Day, and RX5Day, it is found that they all showed similar results. The discordancy measurement results show that the proportion of grids that fails the test in the same region is less than 5.44%. It indicates that 3B42-V7 has good data quality assurance, and only a few grids are statistically considered to be "obviously wrong or differ greatly from other sites " in the same region. The proportion in the east is generally low, while that in the west is higher. This may be related to the fact that the terrain in western China is complex and the meteorological station network is sparse. These two are the main factors affecting the quality of 3B42-V7. Complex terrain affects the observation accuracy of satellites, and the sparse meteorological station network implies the lacking of sufficient data for calibration. It should be noted that for the next generation of multi-satellite precipitation products, GPM performs better in complex terrain and is hopeful of providing higher quality precipitation products, but the impact of a sparse station network on product calibration still exists. In any case, from the current results, even under extremely unfavorable conditions, only a very few grids in a region fail the test, which is quite satisfactory. The homogeneity measurement was performed after removing all grids that failed (The proportion of the total grids is less than 3%). The results show that the regions obtained by FCM clustering and adjustment are homogeneous regions, and most of them belong to "acceptably homogeneous". This shows that it is feasible to estimate the precipitation quantile using the same distribution curve in the same region according to the division result. See Table S1 in the Supplementary Material for more details.

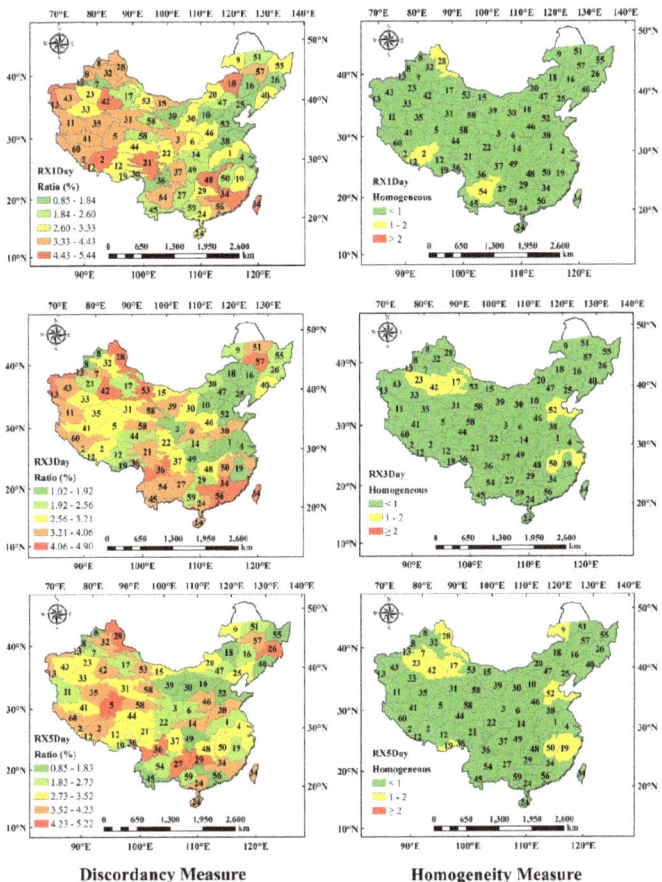

Figure 2. Spatial distribution of discordancy measure and homogeneity measure results based on 3B42-V7.

Table S2 in the Supplementary Material lists the goodness-of-fit measurement results for each region using six alternative distributions, as well as the recommended distribution. The spatial distribution is shown in Figure 3. The results show that the type of selected distribution for each region has a certain spatial continuity. Adjacent regions have a higher probability of selecting the same distribution. Most regions can use a distribution curve with only three parameters (GEV, GLO, GNO, and PE3). GEV and GNO distributions are suitable for most regions in China, followed by PE3, and GPA is not suitable for China. This conclusion is consistent with the results obtained by Wang et al. based on the rain gauge dataset [14]. In RX1Day, GEV is suitable for the southwest, central and northeast, GNO for northwest, and PE3 for the southeast. RX3Day is similar to RX5Day, GNO is more applicable in the north and GEV is more suitable in the south.

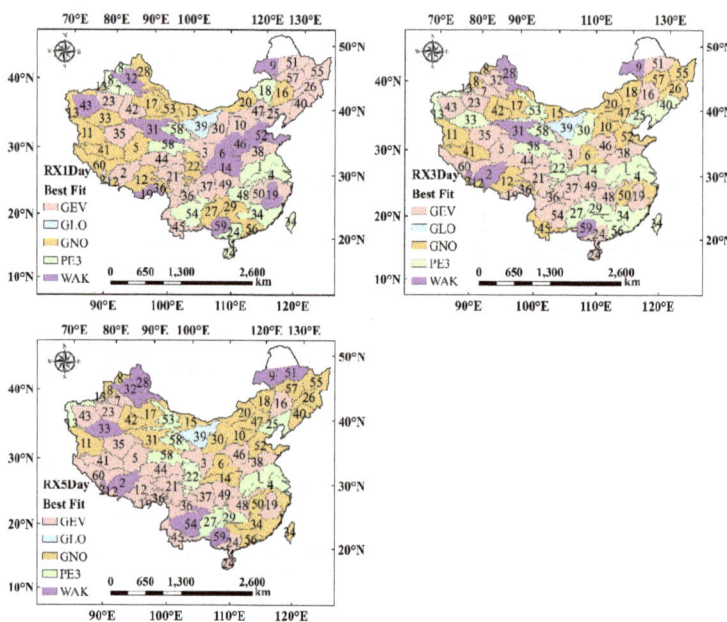

Figure 3. Spatial distribution of best-fit distribution for 60 regions across China.

4.2. Estimation Accuracy

Before comparing extreme precipitation return levels between 3B42-V7 and CGDPA, the steps of region division with FCM, discordancy measurement, homogeneity measurement, and distribution selection on CGDPA are conducted again. This procedure is useful to compare region division results between the two datasets. It is found that the region division results based on CGDPA are similar to those displayed in Figure 2 with a slight difference. The results based on the two products in different return periods (20, 50, 100 years) are shown in Figure 4 (RX1Day), Figure 5 (RX3Day), and Figure 6 (RX5Day). Since CGDPA data have significant errors in western China, we only use the data east of 97.5°E for comparison. No reference data are available west of 97.5°E to compare with the estimation results of 3b42-v7. However, we can still judge whether the results of the western region have reference value by observing the spatial distribution trend and a typical case of the quantile estimation results based on 3B42-V7. The estimation results based on 3B42-V7 show that the precipitation extremes show a decreasing trend from southeast to northwest, which is consistent with the actual spatial distribution of precipitation in China. In the estimation results based on 3B42-V7, there is a region with significantly higher precipitation than the surrounding area in the southwestern part of the Himalayas. It is consistent with the fact that the southwest monsoon from the Indian Ocean is blocked by the Himalayas, and a large amount of water vapor condenses into raindrops here. When the return period becomes longer, the area with less precipitation in the northwest shrinks, while the precipitation in the southeast increases significantly.

Compared with the estimation results based on CGDPA, both of them have a similar spatial distribution pattern of precipitation extremes. In general, when the return period is 20 years, the results of 3B42-V7 are almost the same as those of CGDPA. When using different extreme precipitation indices, 3B42-V7 tends to overestimate the quantile of parts of the southern coast. As the return period becomes longer, there are some differences in the estimation results based on different precipitation inputs. The results of RX1Day show that when the return period is 20 years, 3B42-V7 will overestimate the quantile of parts of the northeast; when the return period is 50 years, the overestimated grids in the northeast is decreased, but the southeast is overestimated; when the return period is 100 years, the results in the northeast are basically the same, and the areas that are mainly overestimated are in the south and southeast. The results of RX3Day and RX5Day indicate that the northeast is not overestimated, and both believe that the southeast has high precipitation extremes. The only divergence is that 3B42-V7 believes that there is a large quantile in the south. In summary, using 3B42-V7 as the precipitation input to estimate precipitation extremes in most regions of China will lead to a similar conclusion with that of using CGDPA, only a few regions are overestimated.

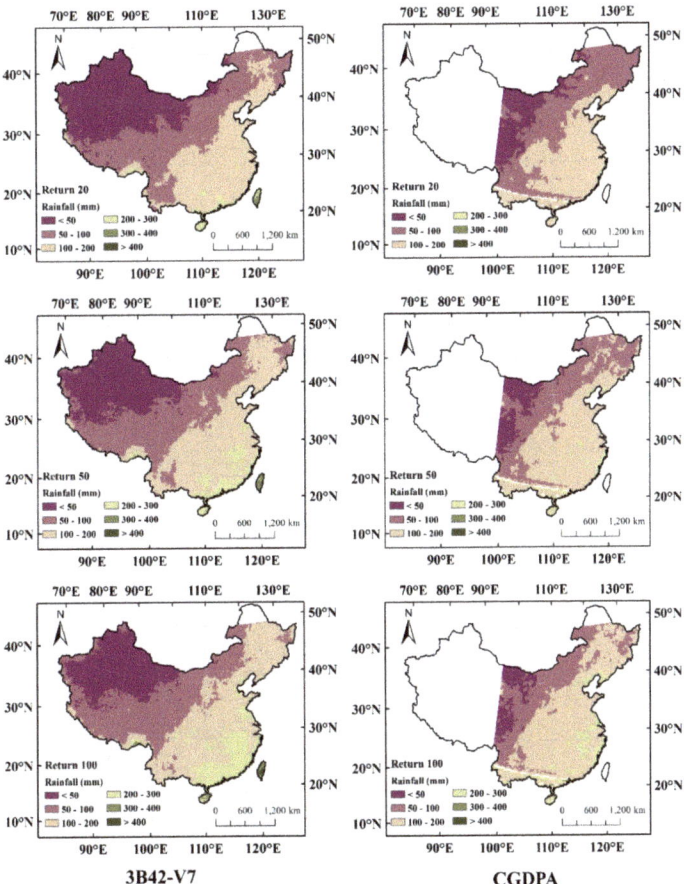

Figure 4. Spatial distribution of 1-day precipitation (RX1DAY) estimated by 3B42-V7 and China Gauge-based Daily Precipitation Analysis (CGDPA) under different return periods.

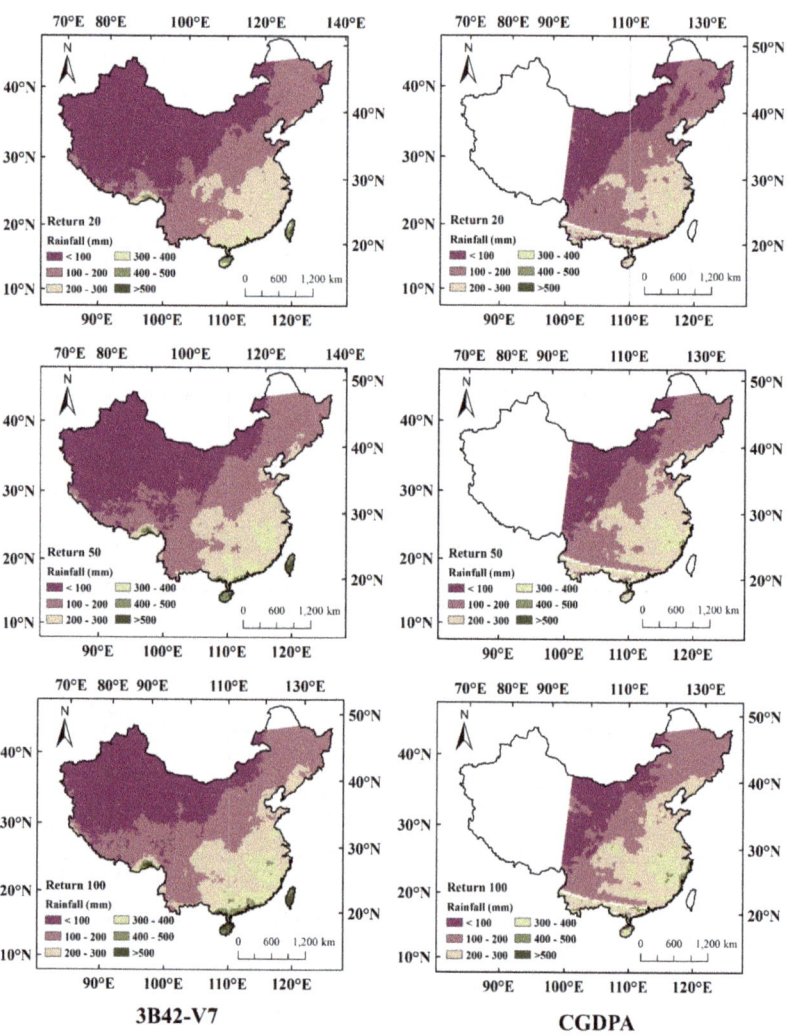

Figure 5. Spatial distribution of RX3DAY estimated by 3B42-V7 and CGDPA under different return periods.

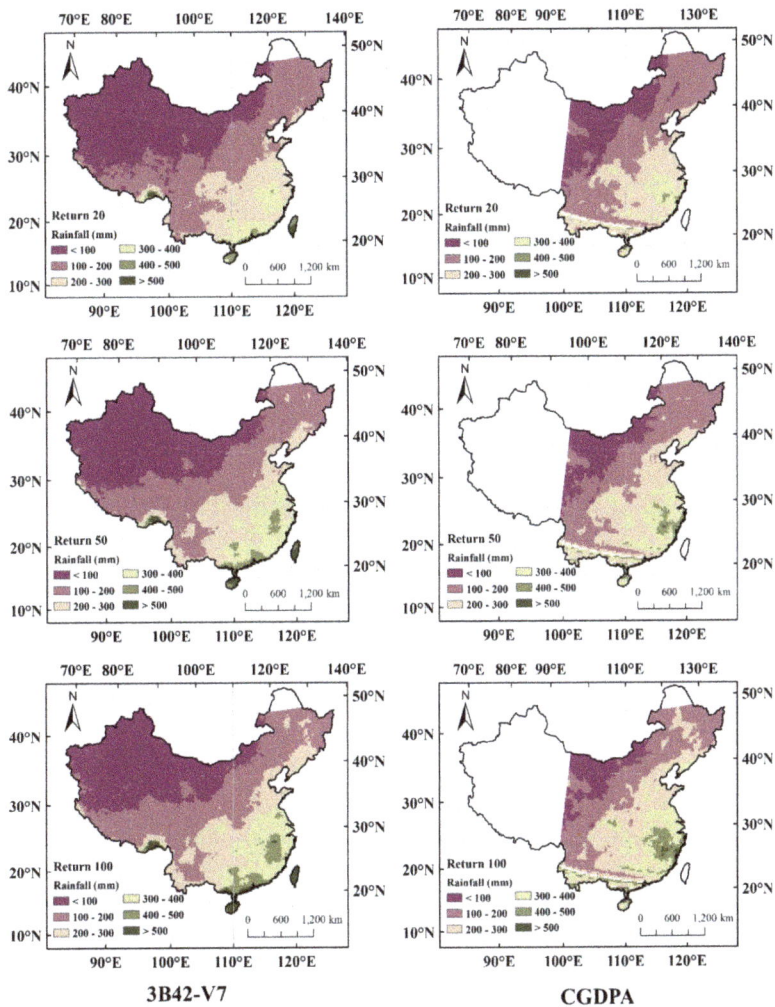

Figure 6. Spatial distribution of RX5DAY estimated by 3B42-V7 and CGDPA under different return periods.

The statistical evaluation results are shown in Table 1. In different return periods, 3B42-V7 and CGDPA estimated RX1Day, RX3Day and RX5Day had high correlations (R > 0.85), of which RX5Day had the strongest correlation. This again shows that the estimation results of 3B42-V7 have a high spatial similarity with that of CGDPA. RMSE measures the deviation between the 3B42-V7 estimate and the CGDPA estimate. As Table 1 shows, RMSE increases slightly with the increase in the return period. It should be noted that RMSE is a dimensioned index, so it is normal to increase with the total rainfall increase. Figure 7 shows the spatial distribution of BIAS. In most areas, the value of BIAS is positive, indicating that the results based on 3B42-V7 tend to overestimate precipitation extremes. The error range of most areas is controlled within ±25%. The results of RX1Day show that precipitation extremes are mainly grossly overestimated in three regions (BIAS >0.5), which are northeast, south, and southwest of China. Among them, the gross overestimation in northeastern China will be alleviated as the return period becomes longer. The results of RX3Day show that the spatial extent of the gross overestimation of extreme precipitation in northeastern and southwestern China is significantly reduced compared with RX1Day. The results of RX5Day indicate that there is only a small portion of the northeastern and southwestern regions that are overestimated (BIAS ranges from 0.25 to 0.5). In summary, using 3B42-V7 to estimate China's precipitation extremes, good results can be achieved in most areas with small errors. When using in southern China, it needs to pay attention to the problem of gross overestimation. When using in the northeast and southwest, it needs to judge the severity of the overestimation according to the selected extreme precipitation index and the return period.

Table 1. Accuracy assessment results of extreme precipitation indicators under different return periods.

Return (year)	RX1DAY		RX3DAY		RX5DAY	
	R	RMSE (mm)	R	RMSE (mm)	R	RMSE (mm)
20	0.86	27.13	0.87	37.96	0.88	41.72
50	0.86	30.73	0.86	45.17	0.87	49.32
100	0.85	33.91	0.85	51.87	0.86	56.54

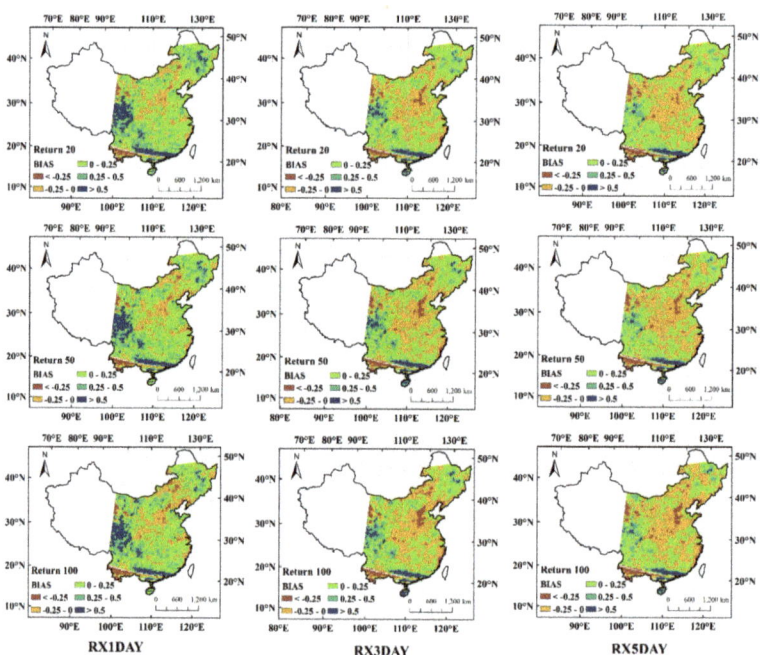

Figure 7. Spatial distribution of relative error (BIAS) under different return periods.

Overall, the estimation of precipitation extremes based on 3B42-V7 can achieve similar results with that based on gauge-based precipitation data. Certainly, it is better to combine, if possible, with gauge-based data to further reduce the error for some regions where 3B42-V7 performs relatively poorly.

5. Discussion

In the precipitation quantile estimation results based on CGDPA, we found some obvious errors in the west. Therefore, only the data east of 97.5°E was used, so as not to affect the final conclusion. This situation may be due to errors in the data recording process, and the sparse rain gauge network makes the impact of a single station larger. This is often difficult to avoid, even if the quality of the data is strictly controlled. For example, for the Historical Climatology Network from the National Climatic Data Center, although its raw data have been checked and preprocessed, 38% of the stations have experienced at least one serious error [49]. In comparison, the advantages of multi-satellite precipitation products are more obvious. On one hand, precision sensors on satellites are less likely to fail than rain gauges. On the other hand, even if one sensor fails, it is possible to minimize the impact by using the data from other sources. This is good news for many developing countries and underdeveloped regions.

In fact, using CGDPA for regional frequency analysis has encountered more problems in practical operations than using 3B42-V7, such as the division of homogeneous regions. In the case of only using longitude, latitude, elevation and annual average rainfall, the FCM algorithm can be used to effectively cluster homogeneous regions based on 3B42-V7 precipitation data. Usually, only a few regional boundaries need to be fine-tuned to pass the homogeneity measurement. However, clustering results based on CGDPA require adjustments to most regions, and some regions need to be subdivided into two regions. Adjustment work is time-consuming and may be an inevitable process if using measured precipitation data. Because regional frequency analysis works on a "regional" scale, and rain gauge station data are "point" scale data, errors will inevitably occur when interpolation. The effects of these errors continue in subsequent clustering (due to the use of annual average rainfall) and homogeneity measurement (extracting RX1Day, RX3Day, and RX5Day from the data). Therefore, it is easy to see that the clustering result does not pass the homogeneity test. Considering the convenience of operation, it is recommended to use 3B42-V7 for regional frequency analysis.

The results of this study were also compared with the results of Wang et al. [14]. Among them, the spatial distribution pattern of precipitation extremes is consistent, the precipitation is basically at the same level, and no abnormal regions are observed. In addition, since the TRMM satellite has only accumulated nearly 20 years of data from the launch, the error is inevitable when using the 3B42-V7 for quantile estimation. However, the dataset is indeed important for areas that lack data, and given the current results, it tends to give an overestimated result, which is not a bad thing to ensure the security of infrastructure design. Additionally, one may consider combining 3B42-V7 with gauge-based precipitation data. In summary, it is possible to use 3B42-V7 providing rainstorm design data for the data-deficient regions.

Extreme precipitation estimation based on 3B42-V7 provides the extreme precipitation spatial distribution under different return periods, which is an important reference when the governments or stakeholders make flood defenses and adaptations. In particular, as our results show, the southeast coastal areas have higher return levels of extreme precipitation, suggesting potential higher flood risk than inland. Additionally, in the southwestern part of the Himalayas (around 25–30°N, 95–100°E), the estimation results based on 3B42-V7 point to potential high flood risk. Therefore, local agencies should pay more attention and make more preparedness regarding flood-related disasters such as flash floods, landslides, and debris flows.

6. Conclusions

In this study, the 3B42-V7 precipitation product was used in combination with the L-moments-based regional frequency analysis to estimate extreme precipitation in China, and the accuracy of the estimation based on 3B42-V7 was evaluated. The main conclusions are summarized below:

The data quality of 3B42-V7 meets the requirements of the L-moments-based regional frequency analysis method, and continuously, China can be divided into 60 homogeneous regions based on the FCM algorithm. For most regions, the GEV and GNO distributions are preferable, followed by PE3 and GLO. In terms of RX1DAY fitting, GEV is suitable for southwest, central and northeast China, while GNO and PE3 are preferable for northwest

China and southeast China, respectively. For RX3Day and RX5Day, GNO and GEV are more applicable over north China and south China, respectively.

The estimation results of 3B42-V7 have a high correlation (R > 0.85) with those of the CGDPA results, with similar spatial distribution patterns of precipitation extremes, and the BIAS of 3B42-V7 is ~25% for most regions of China. In addition, 3B42-V7 tends to overestimate in south China. Overall, however, using the L-moment-based regional frequency analysis method and 3B42-V7, the estimation of extreme precipitation over China is accurate, indicating that the 3B42-V7 product is a reliable way to achieve extreme precipitation estimates.

Supplementary Materials: The following are available online at https://www.mdpi.com/2072-4292/13/2/209/s1.

Author Contributions: Writing—original Draft Preparation, J.C.; Methodology, Z.W.; Supervision, X.W.; Investigation, C.L.; Data Curation, X.C. All authors have read and agreed to the published version of the manuscript.

Funding: This research was funded by the Guangdong Basic and Applied Basic Research Foundation (2019A1515111144), the National Natural Science Foundation of China (51879107, 51709117), the China Postdoctoral Science Foundation (2019M662919), and the Water Resource Science and Technology Innovation Program of Guangdong Province (2020-29).

Acknowledgments: The authors wish to express their gratitude to all authors of the numerous technical reports used for this paper.

Conflicts of Interest: The authors declare no conflict of interest.

References

1. Prakash, S.; Mitra, A.K.; Pai, D.S.; AghaKouchak, A. From TRMM to GPM: How well can heavy rainfall be detected from space? *Adv. Water. Resour.* **2016**, *88*, 1–7. [CrossRef]
2. Pfahl, S.; O'Gorman, A.P.; Fischer, M.E. Understanding the regional pattern of projected future changes in extreme precipitation. *Nat. Clim. Chang.* **2017**, *7*, 423–427. [CrossRef]
3. Swain, L.D.; Langenbrunner, B.; Neelin, D.J.; Hall, A. Increasing precipitation volatility in twenty-first-century California. *Nat. Clim. Chang.* **2018**, *8*, 427–433. [CrossRef]
4. Wu, X.; Guo, S.; Yin, J.; Yang, G.; Zhong, Y.; Liu, D. On the event-based extreme precipitation across China: Time distribution patterns, trends, and return levels. *J. Hydrol.* **2018**, *562*, 305–317. [CrossRef]
5. Hong, Y.; Hsu, K.L.; Sorooshian, S.; Gao, X.G. Improved representation of diurnal variability of rainfall retrieved from the Tropical Rainfall Measurement Mission Microwave Imager adjusted Precipitation Estimation From Remotely Sensed Information Using Artificial Neural Networks (PERSIANN) system. *J. Geophys. Res. Atmos.* **2005**, *110*. [CrossRef]
6. Huffman, G.J.; Adler, R.F.; Bolvin, D.T.; Gu, G.J.; Nelkin, E.J.; Bowman, K.P.; Hong, Y.; Stocker, E.F.; Wolff, D.B. The TRMM multisatellite precipitation analysis (TMPA): Quasi-global, multiyear, combined-sensor precipitation estimates at fine scales. *J. Hydrometeorol.* **2007**, *8*, 38–55. [CrossRef]
7. Joyce, R.J.; Janowiak, J.E.; Arkin, P.A.; Xie, P.P. CMORPH: A method that produces global precipitation estimates from passive microwave and infrared data at high spatial and temporal resolution. *J. Hydrometeorol.* **2004**, *5*, 487–503. [CrossRef]
8. Huffman, G.J.; Adler, R.F.; Bolvin, D.T.; Nelkin, E.J. The TRMM Multi-Satellite Precipitation Analysis (TMPA). In *Satellite Rainfall Applications for Surface Hydrology*; Gebremichael, M., Hossain, F., Eds.; Springer: Dordrecht, The Netherlands, 2010.
9. Kummerow, C.; Barnes, W.; Kozu, T.; Shiue, J.; Simpson, J. The Tropical Rainfall Measuring Mission (TRMM) sensor package. *J. Atmos. Ocean. Tech.* **1998**, *15*, 809–817. [CrossRef]
10. Liu, J.Z.; Duan, Z.; Jiang, J.C.; Zhu, A.X. Evaluation of Three Satellite Precipitation Products TRMM 3B42, CMORPH, and PERSIANN over a Subtropical Watershed in China. *Adv. Meteorol.* **2015**, *2015*, 2731–2738. [CrossRef]
11. Prakash, S.; Mitra, A.K.; Rajagopal, E.N.; Pai, D.S. Assessment of TRMM-based TMPA-3B42 and GSMaP precipitation products over India for the peak southwest monsoon season. *Int. J. Climatol.* **2016**, *36*, 1614–1631. [CrossRef]
12. Worqlul, A.W.; Maathuis, B.; Adem, A.A.; Demissie, S.S.; Langan, S.; Steenhuis, T.S. Comparison of rainfall estimations by TRMM 3B42, MPEG and CFSR with ground-observed data for the Lake Tana basin in Ethiopia. *Hydrol. Earth Syst. Sc.* **2014**, *18*, 4871–4881. [CrossRef]
13. Jung, M.; Reichstein, M.; Ciais, P.; Seneviratne, S.I.; Sheffield, J.; Goulden, M.L.; Bonan, G.; Cescatti, A.; Chen, J.Q.; de Jeu, R.; et al. Recent decline in the global land evapotranspiration trend due to limited moisture supply. *Nature* **2010**, *467*, 951–954. [CrossRef] [PubMed]
14. Wang, Z.L.; Zhong, R.D.; Lai, C.G. Evaluation and hydrologic validation of TMPA satellite precipitation product downstream of the Pearl River Basin, China. *Hydrol. Process.* **2017**, *31*, 4169–4182. [CrossRef]

15. Wang, Z.L.; Chen, J.C.; Lai, C.G.; Zhong, R.D.; Chen, X.H.; Yu, H.J. Hydrologic assessment of the TMPA 3B42-V7 product in a typical alpine and gorge region: The Lancang River basin, China. *Hydrol. Res.* **2018**, *49*, 2002–2015. [CrossRef]
16. Zhong, R.D.; Chen, X.H.; Lai, C.G.; Wang, Z.L.; Lian, Y.Q.; Yu, H.J.; Wu, X.Q. Drought monitoring utility of satellite-based precipitation products across mainland China. *J. Hydrol.* **2019**, *568*, 343–359. [CrossRef]
17. Dalrymple, T. Water Supply Paper. In *Flood-Frequency Analyses, Manual of Hydrology: Part 3*; USGPO: Washington, DC, USA, 1960.
18. Hosking, J.R.M.; Wallis, J.R. *Regional Frequency Analysis: An Approach Based on L-Moments*; Cambridge University Press: Cambridge, UK, 1997.
19. Bharath, R.; Srinivas, V.V. Regionalization of extreme rainfall in India. *Int. J. Climatol.* **2015**, *35*, 1142–1156. [CrossRef]
20. Bhuyan, A.; Borah, M.; Kumar, R. Regional Flood Frequency Analysis of North-Bank of the River Brahmaputra by Using LH-Moments. *Water Resour. Manag.* **2010**, *24*, 1779–1790. [CrossRef]
21. Chen, Y.D.; Zhang, Q.; Xiao, M.Z.; Singh, V.P.; Leung, Y.; Jiang, L.G. Precipitation extremes in the Yangtze River Basin, China: Regional frequency and spatial-temporal patterns. *Theor. Appl. Climatol.* **2014**, *116*, 447–461. [CrossRef]
22. Feng, J.; Yan, D.H.; Li, C.Z.; Gao, Y.; Liu, J. Regional Frequency Analysis of Extreme Precipitation after Drought Events in the Heihe River Basin, Northwest China. *J. Hydrol. Eng.* **2014**, *19*, 1101–1112. [CrossRef]
23. Hussain, Z. Application of the Regional Flood Frequency Analysis to the Upper and Lower Basins of the Indus River, Pakistan. *Water Resour. Manag.* **2011**, *25*, 2797–2822.
24. Nam, W.; Shin, H.; Jung, Y.; Joo, K.; Heo, J.H. Delineation of the climatic rainfall regions of South Korea based on a multivariate analysis and regional rainfall frequency analyses. *Int. J. Climatol.* **2015**, *35*, 777–793. [CrossRef]
25. Smithers, J.C.; Schulze, R.E. A methodology for the estimation of short duration design storms in South Africa using a regional approach based on L-moments. *J. Hydrol.* **2001**, *241*, 42–52. [CrossRef]
26. Wu, X.; Wang, Z.; Zhou, X.; Lai, C.; Lin, W.; Chen, X. Observed changes in precipitation extremes across 11 basins in China during 1961–2013. *Int. J. Climatol.* **2016**, *36*, 2866–2885. [CrossRef]
27. Li, J.; Wang, Z.; Wu, X.; Xu, C.-Y.; Guo, S.; Chen, X. Toward monitoring short-term droughts using a novel daily scale, standardized antecedent precipitation evapotranspiration index. *J. Hydrometeorol.* **2020**, *21*, 891–908. [CrossRef]
28. Li, J.; Wang, Z.; Wu, X.; Chen, J.; Guo, S.; Zhang, Z. A new framework for tracking flash drought events in space and time. *Catena* **2020**, *194*, 104763. [CrossRef]
29. Wang, J.A.; Xiao, H.; Hartmann, R.; Yue, Y. Physical Geography of China and the U.S. In *A Comparative Geography of China and the U.S.*; Hartmann, R., Wang, J.A., Ye, T., Eds.; Springer: Dordrecht, The Netherlands, 2014; pp. 23–81.
30. Wu, X.; Wang, Z.; Guo, S.; Liao, W.; Zeng, Z.; Chen, X. Scenario-based projections of future urban inundation within a coupled hydrodynamic model framework: A case study in Dongguan City, China. *J. Hydrol.* **2017**, *547*, 428–442. [CrossRef]
31. Salio, P.; Hobouchian, M.P.; Skabar, Y.G.; Vila, D. Evaluation of high-resolution satellite precipitation estimates over southern South America using a dense rain gauge network. *Atmos. Res.* **2015**, *163*, 146–161. [CrossRef]
32. Yang, Y.; Cheng, G.; Fan, J.; Sun, J.; Weipeng, L.I. Representativeness and reliability of satellite rainfall dataset in alpine and gorge region. *Adv. Water Sci.* **2013**, *24*, 24–33.
33. Shen, Y.; Xiong, A.Y.; Wang, Y.; Xie, P.P. Performance of high-resolution satellite precipitation products over China. *J. Geophys. Res. Atmos* **2010**, *115*. [CrossRef]
34. Prakash, S.; Mitra, A.K.; Momin, I.M.; Pai, D.S.; Rajagopal, E.N.; Basu, S. Comparison of TMPA-3B42 Versions 6 and 7 Precipitation Products with Gauge-Based Data over India for the Southwest Monsoon Period. *J. Hydrometeorol.* **2015**, *16*, 346–362. [CrossRef]
35. Xue, X.W.; Hong, Y.; Limaye, A.S.; Gourley, J.J.; Huffman, G.J.; Khan, S.I.; Dorji, C.; Chen, S. Statistical and hydrological evaluation of TRMM-based Multi-satellite Precipitation Analysis over the Wangchu Basin of Bhutan: Are the latest satellite precipitation products 3B42V7 ready for use in ungauged basins? *J. Hydrol.* **2013**, *499*, 91–99. [CrossRef]
36. Li, C.M.; Tang, G.Q.; Hong, Y. Cross-evaluation of ground-based, multi-satellite and reanalysis precipitation products: Applicability of the Triple Collocation method across Mainland China. *J. Hydrol.* **2018**, *562*, 71–83. [CrossRef]
37. Xie, P.P.; Yatagai, A.; Chen, M.Y.; Hayasaka, T.; Fukushima, Y.; Liu, C.M.; Yang, S. A Gauge-based analysis of daily precipitation over East Asia. *J. Hydrometeorol.* **2007**, *8*, 607–626. [CrossRef]
38. Shen, Y.; Xiong, A.Y. Validation and comparison of a new gauge-based precipitation analysis over mainland China. *Int. J. Climatol.* **2016**, *36*, 252–265. [CrossRef]
39. Ma, J.; Sun, W.W.; Yang, G.; Zhang, D.F. Hydrological Analysis Using Satellite Remote Sensing Big Data and CREST Model. *IEEE Access* **2018**, *6*, 9006–9016. [CrossRef]
40. Sun, R.C.; Yuan, H.L.; Liu, X.L.; Jiang, X.M. Evaluation of the latest satellite-gauge precipitation products and their hydrologic applications over the Huaihe River basin. *J. Hydrol.* **2016**, *536*, 302–319. [CrossRef]
41. Jarvis, A.; Reuter, H.I.; Nelson, A.; Guevara, E. *Hole-Filled Seamless SRTM Data V4*; Technical Report No.; International Centre for Tropical Agriculture (CIAT): Bogota, Colombia, 2008; Available online: https://srtm.csi.cgiar.org (accessed on 28 November 2020).
42. Bezdek, J.C. *Pattern Recognition with Fuzzy Objective Function Algorithms*; Springer: Boston, MA, USA, 1981; p. 256.
43. Dunn, J.C. A Fuzzy Relative of the ISODATA Process and Its Use in Detecting Compact Well-Separated Clusters. *J. Cybern.* **1973**, *3*, 32–57. [CrossRef]
44. Dikbas, F.; Firat, M.; Koc, A.C.; Gungor, M. Classification of precipitation series using fuzzy cluster method. *Int. J. Climatol.* **2012**, *32*, 1596–1603. [CrossRef]

45. Shu, C.; Burn, D.H. Homogeneous pooling group delineation for flood frequency analysis using a fuzzy expert system with genetic enhancement. *J. Hydrol.* **2004**, *291*, 132–149. [CrossRef]
46. Guttman, N.B. The Use of L-Moments in the Determination of Regional Precipitation Climates. *J. Clim.* **1993**, *6*, 2309–2325. [CrossRef]
47. Hosking, J.R.M.; Wallis, J.R. Some statistics useful in regional frequency analysis. *Water Resour. Res.* **1993**, *29*, 271–281. [CrossRef]
48. Gubareva, T.S.; Gartsman, B.I. Estimating distribution parameters of extreme hydrometeorological characteristics by L-moments method. *Water Resour.* **2010**, *37*, 437–445. [CrossRef]
49. Wallis, J.R.; Lettenmaier, D.P.; Wood, E.F. A daily hydroclimatological data set for the continental United States. *Water Resour. Res.* **1991**, *27*, 1657–1663. [CrossRef]

Article

Spatiotemporal Analysis of Hydrological Variations and Their Impacts on Vegetation in Semiarid Areas from Multiple Satellite Data

Yonghua Zhu [1,2,3], Pingping Luo [2,3,*], Sheng Zhang [4] and Biao Sun [4]

1. College of Civil Engineering and Architecture, Yan'an University, Yan'an 716000, China; zhuyonghua@yau.edu.cn
2. Key Laboratory of Subsurface Hydrology and Ecological Effects in Arid Region, Ministry of Education, Chang'an University, Xi'an 710054, China
3. School of Water and Environment, Chang'an University, Xi'an 710054, China
4. Water Conservancy and Civil Engineering College, Inner Mongolia Agricultural University, Hohhot 001018, China; Yh_Z@emails.imau.edu (S.Z.); sunbiao@imau.edu.cn (B.S.)
* Correspondence: lpp@chd.edu.cn

Received: 7 November 2020; Accepted: 16 December 2020; Published: 20 December 2020

Abstract: Understanding the spatiotemporal characteristics of hydrological components and their impacts on vegetation are critical for comprehending hydrological, climatological, and ecological processes under environmental change and solving future water management challenges. Innovative methods need to be developed in semiarid areas to analyze the special hydrological factors in the water resource systems of these areas. Gravity Recovery and Climate Experiment (GRACE) and Global Land Data Assimilation System (GLDAS) were applied with the normalized difference vegetation index (NDVI) data in this paper to analyze spatiotemporal changes of hydrological factors in the Xiliaohe River Basin (XRB). The results showed that precipitation (P), evapotranspiration (ET) and temperature (T) had similar seasonal change patterns at rates of 0.05 cm/yr., 0.01 cm/yr. and −0.05 °C/yr., respectively. Total water storage change (TWSC) was consistent with the change trend of soil moisture change (SMC) and showed a fluctuating trend. Groundwater change (GWC) showed a decreasing trend at a rate of −0.43 cm/yr. P and ET had a greater impact on GLDAS data ($R = 0.634$, $P < 0.05$ and $R = 0.686$, $P < 0.01$, respectively) than on other factors. GWC was more sensitive to changes in T ($R = 0.570$, $P < 0.05$). Furthermore, a lag period of 0 to 1 months was observed for the effects of P and ET on TWSC and GLDAS. NDVI showed an upward trend at a rate of 0.001 yr^{-1} between 2002 and 2014. A spatial distribution of NDVI was heterogeneous in the study area. ET, GLDAS and GWC in growing season limited vegetation growth and were more important than other factors in XRB. The results may contribute to an understanding of the relationships between the hydrological cycle and climate change and provide scientific support for local environmental management.

Keywords: semiarid area; hydrological variations; normalized difference vegetation index; total water storage change; groundwater change

1. Introduction

Approximately 30% of continental land area is characterized as arid and semiarid [1]. Water cycle conditions and vegetation ecosystems are fragile and sensitive in these areas [2], and increasing water demand from all kinds of water users has seriously impacted vegetation ecosystems. Along with climate change, this increase in demand has greatly changed hydrological factors and water balance in semiarid areas [3,4]. The Xiliaohe River Basin (XRB) has experienced dramatic changes in its hydrological cycle and water balance [5–7]. Distribution and growth of vegetation have changed observably in the area, which may be related to the important role of hydrological factors in a vegetation ecological environment [8,9]. It is within this context that hydrological changes and their impact on vegetation are among the key issues in semiarid areas, where water resources are scarce and the ecological environment is fragile.

XRB is an agropastoral ecotone in a semiarid area that has experienced significant climate change [10], with an average annual temperature (T) increase of 0.5–0.7 °C. At the same time, the groundwater level in the study area decreased notably from 2 m to 6 m over the past 30 years due to unrestrained development and water resource use [11]. Water cycle at the regional scale has been directly affected by regional climate change and groundwater overexploitation, which had impacts in local vegetation change and distribution [12]. Precipitation (P) at 46 meteorological stations decreased from 1960–2012 in Inner Mongolia [13], and P in XRB showed a similar decreasing trend, which led to decreased discharge from the four inbound rivers and a significant decrease in the groundwater level from 1951 to 2007 [14]. Meanwhile, evapotranspiration (ET) (determined by the Penman-Monteith method) varied significantly in both time and space in the area [15,16]. Relationships between hydrological factors were analyzed based on the results at a few observation points [17,18] and thus could not well represent the spatial heterogeneity because of the limited observation points. Some conventional hydrological and climatic indicators have also been used to analyze the hydrological variations and their impacts on the vegetation [13,19]. However, these results could not reveal the balance of regional water resources and its impact on vegetation change, especially on a different scale. Moreover, due to the limited number of observations at the regional scale and in remote areas, data on certain hydrological and meteorological factors may not be available, such as the change in the total water storage change (TWSC), soil moisture change (SMC), groundwater storage change (GWC), etc. Thus, determining the hydrological, climatological and ecological processes may be difficult. Now, this is possible using satellite techniques for monitoring land meteorological and hydrological characteristics [11,20]. Since Gravity Recovery and Climate Experiment (GRACE) satellite launch in March 2002, it has provided a unique way to monitor changes in the earth's gravitational field, especially terrestrial water reserve changes at a regional scale [21]. At present, many achievements have been harvested in related fields, such as hydrological characteristics of TWSC, which were estimated in many regional basins, e.g., China [22], Tarim River basin [23] and so on. Moreover, GWC could also be detected on different spatial scales [24,25]. Zhong et al. [11] found that the GWS showed a prolonged declining rate of −17.8 ± 0.1 mm/yr. during 1971–2015 in the North China Plain, based on in situ groundwater-level measurements and satellite observations. Han et al. [26] discussed the GWC dynamic at multi-timescales in Yunnan Province and the correlations with extreme meteorological factors. Lv et al. [27] found that human factors were the main influencing factors of regional hydrological characteristics, through analyzing the quantitative attribution of terrestrial water storage (TWS) variation from hydroclimatic and anthropogenic factors. In addition, the accuracy of GWC retrieved from the GRACE satellite data in a semiarid area was verified by a comparison with in-situ data [11]. In these analyses, regional hydrological characteristics dynamics and their correlation with meteorological factors are analyzed, thus ignoring the lag time between them. In contrast, more attention to the correlations between regional hydrological characteristics and meteorological factors at multiple-time scales were paid in this paper. Furthermore, in order to fully reveal the impacts of hydrological variations on vegetation in semiarid areas, their evolution characteristics at multiple-time scales and spatial scale were examined.

In summary, spatiotemporal changes of hydrological and meteorological factors, such as P, ET, T, TWSC, SMC and GWC, etc., especially their impacts on vegetation in XRB, have rarely been comprehensively discussed. This paper analyzed regional water balance and vegetation factors based on data from multiple satellite observations. Comprehensive correlations among climatological, hydrological, and vegetation factors in XRB were simultaneously analyzed using time-series data from 2002 to 2014 at the regional scale. The main objectives of this study are (1) to analyze spatiotemporal dynamic of hydrological factors and normalized difference vegetation index (NDVI) in agropastoral ecotone of semiarid region based on multiple satellite data, (2) analyze response relationships among regional hydrometeorological factors at multiple-time scales, and (3) evaluate the impacts of hydrological variations on vegetation.

2. Materials and Methods

2.1. Study Area

Xiliaohe River Basin (XRB), Inner Mongolia Autonomous Region, northeast part of China, lies between the latitude 42°30′ to 45°00′ N and longitude 120°00′ to 123°30′ E (Figure 1), which has an area of 3.2×10^4 km^2 and an average elevation of 800 m (400~1300 m). Three main rivers once flowed through the study area: Xiliao River, Jiaolai River, and Xinkai River. However, the rivers' discharge has been reduced and may even dry up either seasonally or perennially [28]. Moreover, the increased water demand from irrigation in recent decades has led groundwater to become the main total terrestrial water storage source supplied to meet agricultural, industrial, and domestic water demand. The overexploitation and utilization of groundwater has caused various environmental problems [29], e.g., regional groundwater table and pollution, land subsidence, and ecological environment deterioration.

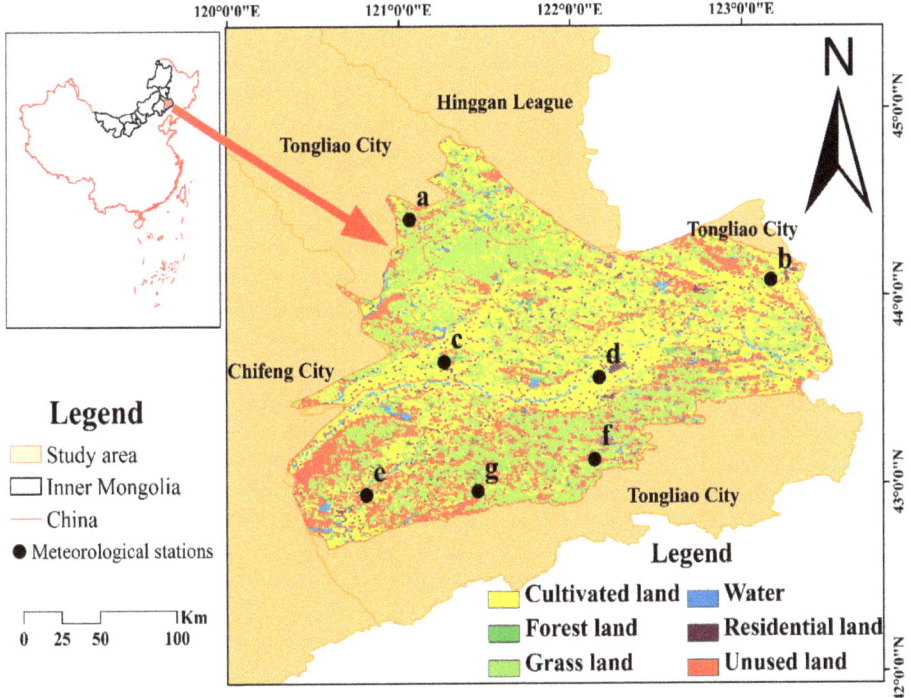

Figure 1. Location of the study area (Land-Use and Land-Cover Change data from 2015 are from Data Center for Resources and Environmental Sciences, Chinese Academy of Sciences (http://www.resdc.cn)).

In XRB, annual precipitation ranged from 350 mm to 450 mm, the annual average temperature was 6.0 °C, and the pan evaporation (Φ 20 cm) was 1817 mm [30]. Approximately 80% of the total precipitation occurs in summer (June to September). As a typical ecotone between Farming and Animal Husbandry in semi-arid area, crops, trees and grassland constitute the main vegetation types [31].

In order to better understand land distribution and vegetation types, the field survey in July of 2015 and August of 2016, were launched.

2.2. Data

2.2.1. Meteorological Data

There are seven meteorological stations in XRB (Figure 1). Monthly meteorological data, e.g., temperature (T) and precipitation(P) are provided by the National Climate Center of China (http://ncc.cma.gov.cn) for time series analysis. As most of these stations were built in the late 1950s, only monthly data from 2002 to 2014 were used in this study to ensure that the lengths of the data from multiple satellite data (GRACE) were consistent. Detailed meteorological stations' information is shown in Table 1.

Table 1. Meteorological stations' information in XRB.

Meteorological Stations (2002–2014)	Code	Location (Lat & Lon)		Data					
				P (cm)			T (°C)		
		X	Y	Max	Min	Avg	Max	Min	Avg
Zhalute	a	120.54	44.34	55.0	22.1	35.8	8.9	6.4	7.5
Kezuozhongqi	b	123.17	44.08	55.6	18.7	34.3	8.0	6.4	7.4
Kailu	c	121.17	43.36	49.6	21.3	31.9	8.5	6.0	7.3
Tongliao	d	122.16	43.36	45.2	22.7	31.0	8.6	6.4	7.5
Naiman	e	120.39	42.51	39.1	21.3	30.1	9.0	6.6	7.5
Kezuohouqi	f	122.21	42..58	51.5	21.7	41.8	8.6	6.4	7.3
Kulun	g	121.45	42.44	56.2	29.3	34.7	9.0	6.6	7.5

China's annual average temperature and annual precipitation spatial interpolation data set (1980–2015) was based on daily observation data of more than 2400 meteorological stations across the country, and generated through sorting, calculation and spatial interpolation processing, which is provided by Data Center of Resources and Environmental Sciences, Chinese Academy of Sciences (http://www.resdc.cn) for analyzing spatial distribution and variation characteristics. The annual average temperature and precipitation units were 0.1 degrees Celsius and 0.1 mm, respectively. Interpolation of climatic factors such as temperature and precipitation use the Australian ANUSPLIN interpolation software. ANUSPLIN is a tool that uses smoothing spline functions to analyze and interpolate multivariate data, e.g., a method of approximating a curved surface using a function [32]. Mask extraction and resampling analysis of spatial interpolation data sets from 2002 to 2014 were carried out by ArcGIS software, and then to extract annual precipitation and temperature average values of the study area for analysis of spatial distribution and change trends.

2.2.2. Actual Evapotranspiration (ET)

Monthly MOD16A2 ET was used to analyze the change of hydrological factors. ET data from 2002 to 2014 were obtained from NASA/EOS (http://www.ntsg.umt.edu/) have a spatial resolution of 1.0 km^2. The improved ET model [33] was applied rather than the Penman-Monteith equation to validate ET. The model has been widely used to calibrate the water cycle factors and their interactions with environmental change [33,34], although differences were observed between measurements and MOD16 data.

2.2.3. Terrestrial Water Storage

As a key variable in hydrological cycle, terrestrial water storage value was obtained from the GRACE satellite and designed mainly to observe gravity field changes with time [35]. Monthly terrestrial water storage data (Unit: cm) from 2002 to 2014 was produced by the Jet Propulsion Laboratory (JPL) with a spatial resolution of 0.5 degrees (https://grace.jpl.nasa.gov).

JPL data were the results computed from GRACE Level-1 data by mascon method [36]. In the process of calculation, the parameters required in the model, e.g., C20 term, geocentric first-order term value and post-ice rebound correction coefficien were calculated by relevant methods [37] and models [38,39]. Furthermore, based on constraints, the area of the world is divided into 4551 spherical caps with equal areas for calculation to reduce the measurement error. In order to improve spatial resolution of the data, CLM 4.0 hydrological model and coastline resolution refinement (CRI) filtering method are used to recover the signal of the solution from the mascon model and separate data from the land and sea for producing spatial distribution grid with 0.5° resolution [40]. While previous studies showed that GRACE satellite data still have some shortcomings, such as lower spatial and temporal resolutions, these data are still widely used because they can simulate water resources, including groundwater, for different land types [35].

2.2.4. Auxiliary Data from Global Land Data Assimilation System (GLDAS)

Due to the lack of continuous monitoring of hydrological data, GLDAS-NOAH data was used to analyze hydrological factor data [41]. Soil moisture, snow water equivalent and total canopy water storage data in this paper were selected from NOAH data products in GLDAS (https://giovanni.gsfc.nasa.gov and https://disc.gsfc.nasa.gov/). GLDAS-NOAH data time series was from January 2002 to December 2014, with monthly temporal resolution and 0.25° spatial resolution. In order to be consistent with GRACE satellite data for calculating groundwater change, ArcGIS software was used to resample and transform its spatial scale to 0.5°. The above data units are kg/m^2, and unit for soil moisture data was depth, and it was recorded at depths of 10 cm, 40 cm, 100 cm, and 200 cm, respectively.

2.2.5. Normalized Difference Vegetation index (NDVI)

Third-generation global inventory modeling and Mapping Research (GIMMS ndvi3g) NDVI data set from NASA Goddard Space Center (https://ecocast.arc.nasa.gov/data/pub/gimms/3g.v1/), which was used in this paper. The changes in vegetation were modified by NDVI data, which had spatial resolution of 0.083° in 15-day intervals from 2002 to 2014. The changing trend of vegetation was analyzed through the seasonal data (winter: Dec, Jan-Feb; spring: Mar-May; summer: Jun-Aug; autumn: Sep-Nov; and growing season: May-Oct).

2.3. Methods

2.3.1. Determination of Terrestrial Water Storage Change (TWSC)

Data of GRACE satellite's monthly gravity model reflects the components related to the earth's static structure, which is the difference between the cell's monthly water storage and the multi-year average of the cell's water storage [42,43]. Thus, the TWSC value was determined through JPL data for a total of 153 months from April 2002 to December 2014 and subtract the average in the period between 2002 and 2010. The numerical value was used to reflect the TWSC, and the positive and negative signs are used to reflect the direction of change, representing the accumulation or loss of TWSC, respectively. However, during the commissioning and operation phase of the GRACE satellite, there were problems such as sensor performance degradation and insufficient energy supply, which resulted in poor quality of the observation data of the GRACE satellite during these two periods [44,45]. The data in thirteen months were not available during the study period, June 2002, July 2002, June 2003, January 2011, June 2011, May 2012, October 2012, March 2013, August 2013, September 2013, February 2014, July 2014 and December 2014. In this paper, in order to maintain the average seasonal cycle well, interpolation, which was the average of the values for each cell from the months either side of the missing data, was used to fill in missing data [46].

2.3.2. Estimation of Groundwater Change (GWC)

Spatiotemporal changes of the GWC were obtained from the GRACE (TWSC) and GLDAS data among, monthly data represented by the monthly average values [35]. The GWC had a significant correlation coefficient with the in situ observed groundwater changes, which can be used to characterize regional groundwater conditions [45]. The water balance equation was expressed as follows:

$$GWC = TWSC - GLDAS_{(SMC+SWEC+TCWSC)} \tag{1}$$

where GWC is groundwater storage change, TWSC is terrestrial water storage change, SMC is soil moisture change, SWEC is snow water equivalent change, TCWSC is total canopy water storage change, and GLDAS$_{(SM+SWE+TCWS)}$ is the sum of SMC, SWEC and TCWSC. The above data units are cm.

2.3.3. Analysis of NDVI

Maximum-Value Composite

Changes in vegetation was analyzed through the method of maximum-value composite [47]. Based on the pixel-by-pixel data of the NDVI image from January to December each year, maximum value of a pixel was determined by the calculation, and the MVC image was then generated.

Analysis of Spatial Trend

The spatial trend of NDVI was analyzed through the unitary linear regression method, in which time and effect factors were independent and dependent variables, respectively. Slope of the straight line from linear regression was applied to illustrate the spatial trend of NDVI [48].

$$slope = \frac{n \times \sum_{i=1}^{n} i \times a_i - \left(\sum_{i=1}^{n} i\right)\left(\sum_{i=1}^{n} a_i\right)}{n \times \sum_{i=1}^{n} i^2 - \left(\sum_{i=1}^{n} i\right)^2} \tag{2}$$

where slope is the linear tendency index, a_i is the annual NDVI in each grid, $n = 13$ is the number of years, and i is the year, e.g., 2003 was the 1st year, 2004 was 2nd year, etc. Value of slope > 0 represents increasing trend; and value of slope < 0 represents decreasing trend.

Analysis of Hurst Index

The hurst index is an effective method for quantitatively representing the long-range correlation of time series, and it has been widely used in hydrology, economics, climatology, geology, and geochemistry [49,50]. Its basic principle is to define the mean sequence for a time sequence $\{NDVI(t)\}, t = 1, 2, \ldots, n$:

$$\overline{NDVI_\tau} = \frac{1}{\tau} \sum_{t=1}^{\tau} NDVI_\tau \quad \tau = 1, 2, \cdots, n \tag{3}$$

$$X_{(t,\tau)} = \sum_{t=1}^{\tau} \left(NDVI_t - \overline{NDVI_\tau}\right) \quad 1 \leq t \leq \tau \tag{4}$$

$$R_\tau = \max_{1 \leq t \leq \tau} X_{(t,\tau)} - \min_{1 \leq t \leq \tau} X_{(t,\tau)} \quad \tau = 1, 2, \cdots, n \tag{5}$$

$$S_\tau = \left[\frac{1}{\tau} \sum_{t=1}^{\tau} \left(NDVI_t - \overline{NDVI_\tau}\right)^2\right]^{\frac{1}{2}} \quad \tau = 1, 2, \cdots, n \tag{6}$$

where τ is the number of elements, t is the time step the year, $\overline{NDVI_\tau}$ is the time series of NDVI, $X(t, \tau)$ is the cumulative deviation, $R\tau$ is the extreme deviation sequence, and $S\tau$ is the standard deviation.

Taking the ratio of $R(\tau)$ and $S(\tau)$, we arrive at the following:

$$\log\left(\frac{R}{S}\right)_n = H \times \log(n) \tag{7}$$

The H value determines whether the NDVI sequence is completely random or persistent. There are three indications according to the value of H index. Value of $0.5 < H < 1$ indicates that NDVI time series is a continuous sequence, i.e., the change in the future would maintain the same trend with the past change trend, and the closer the H is to 1, the stronger the persistence. Value of $H = 0.5$ indicates that time series is a random sequence and there would not be long-term correlation. Value of $0 < H < 0.5$ indicates that future change trend would be opposite to the past change trend, and the closer the H is to 0, the stronger the anti-persistence.

Analysis of Trend Test

The F test was used for the trend significance test. Significance test only represented the confidence level of the changing trend, regardless of change speed. Statistic calculation formula is as follows:

$$F = U \times \frac{n-2}{Q} \tag{8}$$

where U is the error sum of the squares, Q is the regression sum of squares, and n is the number of years. According to the test results, the trend was divided into five levels, i.e., extremely significant decrease (slope < 0, $P < 0.01$), significant decrease (slope < 0, $0.01 < P < 0.05$), non-significant change ($P > 0.05$), significant increase (slope > 0, $0.01 < P < 0.05$), and extremely significant increase (slope > 0, $P < 0.01$).

3. Results

3.1. Changes of Hydrological Factors over Time

The observations in Figure 2 showed that hydrological factors changed annually. In general, total annual precipitation (P) and evapotranspiration (ET) increased by 0.05 cm/yr. and 0.01 cm/yr., respectively, on average from 2002 to 2014. P showed a slightly greater rate of increase after 2008 than before. During 2002 and 2014, mean annual temperature (T) decreased on average by −0.05 °C/yr. Terrestrial water storage change (TWSC) increased from 2002 to 2005 and decreased significantly from 2006 to the beginning of 2012, which led to a value that was approximately 6 cm less than the mean of the whole obtained TWSC series (Figure 2b). TWSC increased again after 2012 and reached a value equivalent to the average of the whole series' mean. Monthly soil moisture change (SMC), snow water equivalent change (SWEC), total canopy water storage change (TCWSC) obtained from GLDAS, fluctuated from 2002 to 2014 (Figure 2c). SMC in the study area had obvious seasonal variations and ranged from −5.24~8.26 cm, and this parameter was sensitive to changes in regional water resources. There were significant differences in the time series of SWEC, which generally reached a peak in December. In the process of freezing and thawing in spring, the value decreased slowly until the end of April. The maximum value of SWEC was 1.36 cm in December 2012. The TCWSC value reached the maximum in July or August in summer and showed an upward trend in the study area, with a range from −0.01 to 0.01 mm. Analysis of TWSC and auxiliary data-Global Land Data Assimilation System (GLDAS = SMC + SWEC + TCWSC) showed that temporal patterns groundwater change (GWC) could be described based on time series analysis (Figure 2d). GWC in the study time showed a clear decreasing trend at rate of −0.43 cm/yr., and especially experienced a significant decrease from 2007 to 2012 at a rate of −0.99 cm/yr. Furthermore, TWSC lagged behind other factors at the time scale, e.g., P, SMC and so on in Figure 2.

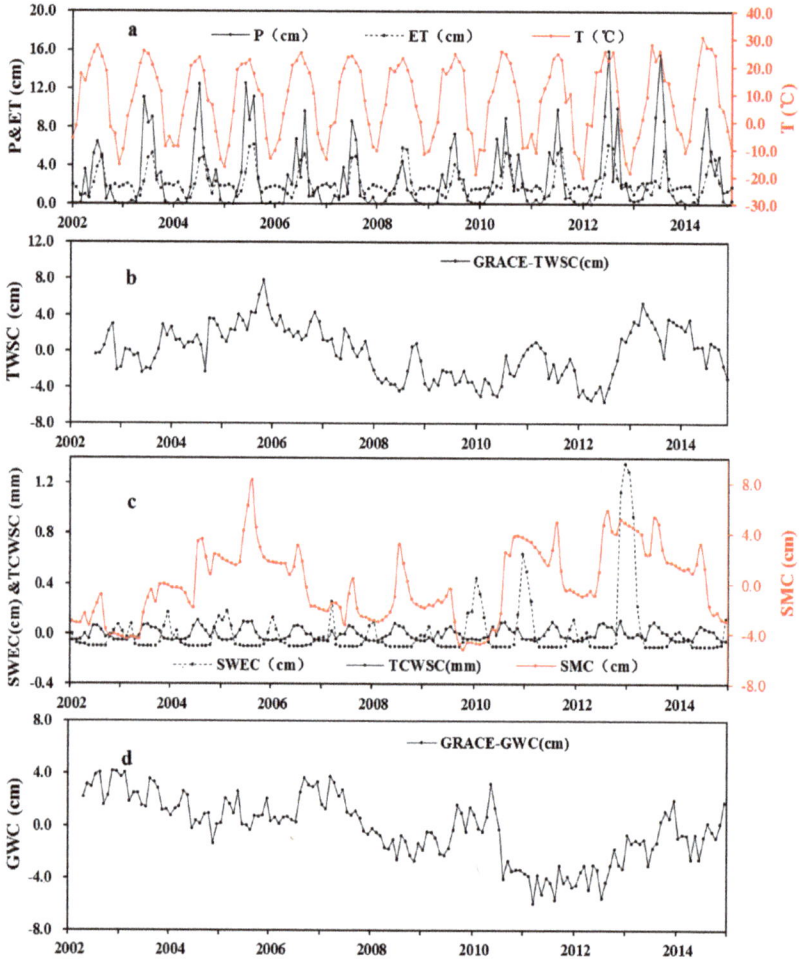

Figure 2. Changes of hydrometeorological factors in XRB from 2002~2014 (**a**) P, T & ET; (**b**) TWSC; (**c**) SWEC, TCWSC & SMC; (**d**) GWC.

In addition to annual changes, hydrological factors changed seasonally as well in Figure 3. P, ET and T values showed similar seasonal change patterns, with a high peak values occurring in summer and a low peak value occurring in winter. TWSC and GLDAS increased from spring to summer and then decreased from autumn to winter. A similar change trend was observed for TWSC and GLDAS with P, T and ET, indicating that there was an interaction between these factors. However, a lag effect was also observed, which required further quantitative analysis. The inter-annual trend of GWC showed a continuous downward trend, gradually decreasing from the maximum value in spring to the minimum value in winter, indicating that the annual groundwater storage gradually decreased, especially in summer where this phenomenon was more obvious.

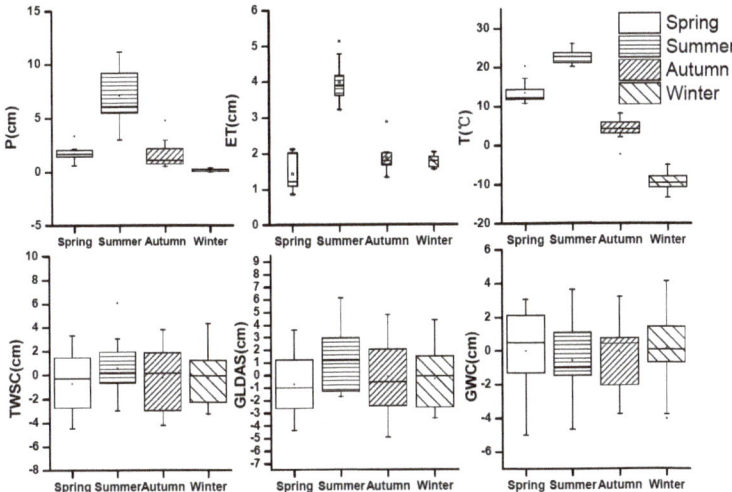

Figure 3. Seasonal P, T, ET, TWSC, GLDAS$_{(SWEC\&SMC\&TCWSC)}$ and GWC from 2002–2014.

3.2. Spatial Distribution of Annual Hydrological Factors

3.2.1. Situ Observation of Hydrological Factors

The in-situ observations of hydrological factors between 2002 and 2014 showed that the mean annual P increased from northwestern to southeastern areas, with an average value of 37.2 cm, while the mean annual T decreased from southwestern to northeastern areas, with an average value of 7.5 °C (Figure 4a,b). This result indicated that hydrological and meteorological factors in the study area have significant spatial heterogeneity, while spatial distributions between P and T were different.

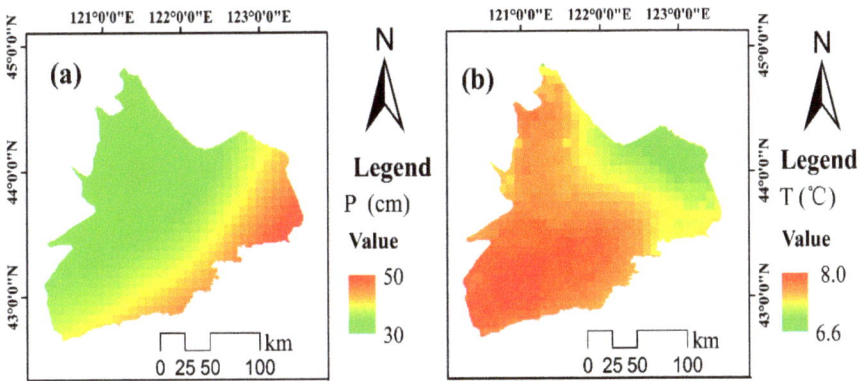

Figure 4. Spatial distribution of P (**a**) and T (**b**) from 2002~2014 (Grid cell size: 1 km).

3.2.2. Hydrological Factors from Satellite Data

TWSC data observed from the satellite in the study area showed a decreasing trend from northeast to southwest on the whole in Figure 5. The northern and central region were in a state of accumulation, while the southern region was in a state of deficit with a relatively obvious decreasing trend (Figure 5a). Based on GLDAS data, SMC, SWEC and TCWS showed equivalent water increases, as shown in Figure 5b. Spatial distribution of GLDAS was similar to that of TWSC, indicating that there was an interaction between these parameters. The GWC value showed a slight surplus state in most area of XRB, which ranged from −1.2 to 1.0 cm (Figure 5c).

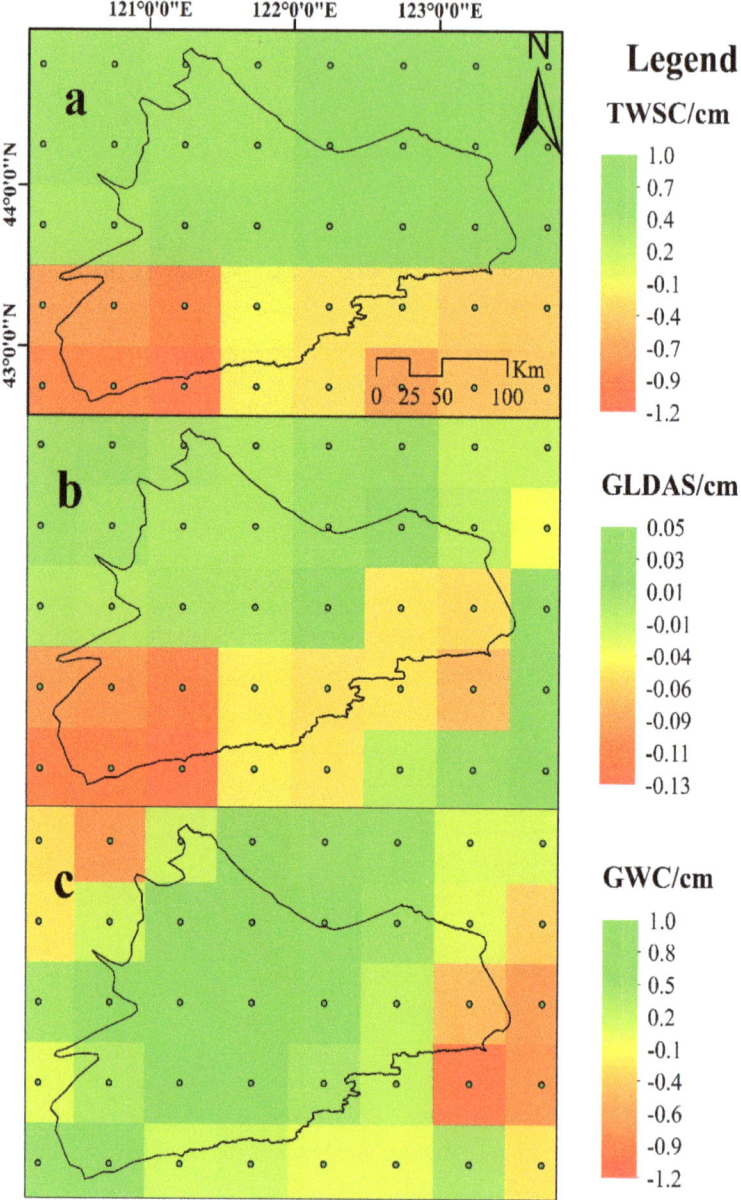

Figure 5. Spatial distribution of TWSC (**a**), GLDAS (**b**) and GWC (**c**) from 2002~2014 (Grid cell size: 0.5°).

3.3. Spatiotemporal Variations of Normalized Difference Vegetation Index (NDVI)

3.3.1. Temporal Variation

Similar to other areas in northern China, NDVI had apparent seasonal variation, and showed an upward trend during 2002 and 2014 in Figure 6. The NDVI increased from May to the most vigorous growing season in July and August, and it then decreased to the minimum level in November and remained almost unchanged until the following April, with a range of the NDVI of 0.16~0.19 (Figure 6a). In addition, NDVI showed a single peak change during the year in Figure 6a as well as hydrological factors in the study area in Figure 2a, meaning that there were relationships between them. Furthermore, the annual value of NDVI increased slightly at a rate of 0.001 yr^{-1} from 2002 to 2014 in Figure 6b. The analysis of the cumulative anomaly method showed that NDVI decreased from 2002 to 2009 and then increased again (Figure 6b). This change trend was also consistent with the change trend of hydrological factors in the study area (Figure 2b,d), indicating that water factors in semiarid areas, especially groundwater, may be the main factors affecting vegetation changes.

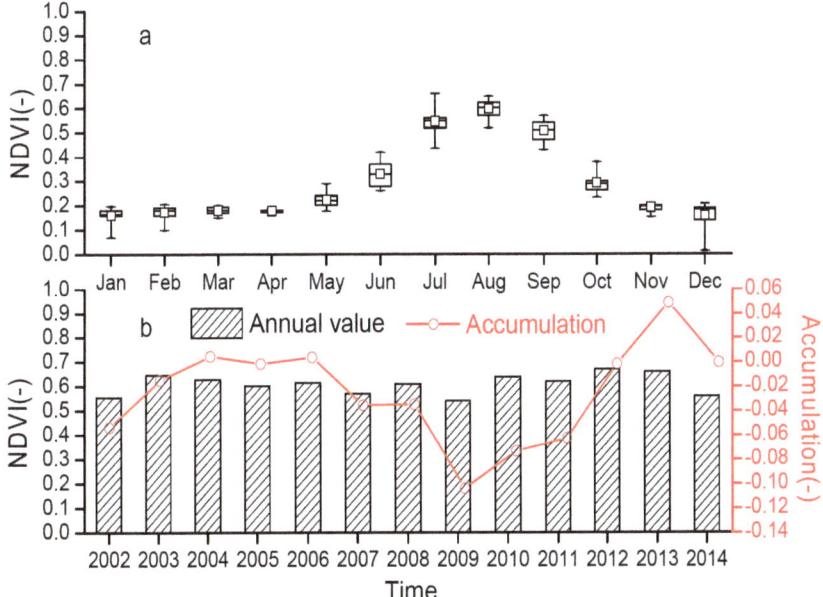

Figure 6. Variations (**a**) and curve of accumulation (**b**) of NDVI with time.

3.3.2. Spatial Variations

Figure 7 showed that the spatial distribution of NDVI was heterogeneous in the area. NDVI value was lower in the southwestern area, with a range of 0.0~0.33, than in the central belt from west to east (Figure 7a). NDVI was greater along the rivers, which ranged from 0.63~1.00, than other parts, which ranged from 0.0~0.63. Furthermore, the annual value of NDVI decreased over 63.2% of the area at a rate of −0.008~0.0 a^{-1}, while 36.8% of the area showed an increase in the NDVI at rate of 0~0.02 a^{-1} (Figure 7b). F test results suggested that significant variance area accounted for 89.3% of the whole area and was mostly distributed in areas with a relatively lower NDVI value (Figure 7c). Significant variance was not observed in cultivated areas along rivers. The Hurst index (0.06~0.97) indicated that in 55.7% of the area, particularly in areas with a decreasing NDVI trend, the variance trend was continuously maintained for a long time while the remaining 44.3% of the area would show fluctuations (Figure 7d).

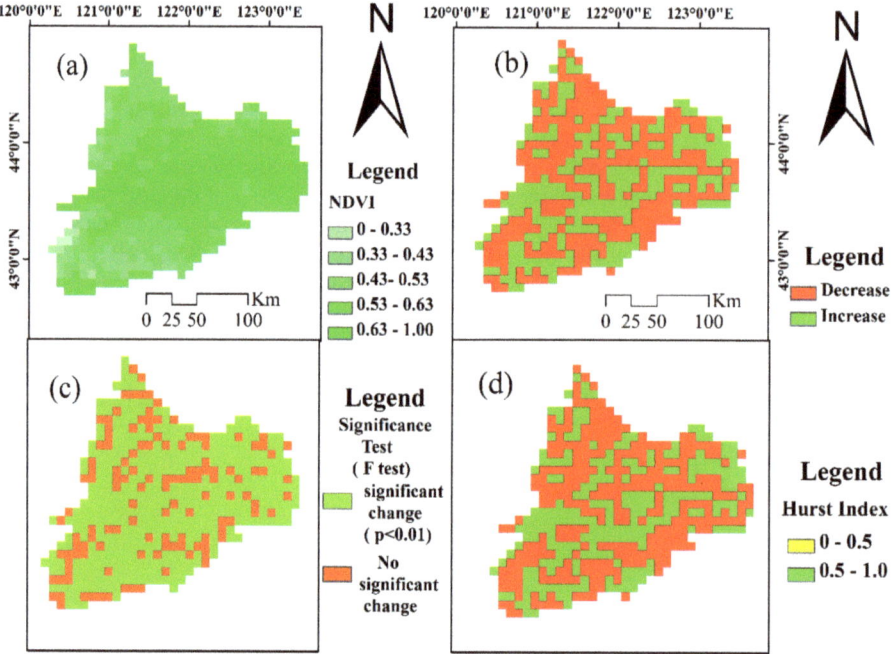

Figure 7. Spatial distribution of annual NDVI value (**a**), annual variation trend (**b**), F test results (**c**), and Hurst index (**d**) from 2002~2014 (Grid cell size: 8 km).

3.4. Relationships between Hydrological Factors

The value of P was highly related to ET, with a correlation coefficient of 0.685 ($P < 0.01$). TWSC had a significant correlation ($R = 0.680$, $P < 0.05$) with GWC, as compared to GLDAS in Table 2, indicating that TWSC mainly consisted of GWC in the area. GLDAS had a significant and positive correlation with P ($R = 0.634$, $P < 0.05$) and ET ($R = 0.686$, $P < 0.01$), meaning that a higher P rate corresponded to more SMC. Significant positive correlation was observed between T and GWC ($R = 0.680$, $P < 0.05$), which indicated that water demand of plants increased as T increased in XRB and groundwater was the main source of water in semiarid area and changed significantly. In summary, there are complex correlations among regional hydrological elements.

Table 2. Correlation between hydrological factors at annual scale.

	P	ET	T	TWSC	GLDAS	GWC
P	1	0.685 **	−0.470	0.472	0.634 *	0.065
ET		1	−0.245	0.387	0.686 **	−0.096
T			1	0.021	−0.481	0.570 *
TWSC				1	0.518	0.680 *
GLDAS					1	−0.193
GWC						1

**. Correlation is significant at the 0.01 level (2-tailed). *. Correlation is significant at the 0.05 level (2-tailed).

P is not the only one of the most important factors underlying the interactions among the hydrological cycle but also the key variable in the water resources in semiarid areas. It accounted for approximately 75% of total water resource recharge in normal years and approximately 57% of recharge in dry years in study area [11]. A lag time of P infiltration is observed based on GLDAS and GWC. Lag time is usually represented in hydrological forecasts based on P in the hydrological equilibrium [51]. Table 3 showed their relationships between TWSC, GLDAS and GWC and P and ET, which were analyzed to determine lag times. The results showed that TWSC and GLDAS had significant correlation ($P < 0.05$) with P at different lag periods, i.e., a one-month time lag occurred ($R = 0.312$, $P < 0.05$). Moreover, these parameters were significantly and positively correlated with the ET in the current month ($R = 0.276$, $P < 0.05$) and one month prior ($R = 0.276$, $P < 0.05$) but not positively correlated ($p > 0.01$) with the ET two to three months later. This finding suggested that the P and ET in the current month and one-month prior could significantly affect the TWSC and GLDAS. Meanwhile, P and ET did not have correlation ($P > 0.05$) with GWC, which might be related to the considerable depth of groundwater table due to overexploitation, long time period for infiltration to recharge groundwater or lack of infiltration recharge of groundwater due to ET. As a result of this, P and ET were increasingly less sensitive to GWC.

Table 3. Correlation between hydrological factors at monthly scale in a different lag period.

	P				ET			
	Lag0-M	Lag1-M	Lag2-M	Lag3-M	Lag0-M	Lag1-M	Lag2-M	Lag3-M
TWSC	0.230 **	0.312 **	0.275 **	0.195 *	0.276 **	0.276 **	0.131	0.025
GLDAS	0.324 **	0.382 **	0.281 **	0.181 *	0.378 **	0.331 **	0.142	0.045
GWC	0.007	0.03	0.067	0.049	−0.012	0.005	0.011	−0.017

**. Correlation is significant at the 0.01 level (2-tailed). *. Correlation is significant at the 0.05 level (2-tailed).

3.5. Relationships between Hydrological Factors and NDVI

To obtain better understanding of relationships between hydrological factors and NDVI, correlation analysis was carried out. As shown in Table 4, the results showed that ET and GLDAS were significantly and positively correlated with NDVI ($R = 0.747$ and 0.704; $P < 0.01$) in the growing season (May-Oct) of vegetation. Correlation coefficient between NDVI and GWC in the growing season was significantly negatively correlated ($R = -0.64$, $P < 0.05$), indicating that groundwater represents an important water source for vegetation growth in the XRB. T and TWSC had no effect on the growth of vegetation in each season, and their correlation coefficients did not pass the significance test. In summary, ET, GLDAS and GWC could be the major hydrological parameters that affect vegetation dynamics in the growing season.

Table 4. Correlation between NDVI and hydrological factors in different seasons.

		P	ET	T	TWSC	GLDAS	GWC
	Spring	−0.075	0.024	−0.226	0.151	0.226	−0.123
	Summer	0.425	0.418	−0.073	−0.162	0.446	−0.357
NDVI	Autumn	−0.140	0.063	0.220	0.370	0.411	0.046
	Winter	−0.585 *	0.407	−0.250	−0.151	−0.484	0.074
	Growing season	0.542	0.747 **	−0.242	0.154	0.704 **	−0.640 *

**. Correlation is significant at the 0.01 level (2-tailed). *. Correlation is significant at the 0.05 level (2-tailed).

4. Discussion

4.1. Dynamics of Hydroecological Elements in Semiarid Area

Water for agricultural irrigation relies heavily on groundwater due to shortages of surface water in study area [47], and reports have indicated that more than 80% of the water use was from groundwater [11]. Spatiotemporal variation and distribution characteristics of the hydrological factors were determined by multiple satellite data, which have been widely used as effective approaches for detecting the hydrological dynamics in ecotones in semiarid areas. Combined with the analysis results of climate change in the study area, spatial distribution of precipitation (P) and temperature (T) was uneven, as shown in Figure 4 and presented a decreasing and increasing trend from the southeast to the northwest, respectively. The results showed that, on the spatial scale, terrestrial water storage change (TWSC) results based on JPL satellite data and auxiliary data-Global Land Data Assimilation System (GLDAS = SMC + SWEC + TCWSC) values were also closely related to P and T. Application of the multi-satellite data would represent an effective approach to monitoring and modeling the dynamics of groundwater change (GWC). Our result confirms that the environment changes have occurred in the semiarid area and that they can be related to dynamics of hydrogeological elements were easily observable and measurable through satellite images. Through correlation analysis, there are complex relationships among regional hydrological elements in Table 2. Furthermore, soil moisture change (SMC) and snow water equivalent change (SWEC) obtained from the GLDAS data played an important role in the hydrological cycle in the area. The SWEC not only affected the surface runoff, groundwater recharge, and SMC in the spring [52–54], but also indicated the snow cover in the winter and the accumulated air temperature in the spring (Figure 2). SMC could even directly reflect the vegetation growth in growing season, and predict the GWC [55,56]. SMC was higher for irrigated land than grassland in the vegetation growth season (Figure 5). In particular, SMC and GWC directly affected the local hydrologic cycle, which could be used to optimize the management and utilization of water resources and improve vegetation growth. However, compared to other areas [57], the response between P and GWC was increasingly insensitive in the study area with a lag period of one month. There are many reasons for this phenomenon, but anthropic factors [58], e.g., irrigation and overexploitation of groundwater, have thickened the vadose zone of soil, and the relationship between P and groundwater, which is is becoming more and more complex.

While the application of the satellite data can significantly improve the level of water resources assessment in this area, there are still some deficiencies, such as time scale (monthly), spatial resolution (0.5 degree) and data accuracy verification, which have a significant impact on the research results of small and medium-sized watersheds, e.g., the correlation strength between them and other environmental molecules is very poor, only the correlation can be considered. Therefore, how to combine the limited water resources monitoring information with new methods, new technologies and new achievements to carry out effective regional water resources management has become a practical problem demand of relevant departments of water resource management.

4.2. Impacts of Hydrological Factors on Vegetation

Hydrological factors played a controlling role in terrestrial ecosystems [59,60]. Field survey in July of 2015 and August of 2016 and normalized difference vegetation index (NDVI) value showed that the northern part of the study was dominated by grassland with medium vegetation coverage, the middle part was dominated by cultivated land with high vegetation coverage, and the southern part was dominated by sandy land with low vegetation coverage. The results in the paper showed that ET, GLDAS and GWC could be the major hydrological parameters that affect the vegetation dynamics in the growing season. The northeastern XRB included grassland with some small rivers with high terrain and vegetation coverage, and it was affected by snow melt water in the spring as well as rainfall. As a result, GLDAS and TWSC showed a cumulative trend with higher values in the northern piedmont plain than in other areas, indicating that the piedmont plain had basically maintained its original ecology and was rarely affected by human activities. Change of GWC in the central part of the study area, which is a flat, wide area of cultivated land, was relatively reduced as shown in Figure 5c. Therefore, irrigation had a significant impact on regional SMC and even affected regional water resource reserves [61], which indicated that the GWC might be affected by human activities, such as the expansion of cultivated land area [62] and overexploitation and utilization of groundwater [63]. TWSC in the southwest of the study area showed a decreasing trend (Figure 5a) [64]. Major land use types in the area consisted of typical steppe, meadow steppe, and cropland in Figure 1, and vegetation growth in these areas was dependent on ET ($R = 0.747$, $P < 0.05$) and GLDAS ($R = 0.704$,

$P < 0.05$) which was mainly consisted of SMC. However, in summer and the vegetation growing season, P and NDVI were positively correlated, although the occurrence of this phenomenon is mainly related to the serious desertification of the local surface and the inability of the soil to store water correlation coefficient failed the significance test (Table 3) which indicated that lower P did not have a significant effect on the vegetation in the semiarid area [65]. NDVI, as an intuitive indicator of vegetation growth, plays a fundamental role in reflecting the characteristics of vegetation growth and distribution. However, NDVI is hardly enough to explain vegetation changes, such as vegetation height, density and so on. Furthermore, grassland will often give a stronger NDVI value than forest but the effect on ET and GWC are very different. In future, more indexes related to water should be used in the study of regional hydroecological processes.

5. Conclusions

Using multiple satellite data, the spatiotemporal changes of hydrological elements in semiarid areas from 2002 to 2014 and their effects on vegetation were analyzed in this paper. The main conclusions are as follows.

(1) Analysis of spatiotemporal characteristics of hydrological elements showed that the hydrological process in the study area has changed significantly. Annual variations of precipitation (P), evapotranspiration (ET) and temperature (T) were all in the form of a single peak, and the interannual variation law was slightly different, with rates of 0.05 cm/yr., 0.01 cm/yr. and −0.05°C/yr., respectively. Terrestrial water storage change (TWSC) showed a fluctuating trend, with initial increase, then decrease, and finally an increase, which was consistent with the change trend of soil moisture change (SMC). Groundwater change (GWC) showed a decreasing trend at a rate of −0.17 cm/yr.

(2) Complex correlations occurred among regional hydrological elements. P and ET were significantly correlated ($R = 0.685$, $P < 0.01$) and had a greater impact on the GLDAS ($R = 0.634$, $P < 0.05$ and $R = 0.686$, $P < 0.01$) than on the TWSC and GWC. GWC is an important component of the TWSC in the region ($R = 0.680$, $P < 0.05$), and it was more sensitive to the T response ($R = 0.570$, $P < 0.05$). Furthermore, P would lead to greater TWSC and GLDAS values when P preceded the TWSC by one month, whereas smaller changes would be observed when P preceded these parameters by two months. The time lag of the GLDAS that was influenced by P was more obvious than that of the TWSC. The TWSC and GLDAS were significantly and positively correlated with the ET in the current month and one month prior and not positively correlated with the ET two to three months later. Due to overexploitation, P and ET did not have any effect on the GWC.

(3) Normalized difference vegetation index (NDVI) had obvious seasonal variations and showed an upward trend at a rate of 0.001 yr^{-1} during 2002 and 2014. Spatial distribution of NDVI was heterogeneous in study area. NDVI decreased by 63.2% of the area at rate of $-0.008 \sim 0.0$ yr^{-1}. The area showing significant variance accounted for 89.3% of the whole area, and 55.7% of the area would maintain the variance trend continuously for a long time, with these areas mainly showing decreasing NDVI change trends. However, other 44.3% area would show fluctuations. Hydrological factors play a controlling role on terrestrial ecosystems. In growing season, ET, GLDAS and GWC were the parameters that limited vegetation growth, and they were more important than other factors in XRB.

Application of satellite data could significantly improve the water assessment capability in semiarid areas and could be used for regional water resource and eco-environment management in semiarid areas. Hydrological factors, such as TWSC, SM and GWC, spatiotemporal dynamic and their correlations were successfully determined and analyzed in this study. Impacts of hydrological change on NDVI were also identified based on the analysis. These results will help to understand regional hydroecological processes, and also provide a scientific basis for local environmental management.

Author Contributions: Methodology, S.Z.; software, B.S.; data curation, Y.Z.; writing—original draft preparation, Y.Z.; writing—review and editing, P.L. All authors have read and agreed to the published version of the manuscript.

Funding: This research was funded by National Key R&D Program of China (2018YFE0103800), Ministry of Water Resources Public Welfare Special Scientific Research Project—Semi-arid Zone Water Cycle and Water Ecological Security Key Technology Research (number 201501031), National Natural Science Foundation of China (51779118, 51669021, 51869020), Fundamental Research Funds for the Central Universities, CHD (300102299302, 300102299102, 300102299104), One Hundred Talent Plan of Shaanxi Province, International Collaborative Research of Disaster Prevention Research Institute of Kyoto University(2019W-02), Excellent Projects for Science and Technology Activities of Overseas Staff in Shaanxi Province (2018038), Yan'an University Project (YDBK2017-19, YDBK2019-35, YDBK2019-36), and Yan'an Science and Technology Bureau (project 2018KS-02).

Acknowledgments: The authors would like to thank the Editors and the anonymous reviewers for their crucial comments, which improved the quality of this paper.

Conflicts of Interest: The authors declare no conflict of interest.

References

1. Huang, J.; Zhang, W.; Zuo, J.; Fu, C.; Chou, J.; Feng, G.; Yuan, J.; Zhang, L.; Wang, S.; Zuo, H.; et al. Development of the semi-arid climate and environment research observatory over Loess Plateau. *Adv. Atmos. Sci.* **2008**, *25*, 906–921. [CrossRef]
2. Nielsen, U.N.; Ball, B.A. Impacts of altered precipitation regimes on soil communities and biogeochemistry in arid and semi-arid ecosystems. *Glob. Chang. Biol.* **2015**, *21*, 1407–1421. [CrossRef] [PubMed]
3. Wu, Y.; Liu, T.X.; Paredes, P.; Duan, L.M.; Wang, H.Y.; Wang, T.S.; Pereira, L.S. Ecohydrology of groundwater-dependent grasslands of the semi-arid Horqin sandy land of inner Mongolia focusing on evapotranspiration partition. *Ecohydrology* **2016**, *9*, 1052–1067. [CrossRef]
4. Villalobos-Vega, R.; Salazar, A.; Miralles-Wilhelm, F.; Haridasan, M.; Franco, A.C.; Goldstein, G.; Wesche, K. Do groundwater dynamics drive spatial patterns of tree density and diversity in Neotropical savannas? *J. Veg. Sci.* **2014**, *25*, 1465–1473. [CrossRef]
5. Dai, X.Q.; Shi, H.B.; Li, Y.S.; Ouyang, Z.; Huo, Z.L. Artificial neural network models for estimating regional reference evapotranspiration based on climate factors. *Hydrol. Process.* **2009**, *23*, 442–450. [CrossRef]
6. He, D.; Liu, Y.; Pan, Z.; An, P.; Wang, L.; Dong, Z.; Zhang, J.; Pan, X.; Zhao, P. Climate change and its effect on reference crop evapotranspiration in central and western Inner Mongolia during 1961–2009. *Front. Earth Sci.* **2013**, *7*, 417–428. [CrossRef]
7. Zhang, F.; Zhou, G.S.; Wang, Y.; Yang, F.L.; Nilsson, C. Evapotranspiration and crop coefficient for a temperate desert steppe ecosystem using eddy covariance in Inner Mongolia, China. *Hydrol. Process.* **2012**, *26*, 379–386. [CrossRef]
8. Miao, L.; Jiang, C.; Xue, B.; Liu, Q.; He, B.; Nath, R.; Cui, X. Vegetation dynamics and factor analysis in arid and semi-arid Inner Mongolia. *Environ. Earth Sci.* **2014**, *73*, 2343–2352. [CrossRef]
9. Li, S.G.; He, Z.Y.; Chang, X.L.; Harazono, Y.; Oikawa, T.; Zhao, H.L. Grassland desertification by grazing and the resulting micrometeorological changes in Inner Mongolia. *Agric. For. Meteorol.* **2000**, *102*, 125–137.
10. Zunya, W.; Yihui, D.; Jinhai, H.; Jun, Y. An updating analysis of the climate change in china in recent 50 years. *Acta Meteorol. Sin.* **2004**, *62*, 228–236.
11. Zhong, Y.L.; Zhong, M.; Feng, W.; Zhang, Z.Z.; Shen, Y.C.; Wu, D.C. Groundwater Depletion in the West Liaohe River Basin, China and Its Implications Revealed by GRACE and In Situ Measurements. *Remote Sens.* **2018**, *10*, 493. [CrossRef]
12. Li, B.; Yu, W.; Wang, J. An Analysis of Vegetation Change Trends and Their Causes in Inner Mongolia, China from 1982 to 2006. *Adv. Meteorol.* **2011**, *2011*, 13–30. [CrossRef]
13. Huang, J.; Sun, S.; Xue, Y.; Zhang, J. Changing characteristics of precipitation during 1960–2012 in Inner Mongolia, northern China. *Meteorol. Atmos. Phys.* **2014**, *127*, 257–271. [CrossRef]
14. Yang, H.S.; Liu, J.; Liang, H.Y. Change characteristics of climate and water resources in west Liaohe River Plain. *J. Appl. Ecol.* **2009**, *20*, 84–90. (In Chinese)
15. Gao, Z.; He, J.; Dong, K.; Bian, X.; Li, X. Sensitivity study of reference crop evapotranspiration during growing season in the West Liao River basin, China. *Theor. Appl. Climatol.* **2015**, *124*, 865–881. [CrossRef]
16. Schaffrath, D.; Bernhofer, C. Variability and distribution of spatial evapotranspiration in semiarid Inner Mongolian grasslands from 2002 to 2011. *Springerplus* **2013**, *2*, 547. [CrossRef]

17. Liu, X.P.; He, Y.H.; Zhao, X.Y.; Zhang, T.H.; Li, Y.L.; Yun, J.Y.; Wei, S.L.; Yue, X.F. The response of soil water and deep percolation under Caragana microphylla to rainfall in the Horqin Sand Land, northern China. *Catena* **2016**, *139*, 82–91. [CrossRef]
18. Yao, S.X.; Zhao, C.C.; Zhang, T.H.; Liu, X.P. Response of the soil water content of mobile dunes to precipitation patterns in Inner Mongolia, northern China. *J. Arid. Environ.* **2013**, *97*, 92–98. [CrossRef]
19. Zhu, L.; Gong, H.L.; Dai, Z.X.; Xu, T.B.; Su, X.S. An integrated assessment of the impact of precipitation and groundwater on vegetation growth in arid and semiarid areas. *Environ. Earth Sci.* **2015**, *74*, 5009–5021. [CrossRef]
20. Feng, W.; Zhong, M.; Lemoine, J.M.; Biancale, R.; Hsu, H.T.; Xia, J. Evaluation of groundwater depletion in North China using the Gravity Recovery and Climate Experiment (GRACE) data and ground-based measurements. *Water Resour. Res.* **2013**, *49*, 2110–2118. [CrossRef]
21. Wahr, J.; Molenaar, M.; Bryan, F. Time variability of the Earth's gravity field: Hydrological and oceanic effects and their possible detection using GRACE. *J. Geophys. Res. Solid Earth* **1998**, *103*, 30205–30229. [CrossRef]
22. Chen, Z.; Jiang, W.; Wu, J.; Chen, K.; Deng, Y.; Jia, K.; Mo, X. Detection of the spatial patterns of water storage variation over China in recent 70 years. *Sci. Rep.* **2017**, *7*, 6423. [CrossRef] [PubMed]
23. Yang, P.; Xia, J.; Zhan, C.; Qiao, Y.; Wang, Y. Monitoring the spatio-temporal changes of terrestrial water storage using grace data in the Tarim river basin between 2002 and 2015. *Sci. Total Environ.* **2017**, *595*, 218–228. [CrossRef] [PubMed]
24. Yeh, P.J.-F.; Swenson, S.C.; Famiglietti, J.S.; Rodell, M. Remote sensing of groundwater storage changes in illinois using the gravity recovery and climate experiment (grace). *Water Resour. Res.* **2006**, *42*, 4. [CrossRef]
25. Singh, A.K.; Tripathi, J.N.; Kotlia, B.S.; Singh, K.K.; Kumar, A. Monitoring groundwater fluctuations over india during indian summer monsoon (ism) and northeast monsoon using grace satellite: Impact on agriculture—Sciencedirect. *Quat. Int.* **2019**, *507*, 342–351. [CrossRef]
26. Han, Z.; Huang, S.; Huang, Q.; Leng, G.; Wang, H.; He, L.; Fang, W. Assessing grace-based terrestrial water storage anomalies dynamics at multi-timescales and their correlations with teleconnection factors in yunnan province, china. *J. Hydrol.* **2019**, *574*, 836–850. [CrossRef]
27. Lv, M.; Ma, Z.; Li, M.; Zheng, Z. Quantitative Analysis of Terrestrial Water Storage Changes under the Grain for Green Program in the Yellow River Basin. *J. Geophys. Res. Atmos.* **2019**, *124*, 1336–1351. [CrossRef]
28. Zhang, Q.; Singh, V.P.; Sun, P.; Chen, X.; Zhang, Z.X.; Li, J.F. Precipitation and streamflow changes in China: Changing patterns, causes and implications. *J. Hydrol.* **2011**, *410*, 204–216. [CrossRef]
29. Long, W.H.; Chen, H.H.; Li, Z.; Pan, H.J. Evaluation of the groundwater intrinsic vulnerability in West Liaohe plain, Inner Mongolia, China. Geological Bulletin of China. *Geol. Bull. China* **2010**, *29*, 598–602.
30. Zhu, Y.H.; Zhang, S.; Sun, B.; Yan, L.; Wang, Y. Relationship of dominant herbaceous plant species and groundwater depth in tongliao plain, northwestern china. *Appl. Ecol. Environ. Res.* **2019**, *17*, 15363–15374. [CrossRef]
31. Jia, X.; Lee, H.F.; Zhang, W.C.; Wang, L.; Sun, Y.G.; Zhao, Z.J.; Yi, S.W.; Huang, W.B.; Lu, H.Y. Human-environment interactions within the West Liao River Basin in Northeastern China during the Holocene Optimum. *Quat. Int.* **2016**, *426*, 10–17. [CrossRef]
32. Hutchinson, M.F. Interpolation of rainfall data with thin plate smoothing splines I. two-dimensional smoothing of data with shortrange correlation. *Geogr. Inf. Decis. Anal.* **1998**, *2*, 153–167.
33. Mu, Q.; Heinsch, F.A.; Zhao, M.; Running, S.W. Development of a global evapotranspiration algorithm based on MODIS and global meteorology data. *Remote Sens. Environ.* **2007**, *111*, 519–536. [CrossRef]
34. Mu, Q.; Zhao, M.; Heinsch, F.A.; Liu, M.; Tian, H.; Running, S.W. Evaluating water stress controls on primary production in biogeochemical and remote sensing based models. *J. Geophys. Res. Biogeosci.* **2015**, *112*, 863–866. [CrossRef]
35. Ouyang, W.; Liu, B.; Wu, Y.Y. Satellite-based estimation of watershed groundwater storage dynamics in a freeze-thaw area under intensive agricultural development. *J. Hydrol.* **2016**, *537*, 96–105. [CrossRef]
36. Feixiao, G.; Zhongmiao, S.; Feilong, R.; Yun, X. Comparison and Analysis of Different Mascon Model Results. *J. Geod. Geodyn.* **2019**, *39*, 1022–1026.
37. Cheng, M.; Tapley, B.D. Variations in the Earth's Oblateness during the Past 28 Years. *J. Geophys. Res. Solid Earth* **2004**, *109*. [CrossRef]
38. Swenson, S.; Chambers, D.; Wahr, J. Estimating Geocenter Variations from a Combination of GRACE and Ocean Model Output. *J. Geophys. Res. Solid Earth* **2008**, *113*. [CrossRef]

39. Geruo, A.; Wahr, J.; Zhong, S. Computations of the Viscoelastic Response of a 3-D Compressible Earth to Surface Loading: An Application to Glacial Isostatic Adjustment in Antarctica and Canada. *Geophys. J. Int.* **2013**, *192*, 557–572.
40. Watkins, M.M.; Wiese, D.N.; Yuan, D.-N.; Boening, C.; Landerer, F.W. Improved Methods for Observing Earth's Time Variable Mass Distribution with GRACE Using Spherical Cap Mascons. *J. Geophys. Res. Solid Earth* **2015**, *120*, 2648–2671.
41. Rodell, M.; Chen, J.; Kato, H.; Famiglietti, J.S.; Nigro, J.; Wilson, C.R. Estimating groundwater storage changes in the Mississippi River basin (USA) using GRACE. *Hydrogeol. J.* **2006**, *15*, 159–166. [CrossRef]
42. Save, H.; Bettadpur, S.; Tapley, B.D. High-resolution CSR GRACE RL05 mascons. *J. Geophys. Res. Sol. Earth* **2016**, *121*, 7547–7569. [CrossRef]
43. Holger, S.; Patrick, W.; Hansheng, W. Determination of the Earth's structure in Fennoscandia from GRACE and implications for the optimal post-processing of GRACE data. *Geophys. J. Int.* **2010**, *182*, 1295–1310.
44. Rodell, M.; Velicogna, I.; Famiglietti, J.S. Satellite-based estimates of groundwater depletion in India. *Nature* **2009**, *460*, 999–1002. [CrossRef]
45. Nanteza, J.; De Linage, C.R.; Thomas, B.F.; Famiglietti, J.S. Monitoring groundwater storage changes in complex basement aquifers: An evaluation of the grace satellites over eastafrica. *Water Resour. Res.* **2016**, *52*, 9542–9564.
46. Long, D.; Yang, Y.; Wada, Y.; Hong, Y.; Liang, W.; Chen, Y.; Chen, L. Deriving scaling factors using a global hydrological model to restore. *J. Remote Sens. Environ.* **2015**, *168*, 177–193. [CrossRef]
47. Holben, B.N. Characteristics of maximum-value composite images from temporal AVHRR data. *Int. J. Remote Sens.* **1986**, *7*, 1417–1434. [CrossRef]
48. Jiang, L.; Bao, A.; Guo, H.; Ndayisaba, F. Vegetation dynamics and responses to climate change and human activities in central Asia. *Sci. Total Environ.* **2017**, *599–600*, 967–980. [CrossRef]
49. Hou, W.; Hou, X. Spatial–temporal changes in vegetation coverage in the global coastal zone based on gimms ndvi3g data. *Int. J. Remote Sens.* **2019**, *41*, 1118–1138. [CrossRef]
50. Tong, S.; Zhang, J.; Bao, Y.; Lai, Q.; Lian, X.; Li, N.; Bao, Y. Analyzing vegetation dynamic trend on the Mongolian Plateau based on the Hurst exponent and influencing factors from 1982–2013. *J. Geogr. Sci.* **2018**, *28*, 595–610. [CrossRef]
51. Thompson, S.E.; Harman, C.J.; Troch, P.A.; Brooks, P.D.; Sivapalan, M. Spatial scale dependence of ecohydrologically mediated water balance partitioning: A synthesis framework for catchment ecohydrology. *Water Resour. Res.* **2011**, *47*, W00J03. [CrossRef]
52. Abelen, S.; Seitz, F. Relating satellite gravimetry data to global soil moisture products via data harmonization and correlation analysis. *Remote Sens.* **2013**, *136*, 89–98. [CrossRef]
53. Clark, M.P.; Hendrikx, J.; Slater, A.G.; Kavetski, D.; Anderson, B.; Cullen, N.J.; Kerr, T.; Örn Hreinsson, E.; Woods, R.A. Representing spatial variability of snow water equivalent in hydrologic and land-surface models: A review. *Water Resour. Res.* **2011**, *47*, W07539. [CrossRef]
54. Lee, Y.G.; Ho, C.-H.; Kim, J.; Kim, J. Potential impacts of northeastern Eurasian snow cover on generation of dust storms in northwestern China during spring. *Clim. Dyn.* **2012**, *41*, 721–733. [CrossRef]
55. Jackson, T.J. Remote sensing of soil moisture: Implications for groundwater recharge. *Hydrogeol. J.* **2002**, *10*, 40–51. [CrossRef]
56. Zheng, Y.; Rimmington, G.M.; Xie, Z.; Zhang, L.; An, P.; Zhou, G.; Li, X.; Yu, Y.; Chen, L.; Shimizu, H. Responses to air temperature and soil moisture of growth of four dominant species on sand dunes of central Inner Mongolia. *J. Plant Res.* **2008**, *121*, 473–482. [CrossRef]
57. Rabelo, J.L.; Wendland, E. Assessment of groundwater recharge and water fluxes of the guarani aquifer system, brazil. *Hdrgeol. J.* **2009**, *17*, 1733–1748. [CrossRef]
58. Zhang, G.H.; Fei, Y.H.; Shen, J.M.; Yang, L.Z. Influence of unsaturated zone thickness on precipitation infiltration for recharge of groundwater. *J. Hydraul. Eng.* **2007**, *38*, 611–617. (In Chinese)
59. Morris, C.; Badik, K.J.; Morris, L.R.; Weltz, M.A. Integrating precipitation, grazing, past effects and interactions in long-term vegetation change. *J. Arid. Environ.* **2016**, *124*, 111–117. [CrossRef]
60. Zhou, J.; Fu, B.; Gao, G.; Lü, Y.; Liu, Y.; Lü, N.; Wang, S. Effects of precipitation and restoration vegetation on soil erosion in a semi-arid environment in the Loess Plateau, China. *Catena* **2016**, *137*, 1–11. [CrossRef]

61. Irmak, S.; Burgert, M.J.; Yang, H.S.; Cassman, K.G.; Walters, D.T.; Rathje, W.R.; Payero, J.O.; Grassini, P.; Kuzila, M.S.; Brunkhorst, K.J.; et al. Large-Scale On-Farm Implementation of Soil Moisture-Based Irrigation Management Strategies for Increasing Maize Water Productivity. *Trans. Asabe* **2012**, *55*, 881–894. [CrossRef]
62. Ibrahim, M.; Favreau, G.; Scanlon, B.R.; Seidel, J.L.; Le Coz, M.; Demarty, J.; Cappelaere, B. Long-term increase in diffuse groundwater recharge following expansion of rainfed cultivation in the Sahel, West Africa. *Hydrogeol. J.* **2014**, *22*, 1293–1305. [CrossRef]
63. Li, P.; Ren, L. Evaluating the effects of limited irrigation on crop water productivity and reducing deep groundwater exploitation in the North China Plain using an agro-hydrological model: II. Scenario simulation and analysis. *J. Hydrol.* **2019**, *574*, 715–732.
64. Liu, X.; He, Y.; Zhang, T.H.; Zhao, X.; Li, Y.; Zhang, L.; Wei, S.; Yun, J.; Yue, X. The response of infiltration depth, evaporation, and soil water replenishment to rainfall in mobile dunes in the horqin sandy land, northern china. *Environ. Earth Sci.* **2015**, *73*, 8699–8708. [CrossRef]
65. Baldocchi, D.; Knox, S.; Dronova, I.; Verfaillie, J.; Oikawa, P.; Sturtevant, C.; Matthes, J.H.; Detto, M. The impact of expanding flooded land area on the annual evaporation of rice. *Agric. For. Meteorol.* **2016**, *223*, 181–193. [CrossRef]

Publisher's Note: MDPI stays neutral with regard to jurisdictional claims in published maps and institutional affiliations.

© 2020 by the authors. Licensee MDPI, Basel, Switzerland. This article is an open access article distributed under the terms and conditions of the Creative Commons Attribution (CC BY) license (http://creativecommons.org/licenses/by/4.0/).

Article
Reliability of Gridded Precipitation Products in the Yellow River Basin, China

Yanfen Yang [1], Jing Wu [2], Lei Bai [3] and Bing Wang [1,*]

1 State Key Laboratory of Soil Erosion and Dryland Farming on the Loess Plateau, Institute of Soil and Water Conservation, Northwest A&F University, Yangling 712100, China; yfyang@ms.iswc.ac.cn
2 Lanzhou Central Meteorological Observatory, Lanzhou 730020, China; Wujing10@mails.ucas.ac.cn
3 School of Navigation, Wuhan University of Technology, Wuhan 430070, China; bailei09@mails.ucas.ac.cn
* Correspondence: bwang@ms.iswc.ac.cn; Tel.: +86-1582-991-2700

Received: 25 December 2019; Accepted: 20 January 2020; Published: 23 January 2020

Abstract: Gridded precipitation products are the de facto standard in hydrological studies, and the evaluation of their accuracy and potential use is very important for reliable simulations. The objective of this study was to investigate the applicability of gridded precipitation products in the Yellow River Basin of China. Five gridded precipitation products, i.e., Multi-Source Weighted-Ensemble Precipitation (MSWEP), CPC Morphing Technique (CMORPH), Global Satellite Mapping of Precipitation (GSMaP), Tropical Rainfall Measuring Mission (TRMM) Multi-Satellite Precipitation Analysis 3B42, and Precipitation Estimation from Remotely Sensed Information using Artificial Neural Networks (PERSIANN), were evaluated against observations made during 2001–2014 at daily, monthly, and annual scales. The results showed that MSWEP had a higher correlation and lower percent bias and root mean square error, while CMORPH and GSMaP made overestimations compared to the observations. All the datasets underestimated the frequency of dry days, and overestimated the frequency and the intensity of wet days (0–5 mm/day). MSWEP and TRMM showed consistent interannual variations and spatial patterns while CMORPH and GSMaP had larger discrepancies with the observations. At the sub-basin scale, all the datasets performed poorly in the Beiluo River and Qingjian River, whereas they were applicable in other sub-basins. Based on its superior performance, MSWEP was identified as more suitable for hydrological applications.

Keywords: precipitation datasets; evaluation; spatial scale; temporal scale; climate; Yellow River Basin

1. Introduction

Precipitation is the main link in the hydrological cycle and one of the most important meteorological input elements of hydrological models. Accurate precipitation input is the basic condition for obtaining reliable land surface hydrological simulations [1]. Choosing precipitation data is more important than choosing hydrological models [2]. The use of ground rainfall observation stations is the most direct way to measure precipitation. However, rain gauge density in complex terrain is low and unevenly distributed, thereby resulting in scarce or even a lack of observed precipitation, which cannot meet the needs of hydrological simulations. Remote sensing products based on microwave (MW) and infrared (IR) measurements have become a potential and valuable data source owing to their wide coverage and high spatiotemporal resolution. Affected by sampling error, algorithm uncertainty [3–6], the number of stations [7,8], and topographical factors [9,10], gridded precipitation data have errors when comparing to gauge observations, and rigorous quality assessment is required before use.

In recent years, precipitation products have been evaluated at multi-regional, multi-temporal, and multi-spatial scales; the results showed that there are large differences among precipitation products. Beck and Vergopolan [11] evaluated 22 precipitation products at the global scale. Among the uncorrected datasets, the satellite and reanalysis-based Multi-Source Weighted-Ensemble Precipitation (MSWEP)-ng (the full name of the abbreviation was listed in Table 1, the same below) showed the greatest correlation with the observations, followed by the reanalysis data (ERA-Interim, JRA-55, and NCEP CFSR), satellite-reanalysis data (CHIRP), passive MW-based data (CPC Morphing Technique (CMORPH), Global Satellite Mapping of Precipitation (GSMaP_MVK), and Tropical Rainfall Measuring Mission (TRMM) 3B42RT), and products based on IR imagery (GridSat, Precipitation Estimation from Remotely Sensed Information using Artificial Neural Networks (PERSIANN), and PERSIANN-CCS). Among the corrected datasets, the products that directly merge daily gauge observations perform best (CPC Unified and MSWEP), followed by those that incorporate temporally coarser gauge data (CHIRPS, GPCP-1DD, TRMM 3B42, and WFDEI-CRU) and products that indirectly incorporate gauge data through other multi-source datasets (PERSIANN-CDR). In China, the spatial distribution of daily mean precipitation of CMORPH and TRMM 3B42 shows good similarity with the ground station data, and both can describe the diurnal variation in summer precipitation in most regions of China [12]. CMORPH outperforms GSMaP_MVK and PERSIANN [13], and IMERG performs better than TRMM 3B42, CMORPH_CRT, and PERSIANN_CDR [14]. The corrected GSMaP_Gauge is superior to GSMaP_NRT and GSMaP_MVK [15]. In humid regions of China, TRMM 3B42 shows the lowest error and deviation and the highest correlation coefficient (CC) at a monthly scale, but its accuracy is lower at a daily scale compared with that of PERSIANN_CDR and NCEP-CFSR [16]. In the Qinghai-Tibet Plateau, IMERG is superior to TRMM 3B42V7 at multiple time scales, but with discrepancies in the timing of the greatest precipitation intensity and overestimation of the maximum rainfall intensity [17]. The error and deviation of TRMM 3B42 are lower than those of PERSIANN and CMORPH [18,19]. In addition, TRMM 3B42 and CMORPH_BLD outperform CMORPH and TRMM 34B2RT in the Huifa River Basin [20]. IMERG is superior to TRMM 3B42 in detecting precipitation events and precipitation in the Huai River Basin [21]. Compared with PERSIANN_CDR, CHIRPS shows a lower bias (PBIAS) and error, and can describe the spatial pattern of precipitation at a monthly and annual scale more accurately in Xinjiang Province [22].

Table 1. Full name of the abbreviation in the text.

Abbreviation	Full Name
CFSR	Climate Forecast System Reanalysis
CHIRP	Climate Hazards group Infrared Precipitation
CHIRPS	Climate Hazards group Infrared Precipitation with Stations
CMORPH	Climate Prediction Center MORPHing technique
CMORPH_BLD	CMORPH satellite-gauge blended product
CMORPH_CRT	CMORPH bias corrected
CPC	Climate Prediction Center
ERA-Interim	European Centre for Medium-range Weather Forecasts ReAnalysis Interim
GPCP	Global Precipitation Climatology Project
GPCP-1DD	GPCP 1-Degree Daily
GridSat	P derived from the Gridded Satellite
GSMaP_MVK	Global Satellite Mapping of Precipitation (GSMaP) Moving Vector with Kalman (MVK)
GSMaP_NRT	GSMaP Near Real Time
IMERG	Integrated Multi-satellite Retrievals for Global Precipitation Measurement
JRA-55	Japanese 55-year ReAnalysis
MSWEP	Multi-Source Weighted-Ensemble Precipitation
NCEP	National Centers for Environmental Prediction
PERSIANN	Precipitation Estimation from Remotely Sensed Information using Artificial Neural Networks
PERSIANN-CCS	PERSIANN Cloud Classification System

Table 1. *Cont.*

Abbreviation	Full Name
PERSIANN-CDR	PERSIANN Climate Data Record
TMPA	TRMM Multi-satellite Precipitation Analysis
TRMM	Tropical Rainfall Measuring Mission
WFDEI-CRU	WATCH Forcing Data ERA-Interim Climatic Research Unit

The same type of precipitation product shows different performance in different regions and at different temporal scales. MSWEP generally overestimates the precipitation in China but underestimates it in North China. MSWEP overestimates light precipitation but underestimates the heavy precipitation events. It shows the highest accuracy at a monthly scale and the lowest accuracy at a daily scale. There is a significant difference in the annual trend of precipitation between MSWEP and the observations [23]. PERSIANN_CDR can capture the spatiotemporal characteristics of extreme precipitation events at a daily scale in the southeast monsoon region of China [24]. It is also a reliable alternative dataset in the Qinghai-Tibet Plateau, upper Yellow River (UYR) [25], and Xiang River Basin [16]. CMORPH shows a large error in the southeast and poor time correlation of seasonal precipitation in the west and northwest of China [14]. IMERG and TRMM 3B42V7 mostly show high correlation and low relative error in the eastern river basins while showing low correlation and high relative error in the western region of China [26]. The performance of precipitation products also varies seasonally. In China, satellite precipitation and site-corrected products have poor ability to detect precipitation events in winter [12]. IMERG has a stronger ability for light and solid precipitation, and its accuracy for winter precipitation is significantly higher than that of TRMM 3B42, but its accuracy in detecting heavy precipitation needs to be strengthened [26].

The Yellow River Basin in China is characterized by a wide area, complex topography and landforms, diverse climate types, and vegetation coverage. The rain gauges here are unevenly distributed with low density, and have poorly representative and discontinuous data sequences, which cannot meet the needs of hydrological simulations. Sometimes, there is only one to two or even no stations in a target research basin, which often leads to great uncertainty in the input precipitation and poor prediction of the rainfall–runoff model [27]. Gridded precipitation datasets have the potential to improve the quality of precipitation and runoff prediction results. In the Yellow River Basin, it had been reported that there is a good linear relationship between IMERG, TRMM 3B42V7, and ground-based rain gauge data, but the annual precipitation is overestimated by 2.46% and 2.19%, respectively. The CC is relatively high in the southern part of the basin, while the correlation is relatively low in the Ordos Plateau and its north [28]. Seasonally, IMERG and TRMM 3B42 show higher reproducibility in spring and autumn than in winter and summer. The precipitation is underestimated in July and August but is overestimated to different degrees in other months. Among them, the relative error is the largest in December, and the absolute deviation is the largest in September [28].

In summary, the accuracy of precipitation products varies with regions, seasons, and spatiotemporal scales. However, studies in the Yellow River Basin are generally conducted at a large spatial scale, thereby masking the error distribution of small-scale watersheds. In addition, it was common that only one or two products were evaluated, which may result in lacking comprehensive cognition of other kinds of products. The objective of this study was to evaluate of the applicability of multiple precipitation datasets at multiple spatiotemporal scales in the Yellow River Basin in order to provide a reliable source of precipitation data for hydrological simulation and water resources management.

2. Study Area and Methods

2.1. Study Area

The Yellow River Basin, which is located between 96–119°E and 32–42°N, has a drainage area of 7,950,000 km^2 [29]. The basin traverses the Qinghai-Tibet Plateau, Inner Mongolia Plateau,

Loess Plateau, and Huanghuaihai Plain from the west to the east [30]. The Yellow River Basin belongs to the continental monsoon climate and can be roughly divided into arid, semi-arid, semi-humid, and humid climates [29]. The west is arid, while the east is humid. It is dry in winter with drought in spring, and is rainy in summer and autumn [31]. Affected by the topography and atmospheric circulation, the precipitation is unevenly distributed across the seasons, with large interannual and regional variations. Nine provinces are involved in the Yellow River Basin: Qinghai, Sichuan, Gansu, Ningxia, Inner Mongolia, Shaanxi, Shanxi, Henan and Shandong. The drainage area in Shandong Province is long and narrow, and will not be discussed in this study. The basin was divided into seven water systems to facilitate the evaluation at different regional and spatial scales, including the UYR, Gansu–Ningxia water system (GN), Inner Mongolia water system (IM), Northern Shaanxi water system (NSH), Wei River Basin (WR), Fen River Basin (FR), and Western Henan water system (WH). Furthermore, the basin was divided into 24 sub-basins, including the Yellow River source region, Wuding River, and Jing River. The geographical location, elevation, rain gauge distribution, water system division, and sub-basin division of the Yellow River Basin are shown in Figure 1.

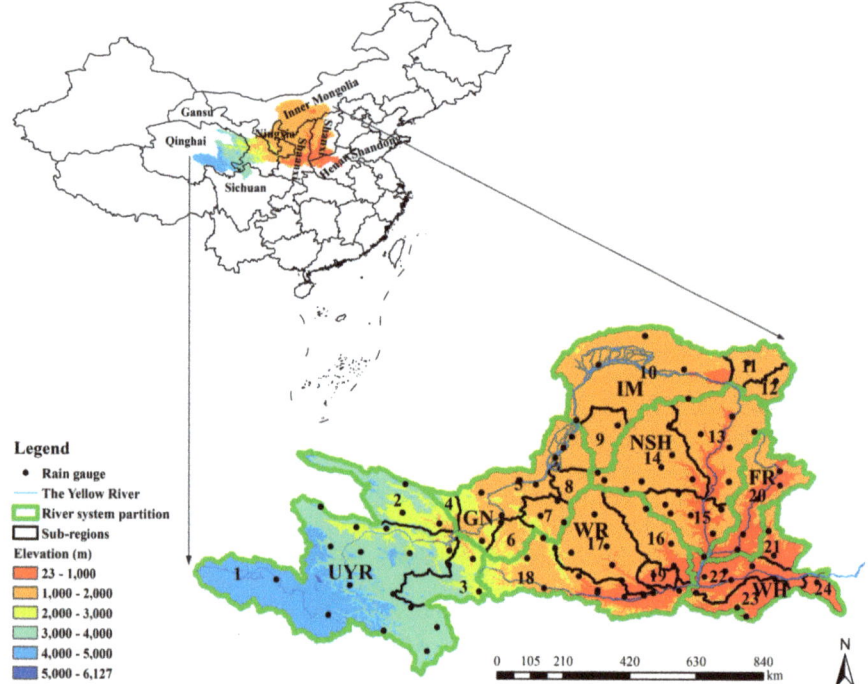

Figure 1. Geographical location, elevation, rain gauge distribution, water system partition and sub-basins of the Yellow River basin. UYR, GN, IM, NSH, WR, FR, and WH are the upper Yellow River, Gansu–Ningxia water system, Inner Mongolia water system, Northern Shaanxi water system, Wei River Basin, Fen River Basin, Western Henan water system, respectively. The No and the name of the sub-basins: 1, Yellow River Source. 2, Huang River. 3, Tao River. 4, Zhuanglang River. 5, Tributary 1 in upper Yellow River. 6, Zuli River. 7, Qingshui River. 8, Kushui River. 9, Dusitu River. 10, Tributary 2 in upper Yellow River. 11, Dahei River. 12, Hun River. 13, Tuwei River, Kuye River, etc. 14, Wuding River. 15, Qingjian River and Yanshui River. 16, Beiluo River. 17, Jing River. 18, Wei River. 19, Shichuan River. 20, Fen River. 21, Qin River. 22, Tributary in middle Yellow River. 23, Yiluo River. 24, Ying River.

2.2. Datasets

The gauge observed daily precipitation (OBS) was used to conduct the point to point evaluation. The gridded dataset CN05.1 interpolated from OBS was used as a reference to evaluate the spatial patterns of the precipitation products. Both OBS and CN05.1 were obtained from the National Meteorological Information Center of China. CN05.1 was interpolated based on the daily precipitation of 2416 rain gauges around China. The spatial resolution was 0.25° × 0.25° and the temporal resolution was daily [32]. The interpolation of CN05.1 was realized by the "anomaly approach". The climatology was first interpolated by thin-plate smoothing splines and then a gridded daily anomaly derived from angular distance weighting method was added to climatology to obtain the final dataset. New and Lister [33] compared several kinds of interpolation methods and indicated that these two methods performed better. Gridded precipitation such as CN05, EA05 and APHRO were all interpolated by using these methods [34–36]. The climatology was first interpolated due to climatic factors, especially precipitation, have great discontinuity in spatial distribution. The climatology is relatively continuous; interpolating it first is beneficial to reduce errors caused by discontinuity. Specifically, thin-plate smoothing splines is used to interpolate climatology by a software named ANUSPLIN. ANUSPLIN was a interpolation package that was widely used to produce climatic elements with high resolution in geography and ecology [37]. CN05.1 was interpolated using ANUSPLIN, taking longitude and latitude as independent variable, and taking elevation as covariable. Then, the anomaly was derived from an angular distance weighting method. The grid value was obtained on the consideration of the weight of angle and distance between the gauge station and the grid.

Five precipitation products were evaluated, including CMORPH_blended, PERSIANN_CDR, GSMaP_MVK, MSWEP V1.1 and TRMM 3B42 V7. Basic information about the products is shown in Table 2. The coarsest temporal and spatial resolutions were daily and 0.25°, and the highest resolutions were 1 h and 0.1°, respectively. To facilitate the point to point evaluation, all the products were downscaled to the spatial location of rain gauge using the bilinear interpolation method. In this method, four nearest grid values are used for calculating the value of a particular point. The weights are derived from the spatial locations in a two-dimensional space. The closer the grid is to the point, the more influence (weight) it will have. The algorithm obtains the pixel value by taking a weighted sum of the pixel values of the four nearest neighbors surrounding the calculated location [38,39]. The raw CMORPH, which was retrieved from MW and IR signals, is a pure satellite precipitation product with a spatial resolution of 8 km and temporal resolution of 30 min [40]. CMORPH_blended is a dataset incorporating raw CMORPH and 30,000 automatic meteorological stations around China, and the spatial and temporal resolutions were 0.1° and 1 h, respectively [41].

Table 2. Brief introduction of precipitation datasets in this study.

Dataset		Time Span	Temporal Resolution	Spatial Resolution	Data Source
Full Name	Abbreviation				
CMORPH_blended	CMORPH	1998–present	Hourly	0.10°	IR, SSM/I, TRMM, AMSU-B, AMSR-E, Automatic weather station in China
PERSIANN_CDR	PERSIANN	1983–2017	Daily	0.25°	IR, TRMM 2A12, NCEP IV, GPCP
TMPA 3B42 V7	TRMM	1998–2016	3 hourly	0.25°	IR, SSMIS, TMI, AMSU-B, MHS, AMSR-E, GPCP
GSMaP_MVK	GSMaP	2000–2014	Hourly	0.10°	IR, TMI, AMSR-E, AMSR, SSMI
MSWEP V1.1	MSWEP	1979–2015	3 hourly	0.25°	CPC Unified, GPCC, CMORPH, GSMaP-MVK, TRMM 3B42RT, ERA-Interim, JRA-55, CHPclim
CN05.1	CN05.1	1961–2015	Daily	0.25°	Gauge

Websites for downloading the datasets. CMORPH_blended: http://data.cma.cn/data/cdcdetail/dataCode/SEVP_CLI_CHN_MERGE_CMP_PRE_HOUR_GRID_0.10.html. PERSIANN_CDR: https://climatedataguide.ucar.edu/climate-data/persiann-cdr-precipitation-estimation-remotely-sensed-information-using-artificial. TMPA 3B42 V7: https://pmm.nasa.gov/data-access/downloads/trmm. GSMAP_MVK: https://sharaku.eorc.jaxa.jp/GSMAP/index.htm. MSWEP V1.1: http://www.gloh2o.org. CN05.1: http://data.cma.cn/data/cdcdetail/dataCode/SEVP_CLI_CHN_PRE_DAY_GRID_0.25.html.

PERSIANN-CDR is generated from the PERSIANN algorithm using GridSat-B1 IR data and adjusted using the GPCP monthly product. The dataset with spatial and temporal resolutions of 0.25° and daily, respectively, was used in this study [42].

TRMM 3B42 V7 was retrieved by MW and IR signals, and corrected by the gauge data. Passive MW data were first corrected by TMI and PR, and then used to correct the IR data. After combining the MW and IR data, the TRMM Multi-Satellite Precipitation Analysis (TMPA) 3B42 V7 was obtained through the correction of global precipitation data (GPCP). The spatial and temporal resolutions were 0.25° and 3 h, respectively [43].

GSMaP_MVK predicted the precipitation rate from the MW data using the Kalman filtering method, and then the rate was improved based on the relationship between the brightness temperature data and the ground precipitation rate. The spatial and temporal resolutions were 0.1° and 1 h, respectively [44].

MSWEP V1.1 blended multiple data sources, including gauge data, satellite data, and reanalysis data. CHPclim was used as the average of long-term precipitation, and the deviation was corrected. The long-term mean of MSWEP was based on Climate Hazards Group Precipitation Climatology (CHPclim) dataset, which was bias corrected using catch-ratio equations and observation-based estimates of long-term streamflow and potential evaporation. Then, the precipitation anomalies of the gauges, satellites, and reanalysis data were combined using the weighted average method. Finally, CHPclim was temporally downscaled through the precipitation anomaly. The spatial and temporal resolutions were 0.25° and 3 h, respectively [45].

2.3. Methods

By comparing the downscaled precipitation products with observed precipitation, the indexes, e.g., CC, PBIAS, and root mean square error (RMSE), were used to measure the quantitative accuracy at an annual, monthly, and daily scale. The variables were significantly correlated when the CC was higher than 0.7 [46], and the precision was acceptable when the PBIAS value ranged from −10% to 10% [47]. The frequency bias index (FBI), probability of detection (POD), false alarm ratio (FAR) and threat score (TS) were used to evaluate the accuracy in detecting precipitation occurrence. In addition, the annual distribution, interannual variation, and spatial pattern of precipitation were also used to clarify the detection capability:

$$CC = \frac{\sum_{i=1}^{n}(O_i - \overline{O})(P_i - \overline{P})}{\sqrt{\sum_{i=1}^{n}(O_i - \overline{O})^2}\sqrt{\sum_{i=1}^{n}(P_i - \overline{P})^2}}, \qquad (1)$$

$$PBIAS = \frac{\sum_{i=1}^{n}(P_i - O_i)}{\sum_{i=1}^{n}O_i} \times 100\%, \qquad (2)$$

$$RMSE = \sqrt{\frac{\sum_{i=1}^{n}(P_i - O_i)^2}{n}}, \qquad (3)$$

$$FBI = \frac{a+b}{a+c}, \qquad (4)$$

$$POD = \frac{a}{a+c}, \qquad (5)$$

$$FAR = \frac{b}{a+b}, \qquad (6)$$

$$TS = \frac{a}{a+b+c},\qquad(7)$$

where a is the number of hits, b is the number of false alarms, and c is the number of misses. Perfect values were FBI = 1, POD = 1, TS = 1, and FAR = 0 [48–51].

3. Results

3.1. Annual Precipitation and Spatial Pattern

The applicability of the precipitation products was interpreted by the CC, RMSE, temporal variation, and spatial pattern of precipitation at an annual scale.

The CCs between the five precipitation products and the observed annual precipitation ranged from −0.61 to 0.99 (Figure 2a). MSWEP and TRMM showed the highest CCs and were significantly correlated with the ground-based rain gauge data with mean values of 0.79 and 0.76, respectively. The CCs of CMORPH and GSMaP were relatively lower, with the mean values of 0.29 and 0.33, respectively.

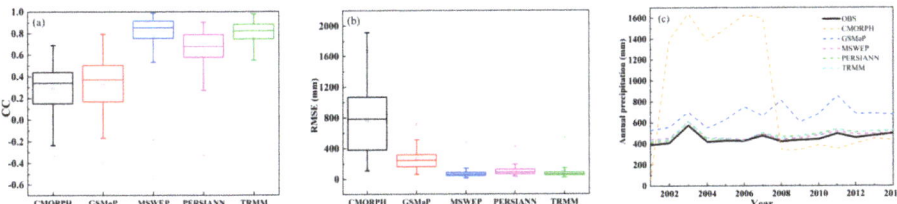

Figure 2. (a) correlation coefficient, (b) root mean square error of annual precipitation between gridded and gauge datasets, and (c) temporal variation of annual precipitation. CC is correlation coefficient, RMSE is root mean square error. (The center line of each boxplot depicts the median value (50th percentile) and the box encompasses the 25th and 75th percentiles of the sample data. The small squares in each boxplot is the arithmetic mean. The whiskers extend from Q1 − 1.5 × (Q3 − Q1) to Q3 + 1.5 × (Q3 − Q1), where Q1 and Q3 are the 25th and 75th percentiles of the sample data, respectively, similarly hereinafter).

The RMSE ranged from 19 mm to 1915 mm (Figure 2b). CMORPH showed the highest RMSE (109–1915 mm) with a mean of 761 mm, followed by GSMaP. The RMSEs of the other three products were relatively lower and ranged from 19 mm to 731 mm. The mean RMSEs of MSWEP and TRMM were the smallest and were similar with values of 88 mm and 89 mm, respectively, thereby indicating that these two products performed better than other datasets, which was consistent with the CC results.

The temporal variation in annual precipitation (Figure 2c) indicated that the observed precipitation showed an upward trend in fluctuation, and the mean ranged from 389 mm to 575 mm. The fluctuating trends and the amount of precipitation estimated by MSWEP, PERSIANN and TRMM were similar to the ground-based rain gauge data, with a range in precipitation between 418 mm and 620 mm. CMORPH and GSMaP demonstrated the largest differences with the ground-based rain gauge data; among them, the precipitation recorded by CMORPH from 2002 to 2007 was above 1380 mm, which was not in line with the actual situation in the Yellow River Basin.

The five products and ground-based rain gauge data showed that the annual precipitation decreased from southeast to northwest (Figure 3), which was consistent with the distribution of climatic conditions in the Yellow River Basin. The contour map of precipitation in the literature showed that the southern part of the Yellow River Basin received the largest amount of precipitation (approximately 700 mm), while the precipitation in the GN in the northwest was reduced to about 200 mm during 1951 to 2001 [52]. The maximum precipitation estimate of CMORPH in the WH was 1400 mm, and there was a 1200 mm high precipitation center in the southern part of the UYR, which was greatly overestimated. The precipitation obtained by GSMaP and PERSIANN in the WH was above 1000 mm, which also overestimated the actual precipitation. The spatial patterns of MSWEP and TRMM were similar to those in the literature. In addition, the results based on the gauge data in the literature indicated that the average annual precipitation in the Yellow River Basin was 483.7 mm [53], and the average annual precipitation estimated by CMORPH, GSMaP, MSWEP, PERSIANN and TRMM was 853 ± 98 mm, 674 ± 91 mm, 483 ± 44 mm, 489 ± 51 mm, and 491 ± 50 mm, respectively. It could be seen that CMORPH and GSMaP overestimated the precipitation to a larger extent, while the precipitation estimates of MSWEP, PERSIANN and TRMM were close to the values in the literature.

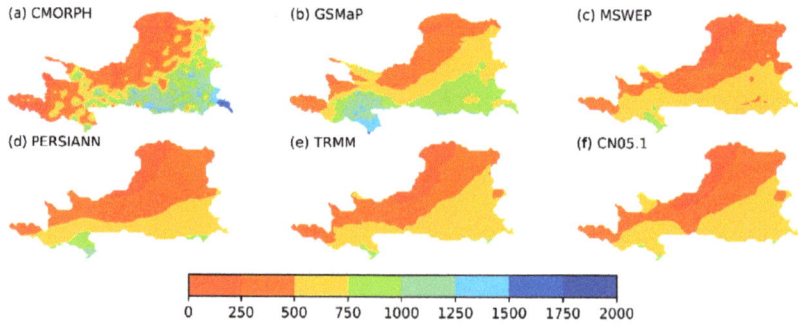

Figure 3. Spatial pattern of annual mean precipitation.

3.2. Monthly Precipitation and Annual Distribution

The CC, RMSE, PBIAS and their distribution during the year were used to interpret the performances of the precipitation products at the monthly scale.

The CCs of the monthly precipitation between the products and the gauge ranged from −0.61 to 0.99 (Figure 4a). More than 97% of the sites from MSWEP, PERSIANN and TRMM were significantly correlated with the ground-based rain gauge data. The CCs of CMORPH and GSMaP were 0.37–0.86 and 0.29–0.86, respectively, and 12.9% and 28.7% of the sites were significantly correlated, respectively. MSWEP showed the largest mean CC (0.93), followed by PERSIANN (0.87) and TRMM (0.91), while CMORPH (0.67) and GSMaP (0.63) showed the smallest mean CCs.

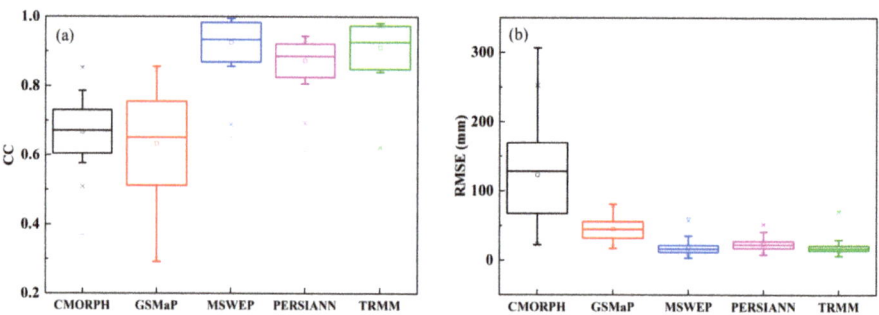

Figure 4. (a) correlation coefficient and (b) root mean square error of monthly precipitation between gridded and gauge datasets. CC is correlation coefficient, RMSE is root mean square error.

CMORPH had the largest RMSE, followed by GSMaP; the ranges were 22.5–307 mm and 17–81 mm and the means were 123 mm and 45.4 mm, respectively. Among the other three precipitation products, MSWEP showed the smallest range and mean RMSE, which were 3.7–62 mm and 18 mm, respectively. PERSIANN showed the largest mean RMSE, which was 23.6 mm (Figure 4b).

The ground-based rain gauge data showed that the monthly precipitation increased from January, reached its peak in July, and then decreased (Figure 5a). The fluctuation trend and amount of precipitation estimated by MSWEP, PERSIANN and TRMM were close to the ground-based rain gauge data, and the differences ranged from −2.3 mm to 10.5 mm. CMORPH underestimated the precipitation slightly from January to April and from October to December, while it overestimated the precipitation to a large extent from May to September. GSMaP overestimated the precipitation, except for in July and August. The annual distributions of PBIAS, CC and RMSE from MSWEP, PERSIANN and TMPA 3B43 showed higher CCs, lower PBIASs and RMSEs. Their variations were relatively smooth and steady without large fluctuation during the year. CMORPH and GSMaP showed lower CCs, higher PBIASs and RMSEs, and a larger fluctuation range (Figure 5b–d).

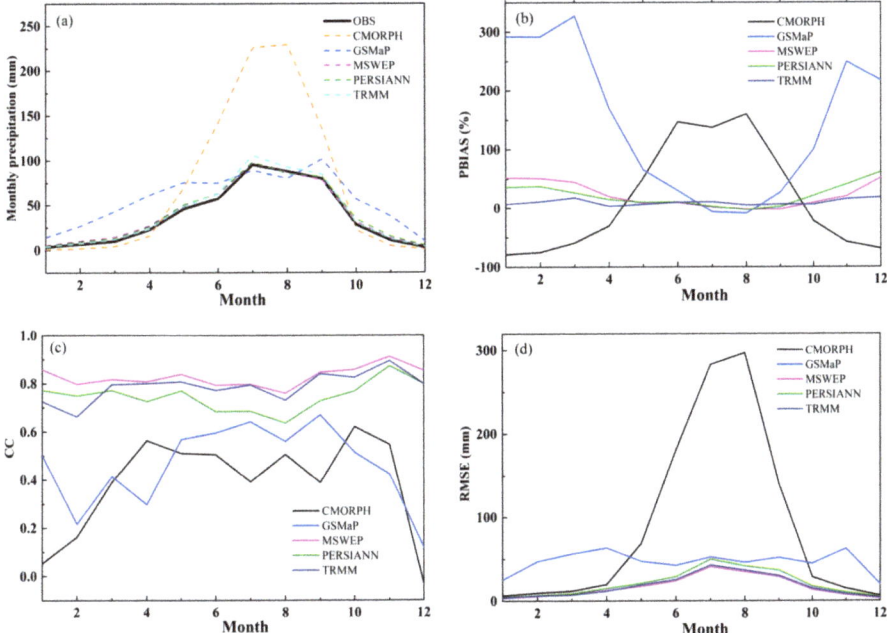

Figure 5. (a) variation of monthly precipitation, (b) PBIAS, (c) correlation coefficient and (d) root mean square error during the year. PBIAS is bias, CC is correlation coefficient, RMSE is root mean square error.

3.3. Daily Precipitation and Precipitation Events

At the daily scale, the assessment was conducted using the precipitation event and the amount of precipitation. The precipitation events were measured by POD, FBI, FAR and TS. The amount of precipitation was quantified by PBIAS, CC and RMSE.

Large differences in the POD among the five precipitation products were observed, as shown in Figure 6a. MSWEP showed the highest POD values, which were all above 0.94 with a mean of 0.97. The POD values of CMORPH, GSMaP and PERSIANN were similar, with a range of 0.6–0.9 and mean of 0.74–0.78. TRMM showed the lowest POD with a range and mean of 0.44–0.85 and 0.64, respectively. From the spatial distribution shown in Figure 7, the PODs of MSWEP at all sites were clearly higher than those of the other four products. CMORPH and GSMaP had higher PODs in the UYR, while PERSIANN showed higher PODs in the FR, WH and WR. TRMM performed relatively poorly in the GN and IM.

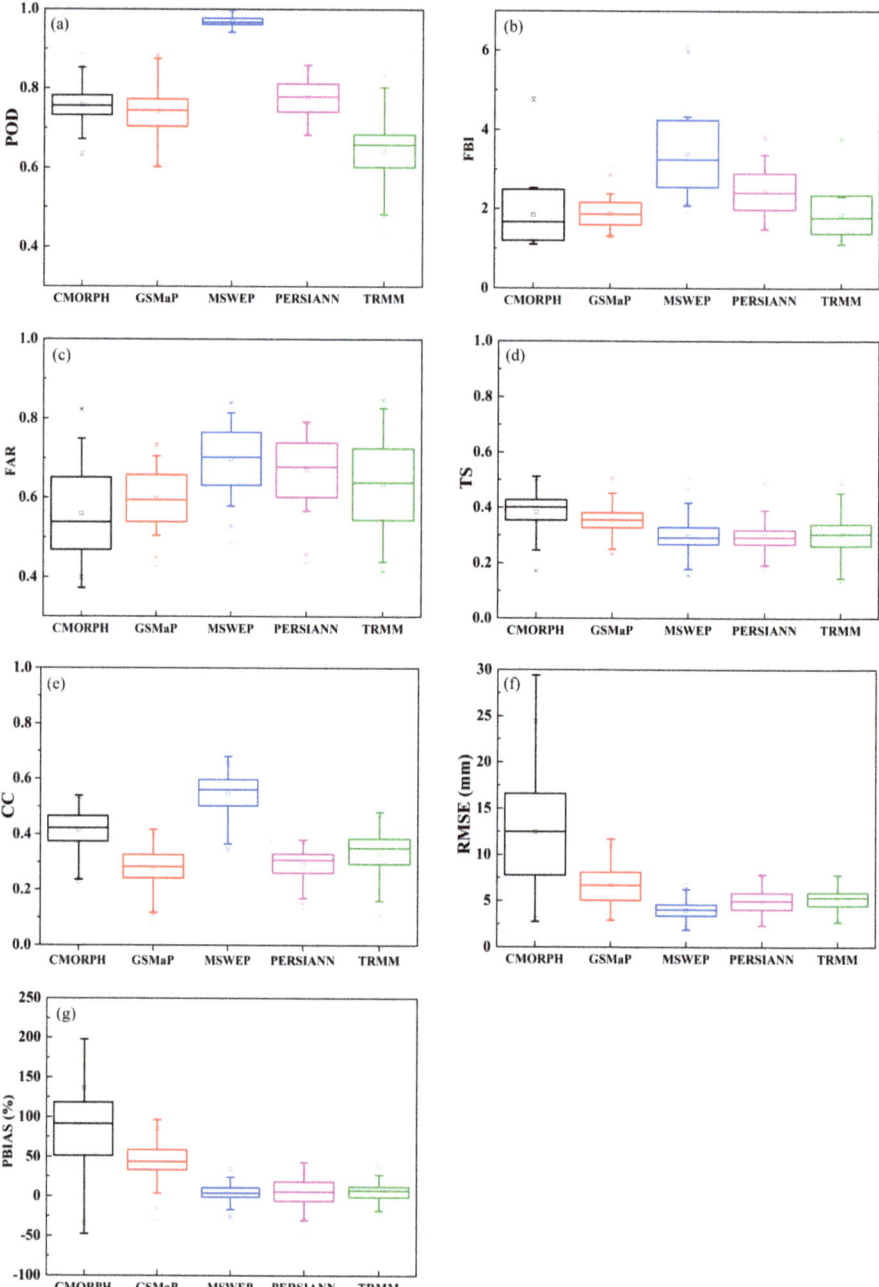

Figure 6. Detective ability of precipitation event: (**a**) POD, (**b**) FBI, (**c**) FAR, (**d**) TS and detective ability of precipitation: (**e**) correlation coefficient, (**f**) root mean square error, (**g**) PBIAS. POD is probability of detection, FBI is the frequency bias index, FAR is false alarm ratio, CC is correlation coefficient, RMSE is root mean square error, PBIAS is bias.

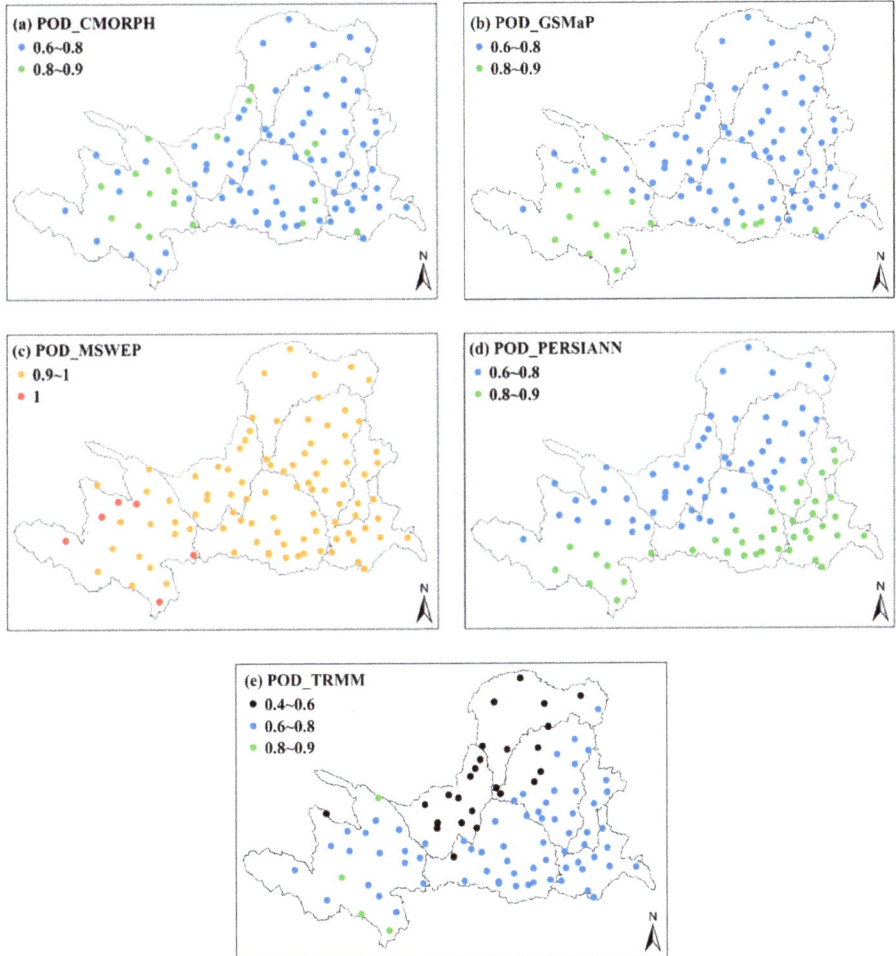

Figure 7. Spatial pattern of POD (probability of detection).

The FBI values of all five products were higher than 1, thereby indicating the overestimation of precipitation occurrence, as shown in Figure 6b. The overestimation degrees of CMORPH, GSMaP and TRMM were relatively lower with FBI values of 1.84, 1.89 and 1.87, respectively. MSWEP showed the highest overestimation degree with a mean FBI value of 3.4. The spatial distribution of the FBI (Figure 8) showed that CMORPH, GSMaP and TRMM had lower overestimation with FBI values ranging from 1 to 2 in the UYR, NSH, FR, WR and WH. MSWEP showed higher overestimation in the GN and IM with FBI values of up to 4 to 6.

Aa shown in Figure 6c, there were high false alarms and the FAR ranged from 0.37 to 0.85. CMORPH showed the lowest FARs, followed by GSMaP, while MSWEP showed the highest false alarms with mean FARs of 0.56, 0.6 and 0.7, respectively. The FARs of the majority sites were higher than 0.5 (Figure 9). All five products demonstrated that the FARs were highest in the GN, which ranged from 0.7 to 0.9. The FARs of MSWEP were clearly higher than those of the other four products in the IM, NSH, FR and WH.

The TSs of all five precipitation products were lower than 0.52. CMORPH showed the highest TS with a mean of 0.38, followed by GSMaP with a mean of 0.35. MSWEP, PERSIANN and TRMM showed similar TS, the means were all valued 0.3 (Figure 6d). The spatial distribution of TS showed that CMORPH performed best, while TSs of MSWEP, PERSIANN and TRMM were lower than CMORPH and GSMaP in almost all water system partitions. The TSs also differed in different water system partition. All five precipitation products showed the highest TS in the UYR with the TSs ranged from 0.36 to 0.42, followed by the WR and the WH. The TSs in the GN and IM were the lowest, with the ranges were 0.18–0.27 and 0.22–0.31, respectively (Figure 10).

At the daily scale, all five products were insignificantly correlated with the ground-based rain gauge data (Figure 6e). MSWEP had the highest CCs between 0.34 and 0.68 with a mean of 0.55. The CCs of GSMaP and PERSIANN were the smallest with mean values of 0.28 and 0.29, respectively. As shown in Figure 6f, MSWEP had the smallest RMSE between 1.9 mm and 6.9 mm with a mean of 4 mm. The RMSEs of PERSIANN and TRMM were similar with mean values of 4.9 mm and 5.2 mm, respectively. CMORPH showed the highest RMSE with a mean of 12.5 mm. MSWEP, PERSIANN and TRMM had smaller and more similar PBIASs, with mean values of 34.9%, 31.3% and 38.5%, respectively. CMORPH showed the largest PBIAS ranging from −48% to 2148% with a mean value of 137% (Figure 6g). From the distribution of PBIAS in Figure 11, CMORPH significantly overestimated the daily precipitation at 80% of the stations, and the PBIAS at less than 1% of the stations was within the acceptable range of ±10%. GSMaP overestimated the daily precipitation at 97% of the stations, but the overestimation degree was less than that of CMORPH. MSWEP, PERSIANN and TRMM overestimated the daily precipitation at 88% to 93% of the stations. The PBIAS of MSWEP was relatively smaller with values at 65% of stations within the acceptable range, followed by TRMM and PERSIANN with acceptable values at 56% and 45% of the stations, respectively.

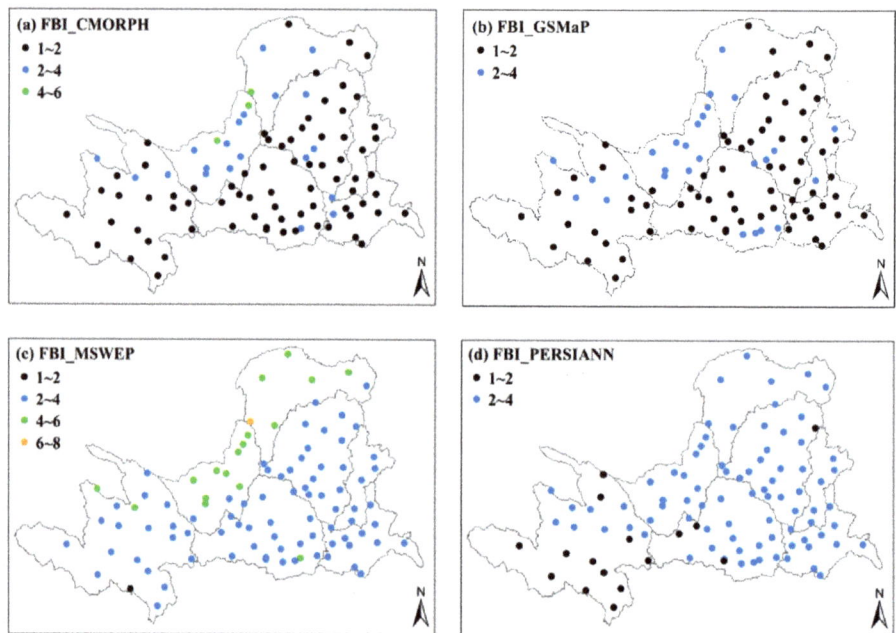

Figure 8. *Cont.*

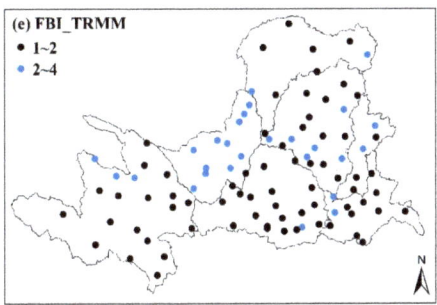

Figure 8. Spatial pattern of FBI (frequency bias index).

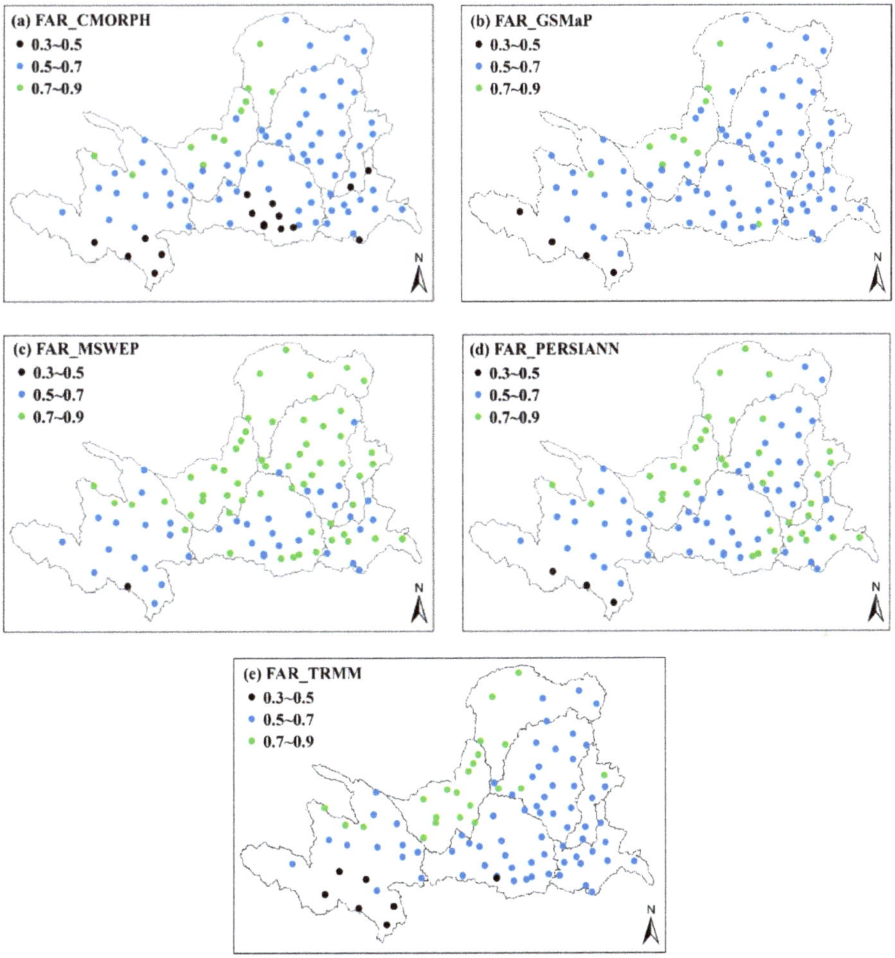

Figure 9. Spatial pattern of FAR (false alarm ratio).

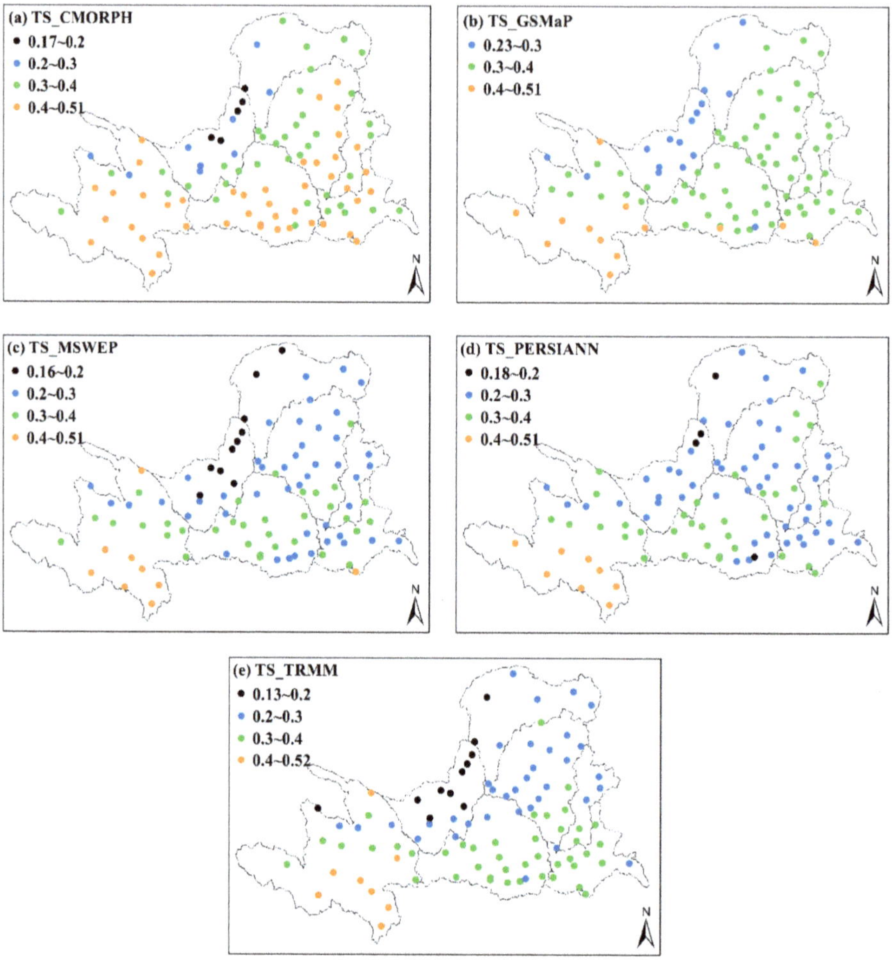

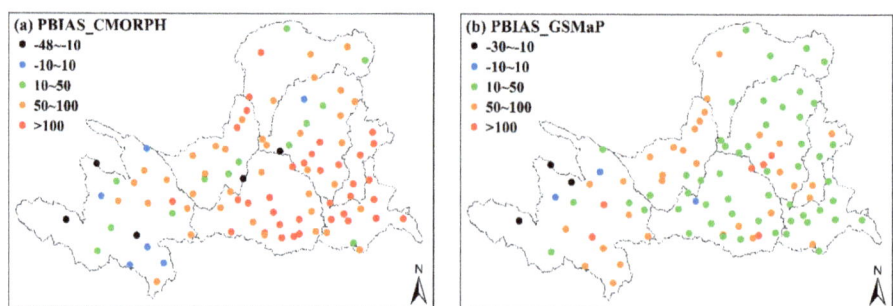

Figure 10. Spatial pattern of TS (threat score).

Figure 11. *Cont.*

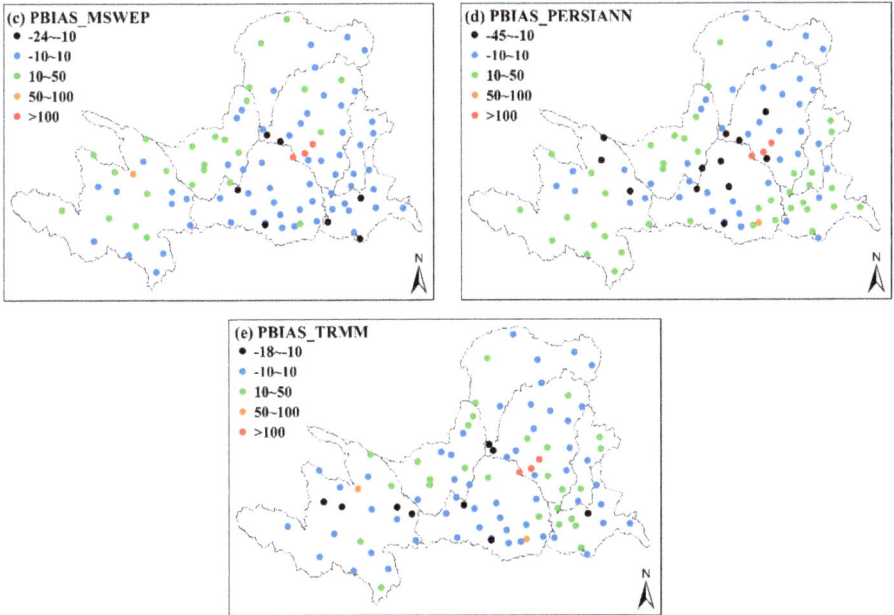

Figure 11. Spatial pattern of PBIAS (bias) on daily scale.

3.4. Frequency Curve of Precipitation

Figure 12a indicates that the precipitation frequency curves of the five products were very similar when the precipitation was above 5 mm/d, and the difference was mainly at precipitation levels below 5 mm/d. All the products underestimated the frequency of dry days, with MSWEP underestimating the frequency to the largest extent (51%), followed by PERSIANN (31%), while CMORPH, GSMaP, and TRMM underestimated the frequency by 17–20%. At the precipitation level of 0–5 mm/d, all the precipitation products overestimated the frequency of precipitation. The degree and rank of overestimation were similar to those of the estimation of dry days.

The PBIAS at different precipitation levels (Figure 12b) showed that all the products overestimated the precipitation at the level of 0–5 mm/d, of which the overestimation was the largest for GSMaP and CMORPH, while it was the smallest for MSWEP. The products underestimated the precipitation above 100 mm/d, with the PBIAS ranging from −85% to −81%, except for CMORPH, which had an acceptable PBIAS. At the level of 5–100 mm/d, all the products underestimated the precipitation, except for the overestimation by CMORPH. The PBIAS increased with the increase in the precipitation level. GSMaP showed the smallest PBIAS at the level of 5–10 mm/d, while CMORPH showed the smallest PBIAS at the level of 10–200 mm/d.

All the precipitation products were insignificantly correlated with the ground-based rain gauge data at each precipitation level, as shown in Figure 12c. There were negative CCs above the precipitation level of 30 mm/d, thereby indicating a decrease in the correlation. The RMSE of the five products showed a gradual increase with the increase in precipitation level. At each precipitation level, CMORPH and MSWEP showed that largest and the smallest RMSE, respectively (Figure 12d).

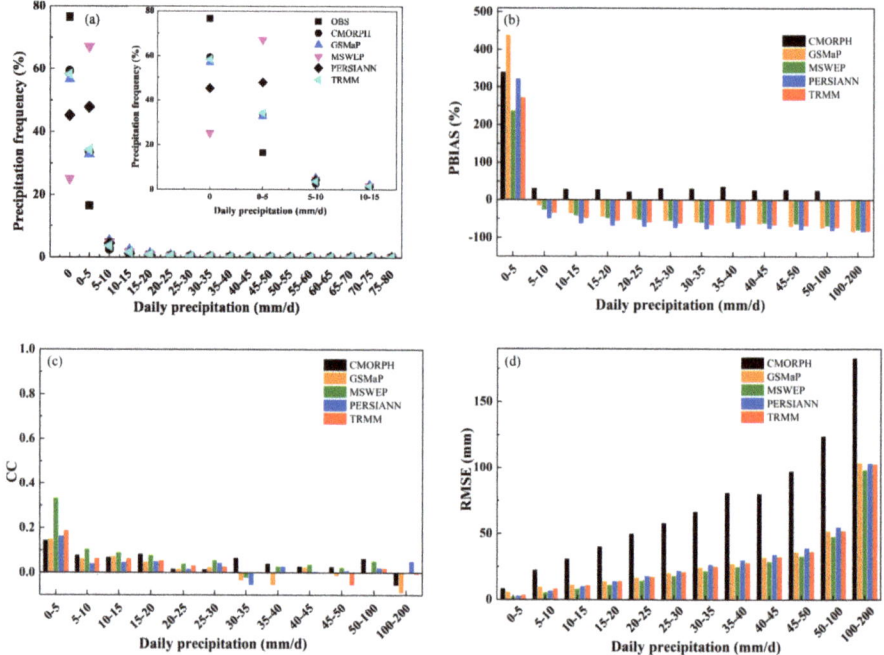

Figure 12. (**a**) precipitation frequency, (**b**) PBIAS, (**c**) correlation coefficient, (**d**) root mean square error for different precipitation level. PBIAS is bias, CC is correlation coefficient, RMSE is root mean square error.

3.5. Applicability in Sub-Regions

In this section, the applicability of the five precipitation products in 24 sub-basins is discussed. There was no gauge in the Kushui River, Zuli River, or Zhuanglang River, which were not analyzed here.

At the annual scale, Figure 13a,b shows that the CCs of MSWEP and TRMM were higher than 0.7 and were significant, except for in the Beiluo River and the Qingjian River Basin. The RMSEs of these two sub-basins were between 200 mm and 1118 mm, which were higher than those of the other sub-basins. The CCs of CMORPH and GSMaP were lower, while the RMSEs were higher than those of other products.

At the monthly scale, the CCs of the products were higher than 0.7, except for CMORPH and GSMaP. MSWEP showed a higher CC, but the CCs of the Beiluo River and Qingjian River were lower than those of the other sub-basins. CMORPH showed the largest RMSE between 40 mm and 307 mm followed by GSMaP with values between 22 mm and 64 mm; MSWEP showed the smallest RMSE with values between 6.7 mm and 34 mm. The products showed the largest RMSE of 33–64 mm in the Beiluo River and Qingjian River, except for CMORPH, as shown in Figure 13c,d.

At the daily scale, the CCs of the five precipitation products in all sub-regions were below 0.7 (Figure 13e). CMORPH showed the largest RMSE, followed by GSMaP and MSWEP with values of 5.1–29 mm, 4.1–9.5 mm, and 2.5–6.3 mm, respectively. The RMSEs of the Ying River, Qin River, and Yiluo River were greater than those of the other sub-basins (Figure 13f). Both CMORPH and GSMaP overestimated the precipitation, with PBIAS values of 4.8–772% and 18–540%, respectively. The PBIAS values of other products were slightly lower, with values between −23% and 400%. The PBIAS values of MSWEP and TRMM were almost within an acceptable range, except for those in the Beiluo River and Qingjian River (Figure 13g).

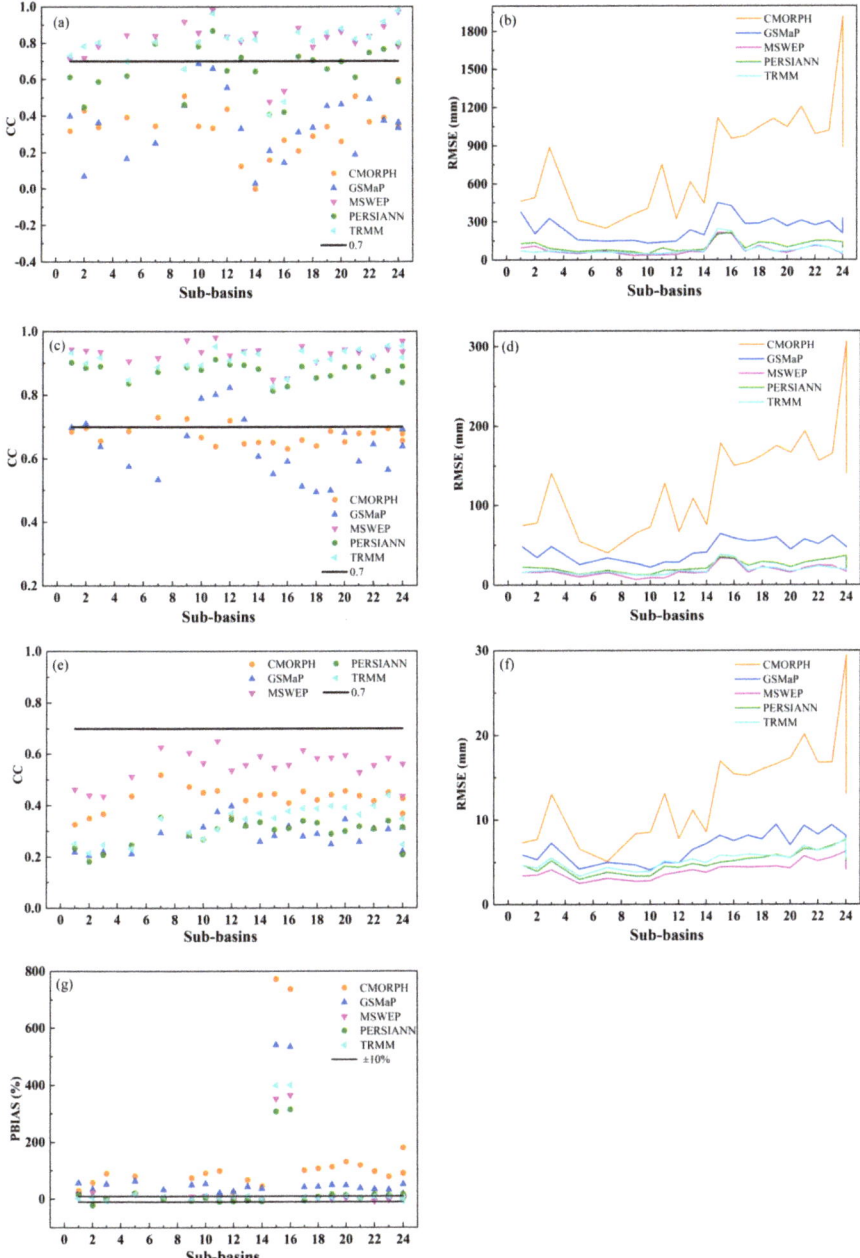

Figure 13. (**a**) annual correlation coefficient and (**b**) root mean square error, (**c**) monthly correlation coefficient and (**d**) root mean square error, (**e**) daily correlation coefficient, (**f**) root mean square error and (**g**) PBIAS in sub-basins. CC is correlation coefficient, RMSE is root mean square error, PBIAS is bias. (The black line in (**a**), (**c**) and (**e**) means the correlation coefficient is equal to 0.7, above which the variables were significantly correlated. The upper black line in (**g**), means PBIAS is equal to 10% and the lower black line means PBIAS is equal to −10%, between which the PBIAS is acceptable.).

www.ingramcontent.com/pod-product-compliance
Lightning Source LLC
LaVergne TN
LVHW070149100526
838202LV00015B/1920